CAMBRIDGE LIBRARY COLLECTION

Books of enduring scholarly value

Physical Sciences

From ancient times, humans have tried to understand the workings of the world around them. The roots of modern physical science go back to the very earliest mechanical devices such as levers and rollers, the mixing of paints and dyes, and the importance of the heavenly bodies in early religious observance and navigation. The physical sciences as we know them today began to emerge as independent academic subjects during the early modern period, in the work of Newton and other 'natural philosophers', and numerous sub-disciplines developed during the centuries that followed. This part of the Cambridge Library Collection is devoted to landmark publications in this area which will be of interest to historians of science concerned with individual scientists, particular discoveries, and advances in scientific method, or with the establishment and development of scientific institutions around the world.

The Scientific Papers of James Prescott Joule

Sir James Prescott Joule (1818–1889) became one of the most significant physicists of the nineteenth century, although his original interest in science was as a hobby and for practical business purposes. The son of a brewer, he began studying heat while investigating how to increase the efficiency of electric motors. His discovery of the relationship between heat and energy contributed to the discovery of the conservation of energy and the first law of thermodynamics. Volume 1 of Joule's scientific papers was published in 1884. It is organised chronologically and reveals the range of Joule's interests and the development of his thought. The topics of the papers include the measurement of heat, voltaic batteries, electromagnets, specific heat, meteorology and thermodynamics. Joule's careful experiments in these areas were fundamental to the development of significant areas of twentieth-century physics, although he was slow to gain recognition from his contemporaries.

Cambridge University Press has long been a pioneer in the reissuing of out-of-print titles from its own backlist, producing digital reprints of books that are still sought after by scholars and students but could not be reprinted economically using traditional technology. The Cambridge Library Collection extends this activity to a wider range of books which are still of importance to researchers and professionals, either for the source material they contain, or as landmarks in the history of their academic discipline.

Drawing from the world-renowned collections in the Cambridge University Library, and guided by the advice of experts in each subject area, Cambridge University Press is using state-of-the-art scanning machines in its own Printing House to capture the content of each book selected for inclusion. The files are processed to give a consistently clear, crisp image, and the books finished to the high quality standard for which the Press is recognised around the world. The latest print-on-demand technology ensures that the books will remain available indefinitely, and that orders for single or multiple copies can quickly be supplied.

The Cambridge Library Collection will bring back to life books of enduring scholarly value (including out-of-copyright works originally issued by other publishers) across a wide range of disciplines in the humanities and social sciences and in science and technology.

The Scientific Papers of James Prescott Joule

Volume 1

James Prescott Joule

CAMBRIDGE UNIVERSITY PRESS

Cambridge, New York, Melbourne, Madrid, Cape Town,
Singapore, São Paolo, Delhi, Tokyo, Mexico City

Published in the United States of America by Cambridge University Press, New York

www.cambridge.org
Information on this title: www.cambridge.org/9781108028820

This edition first published 1884
This digitally printed version 2011

ISBN 978-1-108-02882-0 Paperback

James Prescott Joule

Engraved by C. H. Jeens

London, Published by Macmillan & C.°

THE

SCIENTIFIC PAPERS

OF

JAMES PRESCOTT JOULE,

D.C.L. (OXON.), LL.D. (DUBL. ET EDIN.), F.R.S., HON. F.R.S.E.,

PRES. SOC. LIT. PHIL. MANC., F.C.S., DOC. NAT. PHIL. LUGD. BAT., SOC. PHIL. CANTAB., SOC. PHIL. GLASC., INST. MACH. ET NAUP. SCOT. ET SOC. ANTIQ. PERTH, SOC. HONOR. INST. FR. (ACAD. SCI.), CORRESP. SOC. REG. DAN. HAFN. TAURIN. BOLON., SOC. PHIL. NAT. BASIL., SOC. PHYS. FR. PAR. ET HAL., ET ACAD. AMER. SCI. ET ARTIB. ADSOC. HONOR.

PUBLISHED BY

THE PHYSICAL SOCIETY OF LONDON.

LONDON:

TAYLOR AND FRANCIS, RED LION COURT, FLEET STREET.

1884.

PRINTED BY TAYLOR AND FRANCIS,

RED LION COURT, FLEET STREET.

ADVERTISEMENT.

In issuing this volume, the Council of the Physical Society of London desire to put on record their cordial appreciation of the kindness with which Dr. Joule not only agreed to their request to allow them to publish a collected edition of his Scientific Papers, but undertook personally the labour of getting together and editing the collection.

They desire also to thank the Council of the Philosophical Society of Manchester and the Editors of the 'Philosophical Magazine' for the use of woodcuts, and Messrs. Macmillan and Co. for allowing Jeens's excellent portrait of Dr. Joule, originally published with 'Nature,' to appear as a frontispiece to this volume.

A second volume, containing papers published by Dr. Joule in conjunction with other men of science, is in progress, and will be published as soon as practicable.

February 1884.

PREFACE.

This book appears in consequence of the flattering proposal of the Physical Society of London to collect and reprint the papers on scientific subjects which have appeared in my own name, and those under my own in association with the Rev. Dr. Scoresby, Sir Lyon Playfair, and Sir Wm. Thomson. In this, the first volume, I have endeavoured to fulfil the former part of this design.

The more important papers are literal transcripts from the original; but in some of the earlier a few alterations have been made in the phraseology, not with the intention of altering the meaning, but of making it more clear.

I feel that many imperfections will be found in the compilation, and even that some of the papers may be thought superfluous. However, I have yielded to the wish that the whole should be printed in chronological order as nearly as possible.

J. P. JOULE.

12 Wardle Road, Sale, Cheshire,
November 1883.

TABLE OF CONTENTS.

ON SOME THERMO-DYNAMIC PROPERTIES OF SOLIDS, 413–473.

Introductory.

Thermo-electricity of Iron in different States.

Thermo-electric Intensities of Metals, Alloys, etc.

Experiments on the Thermal Effects of Tension on Solids.

On the Thermal Effects of Longitudinal Compression on Solids.

LIST OF PLATES AND ILLUSTRATIONS.

LIST OF ILLUSTRATIONS.

SCIENTIFIC PAPERS.

Description of an Electro-Magnetic Engine.

[From Sturgeon's 'Annals of Electricity,' vol. ii. p. 122.]

SIR,

I am now making an electro-magnetic engine; and as I imagine that I have succeeded in effecting considerable improvement in the construction of the magnets, and the whole arrangement of the instrument, I hope you will allow me to lay it before the numerous readers of your valuable 'Annals.'

In fig. 1, *e f* represents a side view, and *b b* the poles, of the

Fig. 1. Full size.

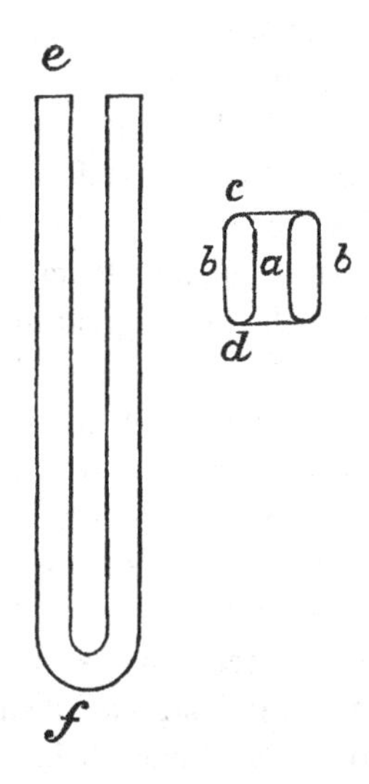

magnet I propose, the distance *a* between the poles being about the fifth part of an inch, and the thickness of each pole or arm of the magnet the same, or perhaps rather less. If the magnets are required to be of greater power, the breadth *c d*, or length *e f*, should be increased, but the thickness left the same. Covered wire is wrapped round until the space between the arms is completely filled.

The advantages obtained by using magnets of the above description may be seen on inspecting fig. 2, where the small arrows indicate the direction that the electricity takes, in passing from P to N.

Fig. 2. Full size.

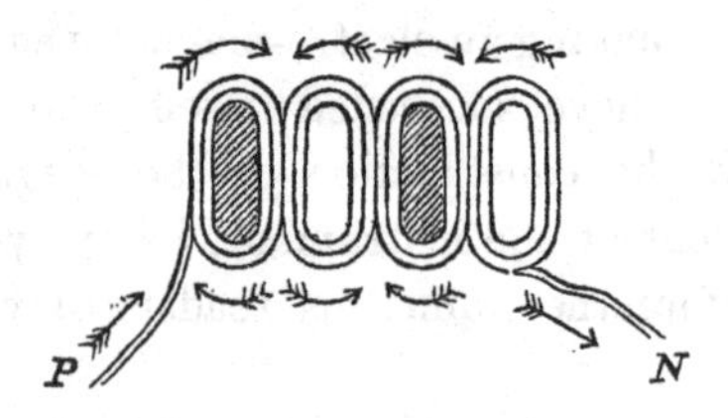

1. The wire round each arm is kept close to the substance of the iron.
2. The greater part of each arm receives the magnetizing effect of the wire wound on its neighbours as well as of that belonging to itself.
3. A great saving of room is effected.
4. The objection which applies to arrangements in which the poles are far distant from one another, viz. that the magnets during a great part of their rotation are almost inactive, is obviated.

I have made several electro-magnets of the above construction. Their lifting power is very good. The spark on breaking battery contact is, however, remarkably brilliant, which may be considered in some respects disadvantageous.

In the engine the electro-magnets I have described are to be arranged in two compact circles, one of which is represented by fig. 3. A stout board *a b c* supports one circle of electro-magnets, the wires of each being attached to those of its neighbours, so that the whole may be magnetized at once by a current of electricity from *h* to *i*. The opposite, or revolving circle is to be magnetized, and the polarity reversed, by means of the commutator *g*, which consists of two cogged

Fig. 3. Half size.

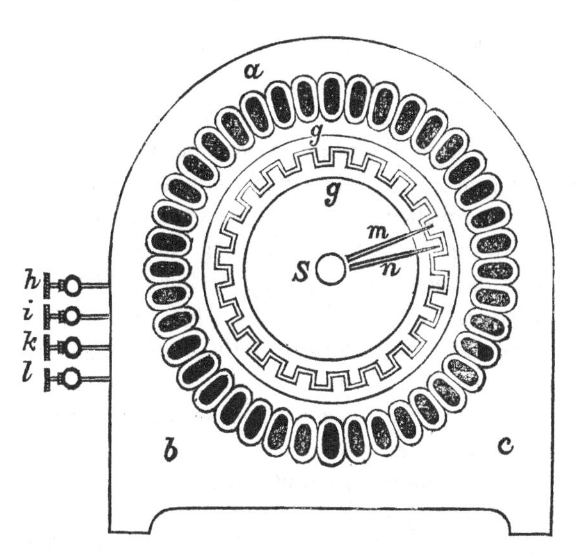

circles of bright brass communicating with the battery by the wires *k l*. The electrodes of the commutator, *m n*, are springs, which, in their revolution, press gently on the circles of brass.

It will be seen that by a proper use of the clamps *h, i, k, l* the current may be made to go consecutively through the fixed and movable circles of electro-magnets, or that a distinct current may traverse each.

The axle carrying the movable circle is supported by the bearing *S*, and may be used to turn any kind of machinery.

I am, yours truly,

J. P. Joule.

Salford, January 8, 1838.

Description of an Electro-magnetic Engine, with Experiments. (In a letter to the Editor of the 'Annals of Electricity'*.)

['Annals of Electricity,' vol. iii. p. 437.]

Dear Sir,

In vol. ii. p. 122 of your interesting work is a communication of mine describing a method of making electro-magnetic engines, which I thought might be adopted with advantage. I finished the one I was working at during last summer. It weighs 7½ lb.; and the greatest power I have been able to develop with a battery of forty-eight Wollaston four-inch plates was to raise 15 lb. a foot high per minute, in which estimate the friction of the working parts, which was very considerable, was reckoned as the load.

The result shows that the advantages of a close arrangement of electro-magnets are not such as I anticipated.

I was desirous, before attempting to make another engine, to satisfy myself by experiment how far it was possible to increase the velocity of rotation, which was only 3½ feet per second in the above trial. Now, of the many things which limit the velocity, the resistance which iron opposes to the instantaneous induction of magnetism is of considerable importance. I think I shall be able to show how this may be obviated in some measure.

The current of electricity produced by a magneto-electric machine is much increased by the insertion of a bundle of iron wires, instead of a solid nucleus of iron, into its coil—a phenomenon evidently occasioned by the peculiar texture of the wires, which allows Mr. Sturgeon's magnetic lines† to collapse with greater suddenness.

With a view to determine to what extent the velocity of rotation could be increased by the use of wire magnets, I

* The experiments were made at Broom Hill, Pendlebury, near Manchester.

† See 'Annals of Electricity,' vol. i. pp. 251, 277.

constructed an apparatus represented by fig. 4, where *a b*, *c d* are steel magnets; *e*, *f*, brass screws with small holes in their

Fig. 4. Scale $\frac{1}{6}$.

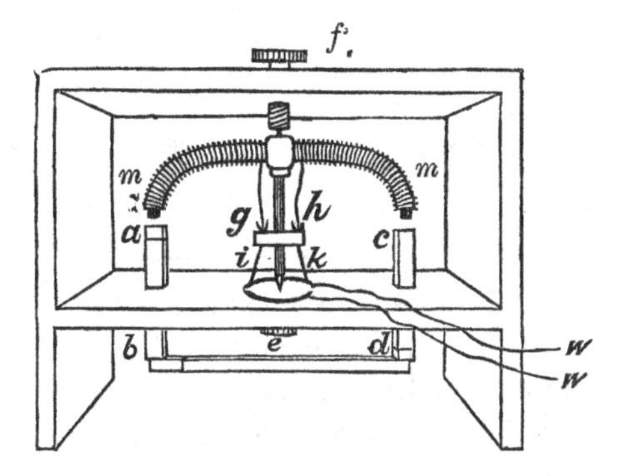

ends to receive the fine points of the steel axle on which the electro-magnet *m m* is fixed; *g*, *h* are mercury-cups to connect the wires of the electro-magnet with the pieces of watch-spring *i*, *k* which dip into two semicircles* of mercury constituting the commutator; *w*, *w* are wires connecting the commutator with the battery. I had four electro-magnets, which, with their axles &c., could, with great expedition, be put off or on the machine by means of the screw *f*.

No. 1 electro-magnet was made of a bar of round iron of 1090 grs. weight; No. 2 was a bundle of nineteen iron wires of about $\frac{1}{20}$ inch diameter: no particular pains were taken to anneal the iron of these magnets. No. 3 and No. 4 were made of iron bar and wires of the same quality and dimensions, but annealed to great softness. Each of the above was first enveloped with a double covering of muslin, and then wound with eleven yards of copper wire $\frac{1}{40}$ inch in diameter covered with silk. Care was taken to make the friction of the pivots equal.

* These semicircles were grooved into the wood. As the convex surface of the mercury stood above the level, the steel springs crossed edgewise over the partition without obstruction or splash.—*Note*, 1881.

The following results, in revolutions per minute, were obtained with the above apparatus :—

	No. 1.	No. 2.	No. 3.	No. 4.
With a single constant cell	146	177	196	192
With two constant cells arranged for intensity	233	274	283	321
Ditto. With weak charge	196	173	224	209
Average	192	208	234	241

The sparks and shocks, on breaking battery circuit, were just sensible in No. 1, twice as great at least in Nos. 2 and 3. In No. 2 they were a little greater than in No. 3, but were by far the most brilliant and powerful in No. 4.

I intend to make another engine presently with magnets made of rectangular wire.

Yours truly,
J. P. Joule.

Salford, December 1, 1838.

On the use of Electro-magnets made of Iron Wire for the Electro-magnetic Engine. (In a letter to the Editor of the 'Annals of Electricity'*)

['Annals of Electricity,' vol. iv. p. 58.]

Dear Sir,

In my last letter I gave you an account of some experiments which were thought to prove that electro-magnets made of iron wire were the most suitable for the electro-magnetic engine. In those experiments the ordinary round wire was used; and it was my opinion that these wire magnets were at a disadvantage in consequence of the interstices between the wires. I therefore arranged fresh experiments as follow :—

* The experiments were made at Broom Hill, Pendlebury, near Manchester.

I constructed two electro-magnets. The one consisted of 16 pieces of square iron wire, each $\frac{1}{11}$ of an inch thick and 7 inches long, bound tightly together so as to form a solid mass whose transverse section was $\frac{4}{11}$ inch square; it was then enveloped by a ribbon of cotton, and wound with 16 feet of covered copper wire of $\frac{1}{16}$ inch diameter. The other was made of a bar of solid iron, but in every other respect was precisely like the first. These electro-magnets were successively fitted to the apparatus used in my last experiments, care being taken to make the friction of the pivots equal in each. On trial, the means of several experiments gave 162 revolutions per minute for the wire magnet, and 130 for the bar magnet.

In the further prosecution of my inquiries, I took six pieces of round bar iron of different diameters and lengths, also a hollow cylinder $\frac{1}{13}$ inch thick in the metal. These were bent in the U-form, so that the shortest distance between the poles of each was half an inch: each was then wound with 10 feet of covered copper wire $\frac{1}{40}$ inch in diameter. Their attractive powers under like currents for a straight steel magnet $1\frac{1}{2}$ inch long, suspended horizontally to the beam of a balance, were, at the distance of half an inch, as follow:—

	No. 1. Hollow.	No. 2. Solid.	No. 3. Solid.	No. 4. Solid.	No. 5. Solid.	No. 6. Solid.	No. 7. Solid.
Length round the bend, in inches ..	6	$5\frac{1}{2}$	$2\frac{2}{3}$	$5\frac{1}{4}$	$2\frac{1}{2}$	$5\frac{1}{4}$	$2\frac{1}{2}$
Diameter, in inches.	$\frac{1}{2}$	$\frac{1}{2}$	$\frac{1}{2}$	$\frac{3}{8}$	$\frac{3}{8}$	$\frac{1}{4}$	$\frac{1}{4}$
Attraction for steel magnet, in grains .	7·5	6·3	5·1	5·0	4·1	4·8	3·6
Weight lifted, in ounces	36	52	92	36	52	20	28

A steel magnet gave an attractive power of 23 grains, while its lifting-power was not greater than 60 ounces.

The above results will not appear surprising if we consider, first, the resistance which iron presents to the induction of magnetism, and, second, how very much the induction is exalted by the completion of the ferruginous circuit.

Nothing can be more striking than the difference between the ratios of lifting to attractive power at a distance in the different magnets. Whilst the steel magnet attracts with a force of 23 grains and lifts 60 oz., the electro-magnet No. 3 attracts with a force of only 5·1 grains, but lifts as much as 92 oz.

To make a good electro-magnet for lifting-purposes:—1st. Its iron, if of considerable bulk, should be compound, of good quality, and well annealed. 2nd. The bulk of the iron should bear a much greater ratio to its length than is generally the case. 3rd. The poles should be ground quite true, and fit flatly and accurately to the armature. 4th. The armature should be equal in thickness to the iron of the magnet.

In studying what form of electro-magnet is best for attraction from a distance, two things must be considered, viz. the length of the iron, and its sectional area.

Now I have always found it disadvantageous to increase the length beyond what is needful for the winding of the covered wire. Then as to the sectional area, you have yourself shown that on placing a hollow and a solid cylinder of iron successively within the same electro-magnetic coil, the hollow piece exerts the greatest influence on the needle. I wished to ascertain whether a hollow magnet could not be represented by a solid one having the same sectional area and girth, but twice the thickness, as in figs. 5 and 6. Two

Fig. 5. Fig. 6.

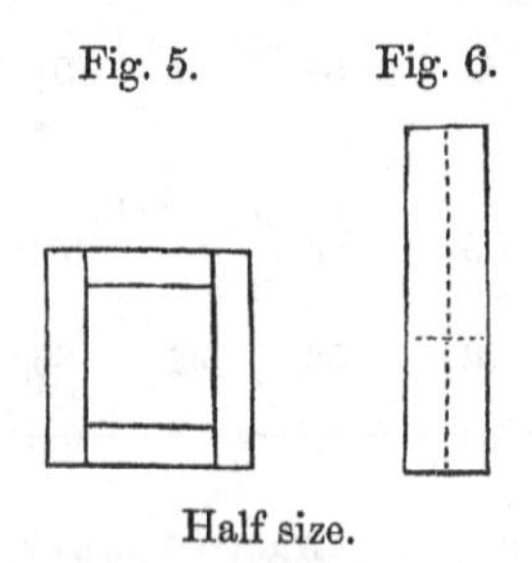

Half size.

electro-magnets were constructed, 7 inches long, and of sections similar to those in the figures, and each wound with 22 feet of covered copper wire $\frac{1}{16}$ inch in diameter.

Their respective attractions, at half an inch distance, for the end of a straight steel magnet were found to be as follow in two trials with different strengths of current :—

Hollow magnet.	Solid magnet.
1·9 grs.	1·7 grs.
4·5 „	4·0 „

It is evident from this that the hollow magnet has the greatest attractive force. But the difference between the two is, I think, hardly sufficient to counterbalance practical advantages which belong to the solid electro-magnet if used in the engine.

Next I made five straight electro-magnets of square iron wire $\frac{1}{11}$ inch thick. Each was 7 inches long, and wound with 22 feet of covered copper wire $\frac{1}{16}$ inch in diameter. No. 1 consisted of nine, No. 2 of sixteen, No. 3 of twenty-five, No. 4 of thirty-six, and No. 5 of forty-nine of the square iron wires. Each was built up into the form of a prism with square base and section. Five other electro-magnets were made of solid iron, but otherwise were exactly similar to the first set. The following attracting-powers (in grains at half an inch distance) for a straight steel magnet were obtained, using three different galvanic forces :—

		No. 1.	No. 2.	No. 3.	No. 4.	No. 5.
1st experiment	Solid magnet ..	1·5	1·9	1·6	2·1	2·0
	Wire magnet ..	2·1	2·1	1·7	2·0	1·9
2nd experiment	Solid magnet ..	2·0	2·5	2·35	2·45	2·2
	Wire magnet ..	2·6	2·8	2·1	2·2	2·05
3rd experiment	Solid magnet ..	2·7	3·6	3·4	3·2	3·1
	Wire magnet ..	3·3	3·8	3·0	2·9	2·65

The iron wire was taken at the same degree of temper as that in which it came from the makers, consequently must have been harder than the bar iron with which it was compared. It will be remarked that while the wire magnets are more powerful in the first Nos., they are less powerful in the last Nos. than the solid magnets. I cannot account for this circumstance, unless by supposing, according to the hypothesis of Dr. Page, that the wires of which the magnets are composed repel one another's magnetism in such a manner as to tend to neutralize the general force of the electro-magnet, and that this effect increases with the number of wires used.

Yours truly,
J. P. JOULE.

Salford, March 27, 1839.

Investigations in Magnetism and Electro-magnetism. (In two letters to the Editor of the 'Annals of Electricity' *.)

['Annals of Electricity,' vol. iv. p. 131.]

DEAR SIR,

I am now able to send you an account of my further investigations on electro-magnetic attraction. It was a matter of importance to use in the research a galvanometer the indications of which could be depended upon.

Fig. 7 represents the form of my galvanometer. The lozenge-shaped needle n is 2 inches long, the wire 10 feet

Fig. 7. Scale $\frac{1}{8}$.

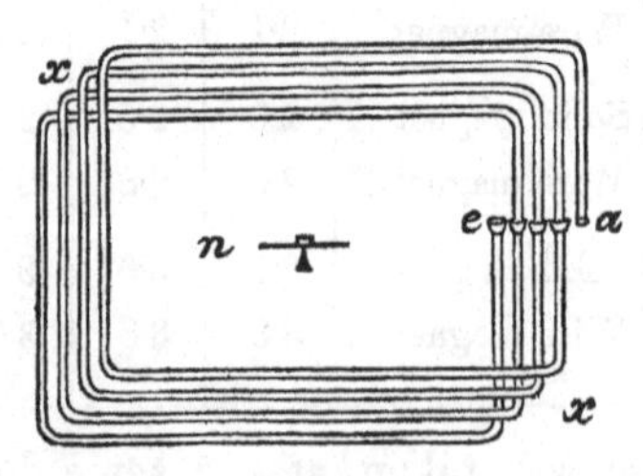

long and $\frac{1}{16}$ of an inch in diameter. It is disposed in four rectangles, mercury-cups being placed at a, b, c, d, e. The

* The experiments were made at Broom Hill, Pendlebury, near Manchester.

wire crosses at x x, but everywhere else is in the same plane.

The process of graduation was performed as follows:—A current of a certain intensity was passed from *a* to *b*, from *a* to *c*, from *a* to *d*, and from *a* to *e*, taking care to decrease the resistance of the battery-wires as the length of the part of the galvanometer-wire through which the current passed was increased: the several deviations of the needle thus obtained were marked 1, 2, 3, and 4 on the card of the instrument. I then increased the power of the battery until the needle stood at 2 when the current passed from *a* to *b*: the former process was then repeated, and I marked on the card 4, 6, and 8; and going on in this manner I obtained the graduations 1, 2, 3, 4, 6, 8, 9, 12, 16, &c. When the galvanometer is used the current is passed from *a* to *b*, and the above numbers represent proportional quantities of electrical current.

In order to give a definite idea of the quantity of electricity indicated by this galvanometer, I took a diluted acid, consisting of 10 parts of water to 1 of sulphuric acid, sp. gr. 1·8, and passed through it by platinum electrodes a current$=$1 of the graduation. In seven minutes 0·62 cubic inch of the mixed gases was evolved.

The electro-magnets used in Table I. were those described in my last letter; they are straight and square, 7 inches long, and wound with 22 feet of covered copper wire $\frac{1}{16}$ inch in diameter. Five of them were of bar iron; and five corresponding ones were bundles of square iron wires. The sides of the square sections of Nos. 1 are $\frac{3}{11}$ of an inch, which dimension is gradually increased by $\frac{1}{11}$ up to Nos. 5, which are $\frac{7}{11}$ of an inch.

The bar-iron electro-magnets were successively suspended from the beam of a balance; and the corresponding wire electro-magnets were brought underneath, so that $\frac{1}{8}$ of an inch intervened, a distance maintained by the interposition of a piece of wood. Currents of the strengths given in the table were passed through the wires of the electro-magnets and the coil of the galvanometer. The attraction was measured in grains.

TABLE I.

Electrical current.	Electro-magnets.				
	Nos. 1.	Nos. 2.	Nos. 3.	Nos. 4.	Nos. 5.
6	76	65	88	62	42
8	133	100	180	103	98
12	258	296	300	286	206
16	500	548	530	550	410
24	1080	1280	1190	1210	1050

To vary the experiments, and with the view of ascertaining the effect of an increase of length, I constructed ten more electro-magnets of the same sectional areas, but of 14 inches (or double the former) length, and wrapped with 22 yards (or three times the length) of covered wire. Nos. 1 and Nos. 2 were made of square iron wire, the others of bar iron.

TABLE II.

Electrical current.	Electro-magnets.				
	Nos. 1.	Nos. 2.	Nos. 3.	Nos. 4.	Nos. 5.
8	410 cor. 675	667 cor. 990	1150	1205	1175
12	690 cor. 1080	1170 cor. 1740	2150	3025	2625
16	1000 cor. 1460	1920 cor. 2710	4575	5687	4675
24	1460 cor. 2080	3500 cor. 4750	9625	11812	10500

Each of the electro-magnets used above, except Nos. 1 and Nos. 2 of the second table, was wound in two layers by the wire, the larger ones being uncovered at intervals. I must mention, however, that Nos. 1 and Nos. 2 of the second table had to be wound in some parts to three layers, on account of their small diameter. On calculation I give the probable corrections stated in Table II.

It does not appear from the above results that any considerable detriment arises from the increase of the length of the electro-magnets. It is plain that, as the magnets in Table II. are wound with three times the length of wire, 24 degrees of electricity in the first table should have the same effect as 8 in the second table. The difference is to be referred to the increased length of the iron; but I do not feel justified in assigning its value, which, however, seems to decrease as the section of the magnets increases.

The experiments, however, appear to indicate an important law, which may be expressed as follows:—*The attractive force of two electro-magnets for one another is directly proportional to the square of the electric force to which the iron is exposed; or, if* E *denote the electric current,* W *the length of wire, and* M *the magnetic attraction,* $M=E^2W^2$.

The discrepancies from this law may, I think, be owing to magnetic inertia and experimental errors. Perhaps the fairest way to compare theory with experiment, is to take the means of all the results of the first table and the means of Nos. 3, 4, and 5 in the second, omitting Nos. 1 and 2 because it is clear that these were at last becoming saturated with magnetism. The result is as follows:—

From the first Table.			From the second Table.	
Currents.	Observed attraction.	Calculated.	Observed attraction.	Calculated.
6	66·4	66·4		
8	123	118	1177	1177
12	269	265	2600	2648
16	508	472	4079	4708
24	1163	1063	10646	10593

Desirous to ascertain whether the law held in the case of lifting as well as in that of distant attraction, I made trial with a horse-shoe electro-magnet constructed of a cylinder of iron 7 inches long, $\frac{5}{8}$ of an inch in diameter, and wound with 5 yards of covered copper wire. The law seems in this case to fail, principally because the iron is sooner saturated with magnetism. Hence the propriety of giving considerable bulk, rather than length, to electro-magnets designed for lifting-purposes.

Current.	Lifting-power in lb.	Calculated.
4	3·5	3·5
6	6·5	8
8	11·5	14
12	21	31·5

I can hardly doubt that electro-magnetism will ultimately be substituted for steam to propel machinery. If the power of the engine is in proportion to the attractive force of its magnets, and if this attraction is as the square of the electric force, the economy will be in the direct ratio of the quantity of electricity, and the cost of working the engine may be reduced *ad infinitum*. It is, however, yet to be determined how far the effects of magnetic electricity may disappoint these expectations.

I find that the plan which I had proposed for a new engine must yield to the views elicited by the above experiments. As far as I see at present, I think it will be best to use only two, and these very large electro-magnets, and to concentrate upon them all the strength of electrical current I can command.

Yours truly,

J. P. Joule.

Broom Hill, near Manchester,
May 28, 1839.

Investigations in Magnetism and Electro-Magnetism. Letter 2nd.

['Annals of Electricity,' vol. iv. p. 135.]

Dear Sir,

The following experiments were made for the purpose of giving a severe test to the law enunciated in my last letter.

I constructed two pairs of straight electro-magnets. Each of the first pair was made of a bar 30 inches long and 1 inch square; each of the second pair was 30 inches long, 2 inches broad, and 1 inch thick. The sharp edges having been ground to avoid cutting the wire, each bar was carefully insulated and then wound with 88 yards of covered copper wire $\frac{1}{16}$ of an inch in diameter.

The attractions were measured in the same manner as before, excepting that a plate of copper was employed instead of wood in order to keep the electro-magnets asunder at the proper distance. The attraction of the suspended electro-magnet for the fixed one was measured in ounces avoirdupois.

		Currents................	6.	8.	12.	16.	24.	32.
First pair.	at $\frac{1}{8}$ inch	Experiment .	18	33	72	124	260	
		Calculated . .	18	32	72	128	288	
	at $\frac{1}{4}$ inch	Experiment .	7	13	28	47	96	
		Calculated . .	7	12·44	28	49·7	112	
	at $\frac{1}{2}$ inch	Experiment .	3	5·25	12	18	38	62
		Calculated . .	3	5·33	12	21·3	48	85·3
Second pair.	at $\frac{1}{8}$ inch	Experiment .	14	27	60	100	240	
		Calculated . .	14	25	56	100	224	
	at $\frac{1}{4}$ inch	Experiment .	6·25	12	25	40	96	
		Calculated . .	6·25	11·1	25	44·4	100	
	at $\frac{1}{2}$ inch	Experiment .	2·5	5	9·5	17·5	36	
		Calculated . .	2·5	4·44	10	17·7	40	

The experimental results are quite as near to the calculated values as could be expected. Those belonging to the first pair are particularly satisfactory, especially if, with regard to the numbers under currents 16, 24, and 32, some allowance is made for the approaching saturation of the iron.

The above electro-magnets, 30 inches long, with 88 yards of wire and 6 degrees of current, sustained at $\frac{1}{8}$ of an inch distance 7000 grains; but the mean attractive power of Nos. 3, 4, and 5 in my last was, with the same electric force, viz. 24 degrees of current traversing 22 yards of wire, 10646 grains. From these figures we may form some idea of the diminution of power arising from the increased length of the bars.

Yours truly,
J. P. JOULE.

Broom Hill, July 10, 1839.

Description of an Electro-magnetic Engine. (In a letter to the Editor of the 'Annals of Electricity.')

['Annals of Electricity,' vol. iv. p. 203.]

DEAR SIR,

I beg to forward to you the description of my new electro-magnetic engine*. It is represented in perspective by fig. 8, where A B C D is a strong wooden frame, the inside measure of which is 4½ feet long, 1½ foot high, and 1 foot broad. F G is the axle of cast steel, three quarters of an inch square, and turned down to half an inch at the journals, which work in the brass bearings *b b*. H H are holders for the revolving electro-magnets *m m*: there is a third holder between these, which is omitted in the figure, but which contributes very materially to the stability of the apparatus. *n n* are the stationary electro-magnets, firmly secured to the top and bottom of the frame by the brass clamps *k k*. *s s* are silver springs which enter through the frame and press gently on the brass semicircles of the commutator *c*. This commutator is shown on a larger scale in fig. 9. *v v* are copper

* Constructed at Broom Hill.

wires soldered to the brass semicircles, and connected with the revolving electro-magnets. *x x* are wires let into the frame, and connected with the stationary electro-magnets. *z z* are wires to form the necessary connections at the other end.

Fig. 8. Scale $\frac{1}{16}$.

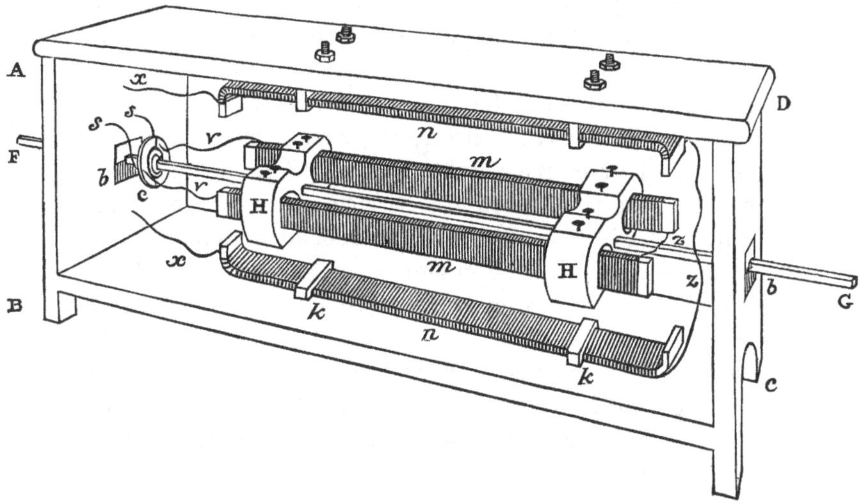

Fig. 9. Scale $\frac{1}{4}$.

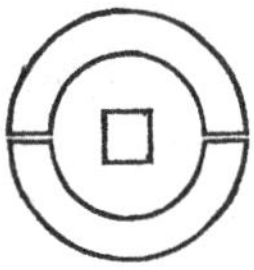

Fig. 10. Scale $\frac{1}{10}$.

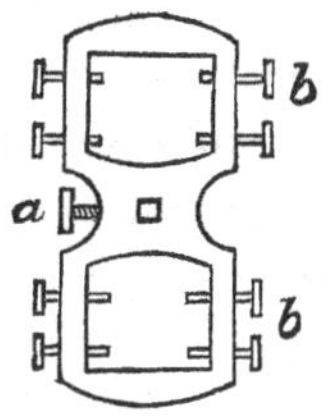

Fig. 11. Full size.

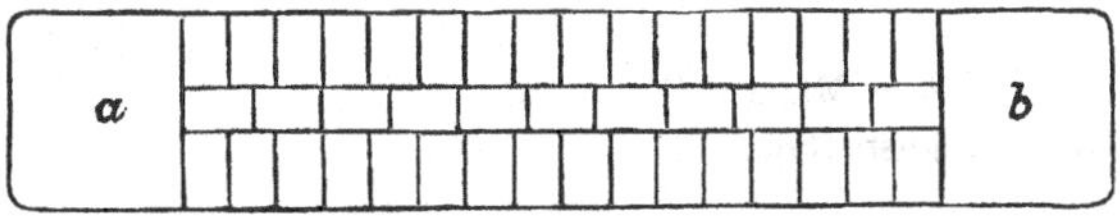

In fig. 10 one of the holders is represented on a larger scale. It is constructed of hard wood strengthened with brass plates. *b b* are brass screws, which act upon wedges, so that I can adjust the electro-magnets at different distances from the axle. *a* is a screw which secures the holder to the axle.

The iron of each revolving electro-magnet is 36 inches long, three inches broad, and half an inch thick: that of the stationary electro-magnets is of the same breadth and thickness; but their length is such that when bent their extremities are $36\frac{3}{4}$ inches asunder; the revolving magnets therefore pass at the distance of $\frac{3}{8}$ of an inch.

One of the revolving electro-magnets is made of a solid bar; the other of rectangular iron wire built up as shown by fig. 11, the two planed iron bars *a b* serving to hold them secure. The iron of these is in the usual condition of hardness as supplied by the trade. I intend to replace them ultimately by electro-magnets constructed of thoroughly annealed iron.

Each electro-magnet is furnished with a strip of thick calico next the iron, and with three separate layers of covered copper wire, each 106 yards long and $\frac{1}{24}$ of an inch in diameter, wound with the utmost regularity.

The engine consists of only four electro-magnets; but it will at once be seen that the principle admits the use of any greater even number. I do not, however, think that the mere increase of number would be attended by advantage.

As the form of the engine enables me with ease to place the electro-magnets in different positions, and as their several coils are insulated, and I am therefore enabled to use the electric current in quantity and intensity arrangements, it offers facilities for experiment. In my preliminary trials I have been much pleased with its performances.

Yours truly,

J. P. Joule.

Broom Hill, near Manchester,
30th August, 1839.

On Electro-magnetic Forces. *By* J. P. Joule*.

['Annals of Electricity,' vol. iv. p. 474.]

About the beginning of last April I made some experiments in electro-magnetism†. I now return to the subject, desirous of placing some of the effects I then witnessed in a more complete and accurate point of view.

I have shown that when a current of voltaic electricity is transmitted through the coils of two electro-magnets, their mutual attraction is in the ratio of the squares of the quantities of electric force; and also that the lifting-power of the "horse-shoe" electro-magnet is governed by the same law.

I have recently made experiments which prove that the attraction of the electro-magnet, for a magnet of constant force, varies in the simple direct ratio of the quantity of electricity passing through the coil of the electro-magnet. In order to succeed in the experiment it is necessary to make the effects of induction insensible by a proper arrangement of the apparatus.

Magnetism appears therefore to be excited in soft iron in the direct ratio of the magnetizing electric force; and electro-magnetic attraction, as well as the attraction of steel magnets, may be considered as proportionate to the product of the intensities of each magnet, or, which is the same thing, to the number of lines which may be drawn between the several magnetic particles of the attracting bodies.

Fig. 12. Fig. 13. Fig. 14.

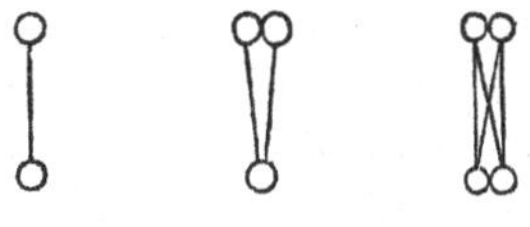

This view is illustrated by figs. 12, 13, and 14, where the

* The experiments were made at Broom Hill, near Manchester.

† 'Annals of Electricity,' iv. pp. 131, 135.

several attractions of the magnetic particles, viz. 1 to 1, 2 to 1, and 2 to 2, are represented by the number of lines drawn in each instance, *i. e.* 1, 2, and 4. Fig. 15 illustrates

Fig. 15.

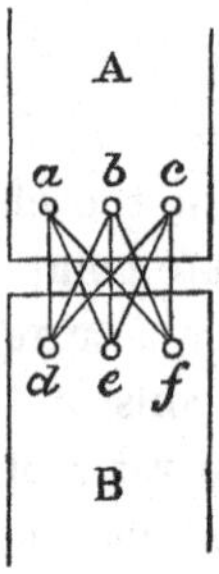

the complex action of forces constituting the aggregate attraction existing between two magnets, A and B. The magnetic particles, of which six only, viz. *a, b, c, d, e, f*, are represented, may be conceived to be an indefinitely large number spread throughout the region of the poles.

If this view be correct, it is obvious that the closer the approximation of the magnetic particles in each system, the greater will be the magnetic attraction; for in that case, for instance, the particle *a* will both be nearer the particle *f*, and the force exerted between them will be in a less oblique direction.

It was in consequence of my entertaining a different view, that I was led to imagine that I had determined the amount of decrease of power arising from the increase of the length of the electro-magnet; for in a comparison of the attractive powers of three pairs of electro-magnets, of the several sizes $\frac{5}{11}$, $\frac{6}{11}$, and $\frac{7}{11}$ inch square, with those of two pairs whose sectional areas were respectively one inch square and two inches by one inch, I found the latter long electro-magnets weaker than the former in the ratio of 7000 to 10646—an effect which now appears to me to be principally owing to diffused polarity.

The loss of attractive power in consequence of length

might be readily ascertained; but neither this nor the diffusion of polarity affects the main conclusion at which I have arrived with regard to the laws under which magnetic attraction, as applicable to the production of motive force, is developed by electricity, viz. *that the attraction of two electro-magnets towards each other is in every case represented by the formula* $M = W^2 E^2$, *where* M *denotes the magnetic attraction,* W *the length of wire, and* E *the quantity of electricity conveyed by that wire in a given period of time,*—a formula *modified* merely by the effects of saturation, inductive power of the iron, and distance of the coils from the surface of the iron.

I have observed that magnetic and electro-magnetic attraction decreases, in certain cases, in the simple ratio of the distances. This was found particularly when the magnets were long and the distances between them small. Mr. Harris has observed a similar effect, which may be accounted for on the principles previously illustrated. It is impossible to doubt the accuracy of Coulomb's law that magnetic attraction obeys the law of the inverse square of the distance.

I will now describe some experiments I have made with the electro-magnetic engine described in page 17. For these I prepared a galvanometer constructed on the plan described in page 10. The coil is rectangular, 12 inches by 6, and of copper wire $\frac{1}{12}$ of an inch in diameter; the length of the needle near 4 inches. A table of the absolute value of its indications was formed in the manner already described.

In the following experiments the engine was fitted up with the unannealed bar and iron-wire electro-magnets. After a few trials it was found undesirable to use the wire united so as to form one length. I therefore soldered the ends of the three coils of each electro-magnet together, and united the combined wires in such a manner that the electric current passed through 424 yards of threefold conducting wire.

In the Tables the first column indicates relative quantities of electrical current; the second, the difference between those quantities; the third, the velocity of the revolving electro-magnets in feet per second; the fourth, the work including friction; the fifth, the duty in pounds raised to the height of one foot by the agency of one pound of zinc.

In calculating the amount of work, I found that current 12·4 was just sufficient to keep the machine in motion, the friction referred to the distance from the axle of the revolving electro-magnets being equal to ten ounces avoirdupois: the same quantity of current was, whatever the velocity might be, always able to overcome exactly the same amount of friction. I therefore felt justified in using it as a basis on which to calculate the force due to other quantities of current electricity. The duty, in the fifth column, is calculated on the basis of the decomposition of water effected by a given current. I must observe that the friction is estimated as part of the work; and that, whenever the motive force was not sufficient by itself to turn the machine, a weight thrown over a pulley on the axis supplied the requisite assistance.

TABLE I.

80 pairs of Wollaston's 4-inch plates of amalgamated zinc and double copper.

Current.	Difference.	Velocity.	Work.	Duty.
24·6	....	0	0	0
	3			
21·6	....	2	3·8	21960
	2			
19·6	....	4	6·25	39740
	1·6			
18·0	....	6	7·89	54800
	1·5			
16·5	....	8	8·85	66950
	1·5			
15·0	....	10	9·15	76140

TABLE II.

40 pairs of Wollaston's plates.

Current.	Difference.	Velocity.	Work.	Duty.
11·8	....	0	0	0
	1·6			
10·2	....	2	0·85	20700
	0·8			
9·4	....	4	1·44	38300
	0·8			
8·6	....	6	1·8	52320
	0·6			
8·0	....	8	2·08	65000

TABLE III.

10 pairs of Wollaston's plates.

Current.	Difference.	Velocity.	Work.	Duty.
5	....	0	0	0
	0·8			
4·2	....	2	0·14	33300
	0·6			
3·6	....	4	0·21	58300
	0·3			
3·3	....	6	0·265	80300
	0·3			
3·0	....	8	0·292	97300

TABLE IV.

Grove's Battery; 10 4-inch plates.

Current.	Difference.	Velocity.	Work.	Duty.
17·6	....	0	0	0
	3·3			
14·3	....	2	1·66	116080
	1·9			
12·4	....	4	2·5	201600
	1·4			
11·0	....	6	2·95	268200
	0·8			
10·2	....	8	3·38	331400

I now united the conductors in such a manner that the fluid was divided between the stationary and revolving electro-magnets; in this case the electricity passed through 212 yards of sixfold wire.

TABLE V.

80 pairs of Wollaston's plates in series of 40.

Current.	Difference.	Velocity.	Work.	Duty.
52	....	0	0	0
	9			
43	....	2	3·76	21800
	5			
38	....	4	5·87	38600
	3·2			
34·8	....	6	7·38	53100
	2·6			
32·2	....	8	8·42	65400
	2·4			
29·8	....	10	9·02	75700

TABLE VI.

40 pairs of Wollaston's plates in series of 20.

Current.	Difference.	Velocity.	Work.	Duty.
28·2	....	0	0	0
	5			
23·2	....	2	1·1	23700
	2·5			
20·7	....	4	1·74	42000
	1·7			
19·0	....	6	2·205	58000
	1·4			
17·6	....	8	2·52	71600

Table VII.

20 pairs of Wollaston's plates in series of 10.

Current.	Difference.	Velocity.	Work.	Duty.
16·8	....	0	0	0
	3			
13·8	....	2	0·387	28000
	1·6			
12·2	....	4	0·605	49600
	1·2			
11·0	....	6	0·738	67100
	1·0			
10·0	....	8	0·813	81300

The above examples will show pretty clearly the effects of magnetic electrical resistance. This resistance is the prime obstacle to the perfection of the electro-magnetic engine; and in proportion as it is overcome will the motive force increase. It therefore claims our first attention.

On comparing the differences with the velocities and currents in each table, the general conclusion is that the magnetic electrical intensity resisting the current is directly proportional to the velocity multiplied by the magnetism, or, which is the same thing, by the electricity which induces that magnetism. It is the latter part of this law which makes the differences decrease in the same ratio with the electrical currents opposite. In forming these conclusions I neglect the first difference, or that which exists between no motion and velocity 2, because this is much augmented by a very trifling inaccuracy in the construction of the commutator.

It appears, moreover, that this law is unaffected by altering the battery-intensity; for on comparing the tables of either system together, it will be seen that the differences are always about one tenth of the currents to which they are opposite.

In the second arrangement the conducting metal was half as long and twice as substantial as in the first: hence it is that half the battery-intensity sufficed to pass twice the quantity of current, and so to produce the same motive power. Compare Table I. with Table V. Also a reference to these tables shows that the differences are twice as great in the second arrangement as in the first, the magnetic force remaining very nearly the same. To understand the reason of this we must observe that—1st, the magnetic electrical intensity has nothing to do with the thickness of the wire upon which it is induced, but exists simply in the direct ratio of the length; consequently that the intensity was only one half as great in the second arrangement as it was in the first in like circumstances; and 2nd, that, as the resistance of the wire to the current in the second arrangement is only one quarter of that in the first, an additional or extraneous resistance will produce four times the effect in the former as in the latter instance. Hence, by compounding the two effects, we have the differences of current due to a given increment of velocity with the same amount of magnetism twice as great in the second as in the first arrangement.

If the intensity of the voltaic battery increases in the ratio of the number of its elements, there will be no variation in economy, whatever the arrangement of the conducting metal may be, or whatever the extent of the battery. For if the battery be doubled in intensity, it must consist of twice the number of pairs in series, which will cause twice the quantity of electricity to pass; and so four times the weight of battery materials will be consumed in the same time; whilst the force of the engine is also increased fourfold, according to the square of the current. See the duty under Tables 1, 2, 5, and 6.

I think my engine might be improved by increasing the conductive power of its coils and the softness of the iron of the electro-magnets; but the augmentation of the intensity of each element of the battery is very important, as it is attended by a proportional increase of duty.

Broom Hill, near Manchester,
March 10, 1840.

On Electro-magnetic Forces. *By* J. P. JOULE*.

['Annals of Electricity,' vol. v. p. 187.]

IN resuming the relation of my researches, I shall dismiss for the present the investigation of electro-magnetic forces acting at a distance, and consider the laws which govern that peculiar condition which is assumed on the completion of the ferruginous circuit—the lifting or sustaining power of the electro-magnet.

Although this wonderful property is known to all, and a variety of forms have been given to the electro-magnet, both as regards the bulk and shape of its iron, and the length and number of its magnetizing spirals, I am not aware that any general rules have been laid down for its manufacture, which is the more to be regretted, as some have been led to imagine that the different capabilities of various arrangements are the consequence of causes too many and recondite to be unravelled. I shall attempt in this paper to throw some light on the subject, and to describe a construction giving greater results than have hitherto been attained. It was my desire to make my experiments as exact as possible; and as I wish the relation of them to be clear and definite, I will begin with some observations on the *measure* of current electricity indicated by the galvanometer, an instrument not only useful but essential in an inquiry like the following.

The great difficulty, if not the impossibility, of understanding experiments and comparing them with one another, arises in general from incomplete descriptions of apparatus and from the arbitrary and vague numbers which are used to characterize electric currents. Such a practice might be tolerated in the infancy of the science; but in its present state of advancement greater precision and propriety are imperatively demanded. I have therefore determined for my own part to abandon my old quantity numbers, and to express my results on the basis of a *unit* which shall be at once scientific and convenient.

* The experiments were made at Broom Hill.

That proposed by Dr. Faraday is, I believe, the only standard of this kind that has been suggested. His discovery of the definite quantity of electricity associated with the atoms or chemical equivalents of bodies, has induced him to use the *voltameter* as a measurer, and to propose that the hundredth part of a cubic inch of the mixed gases forming water should constitute a *degree**. There can be no doubt that this system would offer great advantages to the experimenter in some cases, and when the above instrument is employed. However, as I am not aware that it has been used in the researches of any electrician, not excepting those of Faraday himself, I do not hesitate to advance what I think more appropriate as well as generally advantageous. It is thus simply stated :—

1. *A degree of static electricity is that quantity which is just able to decompose nine grains of water.* 2. *A degree of current electricity is the same amount propagated during each hour of time.* 3. Where both time and length of conductor are elements, as in electro-dynamics, *a degree of electric force, or of electro-momentum, is indicated by that same quantity* (a degree of static electricity) *propagated through the length of one foot in one hour of time.* Whenever in this paper I speak of degrees, I intend those I have just defined.

As 9 is the atomic weight of water, it is obvious how greatly this *degree* would facilitate the calculation of electro-chemical decompositions. I may adduce an illustration from electrotype engraving: here, if a galvanometer graduated according to the proposed scale were included in the circuit, it would only be necessary to multiply the degrees of its indication by 32, the equivalent of copper, and the product by the time in hours during which the work has been carried on, to obtain the weight in grains of the copper which has been precipitated; and there would therefore be no occasion to arrest the process until calculation has shown that the proper quantity of copper was cast.

The galvanometer I described in my last paper was con-

* Experimental Researches, series 7 (736).

nected with an apparatus furnished with fine platinum electrodes. A voltaic current was transmitted through the instruments; and after a few minutes the circuit was broken and the hydrogen measured in a graduated tube. The mean of ten trials showed that 0·76 gr. of water was decomposed per hour by the current indicated by each unit of my former quantity numbers: hence 11·8 of these last are equal to one degree of my present scale. In proof of the accuracy of the indications of the galvanometer thus graduated, I transmitted a current of $0^{\circ}{\cdot}415$ through sulphate of copper by copper electrodes for the interval of an hour and a quarter. The copper deposited amounted to 15·6 grains, while the theoretical result is $0^{\circ}{\cdot}415 \times 32 \times 1{\cdot}25 = 16{\cdot}6$ grs., the slight deficiency being probably owing to the consumption of a part of the current in the decomposition of water, as a few bubbles of hydrogen ascended from the negative electrode.

The dimensions of the single coil of my galvanometer are 12 by 6 inches; and the deviation of its needle for one degree of current is 34° of the graduated card. From these data it is easy to calculate the value of the indications of any similar instrument, bearing in mind that the electro-dynamic force on the needle produced by a constant current is directly as the number of coils and inversely as their linear dimensions.

The quantities of current electricity used in the subsequent experiments were frequently sufficient to bring the needle of the above galvanometer to an almost rectangular position to the plane of the coil. I have, for use in these cases, devised the instrument represented by fig. 16, where *C C* is a rod of

Fig. 16. Scale $\frac{1}{8}$.

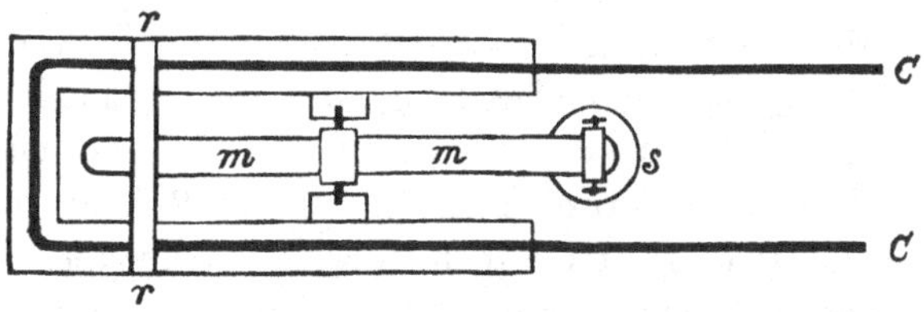

copper, bent and fastened firmly to a wooden frame; *m m*, a magnetized bar of steel, supported, like an ordinary balance-beam, by knife-edges resting on concave surfaces of hard steel; *s* a scale, hung from the nearer end of the magnet for the purpose of receiving the weights by which the strength of the current is measured; and *r r* is a rest, the under surface of which is just touched by the magnet when at zero.

In using this galvanometer it is merely necessary to adjust the magnet to zero, either by means of screws, weights, or by the attraction or repulsion of a small magnet kept for the purpose. Then, on making the necessary battery communications at *C C*, the scale *s* will rise with a force estimated by the weight in grains, tenths, &c. which is required to bring the beam again to zero. In my instrument one degree of current is indicated by 0·69 of a grain in the scale.

The value of the new galvanometer (the sensibility of which may be increased at pleasure by multiplying the number of coils), besides its usefulness in measuring copious currents, consists in its perfect independence of the terrestrial as well as of any other ordinary magnetic influence. In every situation, provided that the intensity of the balance-bar remains constant and no interference is induced after its adjustment to zero, the transmitted current is exactly proportional to the weight lifted by the scale; so that I should have as much confidence in working with it on an iron steamboat as if every particle of neighbouring iron were entirely removed.

I proceed now to describe my electro-magnets, which I constructed of very different sizes in order to develop any curious circumstance which might present itself. A piece of cylindrical wrought iron, 8 inches long, had a hole, one inch in diameter, bored the whole length of its axis; one side was then planed until the hole was exposed sufficiently to separate the thus-formed poles one third of an inch. Another piece of iron, also 8 inches long, was then planed, and being secured with its face in contact with the other planed surface, the whole was turned into a cylinder 8 inches long, $3\frac{3}{4}$ inches in exterior, and 1 inch in interior diameter. The larger piece was then covered with calico and wound with four copper wires covered with silk, each 23 feet long and $\frac{1}{11}$ of an inch in diameter—a quantity which was just sufficient to hide the exterior surface and to fill the interior opened hole. The electro-magnet will perhaps be better understood on reference to fig. 17, where *m* is the "horse-shoe," on which I have drawn some lines to show the position of the conducting wire, *a* is the armature, and *s s* are hooks and eye-holes for the purpose of suspension. The above is designated No. 1; and the rest are numbered in the order of their description.

Fig. 17. Scale $\frac{1}{10}$.

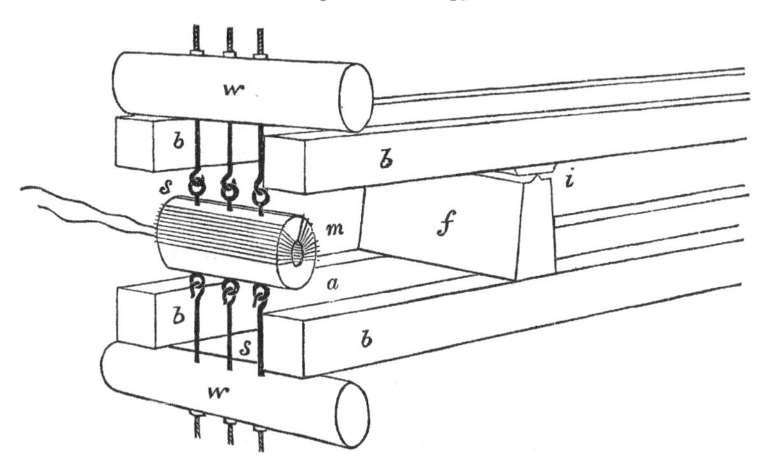

I made No. 2 of a bar of half-inch round iron 2·7 inches long. It was bent into an almost semicircular shape, and then covered with 7 feet of insulated copper wire $\frac{1}{20}$ of an inch thick. The poles are half an inch asunder; and the wire completely fills the space between them.

A third electro-magnet was made of a piece of iron 0·7 inch long, 0·37 inch broad, and 0·15 inch thick. Its edges were reduced to such an extent that the transverse section was elliptical. It was bent into a semicircular shape, and wound with 19 inches of silked copper wire $\frac{1}{40}$ of an inch in diameter.

To procure a still more extensive variety, I constructed what might, from its extreme minuteness, be termed an *elementary electro-magnet*. It is the smallest, I believe, ever made, consisting of a bit of iron wire $\frac{1}{4}$ of an inch long and $\frac{1}{25}$ of an inch in diameter. It was bent into the form of a semicircle, and was wound with three turns of *uninsulated* copper wire $\frac{1}{40}$ of an inch in thickness.

A lever, fig. 17, was employed for measuring the strongest lifting powers of No. 1. *b*, *b*, *b*, *b* are beams of ash, 3 inches square and 10 feet long, and strengthened by iron plates; they are made into two pairs by boards, not shown in the figure, screwed to their upper surfaces; *i*, *i* are movable iron bearings, and *f* is the fulcrum, also movable, and armed with iron; *w*, *w* are strong cylinders of wood, which bear upon the levers and carry the hooks which are affixed to the electro-magnet and its armature.

In Table I. the first column gives degrees of current electricity as already defined. The second gives the products of the length in feet of wire wound on the magnet into the numbers of the currents; it therefore contains degrees of electric force. The last gives the weight lifted in pounds avoirdupois.

TABLE I.

Electro-magnet No. 1. Weight of its iron 15 lb.
Length of wire 23 feet.

Electric Current.	Electric Force.	Lifting Power.
°0·28 (Estimated.)	°6·5 (Estimated.)	2·75
0·64 (Estimated.)	14·6 (Estimated.)	10
0·91 (Estimated.)	21·0 (Estimated.)	23
1·34 (Estimated.)	30·8 (Estimated.)	45
2·85 (Estimated.)	65·5 (Estimated.)	238
3·83 (Estimated.)	88·0 (Estimated.)	540
4·3	99·3	670
5·7	132·5	890
8·6	198·7	1060
14·4	331·0	1400
21·6	497·0	1800
36·0	828·0	2030

On one occasion a weight of 2090 lb. was required to detach the armature, which is, I believe, the greatest weight any magnet has hitherto carried, and is certainly vastly superior to the performance of any other of its size.

The first two columns in the upper half of the table were reduced from the actual indications, as I had detected a certain want of insulation of the coil.

TABLE II.

Electro-magnet No. 2. Weight of its iron 1057 grains. Length of wire 7 feet.

Electric current.	Electric force.	Weight lifted.
$\overset{\circ}{0}$·51	$\overset{\circ}{3}$·57	20
1·53	10·7	38·5
6·1	42·7	49

TABLE III.

Electro-magnet No. 3. Weight of its iron 65·3 grains. Length of wire 1·58 foot.

Electric current.	Electric force.	Weight lifted.
$\overset{\circ}{0}$·42	$\overset{\circ}{0}$·66	5·5
1·0	1·58	9
2·0	3·16	11

With great care this small electro-magnet supported in one instance twelve pounds, or 1286 times its own weight.

No. 4, the weight of which was only half a grain, carried in one instance 1417 grains, or 2834 times its own weight.

It required much patience to work with an arrangement so minute as this last; and it is probable that I might ultimately have obtained a larger figure than the above, which, however, exhibits a power proportioned to its weight far greater than any on record, and is more than eleven times that of the celebrated steel magnet which belonged to Sir Isaac Newton.

It is well known that a steel magnet ought to have a much greater length than breadth or thickness; and Mr. Scoresby has found that when a large number of straight steel magnets are bundled together, the power of each when separated and examined is greatly deteriorated*. All this is easily understood, and finds its cause in the attempt of each part of the system to induce upon the other part a contrary magnetism to its own. Still there is no reason why the principle should in all cases be extended from the steel to the electro-magnet, since in the latter case a great and commanding inductive power is brought into play to sustain what the former has to support by its own unassisted retentive property. All the preceding experiments support this position; and the following table gives proof of the obvious and necessary general consequence, that *the maximum power of the electro-magnet is directly proportional to its least transverse sectional area.* The second column of Table IV. contains the least sectional area in square inches of the entire magnetic circuit. The maximum power in pounds avoirdupois is recorded in the third; and this, reduced to one inch square of sectional area, is given in the fourth column under the title of specific power.

TABLE IV.

		Least sectional area.	Maximum power.	Specific power.
My own electro-magnets .	No. 1..	10	2090	209
	No. 2..	0·196	49	250
	No. 3..	0·0436	12	275
	No. 4..	0·0012	0·202	162
Mr. J. C. Nesbit's. Length round the curve 3 feet. Diameter of iron core $2\frac{3}{4}$ inches. Sectional area 5·7 inches; do. of armature 4·5 inches. Weight of iron about 50 lb.		4·5	1428	317
Prof. Henry's. Length round the curve 20 inches. Section 2 inches square. Sharp edges rounded off. Weight 21 lb.		3·94	750	190
Mr. Sturgeon's original. Length round the curve about 1 foot. Diameter of the round bar half an inch		0·196	50	255

* Scoresby's 'Magnetical Investigations,' p. 37.

The above examples are, I think, sufficient to prove the rule I have advanced. No. 1 was probably not fully saturated; otherwise I have no doubt that its power per square inch would have approached 300. Also the specific power of No. 4 is small, because of the difficulty of making a good experiment with it. The electric force used on Mr. Nesbit's electro-magnet was exceedingly great, consisting of 19 of Daniell's two-feet cells with coils of 14 lengths of copper wire, each 70 feet long and $\frac{1}{14}$ inch thick. On the other hand, Professor Henry used only a pair of voltaic plates, with a much inferior conductor.

The mean of the specific powers of No. 2, No. 3, and Mr. Nesbit's may, I think, be fairly taken for the expression of the maximum magnetic force of iron under ordinary circumstances, which therefore is given by the formula $x=280a$, where a is the sectional area of the magnetic circuit in square inches, and x is the force of adhesion of both poles.

Since the element of length has no place in the above formula, and has in fact only a secondary influence, playing the part of a resistance which it requires a large additional electric force to overcome, it is obvious that in proportion to its reduction the attractions relatively to the weight of iron will increase: hence the large power relative to weight of my short electro-magnets.

The above corroborates what I have already said with regard to the proper construction of the electro-magnet for lifting weights, the condition being well illustrated by fig. 15. When A and B come into contact the oblique forces disappear, and the attraction is in the simple ratio of the number of saturated magnetic particles opposed to each other.

With respect to the magnetizing coils, I may observe that each particle of space through which a certain quantity of electricity is propagated appears to operate in moving the magnetism of the bar with a force proportionate to the inverse square of its distance from the surface of the iron, and that, when the tension of the magnetism is the same, the thickness of the iron on which that particle of conducting space acts has nothing to do with the whole effect. Now it may be shown that, such being the law, if the particle induce upon a large surface, the resulting magnetic induction will not vary very much with the distance, but be a very constant quantity for any distance which bears a very small ratio to the dimensions of that surface. Hence it is that a coil *within* a hollow piece of iron has no power to magnetize it*; for in that case its energy is directed in equal quantities towards opposite directions, the nearness of one surface counterbalancing the size of its opposite. Hence also in my electro-magnet No. 1, where the surface is large, the coil exercises nearly the full extent of its duty even in the places where it does not lie closely to the iron.

Where the magnetization is considerably below the point of saturation, the resistance to induction arising from the *length* of the magnet becomes a very sensible quantity, varying probably in the direct ratio of that element. Some idea of its effect may be formed from the following table, in which I have compared half the maximum powers of three of my electro-magnets with the electric forces which produce them; and by dividing the former by the latter, I have obtained a fourth column which, under the title power per degree, contains the lifting-power for half the maximum magnetization due to a unit of electric force.

TABLE V.

	Length of magnetic circuit in inches.	Electric force.	Half maximum power.	Power per degree of electric force.
		°	lb.	lb.
No. 1	7·46	200	1060	5·3
No. 2	3·7	4·5	25	5·5
No. 3	1·05	0·66	5·5	9·2

It is well known that, after the galvanic circuit is broken, the armature of an electro-magnet is still retained with very considerable force. I was desirous of trying the capability of No. 1 in this respect. The following table contains in its first column the degrees of electric force which were cut off,

* 'Scientific Memoirs,' part 5, p. 14.

the second gives the lifting-powers due to those forces, and the third gives the weight required to detach the armature after the current was broken.

TABLE VI.

Electric force cut off.	Lifting-power due to electric force cut off.	Lifting-power remaining after the electric force was removed.
°	lb.	lb.
88	540	33
29	40	16
14·5	10	10

There was considerable difficulty in making a good experiment with so powerful an electro-magnet as No. 1 when small forces were in question. Nevertheless the above results afford good evidence that nearly all the lifting-power due to feeble currents is retained after those currents are cut off.

When the current is not entirely cut off, but is merely reduced in strength by the interposition of a thin wire, a surprising quantity of lifting-power can be retained by the small electric force left. Having subjected No. 1 to 90° of electric force, a quantity adequate to bring up its power to 560 lb., I reduced the current to a lower intensity, and then found the weight requisite to detach the armature. The following table gives the results of several such experiments.

TABLE VII.

Electric force remaining after 90° was cut off.	Lifting-power due to this force.	Weight required to detach armature.
°	lb.	lb.
31	45	294
21	23	210
14·5	10	112
6·2	2·6	63
4·1	1·1	56

A voltaic cell, the size of a common sewing-thimble, was quite sufficient to produce 31° of electric force in No. 1, and consequently to sustain a magnetic power of about 300 lb.; and it is easy to perceive that, by increasing the size of the electro-magnet and the quantity of conducting-wire, this minute source could support a magnetic virtue of indefinite amount.

Note on Voltaic Batteries.

HAVING had occasion, about a year ago, to construct a battery of great intensity, it became an object to devise an arrangement which should be convenient to use and easily refitted. After trying two or three systems, I succeeded in producing one which answered very well; but as I felt that long experience could be the only strict test of its value, I have hitherto refrained from presenting it to public notice. Now, however, that I have worked with it during nine or ten months, and have found it to possess every quality that could be desired, I hope by describing it to give the same facilities to others which I enjoy myself.

Fig. 18 represents three elements of my battery. A B is

Fig. 18. Scale $\frac{1}{3}$.

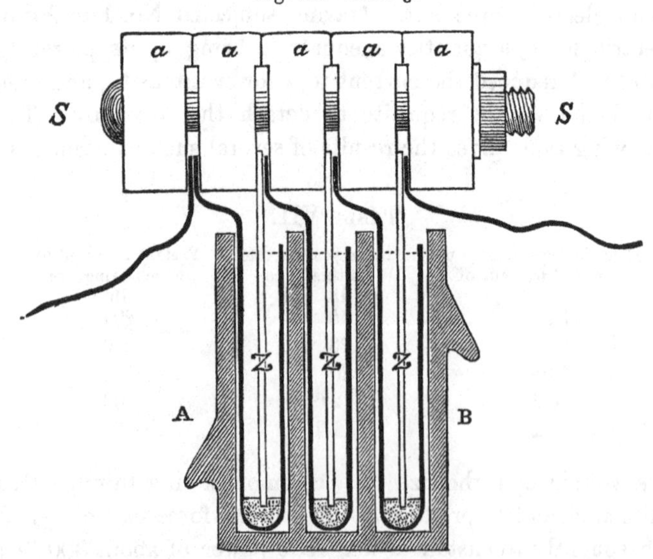

the common divided Wollaston trough with the front side

removed in order to show the interior. The black lines within the cells are rectangular pieces of strong sheet copper, bent on a gauge to the shape seen in the figure. Within these, *z, z, z* represent plates of sheet zinc, amalgamated in those parts which are in contact with the dilute sulphuric acid, and fixed in their places by pieces of hard wood furnished with grooves, and extending the whole breadth of the zinc. Lastly, *a, a, a, a, a* are pieces of wood with holes in their centres to admit the screw bolt *s s*, which secures the whole.

When the battery is worn out, empty the trough and replace it therein; then unscrew the bolt and remove it with the pieces of wood; change the old zinc plates for new ones, taking care in the mean time to see that those parts of the coppers which touch the zincs are bright; then replace the pieces of wood and screw them tightly together.

Mr. Smee's battery may be fitted up on the above plan. I prefer, however, for continued use sheet iron before either copper or platinized silver. In using sheet iron it is well to tin that part which is to touch the zinc, in order to ensure perfect contact.

I have lately constructed a large battery on Mr. Sturgeon's plan; and from my experience with it I am convinced that it presents a very superior arrangement of voltaic elements. It consists of eleven cast-iron cells, each 1 foot square and $1\frac{1}{2}$ inch in interior diameter. With eight of the pairs arranged in a series of four, I can raise to a full red heat 18 inches of copper wire $\frac{1}{10}$ inch thick.

Broom Hill, near Manchester,
August 21st, 1840.

On Electro-magnetic Forces. *By* J. P. Joule*.

['Annals of Electricity,' vol. v. p. 170.]

In my last paper I have described a method of constructing the electro-magnet which gave great lifting-power. The fresh experiments will, I hope, be deemed confirmatory of the principles before advanced.

A piece of "stub" iron was, as in the manufacture of gun-barrels, formed in a spiral, and welded on a mandril into the shape of a thick tube, the process rendering the iron very compact and sound throughout. This, and another piece of iron intended for the armature were planed, turned, and fitted with eyehole screws in the manner I have already described.

In fig. 19, C represents the electro-magnet, D the armature,

Fig. 19. Scale $\frac{1}{8}$.

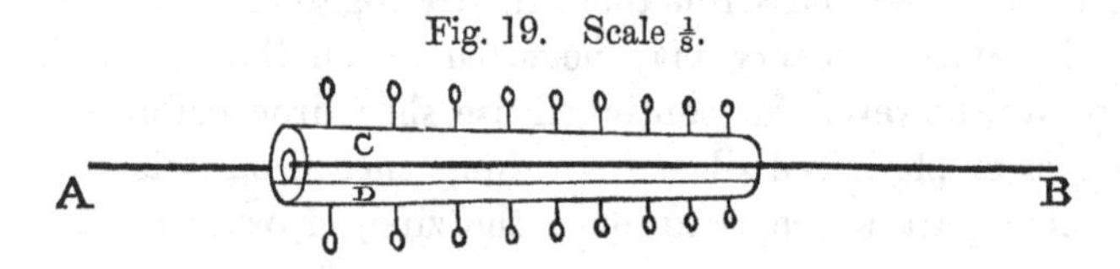

and A B a conductor of copper rod or wire passing along one side, returning by the axis, and then away by the other side, so as to go about the iron once only, and in a shape somewhat similar to that of the letter S. The length of the cylinder formed when the armature is in contact is 2 feet; its external diameter is 1·42 inch, and its internal 0·5 inch. The weight of the iron of the magnet along with the screws is 6 lb. 11 oz., that of the armature 3 lb. 7 oz. The least sectional area of its magnetic circuit is $10\frac{1}{4}$ square inches: I call it No. 5, to distinguish it from those already described.

A copper rod, $\frac{3}{8}$ of an inch in diameter, covered with tape, was bent about the electro-magnet as just described. It and its armature respectively were then secured by means of cords passing through the eyeholes to the lever (fig. 17) used in my former experiments. Eight large cast-iron cells, each of which presented an effective surface of 2 square feet, were

* The experiments were made at Salford Brewery.

arranged as a single pair, and gave a lifting-power of 1350 lb.

Thinking that a bundle of wire was probably a better conductor than a rod of the same length and weight, I removed the copper rod, and substituted for it a bundle of 60 wires, each $\frac{1}{25}$ of an inch thick. With this it was found that sixteen of the cast-iron cells arranged in a series of two produced a lifting-power of 1856 lb., or 183 times the weight of iron employed in the construction of both the electro-magnet and its armature.

By dividing the power thus obtained by the least sectional area, the weight lifted per inch is 181 lb., or about two thirds of that which a comparison with other electro-magnets would lead us to expect. I do not think that the electric force was sufficient to bring the magnetism close to the point of saturation; and, moreover, it was difficult to secure the necessary condition of equal tension along the whole length of the iron.

Suspecting that the greatest power of the large electro-magnet No. 1 had not been reached in my last experiments, I refitted it, using a coil consisting of 21 copper wires, each 23 feet long and $\frac{1}{25}$ of an inch in diameter, bound together by cotton tape. Sixteen of the large cast-iron cells were arranged in a series of four; and when connected by sufficiently good conductors to the electro-magnet, a weight of 2775 lb. was found requisite to detach the armature.

Now, by the formula in my last paper, $280 \times 10\frac{1}{4} = 2870$ is the estimated value for the power of this electro-magnet on its approach to saturation. That this was very nearly attained is manifest from the fact that the electric force used in the last experiment was four times as great as that which was competent to give a lifting-power of 2128 lb.

Although the battery I have used for obtaining maximum results is powerful, a very good effect may be obtained by means of a very small voltaic arrangement. For instance, No. 1 can carry eight hundredweight when the current generated by a single pair of four-inch plates of iron and amalgamated zinc is transmitted through its coils; and with a pair of platinized silver and zinc plates exposing only two square inches of surface, the attraction is such as to render it almost impossible by the hand to slide the armature.

Broom Hill, near Manchester,
Nov. 23rd, 1840.

Description of a new Electro-magnet. By J. P. Joule. (In a letter to Mr. Sturgeon *.)

['Annals of Electricity,' vol. vi. p. 431.]

My dear Sir,

In that part of my researches on Electro-magnetic Forces which was published in the 'Annals' for September last, I showed that the maximum lifting-power of the electro-magnet is proportional to the least transverse sectional area of magnetic circuit; and at the same time I pointed out the method whereby a very great magnetic attraction could be produced between masses of iron of inconsiderable magnitude.

To illustrate my views I made several electro-magnets, two of which have been a long time on exhibition at the "Victoria Gallery."

Stimulated by my success, some gentlemen of Manchester have constructed electro-magnets of a variety of forms, embodying the principle of a large sectional area. They have in consequence carried very heavy weights; Mr. Roberts's, in particular, has sustained the greatest load on record. Mr. Radford's is also very powerful; and his arrangement † of a spiral groove on the face of the magnet to admit the coil involves a new and important principle of magnetic action.

The following is a description of a new electro-magnet which I have recently constructed. Fig. 20 gives its appearance in perspective. B, B are two rings of brass, each 1 foot in exterior diameter, 2 inches broad, and 1 inch thick; to

* The experiments were made at the Salford Brewery.

† 'Annals of Electricity,' vol. vi. p. 231.

each of these flat pieces of iron are fixed by means of the bolt-headed screws *s, s* &c.; twenty-four of them, *m, m* (figs. 20 and 21), have rectangular grooves, and are fixed to the upper

Fig. 20. Scale $\frac{1}{4}$.

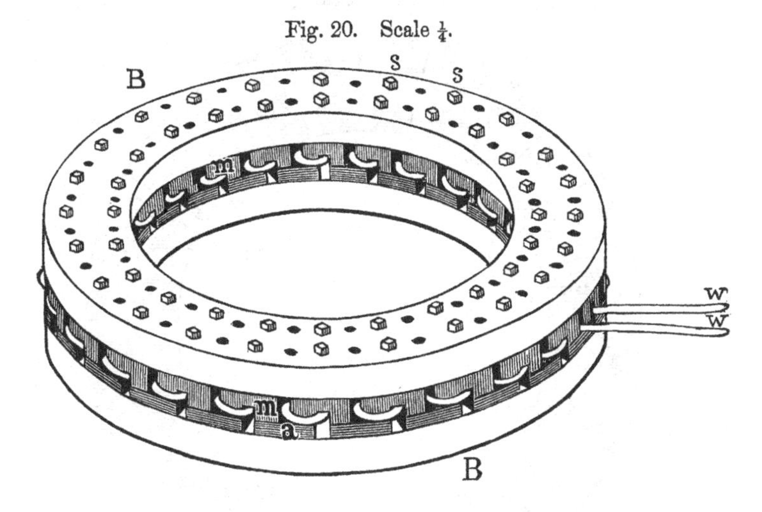

Fig. 21. Scale $\frac{1}{4}$.

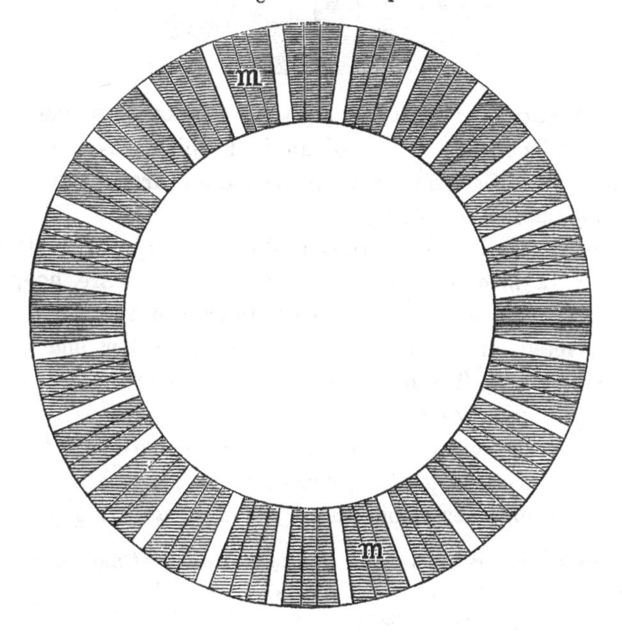

ring; twenty-four, *a, a* &c. (figs. 20 and 22), are plain, and are fastened to the lower ring.

Fig. 22. Scale $\frac{1}{4}$.

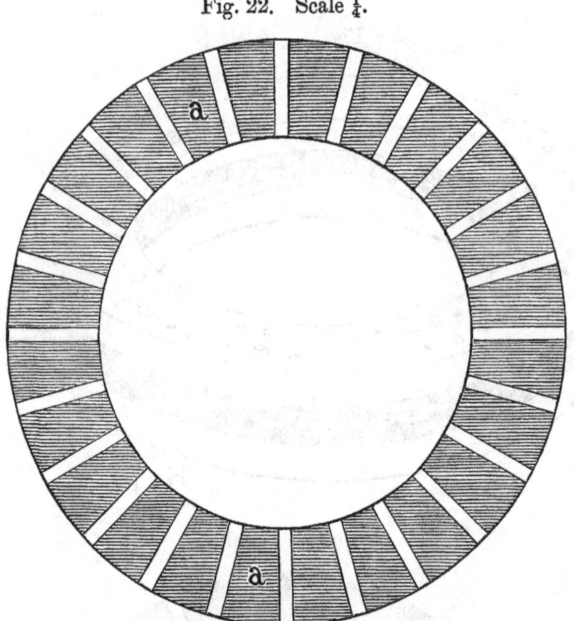

A bundle, *w, w* (fig. 20), consisting of sixteen copper wires, each 16 feet long and $\frac{1}{20}$ of an inch thick, covered with a double fold of cotton tape, was bent to and fro in the grooves of *m m*, as seen in the figure.

Fig. 23 shows the method I adopted to give the electro-magnet a firm and equable suspension: *a a* are hoops of wrought iron, to each of which four bars of the same metal are riveted, and welded together at the other end into a massive hook. The hoops are bound down to the brass rings by copper wires *c c* &c.

On connecting the coil of the above electro-magnet with a battery*, consisting of sixteen of your large cast-iron cells arranged in a series of four, I obtained a lifting-power of

* Each cell was 1 foot square and $1\frac{1}{2}$ inch wide. Arranged as above, the battery was able to give an electric discharge between the poles as much as $\frac{1}{8}$ of an inch in length.

2710 before the armature was detached. Theoretically, the maximum power would be 0·635 (sectional area) $\times 24 \times 280 = 4267$ lb.

Fig. 23. Scale $\frac{1}{8}$.

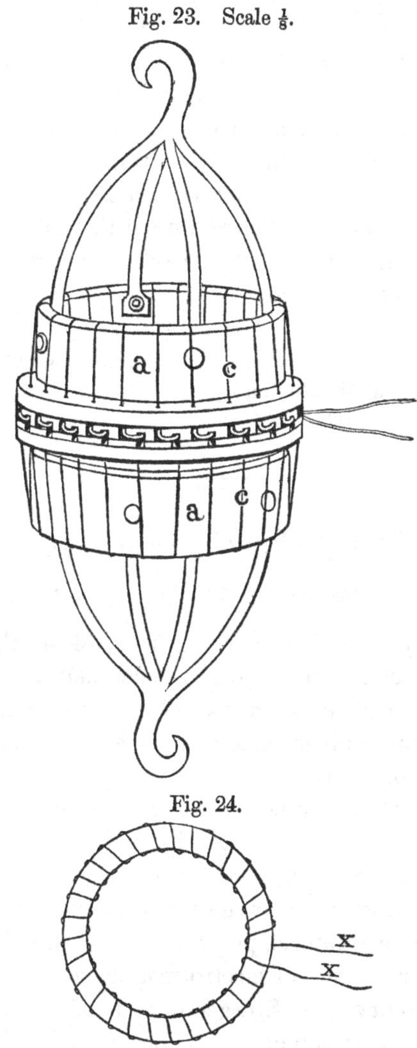

Fig. 24.

The weight of the pieces of grooved iron is 7·025 lb., that of the plain pieces constituting the armature 4·55 lb. The power per lb. of magnetized iron was therefore $\frac{2710}{11·575} = 234$ lb.

When the apparatus is in the position which is represented by figs. 20 and 23, it is evident that the zigzag ring of iron is magnetized by the conducting wire in the same way as the plain ring (fig. 24) would be by the passage of electricity along the wire *x x* which is coiled upon it; wherever such a ring is cut, the display of maximum lifting-power is proportional to the least transverse sectional area of the entire magnetic circuit. In the above position it is impossible to consider the instrument other than a single electro-magnet; but when the armature is turned until the plain pieces affixed to it cover the grooves of the other pieces, a compound electro-magnet is formed.

I remain, dear Sir,
Yours respectfully,
J. P. Joule.

Broom Hill, near Manchester,
April 30, 1841.

On a new Class of Magnetic Forces. By J. P. Joule *.

['Annals of Electricity,' vol. viii. p. 219.]

As it is my intention to bring forward in this paper an electro-magnetic principle in reference partly to the motion of machines, I hope that a few preliminary observations with respect to the ordinary electro-magnetic engines will not be deemed out of place.

The great attractive powers of the electro-magnet, and the extreme rapidity with which its polarity is reversed by changing the direction of the current, very readily present to the reflecting mind an idea that its power may be made available for mechanical purposes. Accordingly, as soon as the general principles of electro-magnetism were understood, Professor Henry, Mr. Sturgeon, and, after them, a great number of ingenious individuals constructed various arrangements of machinery to be set in motion by magnetic attraction and repulsion.

* Lecture at the Victoria Gallery, Manchester, February 16, 1841.

At that period the expectations that electro-magnetism would ultimately supersede steam, as a motive force, were very sanguine. There seemed to be nothing to prevent an enormous velocity of rotation, and consequently an enormous power, except the resistance of the air, which it was easy to remove, the resistance of iron to the induction of magnetism, which I had succeeded in overcoming to a great extent by annealing the iron bars very well, and the inertia of the electric fluid.

We are indebted to Professor Jacobi for the exposition of the principal obstacle to the perfection of the electro-magnetic engine. He has shown that the electric action produced by the motion of the bars operates against the battery current, and in this way reduces the magnetism of the bars, until, at a certain velocity, the forces of attraction become equivalent to the load on the axle, and the motion in consequence ceases to be accelerated. Jacobi had not, however, given precise details concerning the duty of his apparatus; nor had he then determined the laws of the engine. I was therefore induced to construct an engine adapted for experiment, and with it found:—

1st. That the counter electric action, or, in other words, the magneto-electric resistance to the battery-current, is proportional to the velocity of rotation and the magnetism of the bars.

2nd. That the economical duty at a given velocity of rotation is a constant quantity, whatever the number of similar pairs may be in series, provided the resistance of the battery is kept constant.

3rd. That at small velocities great advantage is obtained by reducing as far as possible the resistance of the battery, and by arranging the coils so as to facilitate as far as possible the transmission of the current.

4th. That the economical duty at a given velocity, and for a given resistance of the battery, is proportional to the mean of the intensities of the several pairs of the battery.

With my apparatus every pound of zinc consumed in a Grove's battery produced a mechanical force (friction included) equal to raise a weight of 331,400 lb. to the height of one foot, when the revolving magnets were moving at the velocity of 8 feet per second.

Now the duty of the best Cornish steam-engine is about 1,500,000 lb. raised to the height of 1 foot by the combustion of a lb. of coal, which is nearly five times the extreme duty that I was able to obtain from my electro-magnetic engine by the consumption of a lb. of zinc. This comparison is so very unfavourable that I confess I almost despair of the success of electro-magnetic attractions as an economical source of power; for although my machine is by no means perfect, I do not see how the arrangement of its parts could be improved so far as to make the duty per lb. of zinc superior to the duty of the best steam-engines per lb. of coal. And even if this were attained, the expense of the zinc and exciting fluids of the battery is so great, when compared with the price of coal, as to prevent the ordinary electro-magnetic engine from being useful for any but very peculiar purposes.

New Class of Magnetic Forces.

A few weeks ago an ingenious gentleman of this town suggested to me a novel form of electro-magnetic engine. He was of opinion that a bar of iron was increased in length by receiving the magnetic influence, and that, although the increment was perhaps very small, it still might be found valuable as a source of power on account of the great force with which it would operate. At that gentleman's request I undertook experiments to ascertain whether his opinion was correct, and if so, to ascertain whether the new source of power could be advantageously employed for the movement of machinery.

After some preliminary trials, I adopted the following method of experiment. A length of 30 feet of copper wire $\frac{1}{20}$ of an inch thick, covered with cotton thread, was formed into a coil 22 inches long and $\frac{1}{3}$ of an inch in interior diameter. This coil was secured in a perpendicular position, and the rod of iron, of which I wished to ascertain the increment, was suspended in its axis so as to receive the magnetic influence whenever a current of electricity was passed through the coil. Lastly, the upper extremity of the rod was fixed; and the lower extremity was attached to a system of levers which multiplied its motion three thousand times.

A bar of rectangular iron wire, 2 feet long, $\frac{1}{4}$ of an inch broad, and $\frac{1}{8}$ of an inch thick, caused the index of the multiplying apparatus to spring from its position, and vibrate about a point $\frac{1}{10}$ of an inch in advance, when the coil was made to complete the circuit of a battery capable of magnetizing the iron to saturation, or nearly so. After a short interval of time the index ceased to vibrate, and began to advance very gradually in consequence of the expansion of the bar by the heat which was radiated from the coil. On breaking the circuit, the index immediately began to vibrate about a point exactly $\frac{1}{10}$ of an inch below that to which it had attained.

By dividing the advance of the index by the power of the levers we obtain $\frac{1}{30000}$ of an inch as the increment of the bar, which may be otherwise stated as $\frac{1}{720000}$ of its whole length*.

Similar results were obtained by the use of an iron wire 2 feet long and $\frac{1}{12}$ of an inch in diameter. Five cells of a nitric-acid battery produced an increment of the thirty-thousandth part of an inch; and when only one cell was employed I had an increment very slightly less, viz. the thirtythree-thousandth of an inch.

The increment did not appear to depend upon the thickness of the bar; for an electro-magnet constructed of an iron bar, 3 feet long and 1 inch square, was found to expand under the magnetic influence to nearly the same extent, compared with its length, as the wires did in the previous experiments.

I made some experiments in order to ascertain the law of the increment. Their results proved it to be very nearly proportional to the length of the bar and the intensity of its induced magnetism.

Trial was made whether any effect could be produced by using a copper wire, which was 2 feet long and about $\frac{1}{10}$ of

* The movement of the index was rendered visible to the whole of the audience.

an inch thick; but I need scarcely observe that the attempt was not attended with the slightest success.

A good method of observing the above phenomena, is to examine one end of an electro-magnet with a microscope while the other end is fixed. The increment is then observed to take place suddenly, as if it had been occasioned by a blow struck at the other extremity. The expansion, though very minute, is indeed so very rapid that it may be felt by the touch; and if the electro-magnet be placed perpendicularly on a hard elastic body, such as glass, the ear can readily detect the fact that it makes a slight jump each time that contact is made with the battery.

When one end of the electro-magnet is applied to the ear, a distinct musical note is heard every time that contact with the battery is made or broken—another proof of the suddenness with which the particles of iron are disturbed.

With regard to the application of the new force to the movement of machinery, I have nothing favourable to advance. An easy calculation on the basis of the modulus of elasticity of iron will show that an electro-magnet, consisting of a bar of iron 1 inch square and 3 feet long, would exert a force of about ten pounds through the length of the twenty thousandth part of an inch every time that contact with the voltaic battery is made or broken, provided the transmitted current is capable of saturating the iron. If, therefore, contact be made and broken a hundred times per second, for an hour together, we shall only have fifteen pounds raised to the height of a foot. The force would therefore be far too minute for the movement of machinery; and the duty per lb. of zinc would be vastly less than that of the ordinary electro-magnetic engine.

In examining the bearing of the new property of the electro-magnet on magnetic theory, I would observe, that in the hypothesis of Ampère the phenomena of magnetism are referred to the attraction and repulsion of currents of electricity moving in the same or contrary directions. Fig. 25 represents the section of six particles of a magnetized iron bar, according to a modification of that philosopher's theory.

The black circles represent atoms of iron; and the shadowed ovals around them represent atmospheres of electricity moving in planes at right angles to the axis of the magnet.

Fig. 25. Fig. 26. Fig. 27.

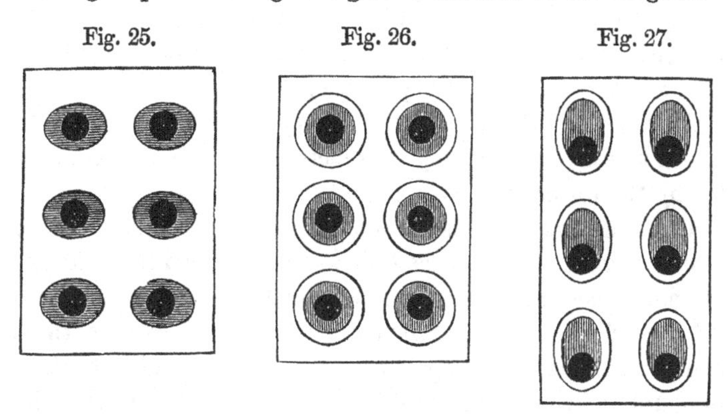

This theory affords a good explanation of most cases of magnetic attraction. But the physical conditions which are demanded by it are impossible, and contrary to the analogy of nature; for it supposes motion, or at least an active force, to be continued against antagonist forces for an indefinite length of time without loss, in order to explain the phenomena exhibited by a hard steel magnet.

The way by which it may be made to account for the fact that iron, after receiving a certain quantity of magnetism, is incapacitated from receiving a further supply, or becomes saturated, is to suppose that the electricity which revolves round each atom of iron has a centrifugal tendency. The velocity of the electric currents around the atoms of iron will tend to be proportional to the influence which urges them; and if the electricity be not endowed with centrifugal force, it is difficult to say why it should refuse to travel beyond a certain velocity; and thus the phenomenon of saturation would be unexplained. If, however, the momentum of electricity, and its consequent centrifugal tendency when rotated, be admitted, the currents will be prevented from going beyond a certain velocity by their interference with one another.

The hypothesis, however, is less successful in accounting

for the increase of length which I have noticed in a bar of iron when under the magnetic influence; for, as the electricity is supposed to revolve in a plane at right angles to the axis of the bar, the divergence of the fluid from each atom of iron by centrifugal force would have the effect of *shortening** the bar, which is opposed to experience.

Turning now to a modified theory of Æpinus, it will be found to explain tolerably well facts which are unaccounted for by the former theory.

Let the black circles in fig. 26 represent in section six atoms of iron, the shadowed circles about them atmospheres of magnetism, and the rings beyond these still rarer atmospheres of electricity. Further, let the space between these compound atoms be supposed to be filled with calorific ether in a state of vibration, or, otherwise, to be occupied with the oscillations of the atoms themselves. Such a state of things may probably give an idea of part of a bar of unmagnetized steel or iron. Now, if an inductive influence be applied to the atoms, the magnetism may be supposed to accumulate on one side of the atoms of iron, as represented by fig. 27, and the bar rendered magnetic.

Such a theory seems to me to afford a natural and complete expression of facts. It supposes nothing which we cannot readily comprehend, except the existence and elementary properties of matter, which are necessarily assumed by every theory, and which the Great Creator has placed utterly beyond the grasp of the human understanding.

When all the magnetism of each atom of iron is accumulated on one side, the bar may be said to be saturated.

It is obvious that when the magnetism is thus being accumulated on one side of the particles, the bar will increase in length in the direction of its polarity, and decrease in a direction at right angles to the length. The former fact I have proved by the foregoing experiments; and there can be little doubt that a very delicate apparatus would exhibit the diminution of the thickness of a bar of iron in consequence of the communication of magnetic virtue.

* Further on it will be seen that a bar under strong tension is shortened by magnetization.—*Note by J. P. J. in* 1882.

The hypothesis will also account for the fact that at a certain degree of heat all the magnetic power of iron is destroyed. I have already observed that the space left between the magnetic particles in fig. 27 represents the room taken up by their vibration. This vibration is called heat, and will of course increase in violence and extent with the increase of the temperature of the bar. Now it is natural to suppose that the atoms of iron have far greater weight and inertia than the atmospheres of magnetism and electricity which surround them—therefore that these atmospheres will be in a state of vibration, while the atoms of iron remain in a state of comparative quiescence; and when the vibration has reached a certain intensity, the inductive influence will not be able to arrange the magnetism in any definite direction with regard to the atoms of iron.

The retentive power may be explained by supposing the magnetism to adhere to the atoms of iron to a certain extent. And if we make another supposition, viz. that an atom of iron, on combining with an atom of carbon, loses its attraction for magnetism in the side which is next the carbon, the superior retentive power of steel, in comparison with that of iron, will be explained.

---

*On Voltaic Apparatus**.

[Proceedings of the London Electrical Society.]

Dear Sir,

Having recently ascertained some circumstances relative to voltaic apparatus, which have, I think, considerable value in a practical point of view, I desire to communicate my experience to you, and, through you, to the Electrical Society.

If a plate of copper be placed in dilute sulphuric acid, it will be gradually dissolved, and, after a certain length of time, the liquid will be found to have acquired a blue tinge owing to the solution of oxide of copper. If now a plate of

* In a letter to C. V. Walker, Sec. Lond. Electr. Soc.

amalgamated zinc be placed in voltaic association with the copper, a powerful current will pass along the connecting wire, equal in intensity* to that which would have been produced by a Daniell's cell. But in the meantime a part of the copper in solution will have precipitated itself on the amalgamated zinc, causing a local action, which will speedily destroy the plate.

But if, instead of allowing the copper to remain alone in the acid before the battery is completed, it be placed in the dilute sulphuric acid along with a piece of amalgamated zinc, connected with it by means of a copper wire, the pair thus formed will not, after the immediate effects of immersion are passed away, be as intense as was the former one; but then it will work without local action.

The due consideration of these facts will enable us to understand why local action is so common an annoyance to those who work with acid batteries in which copper is employed as the negative element. It will also point out the following means for remedying that defect:—

1st. Every part of the copper surface which is immersed in the liquid should be, so to speak, in sight of the amalgamated zinc. Any part *not* so situated is not actively engaged in propagating the current, and is consequently liable to enter into solution, and then to be precipitated on the zinc. If we use the Wollaston's arrangement, we should not (as is common) bend the copper about the zinc, but the zinc about the copper.

2nd. When the copper battery is immersed in its trough, it should be set to work *immediately*, and the current should be allowed to pass with as few intermissions as possible. If we wish to make such a change in our apparatus as will occupy any length of time, we should either take the battery out of its trough, or else we should connect its poles by a wire, the conducting-power of which is sufficient to occasion a slight effervescence on the copper surfaces.

* It will be proper to observe that, throughout this paper, I mean by the word "intensity" electromotive force. It is always proportional to the *quantity* of current multiplied by the resistance of the whole circuit.

By adopting these precautions, I find that I am able to use a copper battery, charged with a strong solution of sulphuric acid, without being annoyed by local action.

I now endeavoured to ascertain the best method of giving the copper a surface adapted for the transmission of a large quantity of electricity. I tried the platinizing process first, and found that, after a plate of copper had been immersed for an instant in an extremely dilute solution of platinum, its surface was closely assimilated in electric character to that of a plate of platinized silver—the quantity of platinum required for the purpose being far too small to be regarded in an economical point of view. Then I tried the plans which had been suggested by Professor Poggendorff and yourself*, particularly that of heating the copper to redness in air—a method which has, I think, some advantage over the other processes.

By this means, and by the precautions against local action of which I have previously spoken, I was able to construct a copper-zinc acid battery, not much inferior in any respect to that of Mr. Smee, while at the same time the stability† of the copper plates gave me the advantage of a superior mechanical arrangement.

Fig. 28.

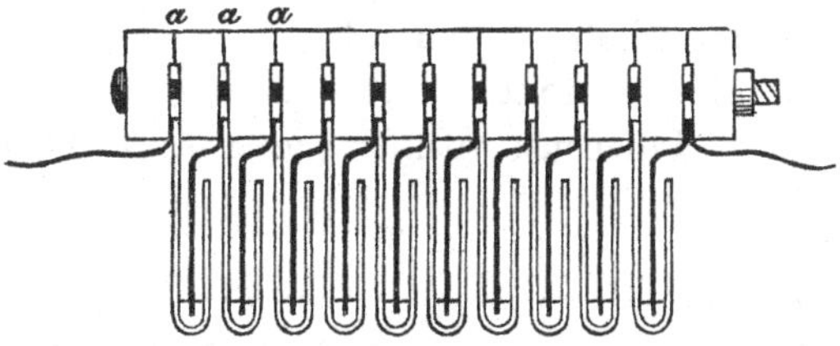

The arrangement to which I am alluding is so convenient in practice that, although it has, with the order of metals reversed, already received publication in the 'Annals of

* Proc. Electr. Soc. part ii. p. 92; also part i. p. 30.

† I have tried the platinized plated copper suggested by Mr. Smee. Its stability renders it much better adapted for the mechanical arrangement of a battery than the platinized silver foil.

Electricity'*, I hope to be permitted to describe it here, connected as it is with the general object of this communication.

Fig. 28 (p. 55) represents my battery. It is adapted for the common divided trough, and is constructed as follows:—I take ten pieces of sheet zinc, each of which is 10 inches long, $3\frac{1}{2}$ inches broad, and about $\frac{1}{30}$ of an inch thick. These I bend over a gauge, into the shape which, for the sake of clear illustration, I have delineated in fig. 29. I amalgamate those parts of them which will have to be immersed in the acid, polish the top of each with glass-paper, and place in the bent part of each a piece of semicircular wood, furnished with a longitudinal saw-cut. Having thus prepared the zincs, I place them in the cells of an empty trough. I then take ten plates of thick sheet copper, bend them into the shape represented by fig. 30, prepare their surfaces by red heat, polish that part which is to come into contact with the zinc, and fix them within the zincs by inserting their bottom edges in the saw-cuts of which I have spoken. Lastly, the pieces of wood, *a a a* &c., are put in their places, a screw-bolt is passed through the whole series by means of a hole in the centre of each, and the whole is secured by screwing on the nut.

Fig. 29.

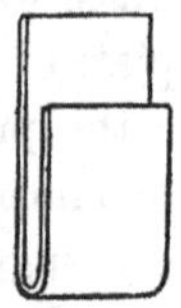

Fig. 30.

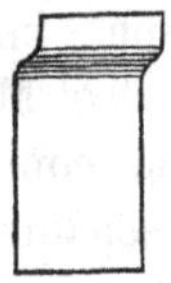

When my object is to construct a battery of a large number of small plates, I dispense with the pieces of wood and screw-

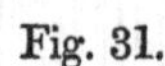

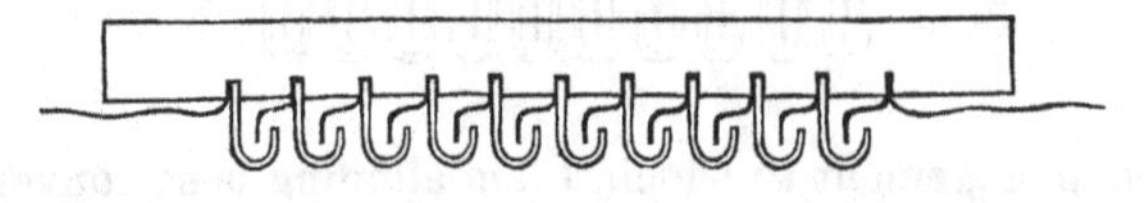

Fig. 31.

bolt, and simply insert the plates in the transverse saw-cuts of a piece of wood. This battery is represented by fig. 31.

With a battery of ten pairs, similar to that represented by fig. 28, furnished with coppers which had been prepared by

* Vol. v. p. 197.

heat, and immersed in a dilute solution of sulphuric acid, I have raised 10 inches of platinum wire $\frac{1}{30}$ of an inch thick, to a white heat. The great intensity, however, which produced such striking calorific effects passed off in a minute or two; gas began to arise from the copper; and the intensity became constant, and about one half of what it was at first. The inferior *constant* intensity was the principal remaining defect of the copper-zinc and acid battery; I therefore endeavoured to devise means for improving its electromotive force. In this inquiry I ascertained the following facts:—

1st. That the intensities of the arrangements of Grove and Daniell are very nearly in the ratio of five to three. My experiments on this point confirm those of Professor Jacobi, who finds these intensities to be in the ratio of 22515 to 13552*.

2nd. That Professor Daniell's battery has the same intensity, whether the fluid in contact with the copper holds in solution the *nitrate* or the *sulphate* of copper.

3rd. That copper-zinc and platinum-zinc pairs have respectively the same intensities as the arrangements of Daniell and Grove, if they are immersed in solutions of nitric acid sufficiently strong to prevent the evolution of hydrogen gas from the negative elements. And

4th. That the intensity of the copper-zinc battery, at the instant of its immersion in dilute sulphuric acid, is equal to that of a Daniell's cell.

With the view of making a practical use of the fourth fact, I caused a circular plate of copper, ten inches in diameter, and dipping to the depth of nearly three inches in dilute sulphuric acid, to be fixed on a horizontal axis, connexion being maintained, through the galvanometer, between it and a plate of amalgamated zinc. On revolving the disk, each part of its circumference was alternately exposed to the atmosphere and immersed in the acid. The electric current generated in these circumstances was equal in intensity to that produced by a Daniell's cell, when it had to traverse 120 yards of copper wire of $\frac{1}{40}$ of an inch diameter. But when a wire of greater

* *Vide* Proc. Electr. Soc. part i. p. 40.

conducting-power was used, the intensity, though considerably greater than it would have been without the revolution of the plate, was not equal to that standard, the quantity of oxygen absorbed from the atmosphere not being equivalent to the quantity of current produced in this latter case. By increasing the size of the disk, I have no doubt that the intensity would be raised as high as that of a Daniell's cell, even when the elements are connected by a very short and thick wire; but in that case the cumbrousness of the apparatus would, I apprehend, preclude it from being practically useful*.

I have also attempted to make a practical use of the third fact. I was aware that you had condemned the use of nitric acid in the charge of the acid battery; but, remembering that it had been used by Dr. Faraday in a very dilute state without producing much local action†, I did not like to relinquish, without another trial, the great intensity which nitric acid gives to the acid battery.

I will not detain you by describing the extensive series of experiments which I have made on this subject. I will only observe that I could work with an acid composed of 16 measures of water, 3 measures of strong oil of vitriol, and 1 measure of nitric acid sp. gr. 1·3, without being annoyed by *much* local action, the intensity, even when the circuit was completed by a short wire, being equal to that of a Daniell's cell, and the *quantity* of electricity transmitted much greater than in a comparative Daniell's cell. Still I was obliged ultimately to revert to your opinion; for the zinc plates were generally acted upon violently after they had been used a few times.

There is also an objection against the use of nitric acid, which has some weight in an economical point of view; for when it is used, there is a portion of nitric acid decomposed by nascent hydrogen from the negative plate, besides the usual equivalent of nitric or sulphuric acid.

I have no further observations to make on batteries without diaphragms, but am not willing to conclude without testifying

* See further on, where the effects of immersion are studied.

† Experimental Researches (1139).

my sense of the value of your method of constructing the acid battery *with* diaphragms for long-continued and constant action, such as is required for electrotype. I have myself used your process, and have witnessed with pleasure and surprise the rapidity with which the copper spread itself over the black lead which I had rubbed on the inner surface of a jar more than two feet deep and three inches in diameter. You state that dilute sulphuric acid alone produces a current of sufficient intensity for electrotype: for other purposes, however, it is desirable to use the cupreous solution mingled with a small quantity of free sulphuric acid, in contact with the copper of a diaphragm battery.

I am, yours truly,
JAMES P. JOULE.

Broom Hill, near Manchester,
March 12th, 1842.

On the Production of Heat by Voltaic Electricity. By J. P. JOULE*.

['Proceedings of the Royal Society,' December 17, 1840.]

THE inquiries of the author are directed to the investigation of the cause of the different degrees of facility with which various kinds of metal, of different sizes, are heated by the passage of voltaic electricity. The apparatus he employed for this purpose consisted of a coil of the wire, which was to be subjected to trial, placed in a jar of water, of which the change of temperature was measured by a very sensible thermometer immersed in it; and of a galvanometer, to indicate the quantity of electricity sent through the wire, which was estimated by the quantity of water decomposed by that electricity. The conclusion he draws from the results of his experiments is, that the calorific effects of equal quantities of transmitted electricity are proportional to the resistance opposed to its passage, whatever may be the length, thickness,

* The experiments were made at Broom Hill, near Manchester.

shape, or kind of metal which closes the circuit; and also that, *cæteris paribus*, these effects are in the duplicate ratio of the quantities of transmitted electricity, and, consequently, also in the duplicate ratio of the velocity of transmission. He also infers from his researches that the heat produced by the combustion of zinc in oxygen is likewise the consequence of resistance to electric conduction.

On the Heat evolved by Metallic Conductors of Electricity, and in the Cells of a Battery during Electrolysis. *By* JAMES PRESCOTT JOULE, Esq.*

['Philosophical Magazine,' vol. xix. p. 260.]

1. THERE are few facts in science more interesting than those which establish a connexion between heat and electricity. Their value, indeed, cannot be estimated rightly, until we obtain a complete knowledge of the grand agents upon which they shed so much light. I hope, therefore, that the results of my careful investigation on the heat produced by voltaic action are of sufficient interest to justify me in laying them before the Royal Society.

CHAP. I.—*Heat evolved by Metallic Conductors.*

2. It is well known that the facility with which a metallic wire is heated by the voltaic current is in inverse proportion to its conducting-power; and it is generally believed that this proportion is exact; nevertheless I wished to ascertain the fact for my own satisfaction, and especially as it was of the utmost importance to know whether resistance to conduction is the *sole* cause of the heating effects. The detail, therefore, of some experiments confirmatory of the law, in addition to those already recorded in the pages of science, will not, I hope, be deemed superfluous.

* The experiments were made at Broom Hill, Pendlebury, near Manchester.

3. It was absolutely essential to work with a *galvanometer* the indications of which could be depended upon as marking definite quantities of electricity. I bent a rod of copper into the shape of a rectangle (A B, fig. 32), 12 inches long and

Fig. 32.

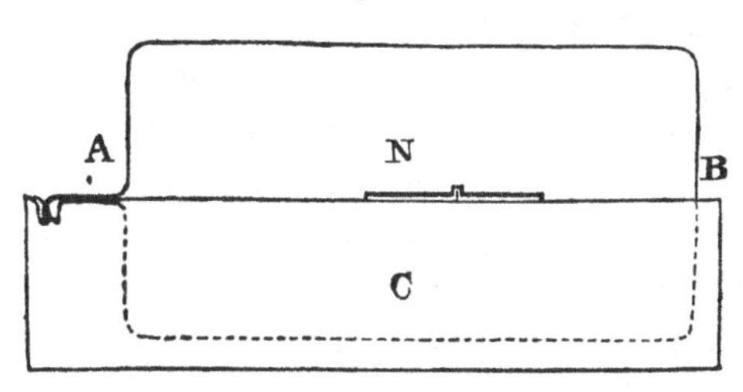

6 inches broad. This I secured in a vertical position by means of the block of wood C; N is the magnetic needle, $3\frac{3}{4}$ inches long, pointed at its extremities, and suspended upon a fine steel pivot over a graduated card placed a little before the centre of the instrument.

4. On account of the large relative size of the rectangular conductor of my galvanometer, the tangents of the deviations of the needle are very nearly proportional to the quantities of current electricity. The small correction which it is necessary to apply to the tangents, I obtained by means of the rigorous experimental process which I have some time ago described in the 'Annals of Electricity'*.

5. I have expressed my quantities of electricity on the basis of Faraday's great discovery of definite electrolysis; and I venture to suggest that that quantity of current electricity which is able to electrolyze a chemical equivalent expressed in grains in one hour of time, be called a *degree*. Now, by a number of experiments I found that the needle of my galvanometer deviated $33^{\circ}\cdot5$ of the graduated card when a current was passing in sufficient quantity to decompose nine grains of water per hour; that deviation, therefore, indicates one *degree of current electricity* on the scale that I propose to

* Vol. iv. pp. 131, 132, and 476.

be adopted. We shall see in the sequel some of the practical advantages which I have had by using this measure.

6. The thermometer which I used had its scale graduated on the glass stem. The divisions were wide, and accurate. In taking temperatures with it, I stir the liquid gently with a feather; and then, suspending the thermometer by the top of its stem, so as to cause it to assume a vertical position, I bring my eye to a level with the top of the mercury. In this way a little practice has enabled me to estimate temperature to the tenth part of Fahrenheit's degree with certainty.

7. In order to ascertain the heating-power of a given metallic wire, it was passed through a thin glass tube, and then closely coiled upon it. The extremities of the coil thus formed were then drawn asunder, so as to leave a small space between each convolution; and if this could not be well done, a piece of cotton thread was interposed. The apparatus thus prepared, when placed in a glass jar containing a given quantity of water, was ready for experiment. Fig. 33 will explain the dispositions: A is the coil of wire; B the glass jar, partly filled with water; T represents the thermometer. When the voltaic electricity is transmitted through the wire, no appreciable quantity passes from it to take the shorter course through the water. No trace of such a current could be detected, either by the evolution of hydrogen, or the oxidation of metal.

Fig. 33.

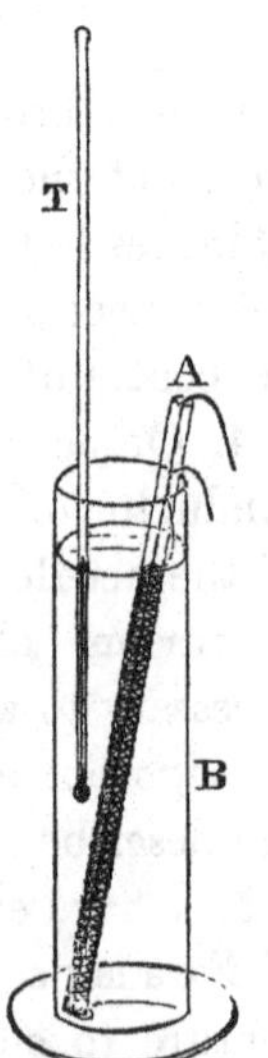

8. Previous to each of the experiments, the necessary precaution was taken of bringing the water in the glass jar and the air of the room to the same temperature. When this is accurately done, the results of the experiments bear the same proportions to one another as if no extraneous cooling agents, such as radiation, were present; for their effects in a given time are proportional to the difference of the temperatures of the cooling and cooled bodies; and hence, although towards the conclusion of some experiments

this cooling effect is very considerable, the *absolute quantities* alone of heat are affected, not the *proportions* that are generated in the same time. [See the table of heats produced during half an hour and one hour, p. 64.]

9. Exp. 1.—I took two copper wires, each two yards long, one of them $\frac{1}{28}$ of an inch, the other $\frac{1}{50}$ of an inch thick, and arranged them in coils in the manner that I have described (7.). These were immersed in two glass jars, each of which contained nine ounces avoirdupois of water. A current of the mean quantity $1^{\circ}\cdot1$ Q* was then passed consecutively through both coils; and at the close of one hour I observed that the water in which the thin wire was immersed had gained $3^{\circ}\cdot4$, whilst the thick wire had produced only $1^{\circ}\cdot3$.

10. Now, by direct experiment, I found that three feet of the thin wire could conduct exactly as well as eight feet of the thick wire; and hence it is evident that the resistances of two yards of each were in the ratio of 3·4 to 1·27, which approximates very closely to the ratio of the heating effects exhibited by the experiment.

11. Exp. 2.—I now substituted a piece of iron wire $\frac{1}{27}$ of an inch thick, and two yards long, for the thick copper wire used in Exp. 1 and placed each coil in half a pound of water. A current of $1^{\circ}\cdot25$ Q was passed through both during one hour, when the augmentation of temperature caused by the iron was 6°, whilst that produced by the copper wire was $5^{\circ}\cdot5$. In this case the resistances of the iron and copper wires were found to be in the ratio of 6 to 5·51.

12. Exp. 3.—A coil of copper wire was then compared with one of mercury, which was accomplished by enclosing the latter in a bent glass tube. In this way I had immersed, each in half a pound of water, $11\frac{1}{4}$ feet of copper wire $\frac{1}{50}$ of an inch thick and $22\frac{3}{4}$ inches of mercury 0·065 of an inch in diameter. At the close of one hour, during which the same current of electricity was passed through both, the former had caused a rise of temperature of $4^{\circ}\cdot4$, the latter of

* I place Q at the end of my *degrees*, to distinguish them from those of the graduated card.

2°·9. The resistances were found by a careful experiment to be in the ratio of 4·4 to 3.

13. Other trials were made, with results of precisely the same character: they all conspire to confirm the fact, that *when a given quantity of voltaic electricity is passed through a metallic conductor for a given length of time, the quantity of heat evolved by it is always proportional to the resistance* which it presents, whatever may be the length, thickness, shape, or kind of that metallic conductor.*

14. On considering the above law, I thought that the effect produced by the increase of the intensity of the electric current would be as the square of that element; for it is evident that in that case the resistance would be augmented in a double ratio, arising from the increase of the *quantity* of electricity passed in a given time, and also from the increase of the *velocity* of the same. We shall immediately see that this view is actually sustained by experiment.

15. I took the coil of copper wire used in Exp. 3, and have found the different quantities of heat gained by half a pound of water in which it was immersed, by the passage of electricities of different degrees of tension. My results are arranged in the following table:—

Mean Deviations of the Needle of the Galvanometer.	Quantities of Current Electricity expressed in Degrees (5).	Quantities of Heat produced in half an hour by the Intensities in Column 2.	Proportional to the Squares of the Intensities in Column 2.	Quantities of Heat produced in one hour by the Intensities in Column 2.	Proportional to the Squares of the Intensities in Column 2.
16°	0·43 Q	....	....	1°·2	1
31½	0·92 Q	3	2·9	4·7	4·55
55	2·35 Q	19·4	18·8		
57⅔	2·61 Q	23	23·2		
58½	2·73 Q	25	25·4	39·6	40

16. The differences between the numbers in columns three

* Mr. Harris, and others, have proved this law very satisfactorily, using common electricity.

and four, and in columns five and six, are very inconsiderable, taking into account the nature of the experiments, and are principally owing to the difficulty which exists in keeping the air of the room in the same state of quiet, of hygrometry, &c. during the different days on which the experiments were made. They are much less when a larger quantity of water is used, so as to reduce the cooling effects (28.).

17. We see, therefore, that *when a current of voltaic electricity is propagated along a metallic conductor, the heat evolved in a given time is proportional to the resistance of the conductor multiplied by the square* * *of the electric intensity.*

18. The above law is of great importance. It teaches us the right use of those instruments which are intended to measure electric currents by the quantities of heat which they evolve. If such instruments be employed (though in their present state they are far inferior in point of accuracy to many other forms of the galvanometer), it is obvious that the *square roots* of their indications are alone proportional to the intensities which they are intended to measure.

19. By another important application of the law, we are now enabled to compare the frictional† and voltaic electricities in such a manner as to determine their elements by the quantity of heat which they evolve in passing along a given conductor; for if a certain quantity of voltaic electricity produce a certain degree of heat by passing along a given conductor, and if the same quantity of heat be generated by the discharge of a certain electrical battery along the same

* The experiments of De la Rive show that the calorific effect of the voltaic current increases in a much greater proportion than the simple ratio of the intensities.—*Ann. de Chimie*, 1836, part i. p. 193. See also Peltier's results, *Ann. de Chimie*, 1836, part ii. p. 249.

† The experiments of Brooke, Cuthbertson, and others prove that the quantity of wire melted by common electricity is as the square of the battery's charge. Harris, however, arrived at the conclusion that the heating-power of electricity is *simply* as the charge (Phil. Trans. 1834, p. 225). Of course the remark in the text is made on the presumption that, when the proper limitations are observed, the calorific effect of electricity is as the square of the charge of any given battery.

conductor, the product of the quantity and velocity of transfer of the *voltaic* electricity will be equal to the product of the quantity and velocity of the *frictional* electicity, or $QV = qv$, whence $\frac{Q}{q} = \frac{v}{V}$.

Chap. II.—*Heat evolved during Electrolysis.*

20. Under the above head, I shall now examine the heat produced in the cells of the battery, and when electrolytes are experiencing the action of the voltaic current. It has been my desire to render these experiments strictly comparable, both with themselves and with those of other philosophers. I have therefore taken care to apply the corrections which either specific heat or other disturbing causes might require, and have by these means been able to express, in every case, the *total* amount of evolved heat.

21. The first of these corrections, which I call Cor. A, arises from the difference between the mean temperature of the liquid used in an experiment and that of the surrounding atmosphere. Its amount is determined by ascertaining the rapidity with which the temperature of the liquid is reduced at the end of each experiment.

22. The second correction (Cor. B) is for the specific heat of the liquids and the vessels which contain them; and when the necessary data could not be found in the tables of specific heat, I have supplied them from my own experiments. The *vessels* were white earthenware jars, 4½ inches deep and 4¼ inches in diameter; their caloric was one twelfth of that contained by two pounds of water, *to which capacity I have reduced the subsequent results* *.

23. As resistance to conduction is the sole cause of the heat produced in the connecting-wire of the voltaic battery, it was natural to expect that it would act an important part in this second class of phenomena also. It was important, therefore, to begin by determining the amount of heat evolved

* In the present paper, however, at p. 79, the capacity of one pound of water appears to have been used.—*Note*, 1881.

by that quantity of conducting metal which I found it convenient to adopt as a *standard of resistance**.

24. Ten feet of copper wire, 0·024 of an inch thick, were formed into a coil in the manner described in (7.); its resistance to conduction was called *unity*. Three experiments were made in order to ascertain its heating-power.

25. 1st. A jar was filled with two pounds of water, and a current, which produced a mean deviation of the needle of the galvanometer (3.) equal to $57\frac{1}{4}^\circ = 2^\circ{\cdot}54$ Q of current electricity, was urged through the coil for twenty-seven minutes, by means of a zinc-iron† battery of ten pairs. The heat thus acquired by the water, after Cor. A and that part of Cor. B which relates to the caloric of the jar had been applied, was $6^\circ{\cdot}22$.

26. 2nd. The battery was now charged with a weaker solution of sulphuric acid. In this case it passed the mean current $2^\circ{\cdot}085$ Q during forty-five minutes. The heat thus produced, when corrected, was $7^\circ{\cdot}04$.

27. 3rd. A battery of five pairs (three of which had platinized silver, one silver, and one copper for their negative plates), passed the mean current $1^\circ{\cdot}88$ Q during one hour, in which time $7^\circ{\cdot}47$ were generated.

28. When the first two experiments are reduced in order to compare them with the third, we have, in accordance with the principles laid down in (17.), $\frac{(1{\cdot}88)^2}{(2{\cdot}54)^2} \times \frac{60'}{27'} \times 6^\circ{\cdot}22$ $= 7^\circ{\cdot}57$, and $\frac{(1{\cdot}88)^2}{(2{\cdot}085)^2} \times \frac{60'}{45'} \times 7^\circ{\cdot}04 = 7^\circ{\cdot}63$. Thus we have $\frac{7^\circ{\cdot}57 + 7^\circ{\cdot}63 + 7^\circ{\cdot}47}{3} = 7^\circ{\cdot}56$, the mean quantity of heat produced per hour by the passage of $1^\circ{\cdot}88$ Q of current electricity, against the above *unit* of resistance.

29. Before I proceed to give an account of some experiments on heat evolved in the cells of voltaic pairs, it is

* This was my first unit of resistance.—*Note*, 1881.

† Whenever an iron battery was used, it was of course placed at a distance from the galvanometer sufficiently great to render its action on the needle inappreciable.

important to observe that every kind of action not essentially electrolytic must be eliminated. For instance, the dissolution of metallic oxides in acid solvents, which has been proved by Dr. Faraday to be no cause of the current, is the occasion of a very considerable quantity of heat, which, if not accounted for in the experiments, would altogether disturb the results. I have taken the oxide of zinc, prepared either by igniting the nitrate, or by burning the metal, and have repeatedly dissolved it in sulphuric acid of various specific gravities; and on taking the mean of many experiments, none of which differed materially from the rest, I have found that the total corrected heat produced by the dissolution of 100 grains of the oxide of zinc in sulphuric acid, is able to raise two pounds of water 3°·44.

30. Exp. 1.—I constructed a voltaic pair consisting of thin plates of amalgamated zinc and platinized silver (Mr. Smee's arrangement. The plates were two inches broad, and were kept one inch asunder by means of a piece of wood, to the opposite sides of which they were bound with string; to the top of each plate a thick copper wire formed a good metallic connexion by means of a brass clamp. The voltaic pair, thus prepared, was immersed in two pounds of sulphuric acid, sp. gr. 1137, contained in one of the earthenware jars (22.). The arrangement is represented by fig. 34.

Fig. 34.

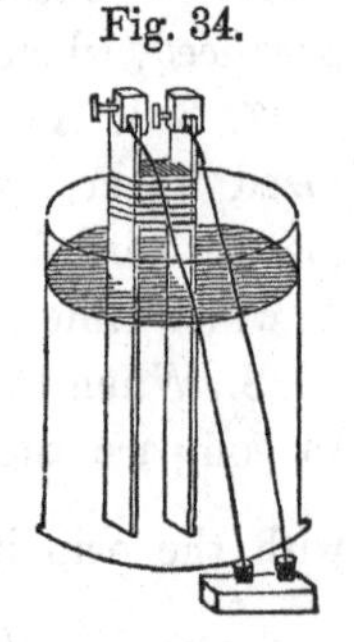

31. When the circuit was completed so as to present to the current the total metallic resistance 0·06, the galvanometer stood at $49\frac{1}{2}°=1°{\cdot}84$ Q, and at $17\frac{1}{2}°=0°{\cdot}453$ Q when the total metallic resistance was increased to 1·16 by the addition to the circuit of ten feet of thin copper wire. Hence, according to the principles laid down by Ohm, $\frac{1{\cdot}84}{r+1{\cdot}16}=\frac{0{\cdot}453}{r+0{\cdot}06}$; from which r, the resistance of the voltaic pair, $=0{\cdot}299$. Immediately after this trial, the temperature of the liquid being exactly 49°, and that of the air 50°·2, the circuit was completed for one hour, during which the needle first advanced

a little from 50°, and then declined to 46°, the average* deviation being 48° 44′=1°·8 Q. The temperature of the liquid was then 53°·7, indicating a rise of 4°·7. Another trial now gave $\frac{1\cdot59}{r'+1\cdot16}=\frac{0\cdot382}{r'+0\cdot06}$; whence r', the resistance of the pair at the close of the experiment, =0·288; the mean resistance of the pair was therefore 0·293.

32. Now in order to obtain the total amount of heat evolved by the pair, reduced to the capacity of two pounds of water, we have 4°·7+0°·4 (on account of Cor. A (21.)) and −0°·5 (on account of Cor. B (22.)) =4°·6. The correction due to the dissolution of oxide of zinc is found by multiplying its quantity by $\frac{3^\circ\cdot44}{100}$ (see (29.)); the quantity of the oxide being obtained by multiplying the equivalent of oxide of zinc by the mean quantity of current electricity. We have then $40\cdot3\times1\cdot8\times\frac{3^\circ\cdot44}{100}=2^\circ\cdot5$; this, when subtracted from 4°·6, leaves 2°·1, the *corrected voltaic heat.*

33. Assuming in this case, as well as in that of a metallic conductor, that the heat evolved is proportional to the resistance multiplied by the square of the electric intensity, we have, from the data in (28.) and (31.), $\frac{(1\cdot8)^2}{(1\cdot88)^2}\times0\cdot293\times7^\circ\cdot56$ $=2^\circ\cdot03$, which is very near 2°·1, the heat deduced from experiment.

34. Exp. 2.—I now constructed another pair, consisting of plates precisely similar to those used in Exp. 1, but half an inch only asunder; it was also immersed in two pounds of sulphuric acid, sp. gr. 1137. The circuit was closed for one hour, during which the needle of the galvanometer advanced gradually from 47½° to 50⅓°, the mean deviation being 49° 35′ =1°·84 Q. The liquid had then gained 4°·8; this +0°·1 (for Cor. A) and −0°·5 (for Cor. B) =4°·4. The heat due to the dissolution of oxide of zinc is in this case

* During each experiment the deflections of the needle were noted at intervals of five minutes, or less. Thence I deduce my averages.

$40{\cdot}3 \times 1{\cdot}84 \times \frac{3^{\circ}{\cdot}44}{100} = 2^{\circ}{\cdot}55$, which, when subtracted from $4^{\circ}{\cdot}4$, leaves the corrected voltaic heat $1^{\circ}{\cdot}85$.

35. The resistance of the pair was ascertained, in this as in every other instance, at the beginning and at the end of the experiment. The equations thus obtained were $\frac{1{\cdot}714}{r+1{\cdot}16} = \frac{0{\cdot}432}{r+0{\cdot}06}$ and $\frac{1{\cdot}91}{r'+1{\cdot}16} = \frac{0{\cdot}446}{r'+0{\cdot}06}$, whence $r=0{\cdot}311$ and $r'=0{\cdot}275$; the mean resistance was therefore $0{\cdot}293$. Now, calculating as before (33.) on the basis of the heat produced by the passage of electricity against the standard of resistance, we have $\frac{(1{\cdot}84)^2}{(1{\cdot}88)^2} \times 0{\cdot}293 \times 7^{\circ}{\cdot}56 = 2^{\circ}{\cdot}12$.

36. Exp. 3.—I formed another pair on Mr. Smee's plan; it was similar to the last, with the exception that the plates were only one inch broad. When the circuit was closed, a current of the mean intensity $1^{\circ}{\cdot}46$ Q passed through the apparatus during one half hour. The heat thereby produced, when corrected, and reduced on account of the dissolution of oxide of zinc, was $0^{\circ}{\cdot}84$.

37. In this instance the mean resistance was $0{\cdot}32$; whence, by a calculation precisely similar to those given under Exps. 1 and 2, we have the theoretical amount of heat $=0^{\circ}{\cdot}74$.

38. The three instances above given are specimens taken from a number of experiments with the platinized silver pairs. The mean of the eight unexceptionable experiments which I have made with them, gives $2^{\circ}{\cdot}08$ of actual and $2^{\circ}{\cdot}13$ of theoretical heat; and not one of the individual experiments presented a greater difference between real and calculated heat than Exp. 2.

39. Exp. 4.—A plate of copper, four inches broad, was bent about a plate of amalgamated zinc three inches and a half broad, so as to form a pair of Wollaston's double battery. It was placed in a jar containing two pounds of dilute sulphuric acid. In this instance, the total voltaic heat that was generated was $1^{\circ}{\cdot}2$, the calculated result being $1^{\circ}{\cdot}0$ only. Repeated experiments with the copper pairs gave similar

results, the real heat being invariably somewhat superior to that which the doctrine of resistance would demand. The cause of this I have found in a slight local action, which it is almost impossible to avoid in the common copper battery.

40. Exp. 5.—I now constructed a single pair on Mr. Grove's plan. The platinum, two inches broad, was immersed in an ounce and a half of strong nitric acid contained in a 4-inch pipe-clay cell; the amalgamated zinc plate, also two inches broad, was immersed (at the distance of an inch and a half from the platinum) in thirty ounces of sulphuric acid, sp. gr. 1156. The whole was contained in one of the jars (22.).

41. A trial, made first as usual, in order to ascertain the resistance of the pair, gave $\frac{4{\cdot}4}{r+2{\cdot}26}=\frac{0{\cdot}816}{r+0{\cdot}06}$, whence $r=0{\cdot}441$. As soon as the slight heat acquired during the above trial was equably diffused through the apparatus, the thermometer placed in the dilute sulphuric acid stood at $51^\circ{\cdot}95$, the temperature of the air being $52^\circ{\cdot}4$. The circuit was then immediately closed for ten minutes, during which time the needle of the galvanometer advanced steadily from $68^\circ\,40'$ to $71^\circ\,20'$, the mean deviation being $70^\circ\,9'=4^\circ{\cdot}77$ Q. As soon as the heat thus generated was equably diffused*, the thermometer immersed in the dilute sulphuric acid stood at $56^\circ{\cdot}7$, indicating a rise of $4^\circ{\cdot}75$. Another trial now gave $\frac{5{\cdot}14}{r'+2{\cdot}26}=\frac{0{\cdot}91}{r'+0{\cdot}06}$, whence $r'=0{\cdot}413$. The mean resistance of the pair was therefore 0·427.

42. $4^\circ{\cdot}75+0^\circ{\cdot}1$ (for Cor. A), and $-0^\circ{\cdot}4$ (for Cor. B, which in this case includes the capacity for heat of the porous cell), $=4^\circ{\cdot}45$. The heat generated by the dissolution of oxide of zinc was in this case $40{\cdot}3\times 4{\cdot}77\times\frac{3^\circ{\cdot}44}{100}\times\frac{10'}{60'}=1^\circ{\cdot}1$, which, subtracted from $4^\circ{\cdot}45$, leaves the corrected voltaic heat $3^\circ{\cdot}35$.

* By gently stirring the dilute sulphuric acid with a feather, so as to bring every part in successive contact with the porous cell, during two minutes.

43. The *theoretical* result is $\frac{(4{\cdot}77)^2}{(1{\cdot}88)^2} \times 0{\cdot}427 \times 7^{\circ}{\cdot}56 \times \frac{10'}{60'}$ $= 3^{\circ}{\cdot}46$.

44. Exp. 6 was made with a pair in every respect similar to the last; the circuit, however, was completed by means of a thin copper wire, in order to reduce the intensity of the current. At the end of one hour, during which the needle of the galvanometer advanced gradually from 41° to 42°, the corrected voltaic heat that was generated was 1°·7. The *theoretical* result was 1°·82.

45. I was desirous of knowing how far the same principles would apply to the heat generated in Prof. Daniell's constant battery. But in this battery the heat is lessened, in consequence of the separation of oxide of copper from the sulphuric acid with which it is combined. This is altogether a secondary effect, and should be eliminated, as was the heat produced by the dissolution of oxide of zinc. I have not yet been able to obtain accurate data for the correction thus needed, and shall therefore content myself with remarking that my results with Mr. Daniell's arrangements are, as far as they go, quite consistent with the theory of resistances.

46. Experiments such as I have related were varied in many ways; and sometimes a number of pairs were arranged so as to form a battery. Still the results were similar, and established the fact that *the heat which is generated in a given time in any pair, by true voltaic action, is proportional to the resistance to conduction of that pair, multiplied by the square of the intensity of the current.*

---

47. I now made some experiments on the heat consequent to the passage of voltaic electricity through electrolytes.

48. Exp. 7.—Two pieces of platinum foil, each of which was an inch long and a quarter of an inch broad, were hermetically sealed, by the wires to which they were attached, into the ends of two pieces of glass tubing: these wires, when the apparatus was in action, terminated in mercury-cups. The tubes thus prepared were bound together by thread, so as to

keep the pieces of platinum foil at the constant distance of half an inch asunder. This apparatus was immersed in two pounds of dilute sulphuric acid, sp. gr. 1154, contained in one of the jars (22.).

49. A battery of twenty (four-inch) double iron-zinc plates was then placed, with its divided troughs (which were charged with a pretty strong solution of sulphuric acid), at a distance from the galvanometer sufficiently great to obviate any disturbing effect on the needle. To the electrodes of this battery thick copper wires were secured, so that by means of one of them connexion could be made to the galvanometer, and by means of the other to the decomposing

Fig. 35.

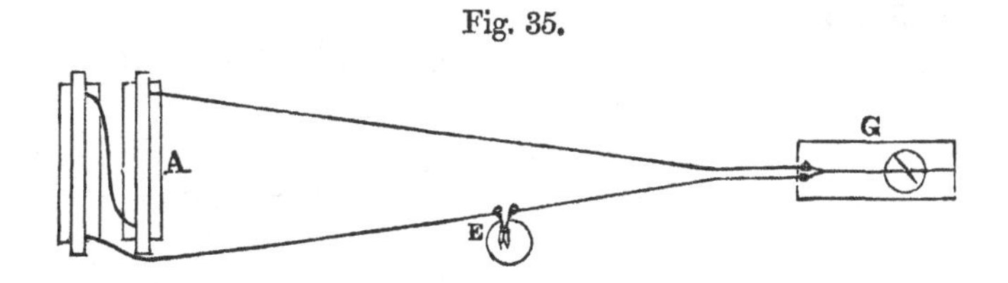

cell. In fig. 35, A represents the battery, G the galvanometer, and E the decomposing apparatus (48.).

50. In order to ascertain the resistances of the battery or of the cell, I provided several coils * of silked copper wire, the resistances of which had been determined by careful experiments. When these were traversed by the current, they were placed in such a position as to prevent any action on the needle; and at the same time they were kept under water, in order to prevent them from becoming hot, which would have had the effect of increasing their resistances.

51. When every thing was duly prepared, the battery was placed in its troughs, and the current from it was urged through the galvanometer and each of three of the coils, which were placed in succession at E (the decomposing ap-

* Two of these coils had been previously employed (31., 41., &c.) in ascertaining the resistances of the voltaic pairs: the resistance 0·06 was that of the galvanometer and connecting wires.

paratus having been removed). The resistances of these coils were 4·4, 5·5, and 7·7; and the currents which they allowed to pass were 1°·88 Q, 1°·65 Q, and 1°·29 Q.

52. The decomposing apparatus was now replaced, and, the proper connexions being made, electrolytic decomposition was allowed to proceed during twenty minutes, in which time the needle of the galvanometer gradually declined from 55° to 48½°, the mean current being 1°·9 Q. The temperature of the liquid had now advanced from 46°·6 to 53°·95, indicating an increase of 7°·35. The temperature of the surrounding atmosphere was 46°·4.

53. The decomposing cell was now removed again, and the several coils, of which the resistances were, as before, 4·4, 5·5, and 7·7, were successively put in its place. The battery now urged through them 1°·73 Q, 1°·48 Q, and 1°·22 Q.

54. In this case, 7°·35 + 0°·55 (for Cor. A) and −0°·64 (for Cor. B) = 7°·26, the heat which was generated in the decomposing jar.

55. The mean intensity of the current when passing through the coil of which the resistance was 4·4, was $\frac{1^{\circ}\cdot88+1^{\circ}\cdot73}{2}$ $=1^{\circ}\cdot805$ Q, but 1°·9 Q when it passed through the decomposing cell. Hence $(4\cdot4+3\cdot15^{*})\frac{1\cdot805}{1\cdot9}=7\cdot17$; this, $-3\cdot15^{*}$, leaves 4·02, the amount of *obstruction* presented by the decomposing cell.

56. Now we must remember that when the electric current was passing through the coils it was urged by the whole intensity of the battery, but that in the case of the decomposing cell a part of the intensity of the zinc-iron battery equal (as I have found by experiment) to 3⅓ pairs, or to one sixth part of the whole, was occupied solely in overcoming the

* From (51., 53.) we have the equations $\frac{1\cdot88}{R+7\cdot7}=\frac{1\cdot29}{R+4\cdot4}$ and $\frac{1\cdot73}{R'+7\cdot7}$ $=\frac{1\cdot22}{R+4\cdot4}$, whence $R=2\cdot81$ and $R'=3\cdot49$: the mean resistance of the battery and connecting wires was therefore 3·15.

*resistance to electrolyzation** of water in the decomposing cell. In order, therefore, to deduce the true *resistance to conduction*, we must subtract $\frac{4{\cdot}02+3{\cdot}15}{6}$ from the *obstruction* 4·02; and thus we have 2·83, the true *resistance to conduction* of the decomposing cell.

57. The latter part of this process is difficult to express clearly; I have therefore drawn a figure to illustrate it. Suppose that (in fig. 36) 6 represents the intensity of the

Fig. 36.

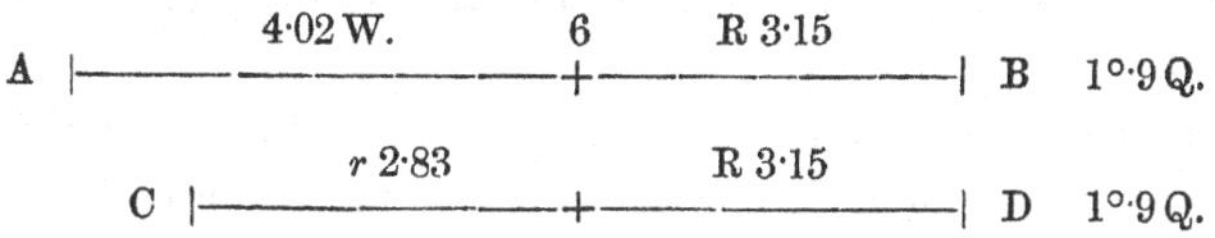

4·02 W. 6 R 3·15
A |——————+——————| B 1°·9 Q.

r 2·83 R 3·15
C |—————+——————| D 1°·9 Q.

battery, the line R 3·15 the resistance of the battery and the connecting wires, and the remainder of the line A B, or 4·02 W, the resistance of wire. I have shown (55.) that the current 1°·9 Q would pass against the resistance A B. But we know that 1°·9 Q was also passed when the cell and the battery formed the sole opposition (52.), and that, on account of the resistance to electrolyzation, the virtual battery-intensity was then one sixth less, and hence that only five sixths of the resistance represented by A B could have been opposed in this case, in order to the passage of the same current. Draw, therefore, another line, C D, one sixth less than A B, and it will represent this resistance, from which, on subtracting R 3·15, we have *r* 2·83, the true resistance to conduction of the decomposing apparatus.

58. From (28.) and the data above given, we have $\frac{(1{\cdot}9)^2}{(1{\cdot}88)^2} \times 2{\cdot}83 \times 7^{\circ}{\cdot}56 \times \frac{20'}{60'} = 7^{\circ}{\cdot}29$, the *theoretical* result.

59. I made three other experiments with the same electrodes, and with the same battery. The results of these, with those of the experiment just given at length, are as follow :—

* Faraday's 'Experimental Researches' (1007.).

	Experimental.		Theoretical.	
	°		°	
Exp. 7......	7·26	......	7·29	
Exp. 8......	8·12	......	8·32	
Exp. 9......	10·2	......	10·2	
Exp. 10 ...	9·64	......	9·75	(Refitted battery.)
Mean ...	8·8	......	8·89	

60. Exp. 11.—The mean current 0°·846 Q from a battery of ten zinc-iron pairs was, by means of the same electrodes, sent through two pounds of dilute sulphuric acid for half an hour, during which the corrected heat that was generated was 3°·09.

61. In order to find the true resistance to conduction of the decomposing cell, it was necessary to remember that in this instance one third of the intensity of the ten pairs was expended in overcoming the resistance to electrolyzation of the water. With this exception the calculations were made precisely as before, and gave 3·76 for the resistance of the cell, whence we have the theoretical heat 2°·88.

62. I now dismissed the narrow electrodes, and substituted for them two pieces of platinum foil, dipping to the bottom of the liquid. They were one inch apart; and each presented to the dilute sulphuric acid a surface of seven square inches. In this case I used twenty pairs of zinc-iron plates arranged in a series of ten.

63. The mean of six experiments with this apparatus gave 4°·42 of actual, and 4°·13 of theoretical heat. I have no doubt that the difference is principally occasioned by the formation of the deutoxide of hydrogen, which is known to occur to a considerable extent when oxygen is evolved from an extended surface. Of this we have another instance in the following experiment.

64. Exp. 12.—Using the same electrodes, and a battery of ten zinc-iron pairs, I now passed a current of the mean intensity 1°·08 Q through two pounds of dilute nitric acid, sp. gr. 1047, for half an hour. The heat that was thus generated, when properly corrected, was 3°.

65. This experiment was, as the others, conducted in the

manner described at length under Exp. 7. Water chiefly* was decomposed; and I ascertained, experimentally, that about $\frac{1}{3{\cdot}5}$ of the intensity of the battery was expended in overcoming resistance to electrolysis. Thus I had $3{\cdot}52-\frac{3{\cdot}52+1{\cdot}68}{3{\cdot}5}$ $=2{\cdot}03$, the resistance to conduction; and hence $\frac{(1{\cdot}08)^2}{(1{\cdot}88)^2}$ $\times 2{\cdot}03\times 7^{\circ}{\cdot}56\times\frac{30'}{60'}=2{\cdot}53$, the theoretical heat

66. Exp. 13.—Two plates of copper, each of which was two inches broad, were secured at the distance of one inch apart, and immersed in two pounds of a saturated solution of sulphate of copper. Through this apparatus, a battery of ten zinc-iron pairs passed the mean current 1° Q during half an hour. The heat thus produced, when properly corrected, was 5°·8.

67. In this case there was no *resistance to electrolysis*; and the action may be regarded simply as a transfer of copper from the positive to the negative electrode. All the *obstruction*, therefore, that was presented to the current, was *resistance to conduction*. Its mean was 5·5; whence we have $\frac{1}{(1{\cdot}88)^2}$ $\times 5{\cdot}5\times 7^{\circ}{\cdot}56\times\frac{30'}{60'}=5^{\circ}{\cdot}88$, the theoretical heat.

68. We have thus arrived at the general conclusion that *the heat which is evolved by the proper action of any voltaic current is proportional to the square of the intensity of that current, multiplied by the resistance to conduction which it experiences.* From this law the following conclusions are directly deduced :—

69. 1st. That *if the electrodes of a galvanic pair of given intensity be connected by any simply conducting body, the total voltaic heat generated by the entire circuit (provided always that no local action occurs in the pair) will, whatever may be the resistance to conduction, be proportional to the number of*

* See Faraday, on the Electrolysis of Nitric Acid, 'Experimental Researches' (752.).

atoms (whether of water or of zinc) concerned in generating the current. For if the resistance to conduction be diminished, the quantity of current will be increased in the same ratio, and hence, according to the law (68.), the quantity of heat which will thus be generated in a given time will be also proportionally increased; whilst of course the unmber of atoms which will be electrolyzed in the pair will be increased in the same proportion.

70. 2nd. That *the total voltaic heat which is produced by any pair is directly proportional to its intensity and the number of atoms which are electrolyzed in it.* For the quantity of current is proportional to the intensity of the pair, and consequently the quantity of heat evolved in a given time is proportional to the square of the intensity of the pair; but the number of atoms electrolyzed is proportional, in the same time, to the simple ratio only of the current, or of the intensity of the pair.

71. And 3rd. That *when any voltaic arrangement, whether simple or compound, passes a current of electricity through any substance, whether an electrolyte or not, the total voltaic heat which is generated in any time is proportional to the number of atoms which are electrolyzed in each cell of the circuit, multiplied by the virtual* intensity of the battery.*

72. Berzelius thinks that the light and heat produced by combustion are occasioned by the discharge of electricity between the combustible and the oxygen with which it is in the act of combination; and I am of opinion that the heat arising from this, and some other chemical processes, is the consequence of resistance to electric conduction. My experiments on the heat produced by the combustion of zinc turnings in oxygen (which, when sufficiently complete, I hope to make public) strongly confirm this view; and the quantity of heat which Crawford produced by exploding a mixture of

* If a decomposing cell be in the circuit, the *virtual* intensity of the battery is reduced in proportion to its resistance to electrolyzation, the "virtual intensity of the battery" signifying the total electromotive power of the entire circuit, where battery-cells are positive quantities and cells in which decomposition is effected are negative quantities.

hydrogen and oxygen may be considered almost decisive of the question. In his unexceptionable experiments, one grain of hydrogen produced heat sufficient to raise one pound of water 9°·6. Now we know from Exp. 5 that the heat generated in one of Mr. Grove's pairs by the electrolysis of $\frac{4\cdot77 \times 32\cdot3}{6} = 25\cdot7$ grains of zinc, is theoretically 3°·46; and the heat which must in the same time have been generated by the *metallic* part of the circuit, which presented the resistance 0·06, is $\frac{0\cdot06}{0\cdot427} \times 3^\circ\cdot46 = 0^\circ\cdot48$; the total voltaic heat was therefore 3°·94. Hence the total heat which would have been evolved by the electrolysis of an equivalent, or 32·3 grains of zinc, is $\frac{32\cdot3}{25\cdot7} \times 3^\circ\cdot94 = 4^\circ\cdot95$; which, when reduced to the capacity of one pound of water, is 9°·9. But, from the table of the intensities of voltaic arrangements (74.), the intensity of Mr. Grove's pair, compared with the affinity of hydrogen for oxygen, is $\frac{1}{0\cdot93}$; whence, from (70.), we have $9^\circ\cdot9 \times 0\cdot93 = 9^\circ\cdot2$, the heat which should be generated by the combustion of one grain of hydrogen, according to the doctrine of resistances; the result of Crawford is only 0°·4 more.

73. I am aware that there are some anomalous conditions of the current which seem to militate against the general law (68.), particularly when in the hands of Peltier it actually produces *cold**. I have little doubt, however, that the explanations of these will be ultimately found in actions of a secondary character.

Note on Voltaic Batteries.

74. In the foregoing investigation I have had occasion to work very extensively with different voltaic arrangements, and have repeatedly ascertained their relative intensities by

* If antimony and bismuth be soldered together, cold will be produced at the point of junction by the passage of the current from the bismuth to the antimony (Peltier, *Annales de Chimie*, vol. lvi. p. 371). In his paper, however, a misprint has inverted the direction of the current.

the mathematical theory of Ohm. It will not, therefore, I hope, be deemed out of place to subjoin a table, in which the intensities of the batteries which are most generally used are inversely as the number of pairs which would be just requisite in order to overcome the resistance of water to electrolyzation.

Mr. Grove's	Platinum Nitric acid	Amalgamated zinc Dilute sulphuric acid	$\frac{1}{0 \cdot 93}$	Constant Intensities.
Prof. Daniell's	Copper Sulphate of copper	Amalgamated zinc Dilute sulphuric acid	$\frac{1}{1 \cdot 54}$	
Mr. Sturgeon's	Iron. Amalgamated zinc Dilute sulphuric acid		$\frac{1}{3 \cdot 33}$	
Mr. Smee's	Platinized silver. Amalgamated zinc Dilute sulphuric acid		$\frac{1}{3 \cdot 58}$	
	Copper. Amalgamated zinc Dilute sulphuric acid		$\frac{1}{5 \cdot 40}$	

75. Without entering particularly into the respective merits of these arrangements, I may observe that any one of the first four may be used advantageously, according to the circumstances in which the experimenter is placed, or the particular experiments which he wishes to execute. The zinc-iron battery is somewhat inconvenient on account of local action on the iron; but then it presents great mechanical facilities in its construction. Mr. Smee's and Mr. Grove's are also very good arrangements; but the battery of Daniell is the best instrument for general use, and is, moreover, unquestionably the most economical.

Broom Hill, Pendlebury, near Manchester,
March 25, 1841.

P.S. In the above table of galvanic intensities, that of zinc-iron immersed in dilute sulphuric acid is somewhat overstated. Recent experiments convince me that when the iron is in its best condition it possesses the same powers as the platinized silver. I attributed the iron battery to Mr. Sturgeon, who constructed one of these excellent instruments early in

1839*; it consisted of twelve cast-iron tubes, furnished with strips of amalgamated zinc. But I find that the experiments of this gentleman were not published as early as those of Mr. Roberts. Prof. Daniell (Phil. Trans. 1836, p. 114) observed that iron is sometimes more efficient than platinum in voltaic association with amalgamated zinc.

J. P. J.

August 11, 1841.

On the Electric Origin of the Heat of Combustion. *By* J. P. Joule, *Esq.*†

[Read before the Literary and Philosophical Society of Manchester, November 2, 1841; Phil. Mag. ser. 3. vol. xx. p. 98.]

1. In the papers which I had some time ago the honour of communicating to the Royal Society, I related an investigation concerning the calorific effects of voltaic electricity, and stated my opinion with regard to the heat evolved by combustion and certain other chemical phenomena. In the present paper I intend to bring forward some experiments in confirmation of my theory, and to prove that the heat of combustion, terminating in the formation of an electrolyte, is the consequence of resistance to electric conduction.

2. We have seen that when those chemical actions which are not the sources of transmitted electricity are allowed for, the heat evolved from any part of the voltaic apparatus is the effect of the resistance which is presented by that part to the electric current; and that hence it necessarily follows that the total voltaic heat generated by the action of any closed galvanic pair is proportional to the number of chemical equivalents which have been consumed in the act of propelling the current, and the intensity of the galvanic arrange-

[* A paper by Dr. A. Fyfe, on the employment of iron in the construction of voltaic batteries, appeared in Phil. Mag. for August 1837, ser. 3. vol. xi. p. 145.—Edit.]

† The experiments were made at Broom Hill, near Manchester.

ment. Now if it can be shown that the quantity of heat which is evolved by ordinary chemical combination is the same as calculation founded on these facts would lead us to expect, no reasonable doubt can be entertained that it also is the product of resistance to electric conduction.

3. In studying the character of the heat of combustion, the first point was to determine the intensities of the affinities of different combustibles for oxygen. For this purpose I have, in accordance with the views which were first stated by Davy, and have since been adopted by the most eminent electricians, made use of the measure of these intensities which is afforded by the electric current.

4. I had not proceeded far before some curious phenomena were observed, which, though not very well understood, have long been known * to electricians. I shall notice these first, because of their important bearing upon subsequent reasonings and conclusions.

5. I was working with an arrangement consisting of iron, platinized silver, and dilute sulphuric acid. The circuit was closed by a galvanometer, the coil of which consisted of 119 turns of thin silked copper wire, forming a rectangle, measuring 1 foot by 6 inches. The needle indicated a pretty constant deviation of 20°; but on moving the platinized silver backwards and forwards, the needle advanced gradually to 40°, where it was kept for some time by continuing the agitation. As soon as the motion of the platinized silver was discontinued, the needle resumed its former position. Similar effects were produced by stirring the liquid, and thus causing it to impinge against the platinized silver.

6. I repeated the above experiment many times with similar results; but I found that whenever a large quantity of

* In 1830, Mr. Sturgeon remarked that when two pieces of iron are placed in dilute muriatic acid, the *agitation* of one of them will make it operate as copper in the copper-zinc battery: also, that if two pieces of iron are immersed in succession in a solution of nitrous acid, the iron *last* immersed will act as copper in the copper-zinc battery.—*Recent Experimental Researches*, pp. 46-49. We shall hereafter see the true cause of these phenomena.

hydrogen had been evolved from the liquid by the action of the pair, or otherwise, the phenomena were not well produced. This circumstance convinced me that the effects were due to atmospheric air held in solution by the liquid, and that the displacement of a part of it by the hydrogen had occasioned their partial prevention. My opinion was confirmed by the following experiment.

7. I filled three quarters of the contents of a glass flask with dilute sulphuric acid, and then placed it over the flame of a spirit-lamp until I judged that all the atmospheric air had been boiled out. I then removed the lamp, and immediately placed in the mouth of the flask a cork, through which a small piece of platinized silver and a stout iron wire had been passed. On connecting the metals with the galvanometer (5.) its needle was deflected to $32\frac{1}{2}°$; and on shaking the flask very briskly, it could not be made to advance further than to 34°. This advance, slight as it is, was probably entirely occasioned by air, which, notwithstanding my precautions, had found its way into the upper part of the flask.

8. The phenomena originated entirely from the platinized silver; and although a slight advance of the needle was sometimes produced by agitating the *iron*, it was not difficult to see that the real cause was the propulsion thereby occasioned of the aërated liquid against the negative * element, for when this was avoided no advance of the needle could be produced by agitating the positive metal.

9. I thought it probable that an increase of the intensity of the current would be produced by directing a stream of oxygen gas against the negative element. On making the experiment, I found that the needle advanced a few degrees, and that the same effect could be produced by a stream of hydrogen. There could be no doubt that the increase of intensity arose rather from the agitation of the liquid than from

* To avoid misconception, it is perhaps as well to observe that I call those elements of the voltaic battery *negative* which attract or combine with those bodies which are called "positively electrical," or "cations."

any specific action of the gases, and that this experiment was essentially the same as that described in (5.).

10. I impregnated some dilute sulphuric acid with a very small quantity of oxygen, according to Thenard's process, and then immersed into it a plate of platinized silver and a rod of iron, both properly connected with the galvanometer. The needle stood for the first few seconds at 68°; in three minutes it declined to 50°, in five minutes more to 49°, and in another five minutes to 48½°. On agitating the platinized silver so as to bring it repeatedly in contact with the yet undecomposed deutoxide of hydrogen, the needle advanced to above 60°. The same pair, immersed in common acid, would have deflected the needle no further than 29° or 30°.

11. The effect of the presence of oxygen at the negative element is well observed by making it, in water, the positive electrode of a voltaic battery. By this means oxygen is deposited on its surface, and is there ready to produce an extraordinary intensity. This deposit of oxygen is in fact the cause of the action of Ritter's secondary piles.

12. The following was also a very convenient method of showing the increase of intensity arising from the presence of oxygen. Some dilute sulphuric acid was agitated with chlorine until the former had taken up as much of the gas as it could. By pouring a solution of sulphate of silver into the liquid, I now precipitated chloride of silver, leaving sulphuric acid and oxygen in solution. When a pair, consisting of platinized silver and iron, was placed in the acid thus prepared, the galvanometer was permanently deflected 50°; and by agitating the platinized silver, the needle advanced as far as 60°. When a piece of amalgamated zinc was substituted for the iron the permanent deflection of the needle was 65°; and by agitating the negative element as before, the needle advanced to 70°. Had the same pairs been immersed in a simple solution of sulphuric acid, the permanent deflections would have been no greater than 30° and 63°.

13. Similar results were obtained with the solution of chlorine, as might have been anticipated from its strong affinity for hydrogen.

14. From the above experiments, we see that the agitation of the negative element is productive of an increase of intensity, simply because it is thereby brought into contact with bodies capable of combining with the hydrogen, which would otherwise have been evolved from it. When those bodies are present in considerable quantities, as in (10.), (12.), and (13.), the intensity of the current is great, even though the pair be left quiet, because then the negative plate can collect them readily upon its surface. Again, by causing the current to encounter great resistance, the effects of agitation which we have noticed are proportionally increased, because then the number of particles required for neutralizing the hydrogen is less. Hence it is that when I have used a resistance of 500 or 600 yards of thin wire, I have frequently found the deviation of the needle (even when the pair was left quiet in a common solution of sulphuric acid) considerably greater than was due to that resistance. This is also the probable reason why De la Rive in one instance * found the intensity of the copper-zinc pile the same, whether charged with water or nitric acid.

---

15. In the course of the preceding experiments I was forcibly struck with the very great intensity of the pairs at the moment of their immersion, compared with that which they were able to maintain permanently. It appeared to me that the theories which had been put forward to explain the first effect of immersion, though seemingly plausible with regard to the zinc battery, were not at all equal to account for the same phenomenon as existing to a far greater extent when iron is used as a positive element.

16. A rod of iron and a small plate of platinized silver were immersed in a dilute solution of sulphuric acid. On connecting them with the galvanometer (5.), the needle was permanently deflected $29\frac{1}{2}°$. After a few preliminary trials to ascertain the proper point, I caused the needle to be maintained by a glass weight at 55°, *beyond* which it was free to travel. I then exposed the platinized silver to the air during

* Ann. de Chimie, 1836, part i. p. 179.

one minute of time. On re-immersing it the needle sprang as far as 60°, and then immediately recoiled to its resting place at 55°, thus indicating a *transitory* current of about 57½°.

17. On exposing the platinized silver for 5 sec. only, the transitory current, ascertained in a manner similar to that just mentioned, was 41°.

18. Greater effects were obtained by washing and drying the platinized silver before it was immersed. In this way the needle, adjusted at 62°, would spring as far as 66°, indicating a transitory current of about 64°. Having now removed the glass weight, the needle took up a permanent position at 29½°, as at the beginning of the experiments.

19. When, instead of the platinized silver, the positive element (iron) was exposed to the air, whether simply or in conjunction with washing and drying, no appreciable increase of intensity was occasioned by its immersion. And although, on the repetition of the experiment, I sometimes observed slight effects, I conceive that they were owing to the power which the negative element seems to possess of collecting upon its surface the air held in solution by the circumambient liquid.

20. With an arrangement of platinized silver and amalgamated zinc, I obtained results of a similar, though less striking character. The galvanometer indicated a permanent deflection of 62°, and after washing and drying the platinized silver I had a transitory deviation of 72°. The immersion of the amalgamated zinc, after washing and drying, produced no effect.

21. The *maximum* effects of immersion were produced in the following manner. A plate of silver was rubbed with a little nitric acid, and then exposed to a red heat, by means of which the film of nitrate of oxide of silver was decomposed and metallic silver reduced *. When the plate prepared in

* By this process all the oxygen is not driven off, but a considerable quantity remains adhering to the silver so tenaciously that it is not entirely removed by making the plate quite bright with glass paper. The oxygen thus deposited (it can hardly, I think, be considered as chemically com-

this way was associated in dilute sulphuric acid with a piece of iron, the needle would deviate 63½° for some time, and then gradually decline until it took up a permanent situation at 29½°. By experimenting in the same way with amalgamated zinc as a positive element, I had a transitory deflection of 76°, and a permanent deflection of 63°.

22. Very trifling transitory effects were obtained by the immersion of iron, when that metal was associated with amalgamated zinc. But this might have been anticipated, because the transitory current is owing to the presence of oxygen on the negative plate, and it is obvious that the hydrogen evolved by the local action of the iron would, whilst in a nascent state, combine with that oxygen, and thus prevent a great part of it from exercising any influence upon the intensity of the current.

23. An experiment was also made with an arrangement of copper, amalgamated zinc, and dilute sulphuric acid. It was able to deflect the needle 51° pretty permanently. On washing and drying the copper, and experimenting as in (16.), I observed a transitory deflection of 72°. This experiment deserves attention, because it shows that the transitory current occasioned by the copper is the same as that exhibited by platinized silver when experimented with in the same way (20.). I take it as an argument, that when copper is in its best state it forms with amalgamated zinc a battery as intense as the platinized silver.

24. That the transitory currents which we are discussing are not occasioned by the diffusion of the salt formed about the positive element during the cessation of voltaic action, is obvious from the fact that (when the proper precautions are observed) they are not produced by the agitation (8.) or by the immersion (19. and 20.) of the metal about which the salt is formed. And if any thing can render this more evident, it is the fact that the immersion of the copper plate of a Daniell's

---

bined with the silver) is the cause of the great intensity of the current immediately after immersion. By simply heating the silver to redness the same general effects can be produced, though not to the same extent.

battery causes the needle to advance little or no higher than its permanent situation, as might have been anticipated from the theory which refers the transitory effects to chemical combination at the negative plate, on account of the slight affinity of copper for oxygen. The following experiments are also decisive of this question.

25. A glass jar, *a*, fig. 37, containing some dilute sulphuric acid, was placed upon the plate *p p* of an air-pump. A small rod of iron, *i*, was immersed in the liquid, and connected by

Fig. 37.

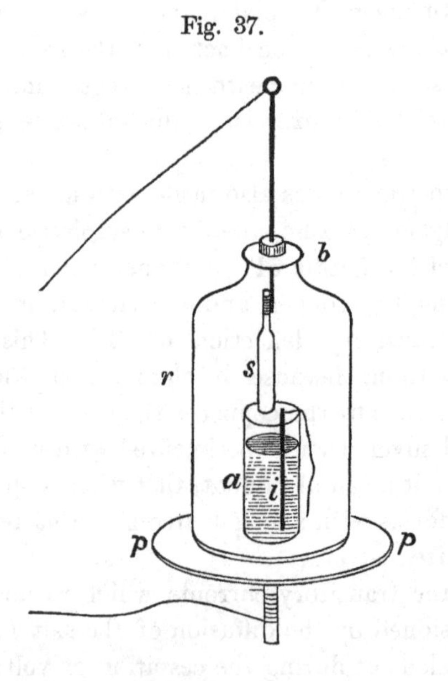

means of the pump-plate to the galvanometer (5.). An open receiver, *r*, was now placed over the jar, and the ground brass plate *b*, with its stuffing-box and sliding rod (the latter having the small piece of platinized silver, *s*, affixed to its extremity), was placed at the top of the receiver. A copper wire, fastened to the ring of the sliding rod, connected the platinized silver with the galvanometer.

26. The sliding rod was now moved until the platinized silver in connexion with it was immersed in the acidulated

water. Then the pump was worked until a very excellent vacuum was obtained; and so tight was every part of the apparatus, that it could be left alone for half an hour without the admission of any appreciable quantity of air. The galvanometer indicated a permanent deflection * of 27°. I now placed a piece of glass so as to prevent the needle from going lower than 27°, and by means of the sliding rod I removed the platinized silver entirely out of the acid. After it had been exposed during a quarter of an hour I re-immersed it, when the needle sprang from 27° to 30° and back, indicating a transitory deflection of about 28½°. Although the effect of immersion exhibited by this experiment is extremely small, it appeared to be almost entirely occasioned by the repose of the electric condition of the iron; for when, instead of entirely withdrawing the platinized silver, its extremity was just allowed to touch the liquid, the transitory deflection was only 27½° after an exposure during a quarter of an hour.

27. On admitting a quantity of air into the receiver sufficient to counterbalance the pressure of one inch of mercury, the effects of immersion were considerable after a very short exposure of the platinized silver. In a quarter of an hour it collected upon its surface sufficient oxygen to cause the needle to spring from 27° to 78°, whether it had or had not remained in contact with the liquid during its exposure.

28. When, instead of the vacuum, I used an atmosphere of hydrogen, the exposure of the platinized silver for any length of time did not render the current more intense at the moment of immersion than it remained permanently. And even when the hydrogen was diluted with one quarter of its bulk of atmospheric air the transitory effects did not appear, on account, no doubt, of the union † of the oxygen with the hydrogen as fast as the former, or both, collected

* No change in the permanent deflection of the needle was occasioned by the removal of atmospheric pressure.

† The phenomenon of Dœbereiner, so fully investigated by Faraday, to whose paper, published in the Phil. Trans. for 1834, I refer the reader for some valuable observations on the power possessed by metals of condensing gases upon their surfaces.

upon the plate. On using a mixture of equal bulks of hydrogen and air, the transitory effects were very small, even after the platinized silver had been exposed for ten minutes.

29. I made several experiments with carbonic acid, but the transitory currents did not entirely disappear as was anticipated. The gas, though prepared carefully and in different ways, could not be obtained perfectly pure, and when exposed to an alkaline solution, $\frac{1}{200}$th of it would remain uncondensed. In order therefore to remove any free oxygen which the gas might contain, I exposed it during two days and two nights to the action of a stick of phosphorus. After this, immersion caused no, or at most very trivial, transitory effects; but on admitting only one per cent. of oxygen they became very considerable—a striking example of the power possessed by metals of collecting and condensing oxygen upon their surfaces. I do not bring forward this experiment as a proof of the *entire* non-action of carbonic acid, because the phosphorus was found to have decomposed it partially.

30. All these phenomena are easily understood, if, with the great body of philosophers, we keep in view the intimate relation which subsists between chemical affinity and the electric current. For let *p*, fig. 38, represent a plate of pla-

Fig. 38.

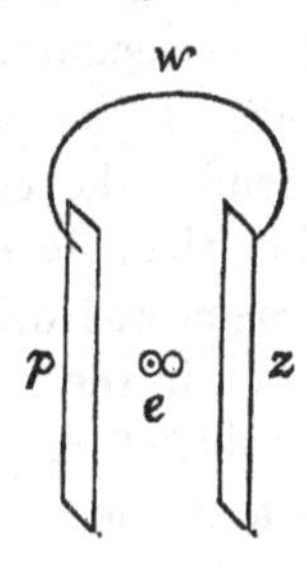

tinum; *z*, a plate of zinc, or other electro-positive metal; and *e*, one of a series of atoms of water extending from *p* to *z*. The intensity of the current along the wire *w* is proportional to the affinity of oxygen for the positive metal, minus the affinity of oxygen for hydrogen. But if *p* be covered with a film of oxygen, the current will be entirely proportional to the affinity of the positive metal for oxygen. In the former case, $c=z-h$; in the latter, $c'=z$.

31. Considering these equations, it is obvious why, as I have observed (15.), the transitory currents are better exhibited with iron than with zinc as a positive element; for in proportion to the smallness of z, provided it remain greater than h, will the difference between c and c' be more manifest. If $c'-c$ be the same for both iron and zinc, we shall have a proof of the accuracy of these principles.

32. Thus from (21.), turning the deflections of the needle into quantities of electricities, we have $63\frac{1}{2}^\circ=0^\circ\cdot034$ Q, and $29\frac{1}{2}^\circ=0^\circ\cdot0072$ Q, of which the difference is $0^\circ\cdot0268$ Q when *iron* is the positive element. We have also $76^\circ=0^\circ\cdot056$ Q, and $63^\circ=0^\circ\cdot027$ Q, of which the difference is $0^\circ\cdot029$ Q when *zinc* is the positive element. I consider these differences as nearly equal as could have been expected from the nature of the experiments.

33. I might now proceed to consider in detail several phenomena (such as the very rapid corrosion of metals when they are exposed to the joint action of air and moisture, &c.) which are occasioned by the great intensity of galvanic action in consequence of the mixture of oxygen with the liquid. But I hasten to fulfil my principal design.

Intensities of the Affinities which unite Bodies with Oxygen.

34. In order to ascertain the intensities of galvanic arrangements, we may either use a galvanometer furnished with a short and thick wire, or with a long and thin wire (within certain limits (14.)). In the former case the calculations must be conducted on the principles of Ohm; in the latter it is only necessary to take care that the resistances of the pairs under comparison are pretty nearly equal, in order that the deviations of the needle may be depended upon in calculating the intensity of the current. I have adopted the latter plan on account of the superior facilities which it presents.

35. *Affinity of zinc for oxygen.*—From (32.) we have the

intensity of the action of zinc$=0°{\cdot}056$ Q; and the intensity required for the electrolysis of water$=0°{\cdot}029$ Q. Hence 29 : 56 : : 1, the affinity of hydrogen, : 1·93, that of zinc for oxygen.

36. *Affinity of iron for oxygen,* likewise obtained from (32.), is 1·27; for 268 : 340 : : 1 : 1·27.

37. *Affinity of potassium for oxygen.*—Twenty grains of potassium were combined with about ten ounces of mercury. The amalgam was poured into a wooden cup, into the bottom of which a copper wire connected with the galvanometer (5.) had been let. At about half an inch above the surface of the amalgam I secured a piece of platinum, also in connexion with the galvanometer. On pouring dilute sulphuric acid into the cup the needle was deflected 74° ($=0°{\cdot}05$ Q) during three successive minutes; but the local action of the amalgam was so vigorous that at the end of this interval of time most of the potassium was dissolved, and the needle declined very fast. On treating 20 grains of zinc in precisely the same manner, I had a deviation of 49° ($=0°{\cdot}0152$ Q). Hence

$$k-h=0{\cdot}05,$$

and

$$z-h=0{\cdot}0152,$$

whence

$$152k=500z-348h.$$

But from (35.), $z=1{\cdot}93$, and $h=1$; therefore k, the affinity of potassium, $=4{\cdot}06$.

38. It is necessary, however, to pay attention to the circumstances under which the experiments were made, in order to obtain correct ideas concerning the above intensities of affinity. The increase of the intensity of the voltaic apparatus by heat is by no means great; and as all the experiments were conducted at common temperatures, no regard need be paid to *it*. But, then, the intensities of affinity were obtained by comparing currents which had been produced under peculiar circumstances with regard to the *condition* of the elements of the galvanic arrangements: in one case the hydrogen was evolved in a *gaseous* state; whilst in the other, the hydrogen, by com-

bining with free and condensed oxygen, did not escape. Now we shall see from the following experiments that electric intensity is *expended* in the act of converting a body into the gaseous state.

39. I took ten glass jars (see fig. 39), made them perfectly

Fig. 39.

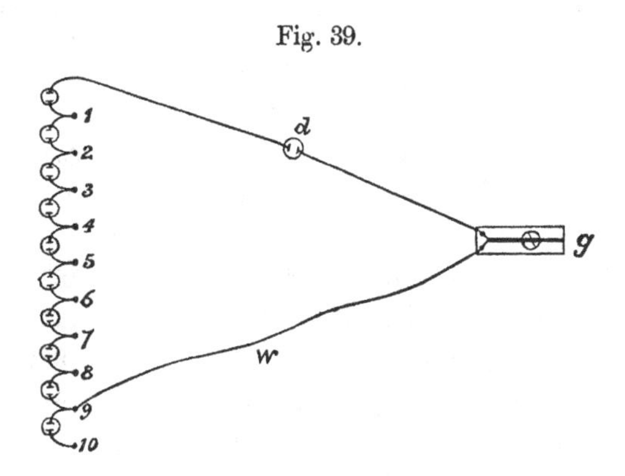

clean and dry*, and placed them in series on a non-conducting substance. Into these I poured a quantity of dilute sulphuric acid, taking care not to wet the glass within an inch of the top of each. Pairs of platinized silver and amalgamated zinc were placed in the jars; and connexions, furnished with the mercury cups 1, 2, 3, &c., were established between them *seriatim*. A decomposing-cell, *d*, furnished with platinum wires, was connected on one hand with the battery, and on the other with the galvanometer (5.). Lastly, I provided a copper wire, *w*, by means of which connexion could be conveniently made between the galvanometer, *g*, and any of the mercury cups, 1, 2, 3, &c.

40. Into *d* I poured a small quantity of dilute sulphuric acid. Then, by placing the wire *w* in each of the mercury

* It is necessary to be very careful in insulating the apparatus, in order to obtain the *maximum* intensity of a battery. The divided porcelain trough has frequently great conducting-powers (particularly when the glaze has been partially destroyed), which render it unfit for accurate experiments.

cups, beginning at 10 and ending at 10, I observed the deviations of the galvanometer contained in the following table:—

No. of pairs in action.	Deflections of galvanometer.				Comparative quantities of Electricity.
	Down.	Up.	Mean.		
10	$72\frac{1}{2}°$	71°	71°	45	341
9	69	68	68	30	284
8	$65\frac{1}{4}$	65	65	7	236
7	60	60	60	0	188
6	53	$53\frac{2}{3}$	53	20	141
5	42	43	42	30	$94\frac{1}{2}$
4	$25\frac{1}{2}$	$27\frac{3}{4}$	26	37	$48\frac{1}{2}$
3	$5\frac{1}{2}$	$5\frac{1}{2}$	5	30	$8\frac{1}{2}$
2	0	0	0	0	0
1	0	0	0	0	0

41. Now if we divide the straight line, AB, fig. 40, into ten equal parts, representing pairs on Mr. Smee's plan, and if at

Fig. 40.

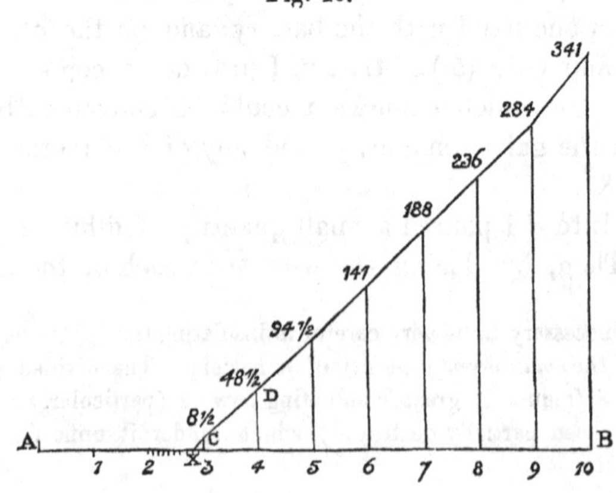

each division we erect straight lines perpendicular to A B, and proportional to the comparative quantities of electricity just given, the principles of electric action demand that the line drawn through the extremities of those perpendiculars should be straight. It is in fact so nearly a straight line, that its slight discrepancies therefrom may be properly referred to unavoidable errors of experiment. Produce the straight line C D so as to meet A B in X, and the straight line A X, equal to 2·8, will indicate the number of pairs necessary to decompose water.

42. Fig. 41 represents an experiment of the same kind,

Fig. 41.

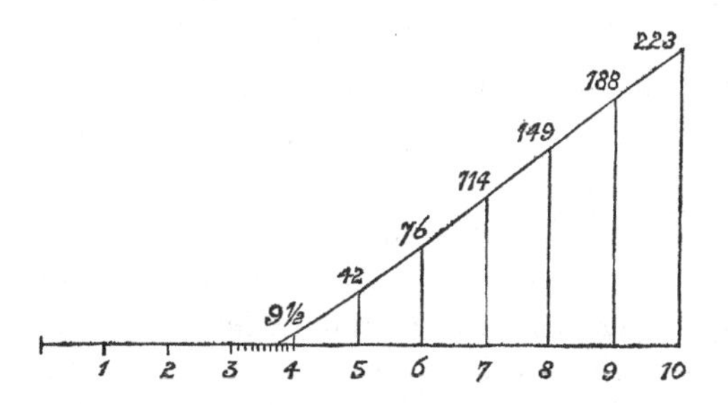

with a solution of sulphate of oxide of zinc in the decomposing cell. Oxide of zinc was decomposed, the oxygen being evolved at the positive, and the zinc being reduced at the negative electrode. The intensity necessary to decompose oxide of zinc is equal to that of 3·7 of Mr. Smee's pairs.

Fig. 42.

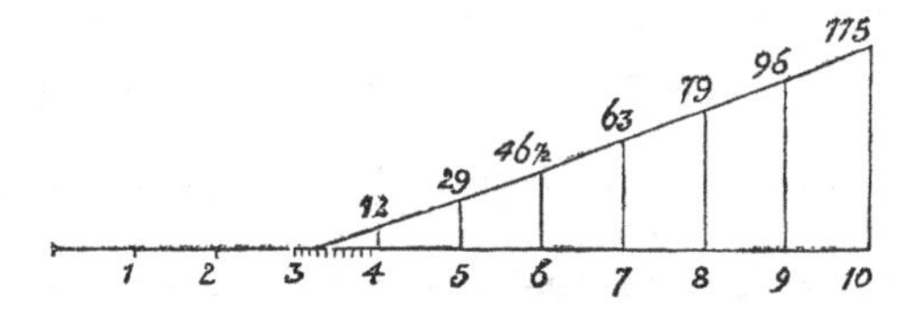

43. With sulphate of protoxide of iron I did not at first succeed, on account of the formation of peroxide at the posi-

tive electrode. However, by placing the negative electrode among some crystals of the salt, pouring water thereon, and suspending the positive electrode in the water, I obviated that difficulty, and obtained the results which are projected in fig. 42, and which indicate 3·3 pairs as the intensity necessary to decompose protoxide of iron.

44. Now from (41.) and (42.) we have (using the same letters as before) 2·8 pairs $=h$*, and 3·7 pairs $=z$; whence $2·8\,z=3·7\,h$, and $z=1·32\,h$; or, in other words, the intensity required to separate oxide of zinc into metal and *gas* is to the intensity required to separate water into its *gaseous* elements as 1·32 : 1. But from (35.), the intensity produced by the union of *non-gaseous* oxygen with zinc is to the intensity necessary to separate water into *non-gaseous* oxygen and *gaseous* hydrogen as 1·93 : 1; and 1·32 : 1 :: 1·93+1·9 : 1+1·9. Wherefore the intensity necessary to give oxygen the gaseous form is to the intensity necessary to separate water into non-gaseous oxygen and gaseous hydrogen as 1·9† : 1.

45. Thus we see that a very great intensity of current is employed in changing the condition of bodies, as well as in separating them from their combinations. The field of investigation here opened is very extensive, but I may not at present enter further upon it. I will only remark, that if the intensity necessary to convert a body into a different state, compared with the heat or cold due to the mechanical or other production of that different state, be such as accords with the relations of intensity and heat which we observe in the voltaic apparatus, we have a proof that some of the effects which are usually referred to "latent heat," are in fact nothing more than the recondite operations of resistance to the electric current ‡.

46. In our investigation into the cause of the heat of com-

* It must, however, be observed that h here means the intensity for the ordinary electrolysis of water; whereas in (37.) it means that required to separate hydrogen gas from non-gaseous oxygen.—*Note*, 1881.

† This is much too high, see p. 118.—*Note*, 1881.

‡ Some experiments, which I have not time to refer to at present, render this hypothesis more than probable.

bustion, it will be necessary to deduce our calculations from the electric intensity which is required in order to reduce the product of combustion to the state in which its elements were prior to combustion. The following is a list of these intensities, reckoning the decomposition of water into its gaseous elements as unity.

47. *Intensity necessary to decompose oxide of zinc into gaseous oxygen and metal,* from (42.) and (44.), is 3·7 pairs of Smee's battery, or 1·32 h.

48. *Intensity necessary to decompose protoxide of iron into gaseous oxygen and metal.*—From (43.), 3·3 of Smee's pairs $=i$; and from (41.), 2·8 pairs$=h$; whence $2{\cdot}8\,i=3{\cdot}3\,h$, or $i=1{\cdot}18$.

49. *Intensity necessary to decompose potassa into potassium and gaseous oxygen.*—From (44.) and (37.) we have $1{\cdot}93+1{\cdot}9 : 4{\cdot}06+1{\cdot}9 :: 1{\cdot}32\,h : 2{\cdot}05\,h$, the intensity required, which may be otherwise expressed by 5·74 of Smee's pairs.

Heat evolved by Combustion, when it terminates in the formation of an Electrolyte.

50. Finding that our information on the quantity of heat evolved by the combustion of metals was not very satisfactory, I have, without wishing to depreciate the labours of Dulong, Despritz, and others, thought it right to bring forward such of my own experiments as were necessary in order to make my investigation complete.

51. I provided two glass jars. The smaller had an internal capacity of 90 cubic inches; and when placed within the other jar, as represented by fig. 43, the space left between the two was sufficient to contain three pounds of water. By means of a scale, *s*, suspended by wire from a thick fold of moistened paper, I was able to introduce a combustible within an atmosphere of oxygen, and by means of a heavy weight I could keep the paper sufficiently close to the top of the jar to prevent the escape of any considerable quantity of heated air, while at the same time it was not so tight as to prevent the admission of air as the oxygen was consumed. The increase

of the temperature of the water was measured by a thermometer of great sensibility.

Fig. 43.

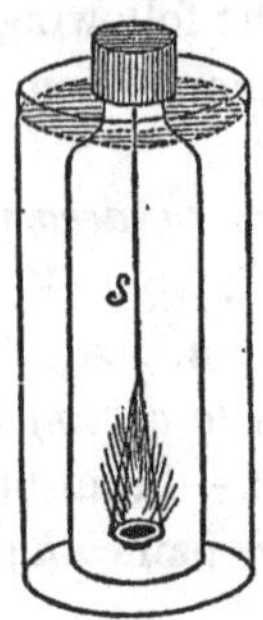

52. *The heat evolved by the combustion of zinc* was ascertained in the following manner. The smaller jar was filled with oxygen, placed in the other jar, and surrounded by three pounds of water, the heat of which was contrived to be as much below the temperature of the surrounding air as it was expected to exceed it at the close of the experiment. A piece of phosphorus, weighing 0·4 grain, was then put into the scale, and over it I placed a heap of fine zinc turnings, weighing 50 grains. I now ignited the phosphorus, and plunged the scale into the inner jar. After the combustion had terminated, and the heat thereby evolved had been evenly distributed throughout the water by stirring, the increase of temperature was noted. The contents of the scale were then thrown into dilute sulphuric acid, and the volume of hydrogen thereby evolved indicated the quantity (generally about 15 grains) which had not been burnt. Two tenths of a degree of heat were deducted from the observed heat, on account of the phosphorus; and an allowance having been made on account of the capacity of glass for heat, the results were reduced to the standard of one pound of water.

53. The mean of several experiments conducted in the above manner showed that the heat evolved by the combustion of 32·3 grains of zinc is able to increase the temperature of a pound of water by 10°·8.

54. *The heat evolved by the combustion of iron* was ascer-

tained in a similar way. The iron was in the state of fine wire, and that portion of it that was not burnt was carefully collected, weighed, and deducted from the original quantity. The mean of several trials indicated that 28 grains could increase the temperature of a pound of water by 9°·48.

55. *Heat evolved by the combustion of potassium.*—This metal, in pretty large lumps, was introduced into an atmosphere composed of equal bulks of oxygen and air. I then introduced a stout iron wire, sharpened at the end, into the jar; and with it I cut the potassium into small pieces. Under this treatment it soon became so soft that every time the rod was lifted it would draw out a string of metal. In this state it often ignited, and the experiment was spoiled on account of the partial formation of peroxide. However, by careful management, I succeeded in making some good experiments, in which nearly all the potassium was converted into *potassa*; and the exact quantity of unoxidized metal was ascertained by observing the volume of hydrogen evolved when the contents of the scale were exposed to the action of water. The mean of these showed that the heat evolved by the conversion of 40 grains of potassium into potassa is able to increase the temperature of a pound of water by 17°·6.

56. *Heat evolved by the combustion of hydrogen.*—The gas was burned in an atmosphere of oxygen, diluted with common air by means of a jet furnished with a very narrow bore. A grain of hydrogen evolved as much heat as is able to increase the temperature of a pound of water by 8°·36.

57. We shall now proceed to examine how far the theory of resistance to electric conduction agrees with the above experimental results.

58. We have seen, (47.), (48.), and (49.), that the intensities of the affinities which unite gaseous oxygen with zinc, iron, potassium, and gaseous hydrogen are as 1·32, 1·18, 2·05, and 1; and the proportional quantities of heat which were generated by the combustion of the equivalents of these bodies are 10°·8, 9°·48, 17°·6, and 8°·36, or 1·29, 1·13, 2·105, and 1—a ratio which is very nearly the same as that of the intensities just given. Hence we see that *the quan-*

tities of heat which are evolved by the combustion of the equivalents of bodies are proportional to the intensities of their affinities for oxygen. Now I proved in my former paper* that a similar law obtains in the voltaic apparatus, in consequence of its heat being produced by resistance to conduction. And hence we have an argument that the heat of combustion has the same origin.

59. But our proof of the real character of the heat of combustion is rendered more complete by regarding *quantities* as well as ratios of heat. From the quantity of heat generated by the motion of a given current along a wire of known resistance, we can deduce the quantities of heat which, according to the theory of resistance to electric conduction, ought to be produced by the combustion of bodies; and then these theoretical deductions may be compared with the results of experiment.

60. The mean of three careful experiments detailed in my former paper† shows that if a wire, the resistance of which is a unit, be traversed by an electric current of $1^{\circ}{\cdot}88$ Q‡ for one hour, the heat evolved by that wire will be able to increase the temperature of a pound of water by $15^{\circ}{\cdot}12$. Now I have ascertained experimentally that a pair consisting of amalgamated zinc and platinized silver, excited by dilute sulphuric acid, is able to propel a current of $0^{\circ}{\cdot}168$ Q against the whole resistance of the circuit when that resistance is 5·2; consequently a similar pair can propel a current of $0^{\circ}{\cdot}168\ \mathrm{Q} \times 5{\cdot}2 = 0^{\circ}{\cdot}874\ \mathrm{Q}$ against the resistance which I have called a unit. But from (42.), the intensity necessary to separate oxide of zinc and gaseous oxygen is to the intensity of one of Smee's pairs as 3·7 : 1. Consequently the electricity produced by the union of zinc and gaseous oxygen

* Philosophical Magazine, October 1841, ser. 3. vol. xix. p. 275 (70:).

† Ibid. p. 266.

‡ I beg to remind the reader that my *degree*, expressed thus (1° Q), indicates that quantity of current electricity which, after passing constantly during one hour, is found to have electrolyzed a chemical equivalent expressed in grains—as 9 grains of water, 36 grains of protoxide of iron, &c.

must be sufficiently intense to propel a current of $0^{\circ}\cdot874$ Q $\times 3\cdot7 = 3^{\circ}\cdot234$ Q against a unit of resistance. Now $1^{\circ}\cdot88$ Q, when urged against a unit of resistance, was able in one hour of time to increase the temperature of a pound of water by $15^{\circ}\cdot12$; therefore $3^{\circ}\cdot234$ Q could, in the same circumstances, produce $\left(\frac{3\cdot234}{1\cdot88}\right)^2 \times 15^{\circ}\cdot12 = 44^{\circ}\cdot74$ of heat. But in (70.) of my former paper I proved that the same quantity of heat should always (according to the theory which refers the whole of the heating power of the voltaic apparatus to resistance to the electric current) be produced by a given quantity and intensity of electrolysis, whether the resistance opposed to the current be small or great. Wherefore the heat which, on these principles, ought to be generated by the combustion of 3·234 equivalents of zinc is $44^{\circ}\cdot74$; or, in other words, *one* equivalent, or 32·3 grains of zinc, should generate heat sufficient to increase the temperature of a pound of water by $13^{\circ}\cdot83$.

61. Now, as I have before stated, the quantities of heat evolved by the combustion of the equivalents of bodies ought, according to the theory of resistance to electric conduction, to be proportional to the intensities of their affinities for gaseous oxygen. These, in the cases of zinc, iron, potassium, and hydrogen, are 1·32, 1·18, 2·05, and 1. Hence $13^{\circ}\cdot83$, $12^{\circ}\cdot30$, $21^{\circ}\cdot47$, and $10^{\circ}\cdot47$ are the quantities of heat which ought, according to our theory, to be produced by the combustion of 32·3 grains of zinc, 28 grains of iron, 40 grains of potassium, and 1 grain of hydrogen.

62. By comparing these results of theory with the quantities of heat, $10^{\circ}\cdot8$, $9^{\circ}\cdot48$, $17^{\circ}\cdot6$, and $8^{\circ}\cdot36$*, which were (53.–56.) obtained from experiment, it will be seen that the former exceed the latter by about one quarter. Considering the difficulty of preventing some loss of heat, in consequence of the escape of air from the mouth of the inner jar (51.) during the first moments of combustion, &c., it will, I think, be

* Crawford, whose method was well adapted to prevent loss of heat, obtained $9^{\circ}\cdot6$. More recently, Dalton observed about $8^{\circ}\cdot5$.

admitted that experiment agrees with the theory as well as could have been expected.

63. I conceive, therefore, that I have proved in a satisfactory manner that the heat of combustion (at least when it terminates in the formation of an electrolyte) is occasioned by resistance to the electricity which passes between oxygen and the combustible at the moment of their union. The amount of this resistance, as well as the manner of its opposition, is immaterial both in theory and in experiment; and if the resistance to conduction be great (as it most probably is when potassium is slowly converted into potassa by the action of a mixture of oxygen and common air) or little (as it probably is when a mixture of oxygen and hydrogen is exploded), still the quantity of heat evolved remains proportional to the number of equivalents which have been consumed and the intensity of their affinity for gaseous oxygen.

64. That the heat evolved by other chemical actions, besides that which is called *combustion*, is caused by resistance to electric conduction, I have no doubt. I cannot, however, enter in the present paper upon the experimental proof of the fact.

On the Electrical Origin of Chemical Heat. *By* JAMES P. JOULE.

['Philosophical Magazine,' ser. 3. vol. xxii. p. 204, having been read before the British Association, at Manchester, on June 25, 1842.]

IN a paper which I read on the 2nd of last November before the Literary and Philosophical Society of Manchester, I endeavoured to account for the heat evolved by the combustion of certain bodies on the hypothesis of its arising from resistance to the conduction of electricity between oxygen and the combustibles at the moment of their union. Taking this view of phenomena, I showed that the heat evolved by the union of two atoms is proportional to the electromotive force of the current passing between them—in other words, to the intensity of their chemical affinity.

In that paper I gave the results of my own experiments, and I apprehended that my numbers might be below the truth, on account of the simplicity of my apparatus. On comparing them, however, with the experiments of Dulong, which were conducted in a manner very well calculated to prevent loss of heat, I now find that they agree so well with the results of that accurate philosopher as to show that the method I adopted of carrying on the combustion in the inner of the two glass jars, whilst the heat evolved was measured in water placed between them, was not unworthy of reliance. In the following table I give the results of Dulong's experiments, reduced to degrees Fahrenheit acquired by a pound of water :—

Quantities converted into Protoxides.	Dulong's results.	My own experiments.	Theoretical results.	Corrected theoretical results.
40 grs. Potassium	......	17·6	21·47	......
33 „ Zinc	10·98	11·03	13·83	11·01
28 „ Iron	9·00	9·48	12·36	8·06
31·6 „ Copper....	5·18	......	9·97	5·97
1 gr. Hydrogen	8·98	8·36	10·47	10·40*

In the above table there is one metal (copper) of which I did not treat in my former paper; it will therefore be well to explain the manner in which the theoretical results for it were obtained.

Platinum wires were immersed in a saturated solution of sulphate of copper. These were successively connected with

* Since Daniell has proved the remarkable fact that, during the electrolysis of dilute sulphuric acid, one quarter of an equivalent of acid goes along with the oxygen to the positive electrode, the true theoretical result will be 9·47.

the poles of different voltaic arrangements, consisting of various numbers of Smee's pairs in series. Using two pairs, I had neither current nor decomposition; but with three there were electrolytic effects, oxygen being evolved from the positive, and copper being deposited on the negative electrode. The ratio of the current passed by three compared with that passed by four pairs, was as nearly as possible 1 : 4. Therefore $2\frac{2}{3}$ pairs are equal to the resistance to electrolysis of sulphate of copper.

Now if I calculate, as I did in my former paper for zinc, iron, and hydrogen, I must argue that electricity equal in intensity to that of $2\frac{2}{3}$ pairs passes between oxygen and copper when they unite by combustion. But one pair of Smee's battery can produce electricity of the intensity that a degree* of it will evolve 3°·74 Fahr. in a pound of water; and multiplying by $2\frac{2}{3}$, we have 9°·97 as the theoretical result for copper, if we suppose that the intensity required to overcome the resistance to electrolysis of sulphate of copper is equal to the intensity of current arising from the union of oxygen and copper in combustion.

There is, however, since the experiments of Daniell, reason to think that this is not quite the case, but that part of the electromotive force engaged in electrolyzing these compound bodies is used in separating the acid from the base prior to or (according to that philosopher's view) simultaneously with the decomposition of the latter. Unfortunately we cannot bring forward a direct experiment to prove the fact, inasmuch as the oxides† are by themselves, and at common temperatures, non-conductors of voltaic electricity, and therefore refuse to yield up their elements. But if, on the principles of our theory, we argue that the heat evolved on the combination of one chemical equivalent with another is a measure of the intensity of the electric current passing between them at the time, we shall have the means of eliminating the electro-

* My degree of electricity is the quantity necessary to electrolyze an equivalent expressed in grains, as 9 grains of water, &c.

† I find that pure water is not at all decomposed by ten pairs of Smee's battery.

motive force employed otherwise than in separating the elements forming the oxides.

I suppose that there are three forces in operation, of which two are against, and one is for, a current engaged in electrolyzing the solution of the sulphate of a metallic oxide. The first two are the affinity of the elements of the oxide for one another and the affinity of the oxide for the acid; and the third (which is in a contrary direction to the two others, and generally less than either) is the affinity of water for sulphuric acid. The latter two forces may be eliminated as follows:—

1st. For *Zinc.*—I find that 41 grains, or an equivalent of oxide of zinc, evolves 2°·82 when dissolved in dilute sulphuric acid. This, which is the quantity of heat due to the intensity of current resulting from the difference of the affinities of sulphuric acid for the oxides of zinc and hydrogen, leaves, when subtracted from 13°·83, 11°·01, the corrected theoretical result which I have given in the 5th column of the table.

2nd. For *Iron.*—The black oxide is dissolved with such difficulty by dilute sulphuric acid that the heat thereby evolved cannot be accurately measured. However, the dissolution of the hydrate of the protoxide is easily effected, the quantity of heat generated thereby being, per equivalent, 2°·74. But we probably arrive nearer the truth by subtracting from the heat evolved by the dissolution of iron in dilute sulphuric acid that portion which is due to the oxidation of the iron. In this way I have $5^{\circ}\cdot 2 - 0^{\circ}\cdot 9 = 4^{\circ}\cdot 3$ as the heat due to the solution of protoxide of iron in dilute sulphuric acid. This, when subtracted from 12°·36, leaves 8°·06, the corrected theoretical result.

3rd. For *Copper.*—The protoxide of copper does not dissolve readily in dilute sulphuric acid. Nevertheless, by keeping the temperature of the surrounding atmosphere equal to that of the liquid, for the long interval of time required by the experiment, I obtained, per equivalent of oxide, 4°·0, which, being subtracted from 9°·97, leaves 5°·97 as the corrected theoretical result.

4th. For *Hydrogen* little correction is needed*. The liquid used in the experiments made to ascertain the resistance to electrolysis of water was mixed with a small quantity only of sulphuric acid. Consequently there were plenty of atoms of water, either uncombined or only slightly attached to the acid, prepared to give up their elements to the current with little or no additional resistance in consequence of its presence.

By inspecting the table it will be seen that the above corrected theoretical results agree very well with the experiments of Dulong and myself. The agreement is accurate in the case of zinc. Iron gives results which are not so satisfactory; but we must remember that this metal is converted by combustion into the magnetic oxide, and that a correction ought therefore to be applied on account of heat evolved by the union of protoxide with oxygen, which it is very difficult entirely to prevent. Potassium gives theoretical and experimental results as accordant as can be well expected, considering the complicated process† by which the former was obtained and the practical difficulty of the latter. In the case of hydrogen, we might have anticipated that theory would give a larger result than experiment; for the resistance to the electrolysis of water appears generally greater than it really is, on account of the peculiar state which the platinum evolving hydrogen is apt to assume, which has, of course, the effect of increasing the theoretical value.

Besides the corrections to the theoretical results which I have supplied, I thought that there might be a slight one needed on account of *light,* which is evolved in such abundance in some instances of combustion. It was of importance to ascertain whether, in the evolution of light, an equivalent of heat was absorbed. With this view I have made an extensive series of experiments with the voltaic apparatus, comparing the heat evolved when no light was exhibited with that evolved when the conducting wire was ignited to whiteness.

* On this question see the note appended to the table. Prof. Daniell's discovery indicated a correction of not less than 1°.

† See p. 97 (49.).

The mean of twelve experiments showed that the heat evolved by a certain quantity of wire immersed in water was, for a given quantity of current and a given length of time, 24°·75; and the mean of sixteen experiments, in which a platinum wire enclosed in a glass tube surrounded with water was ignited so as to give out a quantity of light equal to that of a common tallow candle, gave 24°·4 as the quantity of heat due in this latter case to similar circumstances of resistance and quantity of current. These experiments seem to indicate that heat is lost when light is evolved, but in so slight a degree that my experiments on the heat of combustion need not be corrected for it. The combustion in Dulong's experiments was performed in a box of copper, which, being opaque, would entirely obviate this source of error.

I conceive that the correctness of the idea, entertained, I believe, by Davy, and afterwards more explicitly mentioned by Berzelius, that the heat of combustion is an electrical phenomenon, is now rendered sufficiently evident. I have shown that the heat arises from resistance to the conduction of electricity between the atoms of combustibles and oxygen at the moment of their union. Of the nature of this resistance we are still ignorant.

Some time ago I commenced an investigation on the heat arising from the union of sulphuric acid with potash, soda, and ammonia. This inquiry is more difficult than I expected, and my experiments are not as yet sufficiently complete to lay before the British Association. In a future paper I hope to extend my inquiry, and also to show the relation of latent heat to electrical intensity.

On Sir G. C. Haughton's Experiments. By J. P. JOULE. [In a letter to the Editors of the 'Philosophical Magazine and Journal of Science.']

['Philosophical Magazine,' ser. 3. vol. xxii. p. 265.]

GENTLEMEN,

Allow me to occupy a small portion of your space with a few observations on the subject of an interesting communication by Sir G. C. Haughton inserted in your last Number.

On repeating the experiments on the action of frictional electricity upon the galvanometer, I find that the phenomenon to which Sir Graves has called attention is simply an example of the repulsion of bodies which are in the same electrical state, and is not sensibly owing to the minute currents which are put into play by the electrifying machine. This may be easily shown by the following experiment:—

Take half an inch of copper wire, and suspend it by a filament of silk in the centre of a vertical insulated metallic ring, having an aperture of about an inch and a quarter. Electrify positively both the ring and the copper wire, and the latter will immediately place itself at right angles to the plane of the former. Now cause the copper wire to communicate with the ground for an instant; then, becoming of course negatively electrical by induction, it will be immediately attracted to the plane of the ring, and will vibrate from one side to another of it with the same energy as it did about the angle of 90° when positively electrical.

It is manifest that the unstable equilibrium of the wire, when it is similarly electrical with the ring, and is also exactly in the same plane, may be converted into a stable equilibrium by superadding an attractive force which urges it to remain there. Hence it is that Sir Graves found that a delicately suspended magnetic needle remained steady, though a needle without directive energy was in similar circumstances deflected to 90°.

The interesting and important case of static electrical action observed by Sir Graves ought to be carefully guarded against by those who wish to ascertain the quantity of current frictional electricity by the multiplier. In France this instrument has been employed by M. Peltier for measuring the electricity of the atmosphere; and whilst it cannot be doubted that that able electrician used great precautions, the danger of the apparatus receiving a charge of electricity sufficient to interfere seriously with the results, cannot be too strongly urged upon the attention of those who may repeat his experiments.

Yours truly,
JAMES P. JOULE.

Broom Hill, near Manchester,
March 4, 1843.

On the Heat evolved during the Electrolysis of Water. *By* JAMES P. JOULE*.

['Memoirs of the Manchester Literary and Philosophical Society,' 2nd ser. vol: vii. p. 87. Read January 24th, 1843.]

IN former papers I have endeavoured to prove that the heat evolved by voltaic electricity is proportional to the resistance to conduction and the square of the quantity of current, during electrolysis, and in the cells of the battery as well as in metallic conductors. The heat, however, which is liberated in cases of electrolysis was very uniformly found to exceed the product of resistance and the square of the current; and I attributed this to the solution of oxide of zinc and to other actions which are regarded as secondary. The circumstance is evidently of great importance to the whole subject of voltaic heat. I have therefore undertaken a series of experiments to ascertain its real character, the results of which are, I hope, of sufficient importance to merit the attention of this Society.

It would be quite useless to attempt an inquiry of this kind without the assistance of a *galvanometer* of very considerable

* The experiments were made at Broom Hill, near Manchester.

accuracy. Sensible of this, I have from time to time endeavoured to improve my measures of voltaism. The instrument used in the present investigation is of large dimensions and of great exactness. It is constructed much on the same plan as Pouillet's galvanometer of tangents, the electricity being carried by a thick copper wire bent into a circle of a foot diameter. The magnetic needle, 6 inches in length, traverses a circle of pure brass divided by a machine. Its deflections, appreciable to 5′, are turned into quantities of electric current with the help of a table formed by exact and careful experiments.

Another piece of essential apparatus is a *standard of resistance to conduction.* Mine consists of eight yards of copper wire, $\frac{1}{20}$ of an inch in diameter. It is well insulated, bent double to obviate any action on the needle of the galvanometer, and then arranged into a coil. When used, it is immersed in water, in order, by keeping it cool, to prevent the increase of its resistance.

The *battery* I used is on Prof. Daniell's plan. Each of its cells is 25 inches high and $5\frac{1}{2}$ inches in diameter. They were charged, in the usual way, with acidulated sulphate of copper in contact with the copper; and dilute sulphuric acid, mingled with a variable* quantity of sulphate of zinc, in contact with the amalgamated zinc. Six of these cells in series formed the battery used in all the subsequent experiments; but at the same time I may observe that trials were also made with only three in series, in order to satisfy myself that the results are independent of the extent of the battery.

The *electrolytic apparatus* consisted of a glass jar, containing a pound of liquid, and furnished with electrodes, each of which exposed an active surface of about six square inches.

By ascertaining, immediately before and after each experiment, the current which could pass when the battery and

* As I do not throw away my old solution, but content myself with mixing it occasionally with new, the liquid must have contained considerably more free acid at one time than at another. This cannot, however, interfere with my results, because it did not interfere with the electromotive force of the battery.

connecting-wires (including that of the galvanometer) were only in the circuit, and that which could pass when a unit was added to this resistance by means of the standard of resistance; we have, on the principles of Ohm, the equation $ax=b(x+1)$, or $x=\frac{b}{a-b}$, where a and b are the above currents and x is the resistance of the battery and connecting-wires.

This being known, we have the means of ascertaining the resistance to conduction of the electrolytic cell. It is obtained from the equation $c(x+y)=ax\frac{d-z}{d}$, or $y=\left(a\left(\frac{d-z}{d}\right)-c\right)\frac{x}{c}$, where (the other letters remaining as before) c is the current observed during electrolysis, d is the number of cells in the battery, and z is the *resistance to electrolysis* of the electrolytic apparatus, calling (as I intend to do throughout this paper) the electromotive force of Daniell's cell unity.

This resistance to electrolysis is obtained by observing the currents which pass through the electrolytic cell when different numbers of the battery-cells are employed in propelling them. For this purpose the lowest two numbers adequate to effect electrolysis were chosen; and having reduced the current observed when using the fewest number of cells, on account of the absence in that case of the resistance to conduction of one cell of the battery, the resistance to electrolysis is found by the equation $z=\mathrm{E}-\left(\frac{e}{f-e}\right)$, where E is the smallest number of cells which can effect decomposition, e is the corrected current observed with it, and f is the current passed when one cell more is added to the battery.

The appended table contains the particulars of some experiments on electrolysis in which various electrodes and solutions were used. Each of them is the mean of two experiments, in one of which the needle was deflected towards the right and in the other towards the left hand of the magnetic meridian, so as to neutralize any slight inaccuracy in the position of the galvanometer. The first three numbers of the table, giving the heating-powers of wire, were obtained,

1.	2.		3.	4.	5.	6.	7.	8.	9.	10.	11.	12.	13.
No.	Electrodes. P.	Electrodes. N.	Liquid in Cell.	Current.	Resistance to Electrolysis.	Resistance to Conduction.	Heat evolved.	Heat due to resistance to conduction and square of current.	Difference between Columns 7 and 8 per unit of electrolysis.	Resistance to Electrolysis due to Column 9.	Column 5 minus Column 10.	New Resistance to Conduction.	New theoretic results.
1.	Coil of Copper Wire.		Water.	7°·136	0	0·8876	11°·78	11°·71					
2.	do.		do.	6·770	0	0·8818	10·42	10·47					
3.	do.		do.	6·887	0	0·8810	10·80	10·82					
4.	Platinum.	Amalgamated Zinc.	Dilute Sulphuric Acid.	3·479	2·809	1·025	11·37	6·426	8°·523	1·391	1·418	1·799	11°·281
5.	do.	Platinum.	do.	3·658	2·475	1·049	11·427	7·271	6·817	1·112	1·363	1·635	11·333
6.	Platinized Silver.	Platinized Silver.	do.	6·287	1·748	0·572	13·929	11·715	2·112	0·345	1·403	0·677	13·871
7.	Platinized Platinum.	Platinized Platinum.	do.	4·707	1·760	0·922	12·871	10·581	2·919	0·476	1·284	1·120	12·853
8.	do.	do.	do.	4·918	1·900	0·765	12·628	9·591	3·706	0·605	1·295	1·004	12·584
9.	do.	do.	Solution of Potash, sp. gr. 1·063.	3·111	1·900	1·514	9·370	7·590	3·433	0·560	1·340	1·868	9·366

not in immediate succession, but at considerable intervals of time, during which other experiments were tried. By this means I assured myself that the important condition of uniform conductibility of the standard coil had not suffered by its repeated use. The heat evolved by wire, having been ascertained by these three experiments, furnished the means of calculating the subsequent numbers in column 8, according to the law of resistance to conduction and square of current.

The time during which the electrolysis was allowed to proceed varied from 7½ to 10 minutes. The results are, however, for the sake of uniformity, invariably reduced to 10 minutes. The quantities of heat are reduced to the capacity of a pound avoirdupois of water, by making the requisite corrections (deduced from careful experiments) for specific heat and the influence of the atmosphere. The current is, as in my former papers, expressed in degrees: 1°, passing uniformly during an hour, decomposes a chemical equivalent expressed in grains—as, 9 grs. of water, 41 grs. of oxide of zinc, &c.

By inspecting the table, it will be observed that the heat actually evolved is in every case of electrolysis, greater than that which is due to the product of the resistance to conduction and the square of the current. In a former paper I ascribed this to the solution of oxygen at the positive electrode. Faraday has shown that this occurs sometimes to a considerable extent. In the present experiments I ascertained that the gases were evolved exactly in the right proportions; and this because the small quantity of oxygen dissolved by the solution at the positive electrode was immediately carried by currents to the negative, and there neutralized by hydrogen, as the liquid was not divided by a diaphragm. In this way, however, not more than one sixteenth of the mixed gases was re-formed into water. Nor can this account in any degree for the difference between the numbers in columns 7 and 8; for in proportion to the quantity of gas redissolved is the intensity required for electrolysis undoubtedly diminished, and this, entering the equations, has the effect of increasing the calculated resistances to conduction, and, in the same proportion, the numbers in column 8. We must therefore seek another cause.

And in order to do so with convenience I have enlarged the table. Column 9 contains the difference between columns 7 and 8, reduced to an equivalent of electrolysis. In other words, the differences of the numbers in columns 7 and 8 would have been equal to those in column 9, had the several experiments been continued just long enough to effect the evolution of a grain of hydrogen. Now, I had ascertained by careful experiments given in the first three numbers of the table, that the intensity of a Daniell's cell such as I used is equivalent to 6°·129 in a pound of water per degree of current; and hence I have obtained, by simple proportion, the numbers in column 10, representing, in terms of the intensity of a Daniell's cell, the resistances to electrolysis due to the quantities of heat in column 9. By subtracting these from the numbers in column 5 we obtain those in column 11, which, as we shall presently see, represent the intensity or electromotive force requisite to separate water into its elements and to evolve the constituent gases. Columns 12 and 13, containing new resistances to conduction,* and the heat due to them, are calculated from column 11.

It will be observed that the resistances to electrolysis (contained in column 5) vary according to the nature of the electrodes. With platinized surfaces the resistance is considerably less than with polished platinum· and the latter, again, is exceeded in this respect when the electrode evolving hydrogen is of amalgamated zinc. Now, in all the experiments with dilute acid, the "*chemical*" effect was the same, consisting in the separation of the gaseous elements of water, and the transfer of quarter † of an equivalent of sulphuric acid from the negative to the positive electrode; and in the ex-

* I call these "new resistances to conduction" merely from the want of a more appropriate term to specify what is really a mixture of two distinct sources of resistance. Column 6 contains the ordinary resistances to conduction.

† Phil. Trans. 1840, pt. i. p. 214. I have myself made some rough experiments confirmatory of this remarkable fact.

periment with solution of potassa the same gases were evolved, accompanied, according to Prof. Daniell, by the transfer of the fifth part of an equivalent of alkali from the positive to the negative electrode, which could not fail to require nearly the same intensity as the transfer of acid in the contrary direction. It is quite evident, then, that the resistances in column 5 are not entirely due to *chemical changes*, though I have no doubt that they arise from the joint action of resistance to chemical change and chemical *repulsion*—repulsion principally, owing to the presence of electro-positive materials on the negative electrode. If the resistance to electrolysis which is over and above that due to chemical change were not accounted for elsewhere, it would prove the *annihilation* of part of the power of the circuit, without any corresponding effect. We shall see that this is not the case, but that in the evolution of *heat*, where the excess of resistance takes place, an exact equivalent is restored.

For, by inspecting column 9 we see that the excess of heat is greatest when the resistance to electrolysis is greatest; and least also when the latter is least. Now, when this excess of heat is, in the manner I have before described, turned into the resistance to electrolysis of which it is an equivalent, and then subtracted from the compound resistance in column 5, we obtain the numbers in column 11, which represent the resistance due to the separation of water into its gaseous elements alone, and apart not only from the resistance owing to the state of the electrodes but also from that which is due to the transfer of the fraction of an equivalent of acid or alkali;—for since these are remingled with the liquid as fast as they are determined to their respective electrodes, and the heat evolved by this remingling appears in column 7, the equivalents of resistance due to these transfers appear in column 10, and are subtracted from the compound resistances in column 5.

According to these principles, all the numbers in column 11 should be equal; and, indeed, they do not differ from each other more than might have been anticipated from the nature of the experiments. An error of only the fourth part of a degree in column 7 is sufficient to produce the greatest deviation from the mean in column 11; and the results are likewise dependent upon the accuracy with which the resistance of the battery is ascertained.

1·35, the mean of column 11, will very nearly represent the intensity or electromotive force required for the separation of the elements of water and the assumption by them of the gaseous state. By these means heat becomes "latent," and a reaction on the intensity of the battery takes place without the evolution of free heat. It is most interesting to inquire what part of the whole intensity is due to each action. I have endeavoured to ascertain it in the following manner:—

I removed the thick copper wire of the galvanometer, and substituted for it a coil of a foot diameter, consisting of 200 turns of well-covered copper wire $\frac{1}{40}$ of an inch in thickness. As the resistance to conduction of a voltaic pair of 3-inch-square plates is seldom greater than that of a yard of this thin wire, the galvanometer thus fitted could be depended on as an indicator of the intensity of different arrangements, particularly when care was taken to make their resistances as nearly equal to each other as possible. In this way, calling the intensity of a Daniell's cell unity, I found the intensity of Mr. Grove's arrangement, consisting of platinum in nitric acid associated with amalgamated zinc in dilute sulphuric acid, to be 1·732; while that of Mr. Smee's arrangement of platinized platinum and amalgamated zinc plunged in dilute sulphuric acid (avoiding the first effect of immersion*) was 0·731; also that the intensities of similar arrangements in which *iron* was substituted for zinc as the positive metal were respectively 1·14 and 0·149.

Now, on account of the extreme facility with which oxygen parts from nitric acid, there can be no doubt that the intensites of the above arrangements of Mr. Grove very nearly represent the respective affinities of the positive metals for oxygen which is not in the gaseous state. For the separation of the hydrogen

* Becquerel was, I believe, the first who referred the great intensity of a pair at the moment of its immersion to the reaction of the air adhering to the negative plate.

from the oxygen of the water is simultaneous with the union of hydrogen with the oxygen which may almost be regarded as free at the platinum; and these actions neutralizing each other, it follows that the intensity of the current in the case of either positive metal very nearly represents its affinity for oxygen in a non-gaseous state.

But with the pairs on Mr. Smee's arrangement two things tend to counteract the effect of this affinity. One of them is the force required to separate the elements of water; the other is that required to give hydrogen the gaseous form. Hence we have 1·732−0·731=1·001 with the zinc, and 1·14−0·149=0·991 with the iron. We shall take 1 (which is very nearly the mean of these numbers) as the electromotive force or intensity necessary to separate the elements of water and give hydrogen the gaseous state.

Again, I placed platinum wires as electrodes in a solution of sulphate of zinc, and, connecting them with the long-wire galvanometer, I found the currents which passed when three and then four cells of Daniell were used to be in the ratio of 160 to 615. Hence 2·648 is the resistance to electrolysis of this solution. Of it, part is due to the separation of zinc from oxygen, and part to the transfer of sulphuric acid from the oxide of zinc to the water. We must eliminate the latter; and I know not a better way of doing so than by converting 2°·82, the heat evolved when oxide of zinc is dissolved in dilute sulphuric acid, into 0·46, its equivalent of resistance to electrolysis, and subtracting this from 2·648. Thus we obtain 2·188* as the intensity occupied in separating zinc from oxygen, and giving the latter the gaseous state.

But we have seen that the intensity of the union of non-gaseous oxygen with zinc is represented by 1·732. Therefore

*. I think this number may be rather over the truth, for reasons afterwards stated; but since I have found by an experiment, in which oxy-water was substituted for the nitric acid of a Grove's cell, that the intensity of his arrangement (why, I cannot imagine) is probably greater than is represented by the union of non-gaseous oxygen with zinc, there is probably a compensating error.

2·188−1·732=0·456, the intensity due to the assumption by oxygen of the gaseous state.

Again, the resistance to electrolysis of solution of sulphate of copper is 1·702; and although it is difficult to ascertain with precision the quantity of heat evolved by the solution of oxide of copper in dilute sulphuric acid, I am persuaded that it is not less than 3° nor more than 4° per equivalent. Suppose it to be 3°·5=0·571 of resistance to electrolysis. Then 1·702−0·571=1·131, the intensity due to the separation of oxide of copper into metal and gas. But 0·731 represents the intensity of the union of non-gaseous oxygen with copper; therefore 1·131−0·731=0·4, a result not widely different from what we obtained with zinc. We may, I think, take 0·45 as a near approximation to the intensity due to the gaseous condition of oxygen.

As we have already stated that the intensity necessary to separate oxygen from hydrogen, and give the latter the gaseous state, is almost exactly 1, 1·45 is, according to the above calculations, the intensity required to electrolyze water. This is not very widely different from 1·35*, the mean of the numbers in column 11: 0·45 resistance to electrolysis is equal to 2°·76. It would be curious to ascertain whether the same amount of caloric would be evolved by the mechanical condensation of eight grains of oxygen gas.

We have given 1·35, 1·131, and 2·188 as the intensities due to the separation of gaseous oxygen from hydrogen gas, copper, and zinc. The quantities of heat corresponding with these are respectively 8°·27, 6°·93, and 13°·41; which are therefore the quantities of heat per pound of water which, according to the above data, should be evolved by the combustion of 1 gr. of hydrogen gas, 31·6 grs. of copper, and 33 grs. of zinc. To the first of these, however, about one sixteenth ought to

* If we put hydrogen as requiring an intensity of 0·9 to give it the gaseous condition, the intensity of the union of non-gaseous hydrogen with non-gaseous oxygen would be 1·35−0·9−0·45=0. So that probably in the decomposition of water nearly all the energy is taken up in giving the gases their elastic condition. See Dr. Wright, Phil. Mag. vol. lxvii. p. 178.—*Note*, 1881.

be added on account of the dissolution and recombination of the gases in the electrolytic cell, which has the effect of diminishing the numbers in column 11. On the other hand, the results for copper and zinc are certainly somewhat overstated, on account of that species of resistance to electrolysis which forms column 10. The error thus arising is not indeed great, and in the case of copper is, I have ascertained, not more than 0·13; but nevertheless it ought to be eliminated, for the same reason as we eliminated, in the case of water, the resistances in column 10. After making these corrections, as far as we are able, the above quantities of heat become altered to 8°·79, 6°·14, and 12°·62; which approximate pretty closely to 8°·98, 5°·18, and 10°·98, the quantities of heat of combustion which Dulong obtained by his experiments.

In concluding, I would make a few general observations:—

1st. In an electrolytic cell there are three distinct obstacles to the flow of a voltaic current. The first of these is the ordinary *resistance to conduction*; the second is *resistance to electrolysis without* the necessity of *chemical change*, arising simply from chemical repulsion; and the third is *resistance to electrolysis accompanied by chemical changes*.

2nd. By the first of these heat is evolved, exactly as it is by a wire, according to the amount of the resistance and the square of the current; and it is thus that a part of the heat belonging to the chemical actions of the battery is evolved. By the second, a reaction in the *intensity* of the battery occurs; and wherever it exists, heat is evolved exactly equivalent to the loss of heating-power of the battery arising from its diminished intensity. But the third resistance differs from the second inasmuch as the heat due to its reaction is rendered latent, and is thus lost by the circuit.

3rd. Hence it is that, however we arrange the voltaic apparatus, and whatever cells for electrolysis we include in the circuit, the caloric of the whole circuit is exactly accounted for by the whole of the chemical changes.

4th. Faraday has shown that the *quantity* of current electricity depends upon the number of chemical equivalents which suffer electrolysis in each cell, and that the *intensity* depends on the sum of chemical affinities. Now both the mechanical and heating powers of a current are, per equivalent of electrolysis in any one of the battery-cells, proportional to its intensity or electromotive force. Therefore the mechanical and heating powers of a current are proportional to each other.

5th. The magnetic electrical machine enables us to convert mechanical power into heat by means of the electric currents which are induced by it. And I have little doubt that, by interposing an electromagnetic engine in the circuit of a battery, a diminution of the heat evolved per equivalent of chemical change would be the consequence, and this in proportion to the mechanical power obtained*.

6th. Electricity may be regarded as a grand agent for carrying, arranging, and converting chemical heat. Suppose two of Daniell's cells in series to be connected, by thick wires, with platinized plates immersed in dilute sulphuric acid. Owing to the near balance of affinities, very little free heat will be evolved, per equivalent of chemical action, in any part of the circuit; and that little will be equivalent to the difference of the intensity of the battery and the intensity due to the electrolysis of water, or to $2-1{\cdot}35=0{\cdot}65$, if we do not regard the heat arising from secondary action in the battery. But then a great transfer of latent heat, equal to $8^{\circ}{\cdot}27$ per equivalent, will take place from the battery to the electrolytic cell; and this by the immediate agency of the current. Again, if a large battery be connected by thick conducting-wires with a coil of very thin wire, nearly the whole of the heat due to the chemical changes taking place in the battery will be evolved by that coil, while the battery itself will remain cool.

7th. Pouillet having deduced from his experiments, repeated with great caution by Becquerel, the general conclusion that, during combustion, the oxygen disengages positive and the combustible negative electricity; and Faraday having proved that a constant quantity of electricity is associated with the

* I am preparing for experiments to test the accuracy of this proposition.—*Note*, Feb. 18th, 1843.

combining proportions of all bodies, it only remained to prove that the *intensity* of the electricity passing from the oxygen to the combustible at the moment of their union is just that which is equivalent to the actual heat of combination. This I have attempted; and in so doing have met, I think it will be admitted, with such success as to put the beautiful electrical theory of chemical heat, first suggested by Davy and Berzelius, beyond all question.

Appendix.

[Read February 20, 1844.]

Since the above paper was read I have found that the intensity of Mr. Grove's voltaic arrangement may be increased in the ratio of 100 to 118 by the use of a mixture of peroxide of lead and sulphuric acid instead of nitric acid. Hence it would appear that the intensity of the nitric-acid battery is considerably less than the intensity due to the union of the positive metal with oxygen. Some of the foregoing calculations, based on the intensity of Grove's battery, therefore, require correction*.

I have once or twice made use of the terms "latent heat," "caloric," &c.; but I wish it to be understood that those words were only employed because they conveniently expressed the facts brought forward. I was then as strongly attached to the theory which regards heat as motion among the particles of matter as I am now. In this spirit Proposition 6 was written, the correctness of which I have since established by experiment.

There are many phenomena which cannot be accounted for by the theory which recognizes heat as a substance; and there are several which, though sometimes adduced as triumphant objections to the other theory, tend, when rightly considered, only to confirm it. The heat of fluidity

* I do not think so. For in the peroxide-of-lead arrangement there is an additional source of electromotive force—the affinity of sulphuric acid for oxide of lead.—J. P. J., 1881

may very naturally be regarded as the momentum or mechanical force necessary to overcome the aggregation of particles in the solid state. The heat of vaporization may be regarded partly as the mechanical force requisite to overcome the aggregated condition of atoms in the fluid state, and partly as the force requisite to overcome atmospheric pressure. Again, the heat of combination is only the manifestation, in another form, of the mechanical force with which atoms combine; on the other hand, the phenomena of electrolysis by the voltaic battery give us positive proof that the mechanical force of the current requisite to procure the decomposition of an electrolyte is the equivalent of the heat due to the recombination of the elements. Thus it appears that electricity is a grand agent for converting heat and the other forms of mechanical power into one another.

I have lately been at some pains in framing the following theory of heat. Setting out with the discovery by Faraday, that each atom is associated with the same absolute quantity of electricity, I assume that these atmospheres of electricity revolve with enormous rapidity round their respective atoms; that the momentum of the atmospheres constitutes "caloric," while the velocity of their exterior circumferences determines temperature. This theory may be made to apply to the phenomena of conduction, and satisfies the law of Boyle and Mariotte with respect to elastic fluids. When applied to the doctrine of specific heat, it demands the extension of the law of Dulong and Petit to all bodies, whether compounds or not, and points out the following general law, applicable to all bodies except, perhaps, compound gases, viz. *the specific heat of a body is proportional to the number of atoms in combination divided by the atomic weight*—a law which agrees very well with the results of experiment when some atomic weights (as, for instance, those of hydrogen, nitrogen, and chlorine) are halved, while others (as that of carbon) are doubled. According to this theory, the zero of temperature is only 480° below the freezing-point, indicating that the momentum of the revolving atmospheres of electricity in a pound of water at the freezing-point is equal to a mechanical force able to raise a weight of about 400,000 pounds to the height of one foot.

February 20, 1844.

On the Calorific Effects of Magneto-Electricity, and on the Mechanical Value of Heat. By J. P. JOULE, *Esq.**

[Phil. Mag. ser. 3. vol. xxiii. pp. 263, 347, and 435; read before the Chemical Section of Mathematical and Physical Science of the British-Association meeting at Cork on the 21st of August, 1843.]

IT is pretty generally, I believe, taken for granted that the electric forces which are put into play by the magneto-electrical machine possess, throughout the whole circuit, the same calorific properties as currents arising from other sources. And indeed when we consider heat not as a *substance*, but as a *state of vibration*, there appears to be no reason why it should not be induced by an action of a simply mechanical character, such, for instance, as is presented in the revolution of a coil of wire before the poles of a permanent magnet. At the same time it must be admitted that hitherto no experiments have been made decisive of this very interesting question; for all of them refer to a particular part of the circuit only, leaving it a matter of doubt whether the heat observed was *generated*, or merely *transferred from the coils* in which the magneto-electricity was induced, the coils themselves becoming cold. The latter view did not appear untenable without further experiments, considering the facts which I had already succeeded in proving, viz. that the heat evolved by the voltaic battery is *definite* † for the chemical changes taking place at the same time; and that the heat rendered

* The experiments were made at Broom Hill, near Manchester.

† Phil. Mag. ser. 3. vol. xix. p. 275.

"latent" in the electrolysis of water is at the expense of the heat which would otherwise have been evolved in a free state by the circuit *—facts which, among others, might seem to prove that *arrangement* only, not *generation* of heat, takes place in the voltaic apparatus, the simply conducting parts of the circuit evolving that which was previously latent in the battery. And Peltier, by his discovery that cold is produced by a current passing from bismuth to antimony, had, I conceived, proved to a great extent that the heat evolved by thermo-electricity is transferred † from the heated solder, no heat being *generated.* I resolved therefore to endeavour to clear up the uncertainty with respect to magneto-electrical heat. In this attempt I have met with results which will, I hope, be worthy the attention of the British Association.

Part I.—*On the Calorific Effects of Magneto-Electricity.*

The general plan which I proposed to adopt in my experiments under this head, was to revolve a small compound electro-magnet, immersed in a glass vessel containing water, between the poles of a powerful magnet, to measure the electricity thence arising by an accurate galvanometer, and to ascertain the calorific effect of the coil of the electro-magnet by the change of temperature in the water surrounding it.

The revolving electro-magnet was constructed in the following manner:—Six plates of annealed hoop-iron, each 8 inches long, $1\frac{1}{8}$ inch broad, and $\frac{1}{16}$ inch thick, were insulated from each other by slips of oiled paper, and then bound tightly together by a ribbon of oiled silk. Twenty-one yards of copper wire $\frac{1}{18}$ inch thick, well covered with

* 'Memoirs of the Literary and Philosophical Society of Manchester,' 2nd series, vol. vii. p. 97.

† The quantity of heat thus transferred is, I doubt not, proportional to the square of the difference between the temperatures of the two solders. I have attempted an experimental demonstration of this law, but, owing to the extreme minuteness of the quantities of heat in question, I have not been able to arrive at any satisfactory result.

silk, were wound on the bundle of insulated iron plates, from one end of it to the other and back again, so that both of the terminals were at the same end.

Having next provided a glass tube sealed at one end, the length of which was $8\frac{3}{4}$ inches, the exterior diameter 2·33 inches, and the thickness 0·2 of an inch, I fastened it in a round hole, cut out of the centre of the wooden revolving piece *a*, fig. 44. The glass was then covered with tinfoil,

Fig. 44. Scale $\frac{1}{12}$.

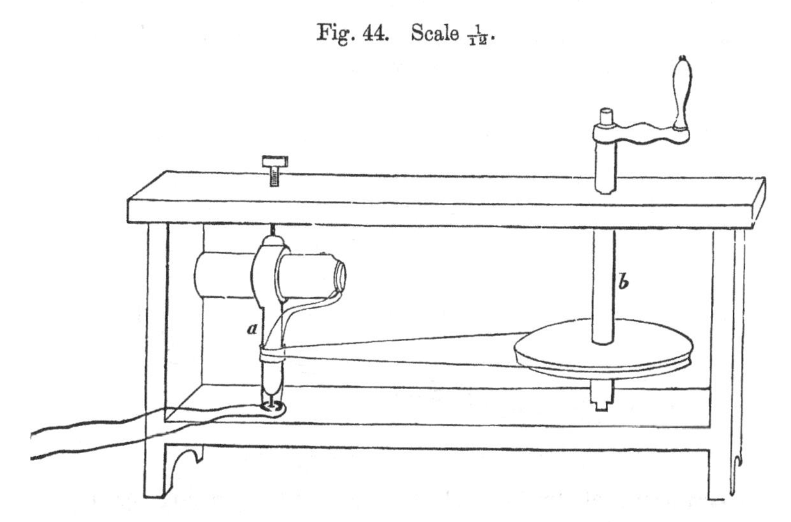

excepting a narrow slip in the direction of its length, which was left in order to interrupt magneto-electrical currents in the tinfoil during the experiments. Over the tinfoil small cylindrical sticks of wood were placed at intervals of about an inch, and over these again a strip of flannel was tightly bound, so as to inclose a stratum of air between it and the tinfoil. Lastly, the flannel was well varnished. By these precautions the injurious effects of radiation, and especially of convection of heat in consequence of the impact of air at great velocities of rotation, were obviated to a great extent.

The small compound electro-magnet was now put into the tube, and the terminals of its wire, tipped with platinum, were arranged so as to dip into the mercury of a commutator *, consisting of two semicircular grooves cut out of the base of the frame, fig. 44. By means of wires connected with the mercury of the commutator, I could connect the revolving electro-magnet with a galvanometer or any other apparatus.

In the first experiments I employed two electro-magnets (formerly belonging to an electro-magnetic engine) for the purpose of inducing the magneto-electricity. They were placed with two of their poles on opposite sides of the revolving electro-magnet, and the other two joining each other beneath the frame. I have drawn fig. 45 representing these

Fig. 45. Scale $\frac{1}{14}$.

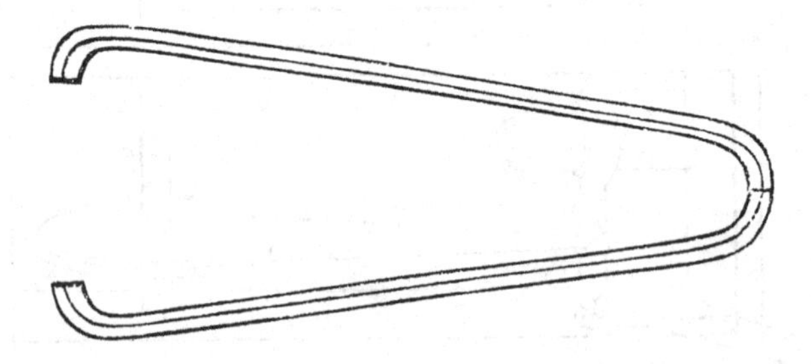

electro-magnets by themselves, to prevent confusing fig. 44. The iron of which they were made was 1 yard 6 inches long, 3 inches broad, and $\frac{1}{2}$ inch thick. The wire which was wound upon them was $\frac{1}{20}$ inch thick; it was arranged so as to form a sixfold conductor a hundred yards long.

The following is the method in which my experiments were made:—Having removed the revolving piece from its place (which is done with great facility by lifting the top of the frame, and with it the brass socket in which the upper steel pivot of the revolving piece works), I filled the

* I had made previous experiments in order to ascertain the best form of commutator, but found none to answer my purpose as well as the above. I found an advantage in covering the mercury with a little water. The steadiness of the needle of the galvanometer during the experiments proved the efficacy of this arrangement.

tube containing the small compound electro-magnet with $9\frac{3}{4}$ oz. of water. After stirring the water until the heat was equally diffused, its temperature was ascertained by a very delicate thermometer, by which I could estimate a change of temperature equal to about $\frac{1}{50}$ of Fahrenheit's degree. A cork covered with several folds of greased paper was then forced into the mouth of the tube, and kept in its place by a wire passing over the whole, and tightened by means of one or two small wooden wedges. The revolving piece was then restored to its place as quickly as possible, and revolved between the poles of the large electro-magnets for a quarter of an hour, during which time the deflections of the galvanometer and the temperature of the room were carefully noted. Finally, another observation with the thermometer detected any change that had taken place in the temperature of the water.

Notwithstanding the precautions taken against the injurious effects of radiation and convection of heat, I was led into error by my first trials: the water had lost heat, even when the temperature of the room was such as led me to anticipate a contrary result. I did not stop to inquire into the cause of the anomaly, but I provided effectually against its interference with the subsequent results by alternating the experiments with others made under the same circumstances, except as regards the communication of the battery with the stationary electro-magnets, which was in these instances broken. And to avoid any objection which might be made with regard to the heat, however trifling, evolved by the wires of the large electro-magnets, the thermometer employed in registering the temperature of the air was situated so as to receive the influence arising from that source equally with the revolving piece.

I will now give a series of experiments in which six Daniell's cells, each 25 inches high and $5\frac{1}{2}$ inches in diameter, were alternately connected and disconnected with the large stationary electro-magnets. The galvanometer, connected through the commutator with the revolving electro-magnet, had a coil a foot in diameter, consisting of five turns of copper wire, and a needle six inches long. Its deflections could be turned into quantities of current by means of a table constructed from previous experiments. The galvanometer was situated so as to be out of the reach of the attractions of the large electro-magnets, and every other precaution was taken to render the experiments worthy of reliance. The rotation was in every instance carried on for exactly a quarter of an hour.

Series No. 1.

		Revolutions of Electro-Magnet per minute.	Deflections of Galvanometer of 5 turns.	Mean Temperature of Room.	Mean Difference.	Temperature of Water. Before.	Temperature of Water. After.	Loss or Gain.
April 15, P.M.	Battery contact broken.	600	0° 0′	54°·69	0°·19+	54°·90	54°·85	0°·05 loss
	Battery in connexion.	600	21 0	54·67	0·20+	54·85	54·88	0·03 gain
	Battery contact broken.	600	0 0	54·61	0·24+	54·88	54·83	0·05 loss
	Battery in connexion.	600	24 0	54·65	0·23+	54·85	54·92	0·07 gain
	Mean, Battery in connexion.	600	22 30	..	0·21+	..	..	0·05 gain
	Mean, Battery contact broken.	600	0 0	..	0·21+	..	..	0·05 loss
	Corrected Result.	600	22° 30′=0·177* of cur. mag.-elect.					0·10 gain

Having thus detected the evolution of heat from the coil of the magneto-electrical apparatus, my next business was to confirm the fact by exposing the revolving electro-magnet to a more powerful magnetic influence; and to do so with the greater convenience, I determined on the construction of a new stationary electro-magnet, by which I might obtain a

* Throughout the paper I have called that quantity of current *unity*, which, passing equably for an hour of time, can decompose a chemical equivalent expressed in grains.

more advantageous employment of the electricity of the battery. Availing myself of previous experience, I succeeded in producing an electro-magnet possessing greater power of attraction from a distance than any other I believe on record. On this account a description of it in greater detail than is absolutely necessary to the subject of this paper will not, I hope, be deemed superfluous.

A piece of half-inch boiler plate-iron was cut into the shape represented by fig. 46. Its length was 32 inches; its

Fig. 46. Scale $\frac{1}{12}$.

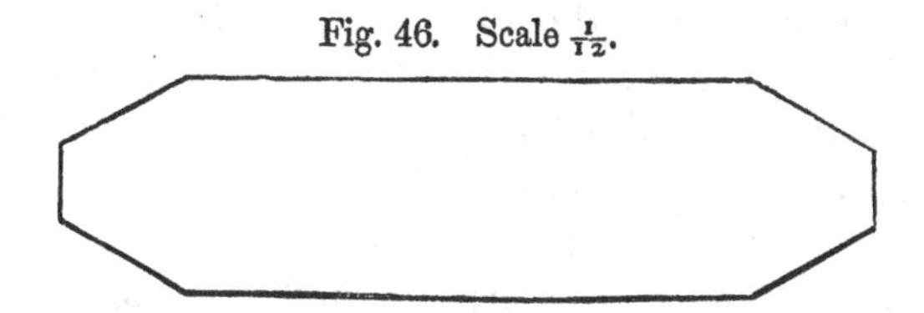

breadth in the middle part 8 inches, at the ends 3 inches. It was bent nearly into the shape of the letter U, so that the shortest distance between the poles was slightly more than 10 inches.

Twenty-two strands of copper wire, each 106 yards long and about one twentieth of an inch in diameter, were now bound tightly together with tape. The insulated bundle of wires, weighing more than sixty pounds, was then wrapped upon the iron, which had itself been previously insulated by a fold of calico. Fig. 47 represents the electro-magnet in its completed state.

Fig. 47. Scale $\frac{1}{12}$.

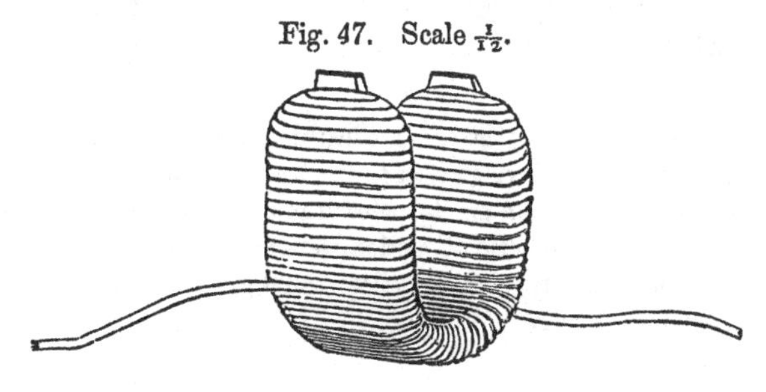

In arranging the voltaic battery for its excitation, care was taken to render the resistance to conduction of the battery equal, as nearly as possible, to that of the coil, Prof. Jacobi having proved that to be the most advantageous arrangement. Ten of my large Daniell's cells, arranged in a series of five double pairs, fulfilled this condition very well, producing a magnetic energy in the iron superior to any thing I had previously witnessed. I will mention the results of a few experiments in order to give some definite idea of it.

1st. The force with which a bar of iron three inches broad and half an inch thick was attracted to the poles was equal, at the distance of $\frac{1}{16}$ of an inch, to 100 lb., at $\frac{1}{4}$ of an inch to 30 lb., at $\frac{1}{2}$ an inch to $10\frac{1}{2}$ lb., and at 1 inch to 4 lb. 13 oz.* 2nd. A small rod of iron three inches long, weighing 148 grains, held vertically under one of the poles, would jump through an interval of $1\frac{3}{4}$ inch: a needle three inches long, weighing 4 grains, would jump from a distance of $3\frac{1}{4}$ inches.

Having fixed the electro-magnet just described with its poles upwards and on opposite sides of the revolving electro-magnet, I arranged to it the battery of ten cells, in a series of five double pairs, and, experimenting as before, I obtained a second series of results. The galvanometer used in the present instance was in every respect similar to that previously described, with the exception of the coil, which now consisted of a single turn of thick copper wire. Great care was taken to prevent, by its distance from and relative position with the electro-magnet, any interference of the latter with its indications.

* The above electro-magnet being constructed for a specific purpose, was not adapted for displaying itself to the best advantage in one respect. On account of the extension of its poles (three inches by half an inch) many of the lines of magnetic attraction were necessarily in very oblique directions. Theoretically, circular poles should give the greatest attraction from small distances.

No. 2.

Date		Revolutions of Electro-magnet per minute.	Deflections of Galvanometer of one turn.	Mean Temperature of Room.	Mean Difference.	Temperature of Water. Before.	Temperature of Water. After.	Loss or Gain.
May 6, A.M.	Battery in connexion.	600	22° 0′	58°·93	0°·17+	58°·20	60°·00	1°·80 gain
May 6, A.M.	Battery contact broken.	600	0 0	59·60	0·40+	60·02	59·98	0·04 loss
May 6, A.M.	Battery in connexion.	600	24 0	59·55	1·23+	59·90	61·67	1·77 gain
May 6, P.M.	Battery contact broken.	600	0 0	59·45	0·19+	59·78	59·50	0·28 loss
May 6, P.M.	Battery in connexion.	600	24 45	58·30	0·05+	57·35	59·35	2·00 gain
May 8, A.M.	Battery in connexion.	600	22 0	57·74	0·32+	57·28	58·83	1·55 gain
May 8, A.M.	Battery contact broken.	600	0 0	58·35	0·49+	58·83	58·85	0·02 gain
May 8, A.M.	Battery in connexion.	600	21 20	58·73	0·78+	58·83	60·20	1·37 gain
	Mean, Battery in connexion.	600	22 49	..	0·51+	..	..	1·70 gain
	Mean, Battery contact broken.	600	0 0	..	0·36+	..	..	0·10 loss
	Corrected Result.	600	22° 49′=0·902 of cur. mag.-elect.					1·84 gain

The corrected result is obtained, as before, by adding the loss sustained when contact with the battery was broken to the heat gained when the battery was in connexion. I have in the present instance, however, made a further correction of 0°·04 on account of the difference between the *mean differences* 0°·51 and 0°·36. The ground of this correction is the result of a previous experiment, in which, by revolving the apparatus at 94° in an atmosphere of 60°, the water sustained a loss of 7°·6, or about one quarter of the difference between the temperature of the atmosphere and the mean temperature of the water.

With the same electro-magnet, but using a battery of only four cells arranged in a series of two double pairs, by which I expected to obtain about half as much magnetism in the iron, the following results were obtained :—

No. 3.

		Revolutions of Electro-magnet per minute.	Deflections of Galvanometer of 5 turns.	Mean Temperature of Room.	Mean Difference.	Temperature of Water. Before.	Temperature of Water. After.	Loss or Gain.
May 8, P.M.	Battery in connexion.	600	38° 0′	57°·00	0°·02 –	56°·73	57°·23	0°·50 gain
	Battery contact broken.	600	0 0	57·25	0·0	57·23	57·27	0·04 gain
	Battery in connexion.	600	38 30	57·53	0·09 +	57·35	57·90	0·55 gain
May 9, P.M.	Battery in connexion.	600	39 45	56·37	0·45 –	55·60	56·25	0·65 gain
	Battery contact broken.	600	0 0	56·75	0·39 –	56·27	56·45	0·18 gain
	Battery in connexion.	600	38 45	57·14	0·37 –	56·50	57·05	0·55 gain
	Mean, Battery in connexion.	600	38 45	..	0·19 –	..	..	0·56 gain
	Mean, Battery contact broken.	600	0 0	..	0·19 –	..	..	0·11 gain
	Corrected Result.	600	38° 45′ = 0·418 of cur. mag.-elect.					0·45 gain

In the next experiments a battery of ten cells in a series of five double pairs was used for the purpose of exciting the large stationary electro-magnet. But, dismissing the galvanometer and the other extra parts of the circuit, I connected the terminal wires of the revolving electro-magnet together, so as to obtain the whole effect of the magneto-electricity The resistance of the coil of the revolving electro-magnet being to that of the whole circuit employed in the experiments No. 2 as 1 : 1·13, and 0·902 of current being obtained in those experiments, I expected to obtain the calorific effect of 1·019 in the new series.

No. 4.

		Revolutions of Electro-magnet per minute.	Mean Temperature of Room.	Mean Difference.	Temperature of Water, Before.	Temperature of Water, After.	Loss or Gain.
May 10, P.M.	Battery in connexion.	600	56°·85	0°·61−	54°·98	57°·50	2°·52 gain
	Battery contact broken.	600	57·37	0·12+	57·48	57·50	0·02 gain
	Battery in connexion.	600	57·52	1·08+	57·48	59·73	2·25 gain
	Mean, Battery in connexion.	600	..	0·23+	..	..	2·38 gain
	Mean, Battery contact broken.	600	..	0·12+	..	..	0·02 gain
	Corrected Result.	600	1·019 of cur. mag.-elect.				2·39 gain

It seemed to me very desirable to repeat the experiments, substituting *steel* magnets for the stationary electro-magnets hitherto used. With this intention I constructed two magnets, each consisting of a number of thin plates of hard steel, —an arrangement which we owe to Dr. Scoresby. My metal was, unfortunately, not of very good quality; but nevertheless an attractive force was obtained sufficiently powerful to overcome the gravity of a small key weighing 47 grains, placed at the distance of three eighths of an inch. The following results were obtained by revolving the electro-magnet acted on by the poles of the steel magnets.

No. 5.

		Revolutions of Electro-magnet per minute.	Deflections of Galvanometer of 5 turns.		Mean Temperature of Room.	Mean Difference.	Temperature of Water.		Loss or Gain.
							Before.	After.	
May 16, A.M.	Circuit complete.	600	26°	0'	59°·72	0°·0	59°·73	59°·70	0°·03 loss
	Circuit broken.	600	0	0	59·82	0·20 –	59·70	59·55	0·15 loss
	Circuit complete.	600	29	0	59·95	0·41 –	59·55	59·53	0·02 loss
	Circuit broken.	600	0	0	59·58	0·12 –	59·52	59·40	0·12 loss
	Circuit complete.	600	27	0	59·65	0·25 –	59·40	59·40	0
	Mean, Circuit complete.	600	27	20	..	0·22 –	..	..	0·016 loss
	Mean, Circuit broken.	600	0	0	..	0·16 –	..	..	0·135 loss
	Corrected Result.	600	27° 20′=0·236 of cur. mag.-elect.						0·10 gain

In order to obtain the whole calorific effect, I now, as in Series No. 4, connected the terminal wires of the revolving electro-magnet, and alternated the experiments with others in which that connexion was broken. The resistance of the coil of the revolving electro-magnet being to the resistance of the whole circuit used in the experiments marked No. 5 as 1 : 1·44, and 0·236 of current electricity being obtained in those experiments, I expected to obtain in the present series the calorific effect of 0·34 of current magneto-electricity.

No. 6.

		Revolutions of Electro-magnet per minute.	Mean Temperature of Room.	Mean Difference.	Temperature of Water. Before.	Temperature of Water. After.	Loss or Gain.
May 17, A.M.	Terminals joined.	600	59°·07	0°·20–	58°·82	58°·92	0°·10 gain
	Terminals separated.	600	59·07	0·22–	58·92	58·78	0·14 loss
	Terminals joined.	600	58·96	0·20–	58·75	58·78	0·03 gain
	Terminals separated.	600	58·88	0·18–	58·78	58·63	0·15 loss
	Mean, Terminals joined.	600	..	0·20–	..	..	0·065 gain
	Mean, Terminals separated.	600	..	0·20–	..	..	0·145 loss
	Corrected Result.	600	0·34 of cur. mag.-elect.				0·21 gain

Although any considerable development of electrical currents in the iron of the revolving electro-magnet was prevented by its disposition in a number of thin plates insulated from each other, I apprehended that they might, under a powerful inductive influence, exist separately in each plate to such an extent as to produce an appreciable quantity of heat. To ascertain the fact, the terminals of the wire of the revolving electro-magnet were insulated from each other, while the latter was revolved between the poles of the large electro-magnet excited by ten cells in a series of five double pairs. The experiments were alternated with others in which contact with the battery was broken. As we shall hereafter give in detail experiments of the same class, it will not be necessary to do more at present than to state that the mean result of the present series, consisting of eight trials, gave 0°·28 as the quantity of heat evolved by the iron alone.

We are now able to collect the results of the preceding

experiments, so as to discover the laws by which the development of the heat is regulated. The fourth column of the following table, containing the heat due to the currents circulating in the iron alone, is constructed on the basis of a law which we shall subsequently prove, viz. *the heat evolved by a bar of iron revolving between the poles of a magnet is proportional to the square of the inductive force.* Column 5 gives the heat evolved by the coils of the electro-magnet alone. No elimination is required for the results of Series Nos. 5 and 6, because in them the iron of the revolving electro-magnet was subject to the influence of the steel magnets in the alternating as well as in the other experiments.

TABLE I.

Series of Experiments.	Current Magneto-electricity.	Heat actually evolved.	Correction for Currents in the Iron.	Corrected Heat.	Squares of Numbers proportional* to those in column 2.	Heat due to Voltaic Currents of the intensities given in col. 2.	The Numbers of column 7 multiplied by $\frac{4}{3}$.
No. 1.	0·177	$0\overset{\circ}{\cdot}10$	$0\overset{\circ}{\cdot}02$	$0\overset{\circ}{\cdot}08$	$0\overset{\circ}{\cdot}062$	$0\overset{\circ}{\cdot}040$	$0\overset{\circ}{\cdot}053$
No. 2.	0·902	1·84	0·28	1·56	1·614	1·040	1·386
No. 3.	0·418	0·45	0·09	0·36	0·346	0·224	0·299
No. 4.	1·019	2·39	0·28	2·11	2·060	1·327	1·769
No. 5.	0·236	0·10	0	0·10	0·109	0·071	0·091
No. 6.	0·340	0·21	0	0·21	0·229	0·148	0·197
1.	2.	3.	4.	5.	6.	7.	8.

On comparing the corrected results in column 5 with the squares of magneto-electricity given in column 6, it will be

* This proportion is arbitrary, and greater than that in Table 2. Column 6 is superfluous. It will be found that in both tables the numbers in column 7 are to those in column 5 nearly as 2 to 3, which difference, as explained in the text, is owing to the uniform intensity of the current in the former case and its pulsatory character in the latter.—*Note*, 1881.

abundantly manifest that *the heat evolved by the coil of the magneto-electrical machine is proportional* (*cæteris paribus*) *to the square of the current.*

Column 7, containing the heat due to voltaic currents of the quantities stated in column 2, is constructed on the basis of three very careful experiments on the heat evolved by passing currents through the coil of the small compound electro-magnet. I observed an increase in the temperature of the water equal to 5°·3, 5°·46, and 5°·9 respectively, when 2·028, 2·078, and 2·145 of current voltaic electricity were passed, each during a quarter of an hour, through the coil. Reducing the first and second experiments to the electricity of the third according to the squares of the current, we have 5°·93, 5°·82, and 5°·9 for 2·145 of current. The mean of these is 5°·88, a datum from which the theoretical results of the preceding and subsequent tables are calculated.

But in comparing the heat evolved by magneto- with that evolved by voltaic electricity, we must remember that the former is propagated by pulsations, the latter uniformly. Now, since the square of the mean of unequal numbers is always less than the mean of their squares, it is obvious that the magnetic effect at the galvanometer will bear a greater proportion to the heat evolved by the voltaic, than the magneto-electricity; so that it is impossible to institute a strict comparison without ascertaining previously the intensity of the magneto-electricity at every instant of the revolution of the revolving electro-magnet. I have not been able to devise any very accurate means for attaining this object; but judging from the comparative brilliancy of the sparks when the commutator was arranged so as to break contact with the mercury at different positions of the revolving electro-magnet with respect to the poles of the stationary electro-magnet, there appeared to be but little variation in the intensity of the magneto-electricity during $\frac{3}{4}$ of each revolution. The remaining $\frac{1}{4}$ (during which the revolving electro-magnet passes the poles of the stationary electro-magnet) is occupied in the reversal of the direction of the electricity. In the experiments all flow of electricity during this $\frac{1}{4}$ is cut off by the divisions of the commutator. In illustration of this I have drawn fig. 48, in which the direction and intensity of the magneto-electricity are represented by ordinates A x, &c., perpendicular to the straight line A B C D E; the intermediate spaces B C, D E, &c. represent the time during which the electricity is wholly cut off by the divisions of the commutator. Were A x x' B &c. perfect rectangles, it is obvious that the heat due to a given deflection of the galvanometer would be

Fig. 48.

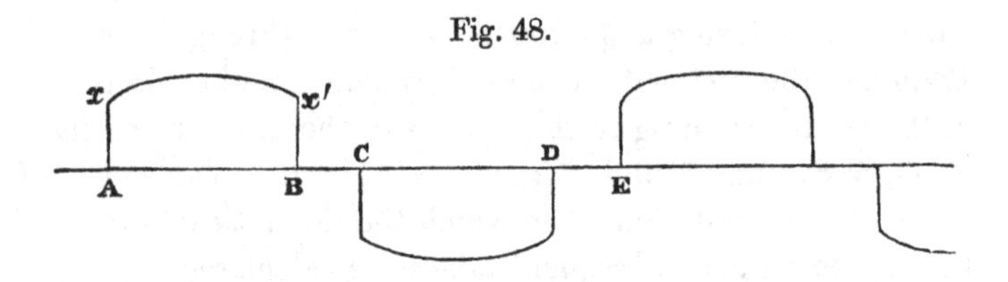

$\frac{4}{3}$ of that due to the same deflection and a uniform current, and column 8 of the table would contain exact theoretical results. But as this is not precisely the case, the numbers in that column are somewhat under the truth.

Bearing this in mind in the comparison of columns 5 and 8, it will, I think, be admitted that the experiments afford decisive evidence that *the heat evolved by the coil of the magneto-electrical machine is governed by the same laws as those which regulate the heat evolved by the voltaic apparatus, and exists also in the same quantity under comparable circumstances.*

Although very little doubt could exist with regard to the heating power of magneto-electricity *beyond* the coil, I thought it would nevertheless be well to follow it there, in order to render the investigation more complete: I am not aware of any previous experiments of the kind.

I immersed five or six yards of insulated copper wire of $\frac{1}{40}$ inch diameter in a flask holding about 12 oz. of water. The terminals of the wire were connected on the one hand with the galvanometer of five turns and on the other with the commutator, and the circuit was completed by a wire extending from the galvanometer to the other compartment of the commutator. The revolving electro-magnet, being now subjected to the inductive influence of the large electro-magnet excited by ten cells in a series of five double pairs, was rotated at the rate of 600 revolutions per minute during a quarter of an hour. The needle of the galvanometer, which remained, as usual, pretty steady, indicated a mean deflection of $32^\circ\ 40'=0{\cdot}31$ of current; and the heat evolved was found to be $0^\circ{\cdot}46$, after the correction on account of the temperature of the surrounding air had been applied. Another experiment gave me $0^\circ{\cdot}4$ for 0·286. The mean of the two is $0^\circ{\cdot}43$ for 0·298 current magneto-electricity.

By passing a voltaic current from four cells in series through the wire, I found that 2·02 of current flowing uniformly evolved $12^\circ{\cdot}0$ in a quarter of an hour. Reducing this to 0·298 of current we have $\left(\frac{0{\cdot}298}{2{\cdot}02}\right)^2 \times 12^\circ = 0^\circ{\cdot}261$. The product of this by $\frac{4}{3}$ (on account of the pulsatory character of the magneto-current) gives $0^\circ{\cdot}348$, which, as theory demands, is somewhat less than the quantity found by experiment.

On the Calorific Effects of Magneto- with Voltaic Electricity.

I now proceeded to consider the heat evolved by voltaic currents when they are counteracted or assisted by magnetic induction. For this purpose it was only necessary to introduce a battery into the magneto-electrical circuit: then by turning the wheel in one direction, I could oppose the voltaic current; or, by turning in the other direction, I could increase the intensity of the voltaic by the assistance of the magneto-electricity. In the former case the apparatus possessed all the properties of the electro-magnetic engine; in the latter it presented the reverse, viz. the *expenditure* of mechanical power.

No. 7.

		Revolutions of Electro-magnet per minute.	Deflections of Galvanometer of 1 turn.	Mean Temperature of Room.	Mean Difference.	Temperature of Water.		Loss or Gain.
						Before.	After.	
May 19.	Circuits complete.	600	22° 40′	57°·43	1°·03 −	55°·62	57°·18	1°·56 gain
May 19.	Circuits broken.	600	0 0	57·45	0·41 −	57·08	57·00	0·08 loss
May 20.	Circuits complete.	600	20 45	59·40	0·08 −	58·65	60·00	1·35 gain
May 20.	Circuits broken.	600	0 0	59·40	0·51 +	60·00	59·83	0·17 loss
May 20.	Circuits complete.	600	23 0	59·10	1·29 +	59·78	61·00	1·22 gain
	Mean, Circuits complete.	600	22 8	....	0·06 +	....	....	1·38 gain
	Mean, Circuits broken.	600	....	....	0·05 +	....	....	0·12 loss
	Corrected Result.	600	22° 8′ = 0·864 of current.					1·50 gain

In the preceding series I used the steel magnets previously described, as the inductive force ; and I had two of the large Daniell's cells in series, arranged so as to pass a current of electricity through the revolving electro-magnet and galvanometer. The wheel was turned in the direction which it would have taken had the friction been sufficiently reduced to allow of the motion of the apparatus without assistance.

I give another series, in which every thing else remaining the same, the direction of revolution was reverse, so as to *increase* the intensity of the voltaic electricity by superadding that of the magneto-electricity.

No. 8.

Date		Revolutions of Electro-magnet per minute.	Deflections of Galvanometer of 1 turn.	Mean Temperature of Room.	Mean Difference.	Temperature of Water. Before.	Temperature of Water. After.	Loss or Gain.
May 23.	Circuits complete.	600	30° 15′	60°·55	0°·19+	59°·30	62°·18	2°·88 gain
May 23.	Circuits broken.	600	0 0	62·28	0·48+	62·92	62·60	0·32 loss
May 24.	Circuits complete.	600	29 40	60·90	0·03−	59·50	62·25	2·75 gain
May 24.	Circuits broken.	600	0 0	59·50	0·0	59·50	59·50	0
May 26.	Circuits complete.	600	29 50	61·85	0·18−	60·33	63·02	2·69 gain
May 26.	Circuits broken.	600	0 0	60·90	0·49−	60·50	60·33	0·17 loss
	Mean, Circuits complete.	600	29 55	....	0·02−	....	....	2·77 gain
	Mean, Circuits broken.	600	....	....	0·01−	....	....	0·16 loss
	Corrected Result.	600	29° 55′=1·346 of current.					2·93 gain

Dismissing the steel magnets, which did not appear to have lost any of the magnetic virtue which they possessed at first, I now substituted for them the large stationary electro-magnet, excited by eight of the Daniell's cells arranged in a series of four double pairs. The revolving electro-magnet completed, as before, a circuit containing the galvanometer and two of Daniell's cells in series. The motive power of the apparatus was now so great that it would revolve rapidly in spite of very considerable friction. In order to give the requisite velocity it was necessary, however, to assist the motion by the hand.

No. 9.

		Revolutions of Electro-magnet per minute.	Deflections of Galvanometer of 1 turn.	Mean Temperature of Room.	Mean Difference.	Temperature of Water.		Loss or Gain.
						Before.	After.	
May 31, P.M.	Circuits complete.	600	16°	62°·50	0°·11 –	62°·00	62°·78	0°·78 gain
	Circuits broken.	600	0	63·00	0·23 –	62·73	62·82	0·09 gain
June 1, A.M.	Circuits complete.	600	14	62·65	0·11 –	62·18	62·90	0·72 gain
	Circuits broken.	600	0	63·15	0·20 –	62·90	63·00	0·10 gain
	Mean, Circuits complete.	600	15	....	0·11 –	....	....	0·75 gain
	Mean, Circuits broken.	600	....	....	0·21 –	....	....	0·095 ga.
	Corrected Result.	600	15° = 0·543 of current.					0·68 gain

The following series of results was obtained with the same apparatus, by turning the wheel in the opposite direction.

No. 10.

		Revolutions of Electro-magnet per minute.	Deflections of Galvanometer of 1 turn.	Mean Temperature of Room.	Mean Difference.	Temperature of Water.		Loss or Gain.
						Before.	After.	
June 2, A.M.	Circuits complete.	600	35° 10′	65°·38	0°·62 +	63°·25	68°·75	5°·50 gain
	Circuits broken.	600	0 0	64·73	0·75 +	65·51	65·45	0·06 loss
June 3, A.M.	Circuits complete.	600	37 10	65·10	1·40 +	63·33	69·66	6·33 gain
	Circuits broken.	600	0 0	64·93	1·23 +	66·28	66·05	0·23 loss
	Mean, Circuits complete.	600	36 10	....	1·01 +	....	....	5·915 ga.
	Mean, Circuits broken.	600	....	....	0·99 +	....	....	0·145 loss
	Corrected Result.	600	36° 10′ = 1·845 of current.					6·06 gain

I give two series more, in which only one cell was connected with the revolving electro-magnet, and the revolution was in the direction of the attractive forces. The magneto-electricity

No. 11.

		Revolutions of Electro-magnet per minute.	Deflections of Galvanometer of 1 turn.	Mean Temperature of Room.	Mean Difference.	Temperature of Water. Before.	Temperature of Water. After.	Loss or Gain.
June 3, P.M.	Circuits complete.	350	0	64°·02	0°·38 –	63°·57	63°·72	0°·15 gain
	Circuits broken.	350	0	63·75	0·02 –	63·73	63·73	0
	Circuits complete.	400	0	63·80	0·02 –	63·70	63·86	0·16 gain
	Circuits broken.	400	0	64·35	0·08 –	64·27	64·27	0
	Mean, Circuits complete.	375	0	....	0·20 –	....	....	0·155 ga.
	Mean, Circuits broken.	375	0	....	0·05 –	....	....	0
	Corrected Result.	375	0					0·12 gain

No. 12.

		Revolutions of Electro-magnet per minute.	Deflections of Galvanometer of 1 turn.	Mean Temperature of Room.	Mean Difference.	Temperature of Water. Before.	Temperature of Water. After.	Loss or Gain.
June 5, A.M.	Circuits complete.	600	9° 40′	60°·40	0°·16 –	60°·02	60°·47	0°·45 gain
	Circuits broken.	600	0	60·64	0·17 –	60·47	60·47	0
	Corrected Result.	600	9° 40′ = 0·34	current in the opposite direction.				0·45 gain

was so intense, at a velocity of 600 per minute, as to overpower the intensity of the single cell, causing the needle to be permanently and steadily deflected to between 9° and 10° in the opposite direction. The command of the magneto-electricity over the voltaic current arising from one cell was beautifully illustrated by the sparks at the commutator. Turning slowly, they were bright and snapping*: increasing the rapidity of revolution, they decreased in brightness; until at a velocity of about 370 per minute they ceased altogether. They were plainly visible again when the velocity reached 600 per minute.

The results of the preceding series of experiments are collected in the following table along with theoretical results calculated in precisely the same manner as those of Table I. The correction for heat evolved by the iron of the revolving electro-magnet is estimated at 0°·18, the product of 0°·28 by $\left(\frac{4}{5}\right)^2$; because in the above experiments the large electro-magnet was excited by $\frac{4}{5}$ of the battery used when 0°·28 was obtained. No correction is needed for the series in which the steel magnets were used, because they remained in their places during the alternating experiments.

TABLE II.

Series of Experiments.	Current Magneto-electricity.	Heat actually evolved.	Correction for Currents in the Iron.	Corrected Heat.	Squares of numbers proportional† to those in col. 2.	Heat due to Voltaic Currents of the intensities given in col. 2.	The Numbers of column 7 multiplied by $\frac{4}{3}$.
No. 7.	0·864	1°·50	0°	1°·50	1°·291	0°·954	1°·272
No. 8.	1·346	2·93	0	2·93	3·133	2·316	3·088
No. 9.	0·543	0·68	0·18	0·50	0·510	0·377	0·503
No. 10.	1·845	6·06	0·18	5·88	5·886	4·351	5·801
No. 11.	0	0·12	0·18	−0·06	0	0	0
No. 12.	0·340	0·45	0·18	0·27	0·200	0·148	0·197
1.	2.	3.	4.	5.	6.	7.	8.

* The most splendid sparks are obtained, when the voltaic is assisted by the magneto-electricity, by turning an electro-magnetic engine in a direction contrary to the attractive forces.

† See note to Table I. (p. 136).

In all these experiments, as well as in those collected in Table I., the time occupied by the platinum wires in crossing the divisions of the commutator was found to be exactly $\frac{1}{4}$ of that occupied by an entire revolution; hence the multiplication by $\frac{4}{3}$ in order to obtain true theoretical results on the supposition that the current flows uniformly during $\frac{3}{4}$ of a revolution. It will be observed that these theoretical results are not so much inferior to the experimental results of column 5 as they were in Table I. The principal reason of this arises from the mixture of the constant effect of the battery with the variable magneto-electrical current, as will be readily seen on inspecting figs. 49 and 50, the former of which represents the currents in series No. 9; the latter, those in series No. 10. The dotted roctangles *a b c d*, &c., represent the constant effect of the battery of two cells, which is in one instance diminished, in the other increased by the magneto-electricity.

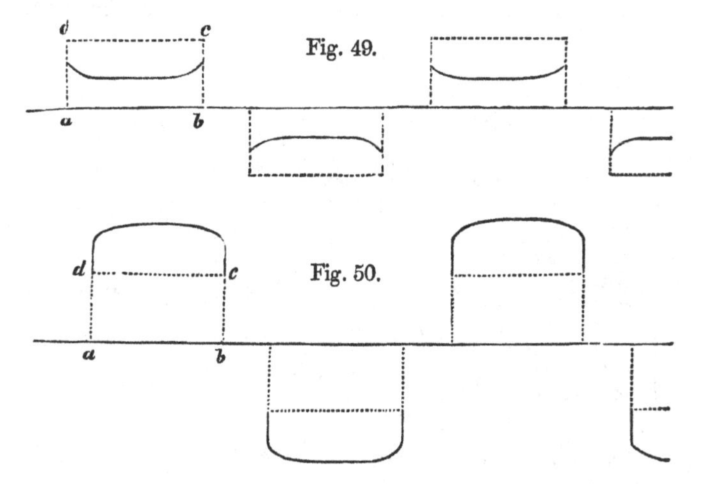

Fig. 49.

Fig. 50.

On comparing columns 6 and 8 with column 5, it is manifest that the law of the square of the electric current still obtains, and is not affected either by the assistance or resistance which the magneto-electricity presents to the voltaic current. Now the increase or diminution of the chemical effects occurring in the battery during a given time is proportional to the magneto-electrical effect, and the heat evolved is always proportional to the square of the current; therefore the heat due to a given chemical action is subject to an increase or to a diminution directly proportional to the intensity of the magneto-electricity assisting or opposing the voltaic current.

We have therefore in magneto-electricity an agent capable by simple mechanical means of destroying or generating heat. In a subsequent part of this paper I shall make an attempt to connect heat with mechanical power in absolute numerical relations. At present we shall turn to a question intimately connected with the previous investigations, and which indeed has already been partly developed.

On the Heat evolved by a Bar of Iron rotating under Magnetic Influence.

Having removed the small electro-magnet from the tube of the revolving piece, I fixed in its stead, in the centre

No. 13.

		Revolutions of the Bar per minute.	Deflections of Galvanometer of 1 turn.	Mean Temperature of Room.	Mean Difference.	Temperature of Water.		Gain or Loss.
						Before.	After.	
June 17, P.M.	Electro-magnet in action.	600	70° 55′	67°·38	0°·35 −	66°·27	67·80	1°·53 gain
	Battery contact broken.	600	..	67·60	0·23 +	67·77	67·90	0·13 gain
	Electro-magnet in action.	600	70 45	67·85	0·67 +	67·85	69·20	1·35 gain
	Battery contact broken.	600	..	68·92	0·30 +	69·18	69·27	0·09 gain
	Mean, Electro-magnet in action.	600	70 50	..	0·16 +	..	..	1·44 gain
	Mean, Battery contact broken.	600	..	..	0·26 +	..	..	0·11 gain
	Corrected Result.	600	70° 50′ = 9·85 current.					1·31 gain

of the tube, a solid cylinder of iron 8 inches long and ¾ inch in diameter. The tube was then, as before, filled with water and rotated for a quarter of an hour between the poles of the large electro-magnet. In the first experiments the electro-magnet was excited by ten cells in a series of five double pairs, a galvanometer being included in the circuit in order to indicate the electric force applied. It was of course placed, as before, so as not to be affected by the powerful attraction of the electro-magnet. The precaution of alternating the experiments was adopted as usual. The results are recorded in No. 13.

Every thing else remaining the same, I now used a battery of six cells arranged in a series of three double pairs to excite the electro-magnet in the experiments of No. 14.

No. 14.

		Revolutions of the Bar per minute.	Deflections of Galvanometer of 1 turn.	Mean Temperature of Room.	Mean Difference.	Temperature of Water.		Gain or Loss.
						Before.	After.	
June 19, A.M.	Electro-magnet in action.	600	64° 10	65°·42	0°·17 −	65°·00	65°·50	0°·50 gain
	Battery contact broken.	600	..	65·55	0·10 −	65·48	65·42	0·06 loss
	Electro-magnet in action.	600	64 10	65·42	0·17 +	65·33	65·86	0·53 gain
	Battery contact broken.	600	..	65·75	0·03 +	65·80	65·77	0·03 loss
	Mean, Electro-magnet in action.	600	64 10	..	0	..	..	0·515 gain
	Mean, Battery contact broken.	600	..	..	0·03 −	..	..	0·045 loss
	Corrected Result.	600	64° 10′ = 6·67 current.					0·56 gain

I give another series obtained with a battery of two cells in series.

No. 15.

		Revolutions of the Bar per minute.	Deflections of Galvanometer of 1 turn.	Mean Temperature of Room.	Mean Difference.	Temperature of Water. Before.	Temperature of Water. After.	Gain or Loss.
June 19, P.M.	Electro-magnet in action.	600	54°	63°65	0°43 –	63°12	63·32	0°20 gain
	Battery contact broken.	600	..	63·80	0·38 –	63·40	63·45	0·05 gain
	Electro-magnet in action.	600	54	63·75	0·20 –	63·45	63·65	0·20 gain
	Battery contact broken.	600	..	64·07	0·37 –	63·68	63·73	0·05 gain
	Mean, Electro-magnet in action.	600	54	..	0·32 –	..	..	0·20 gain
	Mean, Battery contact broken.	600	..	..	0·38 –	..	..	0·05 gain
	Corrected Result.	600	54°=4·17 current.					0·16 gain

The results of the preceding experiments are collected in the following table:—

Table III.

Series of Experiments.	Current employed in exciting the Electro-Magnet.	Heat evolved.	Square of Numbers proportional to those of Column 2.
No. 13.	9·85	1°31	1·2290
No. 14.	6·77	0·56	0·5807
No. 15.	4·17	0·16	0·2203
1	2	3	4

It was discovered by Prof. Jacobi, and by myself also one or two months afterwards*, that the attraction of electro-magnets, either towards one another or for their armatures, is (below the point of saturation †) proportional to the square of the electric force. The *magnetism* in an electro-magnet is therefore simply as the electric force. Consequently the numbers in column 2 are proportional to the magnetic virtue of the electro-magnet. But on comparing columns 3 and 4 together, it will be seen that the heat evolved is as the square of the electricity. Therefore *the heat evolved by a revolving bar of iron is proportional to the square of the magnetic influence to which it is exposed.*

After the preceding experiments there can be no doubt that heat would be evolved by the rotation of non-magnetic substances in proportion to their conducting power, Dr. Faraday having proved the existence of currents in such circumstances, and that their quantity is proportional, *cæteris paribus*, to the conducting power of the body in which they are excited. I have not made any experiments on this subject; but in the next part we shall have occasion to avail ourselves of the good conducting power of copper, in conjunction with the magnetic virtue of the bar of iron, in order to obtain a maximum result from the revolution of a metallic bar.

Part II.—*On the Mechanical Value of Heat.*

Having proved that *heat* is *generated* by the magneto-electrical machine, and that by means of the inductive power of magnetism we can *diminish* or *increase* at pleasure the *heat* due to chemical changes, it became an object of great interest to inquire whether a constant ratio existed between it and the mechanical power gained or lost. For this purpose it was only necessary to repeat some of the previous experiments, and to ascertain, at the same time, the mechanical force necessary in order to turn the apparatus.

* Annals of Electricity, vol. iv. p. 131. Jacobi and Lenz communicated their report to the Petersburg Academy in March 1839—two months previous to the date of my paper.

† I am not aware that Jacobi and Lenz made any limitation to the law.—*Note*, 1881.

To accomplish the latter purpose, I resorted to a very simple device, yet one peculiarly free from error. The axle *b*, fig. 44 (p. 125), was wound with a double strand of fine twine, and the strings (as represented in fig. 51) were carried over very easily-working pulleys, placed on opposite sides of the axle, at a distance from each other of about 30 yards. By means of weights placed in the scales attached to the ends of the strings, I could easily ascertain the force necessary to move the apparatus at any given velocity; for, having given in the first instance the required velocity with the hand, it was easily observed, in the course of about 40 revolutions of the axle, corresponding to

Fig. 51.

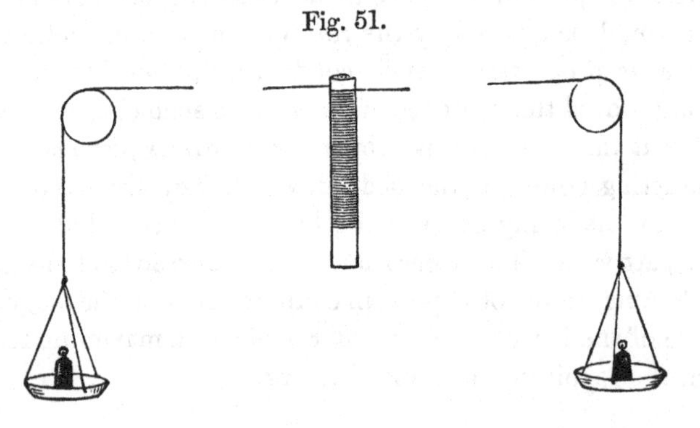

about 270 revolutions of the revolving piece, whether the weights placed in the scales were just able to maintain that velocity.

The experiments selected for repetition first were those of series No. 2. Ten cells, in a series of five double pairs, were connected with the large electro-magnet; and the small compound electro-magnet (restored to its place in the centre of the revolving tube) was connected, through the commutator, with the galvanometer. Under these circumstances a velocity of 600 revolutions per minute was found to produce a steady deflection of the needle to 24° 15′, indicating 0·983 of current magneto-electricity.

To maintain the velocity of 600 per minute, 5 lb. 3 oz. had to be placed in each scale; but when the battery was thrown out of communication with the electro-magnet, and the motion was opposed solely by friction and the resistance of the air, only 2 lb. 13 oz. were required for the same purpose. The difference, 2 lb. 6 oz., represents the force spent during the connexion of the battery with the electro-magnet in overcoming magnetic attractions and repulsions. The perpendicular descent of the weights was at the rate of 517 feet per 15 minutes.

According to series No. 2, Table I., the heat due to 0·983 of current magneto-electricity is $\left(\frac{983}{902}\right)^2 \times 1^\circ{\cdot}56 = 1^\circ{\cdot}85$. But as the resistance of the coil of the revolving electro-magnet was to that of the whole circuit as 1 : 1·13, the heat evolved by the whole conducting circuit was $1^\circ{\cdot}85 \times 1{\cdot}13 = 2^\circ{\cdot}09$. Adding to this, 0°·33 on account of the heat evolved by the iron of the revolving electro-magnet, and 0°·04 on account of the sparks* at the commutator, we have a total of 2°·46. Now in order to refer this to the capacity of a lb. of water, I found:—

	lb.		lb.	
Weight of glass tube	= 1·65 =	capacity for heat of	0·300	of water.
Weight of water........	= 0·61 =	.. ..	0·610	..
Weight of electro-magnet.	= 1·67 =	.. ..	0·204	..
Total weight..	= 3·93 =	.. ..	1·114	..

$2^\circ{\cdot}46 \times 1{\cdot}114 = 2^\circ{\cdot}74$†; and this has been obtained by the power which can raise 4 lb. 12 oz. to the perpendicular height of 517 feet.

1° of heat per lb. of water is therefore equivalent to a mechanical force capable of raising a weight of 896 lb. to the perpendicular height of one foot.

Two other experiments, conducted precisely in the same manner, gave a degree of heat to mechanical forces represented respectively by 1001 lb. and 1040 lb.

* The heat evolved by sparks in the above and subsequent instances had been determined by previous experiments.

† The thermal effects on the large stationary electro-magnet appear to have been neglected in this summary as insignificant. But they were nevertheless in all probability sufficiently great to account for the difference between 838, the equivalent deduced from these experiments, and 772 which was ultimately arrived at.—*Note*, 1881.

I now made an experiment similar to those of series No. 10. Eight cells in a series of four double pairs were connected with the large electro-magnet, and two in series with the small revolving electro-magnet. The velocity of revolution was at the rate of 640 per minute, contrary to the direction of the attractive forces, causing the needle to be deflected to 37° 20′, which indicates a current of 1·955.

A weight of 6 lb. 4 oz. placed in each scale was just able to maintain the above velocity when the circuits were complete; but when they were broken, and friction alone opposed the motion, a weight of 2 lb. 8 oz. only was required, which is less than the former by 3 lb. 12 oz. The fall of the weights was in this instance 551 feet per 15 minutes.

According to series 10, Table II., the heat due to the current observed in the present instance is $\left(\frac{1 \cdot 955}{1 \cdot 845}\right)^2 \times 5^{\circ}{\cdot}88 = 6^{\circ}{\cdot}6$. But I had found by calculations, based as usual upon the laws of Ohm, that in the present experiment the resistance of the coil of the revolving electro-magnet was to that of the whole circuit, including the two cells, as 1 : 1·303. Therefore the heat evolved by the whole circuit, including 0°·18 on account of the iron of the revolving electro-magnet, and 0°·12 on account of sparks at the commutator, was 8°·9, or 9°·92 per capacity of a lb. of water.

Now, when the revolving electro-magnet was stationary, the two cells could pass through it a uniform current of 1·483. The heat evolved from the whole circuit by such a current is $\left(\frac{1 \cdot 483}{2 \cdot 145}\right)^2 \times 5^{\circ}{\cdot}88 \times 1 \cdot 303 \times 1 \cdot 114 = 4^{\circ}{\cdot}08$ per lb. of water per 15 minutes, according to data previously given. Hence the quantity of heat due to the chemical reactions in the experiment is $\frac{1 \cdot 955}{1 \cdot 483} \times 4^{\circ}{\cdot}08 = 5^{\circ}{\cdot}38$, instead of 9°·92, the quantity actually evolved.

Hence 4°·54 were evolved in the experiment over and above the heat due to the chemical changes taking place in the battery, by the agency of a mechanical power capable of raising 7 lb. 8 oz. to the height of 551 feet. In other words, one degree is equivalent to 910 lb. raised to the height of one foot.

An experiment was now made, using the same apparatus as an electro-magnetic engine. The power of the magnetic attractions and repulsions alone, without the assistance of any weights, was able to maintain a velocity of 320 revolutions per minute. But when the circuits were broken, a weight of 1 lb. 2 oz. had to be placed in each scale in order to obtain the same velocity. The deflection of the needle was in this instance $17^\circ\ 15' = 0{\cdot}63$ of current electricity. The perpendicular descent of the weights was 275 feet per 15 minutes.

Now, calculating in a similar manner to that adopted in the last experiment, we have, from series 9, Table II., and other data previously given, $\left(\frac{630}{543}\right)^2 \times 0^\circ{\cdot}50 \times 1{\cdot}303 = 0^\circ{\cdot}877$, which, on applying a correction of $0^\circ{\cdot}012$ on account of sparks at the commutator, and $0^\circ{\cdot}18$ on account of the iron of the revolving electro-magnet, and then reducing to the capacity of a pound of water, gives $1^\circ{\cdot}191$ as the quantity of heat evolved by the whole circuit in 15 minutes.

The quantity of current which the two cells could pass through the revolving electro-magnet when the latter was stationary was in this instance $1{\cdot}538$; and $\left(\frac{1{\cdot}538}{2{\cdot}145}\right)^2 \times 5^\circ{\cdot}88 \times 1{\cdot}303 \times 1{\cdot}114 = 4^\circ{\cdot}38$. Hence, as before, the quantity of heat due to the chemical reactions during the experiment is $\frac{0{\cdot}63}{1{\cdot}538} \times 4^\circ{\cdot}38 = 1^\circ{\cdot}794$, which is $0^\circ{\cdot}603$ more than was obtained during the revolution of the electro-magnet.

Hence $0^\circ{\cdot}603$ has been converted into a mechanical power equal to raise 2 lb. 4 oz. to the height of 275 feet. In other words, one degree per lb. of water may be converted into the mechanical power which can raise 1026 lb. to the height of one foot.

Another experiment, conducted in precisely the same manner as the above, gave, per degree of heat, a mechanical power capable of raising 587 lb. to the height of one foot.

As the preceding experiments are somewhat complicated,

and therefore subject to the accumulation of small errors of observation, I thought it would be desirable to execute some of a more simple character. For this purpose I determined upon an arrangement in which the whole * of the heat would be evolved in the revolving tube.

The iron cylinder used in previous experiments was placed in an electrotype apparatus constructed in such a manner as to render every part of it equally exposed to the voltaic action. In four days 11 oz. of copper were deposited in a hard compact stratum. The ends of the cylinder were then filed until the iron just appeared. Thus I had a cylinder of iron immediately surrounded by a hollow cylinder of pure copper nearly one eighth of an inch thick. This was placed in the centre of a new revolving tube fitted up in precisely the same manner as the former one (which had been accidentally broken), and surrounded with 11¼ oz. of water I give the following series of experiments in which the above was rotated between

No. 16.

		Revolutions of the Bar per minute.	Deflections of Galvanometer of 1 turn.	Mean Temperature of Room.	Mean Difference.	Temperature of Water. Before.	Temperature of Water. After.	Gain or Loss.
July 4, P.M. July 5, A.M.	Battery contact broken.	600	°	67°·50	0°·15−	67°·37	67°·33	0°·04 loss
	Electro-magnet in action.	600	72 35	69·32	0·42−	67·50	70·30	2·80 gain
	Battery contact broken.	600	..	68·80	0·16+	69·00	68·93	0·07 loss
	Electro-magnet in action.	600	72 25	69·70	0·56+	69·00	71·52	2·52 gain
	Mean, Electro-magnet in action.	600	72 30	..	0·07+	..	..	2·66 gain
	Mean, Battery contact broken.	600	..	..	..	..	..	0·05 loss
	Corrected Result.	600	72° 30′=10·93 current.					2·73 gain

* See note p. 151.

the poles of the large electro-magnet excited by ten cells arranged in a series of five double pairs, a galvanometer being included in the circuit to indicate the electric force to which the electro-magnet was exposed.

I now proceeded to ascertain, by means already described, the mechanical power by which the above effects were produced. First, I ascertained the current passing through the coil of the electro-magnet; then the weights necessary to maintain the velocity of 600 revolutions per minute, both when the magnet was in action and when contact with the battery was broken. I have collected the results of my experiments on this subject in the following table. The first five

TABLE IV.

	Deflections of the Galvanometer of one turn completing the circuit of the Electro-magnet.	Weight in each scale, the Electro-magnet being in action.	Weight in each scale, the Electro-magnet being not in action.	Difference.
		lb. oz.	lb. oz.	lb. oz.
	72° 30	4 4	2 5	1 15
	72 30	4 4	2 3	2 1
	72 25	4 2	2 0	2 2
	72 15	5 0	2 10	2 6
	72 5	4 0	2 0	2 0
	68 0	3 14	2 8	1 6
	66 10	3 0	2 0	1 0
Mean of the first 5 experiments	72° 21′=10·82 current.			2·1 lb.
Mean of the last 2 experiments	67° 5′=7·91 current.			1·19 lb.

were obtained with a battery of ten cells in a series of five; the last two with a battery of five pairs in series.

Referring to Series 16, we see that 2°·73 were obtained when the bar was revolved between the poles of the electro-magnet excited by a current of 10·93. Therefore the quantity of heat due to the mean current in the first five experiments of the above table is $\left(\frac{10{\cdot}82}{10{\cdot}93}\right)^2 \times 2^\circ{\cdot}73 = 2^\circ{\cdot}675$.

To reduce this to the capacity of a pound of water, I had in the present instance the following data :—

	lb.		lb.	
Weight of glass tube . =	1·125 =	capacity for heat of	0·205	of water.
Weight of water .... =	0·687 =	.. ..	0·687	..
Weight of metallic bar =	1·688 =	.. ..	0·202	..
Total weight =	3·500 =	.. ..	1·094	..

2°·926, the product of 1·094 and 2°·675, is therefore the heat generated by a mechanical force capable of raising 4·2 lb. to the height of 517 feet.

In other words, one degree of heat per lb. of water may be generated by the expenditure of a mechanical power capable of raising 742 lb. to the height of one foot.

By a similar calculation, I find the result of the last two experiments of the table to be 860 lb.

The foregoing are all the experiments I have hitherto made on the mechanical value of heat. I admit that there is a considerable difference between some of the results, but not, I think, greater than may be referred with propriety to mere errors of experiment. I intend to repeat the experiments with a more powerful and more delicate apparatus. At present we shall adopt the mean result of the thirteen experiments given in this paper, and state generally that,

The quantity of heat capable of increasing the temperature of a pound of water by one degree of Fahrenheit's scale is equal to, and may be converted into, a mechanical force capable of raising 838 *lb. to the perpendicular height of one foot.*

Among the practical conclusions which may be drawn from the convertibility of heat and mechanical power into one another, according to the above absolute numerical relations, I will content myself with selecting two of the more important. The former of these is in reference to the duty of steam-engines; the latter, to the practicability of employing electro-magnetism as an economical motive force.

1. In his excellent treatise on the Steam-engine, Mr. Russell has given a statistical table *, containing the number of

* Encycl. Brit., 7th Edition, vol. xx. part 2, p. 685.

pounds of fuel evaporating one cubic foot of water, from the initial temperature of the water, and likewise from the temperature of 212°. From these facts it appears that in the Cornish boilers at Huel Towan, and the United Mines, the combustion of a lb. of Welsh coal gives 183° to a cubic foot of water, or otherwise 11,437° to a lb. of water. But we have shown that one degree is equal to 838 lb. raised to the height of one foot. Therefore the heat evolved by the combustion of a lb. of coal is equivalent to the mechanical force capable of raising 9,584,206 lb. to the height of one foot, or to about ten times the duty of the best Cornish engines.

2. From my own experiments, I find that a lb. of zinc consumed in Daniell's battery produces a current evolving about 1320°; in Grove's battery, about 2200° per lb. of water. Therefore *the mechanical forces of the chemical affinities which produce the voltaic currents in these arrangements are, per lb. of zinc, equal respectively to* 1,106,160 *lb. and* 1,843,600 *lb. raised to the height of one foot.* But since it will be practically impossible to convert more than about one half of the heat of the voltaic circuit into useful mechanical power, it is evident that the electro-magnetic engine, worked by the voltaic batteries at present used, will never supersede steam in an economical point of view.

Broom Hill, Pendlebury,
near Manchester, July 1843.

P.S.—We shall be obliged to admit that Count Rumford was right in attributing the heat evolved by boring cannon to friction, and not (in any considerable degree) to any change in the capacity of the metal. I have lately proved experimentally that *heat is evolved by the passage of water through narrow tubes.* My apparatus consisted of a piston perforated by a number of small holes, working in a cylindrical glass jar containing about 7 lb. of water. I thus obtained one degree of heat per lb. of water from a mechanical force capable of raising about 770 lb. to the height of one foot, a result which will be allowed to be very strongly confirmatory of our previous deductions. I shall lose no time in repeating and extending these experiments, being satisfied that the grand agents of nature are, by the Creator's fiat, *indestructible*; and that wherever mechanical force is expended, an exact equivalent of heat is *always* obtained.

On conversing a few days ago with my friend Mr. John Davies, he told me that he had himself, a few years ago, attempted to account for that part of animal heat which Crawford's theory had left unexplained. by the friction of the blood in the veins and arteries, but that, finding a similar hypothesis in Haller's 'Physiology'*, he had not pursued the subject further. It is unquestionble that heat is produced by such friction, but it must be understood that the mechanical force expended in the friction is a part of the force of affinity which causes the venous blood to unite with oxygen; so that the whole heat of the system must still be referred to the chemical changes. But if the animal were engaged in turning a piece of machinery, or in ascending a mountain, I apprehend that in proportion to the muscular effort put forth for the purpose, a *diminution* of the heat evolved in the system by a given chemical action would be experienced.

I will observe, in conclusion, that the experiments detailed in the present paper do not militate against, though they certainly somewhat modify, the views I had previously entertained with respect to the electrical origin of chemical heat. I had before endeavoured to prove that when two atoms combine together, the heat evolved is exactly that which would have been evolved by the electrical current due to the chemical action taking place, and is therefore proportional to the intensity of the chemical force causing the atoms to combine. I now venture to state more explicitly, that it is not precisely the attraction of affinity, but rather the mechanical force expended by the atoms in falling towards one another, which determines the intensity of the current, and consequently the quantity of heat evolved; so that we have a simple hypothesis by which we may explain why heat is evolved so freely in the combination of gases, and by

* Haller's 'Physiology,' vol. ii. p. 304.

which, indeed, we may account "latent heat" as a mechanical power prepared for action as a watch-spring is when wound up. Suppose, for the sake of illustration, that 8 lb. of oxygen and 1 lb. of hydrogen were presented to one another in the gaseous state, and then exploded, the heat evolved would be about one degree Fahr. in 60,000 lb. of water, indicating a mechanical force expended in the combination equal to a weight of about 50,000,000 lb. raised to the height of one foot. Now if the oxygen and hydrogen could be presented to each other in a liquid state, the heat of combination would be less than before, because the atoms, in combining, would fall through less space. The hypothesis is, I confess, sufficiently crude at present; but I conceive that ultimately we shall be able to represent the whole phenomena of chemistry by exact numerical expressions, so as to be enabled to predict the existence and properties of new compounds.

August, 1843. J. P. J.

On the Intermittent Character of the Voltaic Current in certain cases of Electrolysis; and on the Intensities of various Voltaic Arrangements. By JAMES P. JOULE*.

[Phil. Mag. ser. 3. vol. xxiv. p. 106. Read before the Manchester Literary and Philosophical Society, December 26, 1843.]

IT can hardly have escaped the notice of electricians that, in some instances of electro-chemical decomposition, the needle of a galvanometer included in the circuit will indicate by its unsteadiness a very irregular flow of electricity. I have not, however, been able to meet with any description of the phenomena, which are generally so trifling in the extent of their manifestation as to induce the belief that they arise from

* The experiments were made at Broom Hill, near Manchester.

accidental and unimportant causes. It is now more than a year since I observed some very striking examples of the phenomena in the course of some experiments on the calorific effects of electrolysis; but I was too much interested in the subject immediately in hand to allow them to occupy much of my attention. They have since, however, appeared to me to have an important bearing on the theory of electrolysis, and on this account to deserve the attention of philosophers. I propose to begin by mentioning the old experiments just referred to, and then to relate the progress I have recently made in the investigation.

The following experiment was made on the 9th of July, 1842. Two plates of iron were immersed in a dilute solution of sulphuric acid, and then connected with the poles of a battery, consisting of six large cells of Daniell in series. After electrolysis had proceeded for a few minutes, I observed that the needle of a galvanometer, which was included in the circuit, indicated by its unsteadiness a very great irregularity in the electrical current. On connecting only one cell of the battery with the iron electrodes, the electrolysis appeared to be carried on with freedom, and the needle was pretty steady.

About the same time I made some experiments with electrodes of copper immersed in a solution consisting of seven parts water and one part strong oil of vitriol. In this case the sudden jerking motion of the needle was not observed; but it invariably happened that the current diminished very rapidly during the first one or two minutes, and then began to increase, and continued to do so until, after a certain interval of time, it arrived nearly at the same degree of intensity that it had at first. I give the following as a fair example, selected out of a number of experiments, which did not differ much from one another.

A vessel containing dilute sulphuric acid was divided into two compartments by a diaphragm of animal membrane. In each of these a bright plate of copper, exposing a surface of about ten square inches, was immersed. The copper plates were then connected with a battery, consisting of six large cells of Daniell in series, a galvanometer furnished with a thick copper wire bent into a circle of a foot diameter being included in the circuit. Immediately after the circuit was closed the current was sufficiently powerful to deflect the needle to 69°. Then, noting the position of the needle at the end of each quarter of a minute, I observed the following deflections, viz. 68°, 60°, 55°, 10°, 20°, 30°, 35°, 41°, 43°. Turning these deflections into quantities of electricity, it appeared that in the short space of one minute the voltaic current declined to $\frac{1}{25}$ of its first intensity, and that at the end of a further interval of one minute and a quarter it had eight times the intensity that it had when at its lowest ebb.

I met with very curious results in using amalgamated zinc as the positive electrode of a battery of six large cells. The needle was pretty steady at first, but after a short time it began to oscillate in the most capricious manner through an arc of about 10°. Sometimes it would remain steady for a few seconds, then it would suddenly spring forwards, and before I had time to make it steady in its new position it would move backwards again.

It was natural enough to suppose that such extraordinary irregularities of the current might be accompanied by a visible change in the character of the electrode. And in this I was not deceived, for on examining the amalgamated zinc, I observed the following very curious phenomenon :—At intervals of one or two seconds a white shade, as of frosted silver, overspread the surface of the amalgamated zinc and then suddenly disappeared, leaving the metal brilliant. The pulsations of the current were evidently simultaneous with these sudden changes in the appearance of the electrode, and the needle received a sudden impulse every time the white film suddenly disappeared.

All the above experiments were made more than a year ago; those which follow are the experiments I have recently made on the subject.

A plate of amalgamated zinc and a stout iron wire were immersed in a solution consisting of one part strong oil of vitriol to six parts of water. The iron was connected with the positive, the zinc with the negative electrode of a battery of five large Daniell's cells, and a galvanometer was included in the circuit. The instant that the circuit was completed a powerful current was transmitted through it, hydrogen being evolved from the negatively electrified zinc, while the positively electrified iron was oxidized and began to dissolve away. In a short time, however, the intensity of the current began to decline very rapidly, and the iron electrode, ceasing to be dissolved, assumed the passive state* described by Schœnbein, and began to evolve oxygen gas, and continued to do so as long as I had patience to watch it. On breaking the circuit and then closing it afresh, the same phenomena were repeated. Having now reduced the battery from five to three cells, the action of the iron became intermittent. First it was dissolved, the needle being at the same time deflected 45°; then it began to evolve oxygen, the needle at the same time declining rapidly until it stood at 15°; and then, again, the oxygen suddenly ceased to be evolved, while at the same moment the needle sprang forward, and began to oscillate about its former resting-place at 45°. The iron remained in each state about half a minute, and a white film was observed to pass over its surface every time that the oxygen was about to rise, and to disappear suddenly when the evolution of oxygen ceased.

Having watched these curious phenomena for some time, it occurred to me to try the effect of dividing the current from the battery between two electrolytic cells. On making the experiment I found that the action of the iron was in both cells intermittent; and, what was very remarkable, the condition of the iron electrodes changed *simultaneously*. They always began to evolve oxygen at about the same time; and when one of them ceased to evolve oxygen and began to be oxidized and dissolved, the same thing happened to the other at the same instant.

When both of the iron electrodes were evolving oxygen, it was only necessary to lift one of them up a little, so as to expose a small portion of its surface to the air, and then to

* Keir appears to have been the first who observed the passive state of iron.—*Note by the Editor of the Phil. Mag.*

plunge it into the acid again, in order to make both irons instantly assume the opposite state. The same effect was also produced by touching the immersed portion of one of the electrodes with a piece of iron or zinc.

Now, as far as regards one electrolytic cell, the above phenomena can be explained, I think, without much difficulty. Adopting the theory of Professor Daniell, which, agreeably to the theory which has been promulgated by Davy and Graham, supposes the positive metal to unite directly with oxysulphion (SO_4), we can readily perceive that oxygen must inevitably rise from the iron whenever the oxysulphion cannot be produced as quickly as is demanded by the intensity of the current. On the other hand, there will not, I think, be much difficulty in admitting that the evolution of oxygen may, by producing currents in the liquid &c., have the effect of restoring to the iron its original aptitude for dissolution; then, if the smallest portion of iron assumes that state, it is evident that it will be positive with regard to the rest of the iron still evolving oxygen; a current, therefore, will be established through the acid from the former to the latter, and the hydrogen thereby liberated immediately uniting with the nascent oxygen of the passive portion of the iron, the whole surface of iron will suddenly become clean and again combine with oxysulphion. On this view the advance of the needle from its smallest to its greatest deflection ought to be very sudden, which accords with my experience.

The simultaneous change of the state of two iron electrodes in separate cells between which the current of the battery is divided, may perhaps be explained by supposing that, when one of the iron electrodes enters into the active state, the sudden increase of the intensity of the current through its cell diverts the current from the other cell to such an extent as to allow its iron electrode also to assume the active state.

In general a current of a certain degree of intensity is requisite in order to produce the intermittent effects. If it be too low, the iron will continue to be dissolved; if too high, the iron will, after the first few moments of action, commence and then continue to evolve oxygen. A great deal seems also to depend on the quality of the iron employed. With some specimens of iron and steel I could not succeed at all, whilst with a piece of rectangular iron wire, a quarter of an inch broad and one eighth thick, I was able to obtain intermittent effects when using a battery consisting of two, three, four, and even five cells of Daniell. In this case the negative electrode was a plate of platinized silver, the solution consisted of six parts of water to one of oil of vitriol, and a diaphragm was used in order to prevent the hydrogen rising from the negative electrode from troubling the liquid in contact with the positive iron. The results of the experiments are given in the Table below, the second and third columns of which give the deflections of the needle during the passive and the active states of the iron, whilst the fourth contains the difference between the currents in the two states.

Number of Daniell's cells in Series.	Deflection in passive state.	Deflection in the active state.	Difference.
2	2°	43°	884
3	23	50	926
4	34	53	837
5	43	56	737

In each of the four instances given in the Table, the different states succeeded each other at intervals of about half a minute; and it was uniformly observed that the active state was assumed with greater suddenness than the passive. It will also be seen, on inspecting the Table, that, as might have been anticipated from theory, the difference between the currents flowing during the different states is nearly a constant quantity.

On repeating my old experiments with a positive electrode of amalgamated zinc, I find that whenever a battery of six or ten large Daniell cells is connected with a plate of amalgamated zinc immersed as a positive electrode in a dilute solution of sulphuric acid, the curious phenomenon already adverted to occurs. It commences at the bottom and edges of the plate, and generally goes on extending until the whole surface is under its influence: the amalgamated zinc loses its brightness in consequence of a white shade overspreading its surface and giving it the appearance of frosted silver: this is hardly formed before it suddenly disappears, and then a new shade overspreads the surface, only to vanish again as suddenly as the one which preceded it. These alternations generally succeed each other very rapidly; but I have sometimes seen them occur at intervals of five seconds or more, and then I have been able to prove, by the motions of the needle of the galvanometer, that the disappearance of the white shade is always accompanied by a sudden increase of the intensity of the current.

On dividing the current between two similar electrolytic cells, I observed that the disappearance of the white shade occurred at the same moment on both the positive electrodes of amalgamated zinc.

It is evident, therefore, that the phenomena obtained with amalgamated zinc are, in a great measure, analogous to those observed with iron; but there is an important distinction between the two, inasmuch as no oxygen is evolved from amalgamated zinc when made positive in dilute sulphuric acid, even when a powerful battery is employed*. So that we see that amalgamated zinc continues to be dissolved even when it has assumed a state analogous to that of passive iron evolving oxygen. I think that this fact might be explained by supposing that, in the active state, the zinc combines immediately with oxysulphion; but that, in consequence of the too tardy arrival of that compound, the zinc sometimes combines with oxygen alone as a proper electrolytic action, depending upon the secondary action of the sulphuric acid for the removal of the film of oxide thus formed. It is easy to see that in the latter case the intensity of the current will

* Oxygen is evolved by zinc when the latter is made positive in a dilute solution of potassa by three or four cells of Daniell in series.

be less than when the metal combines immediately with oxysulphion.

PS. Nearly the whole of the above had been written before I was aware that Schœnbein had already observed the intermitting passivity of iron. As, however, my experiments are considerably different from those of this philosopher, I have not thought it right to suppress them. Schœnbein's experiments were made in the following manner* :—The conducting-wires of a powerful voltaic pair were connected with two mercury-cups; a plate of platinum immersed in dilute sulphuric acid was connected with the negative cup; then a piece of iron wire, previously connected at one of its extremities with the positive mercury-cup, was made to complete the circuit by immersing the other extremity in the dilute acid. Under these circumstances he did not observe any disengagement of hydrogen from the negative platinum, in consequence of the passivity of the iron electrode. He observes that the apparatus may be made to lose this state of inactivity, and so to produce the electrolysis of water, by the following means :—

1st. By putting the negative electrode in contact, for a moment, with the positive electrode of iron. The instant they are separated again a lively disengagement of hydrogen from the negative electrode takes place, which, however, soon begins to diminish, and ceases entirely at the end of some seconds.

2nd. By opening the circuit of the pile for some instants. When it is closed again, a lively disengagement of gas takes place upon the negative electrode, which is soon succeeded by the state of inactivity.

3rd. By putting the immersed portion of the positive electrode of iron in contact with an oxidable metal, as, for example, zinc, tin, copper, or even silver. But in this case the disengagement of hydrogen from the negative electrode does not last longer than some seconds.

4th. By establishing a communication between the two

* De la Rive's Archives de l'Electricité, No. 5, p. 267.

mercury-cups for a few moments, by means of a copper wire three inches long and half a line thick. Then, the moment the wire is removed again, a lively disengagement of hydrogen takes place on the negative electrode, which does not, however, last longer than a few seconds.

5th. By briskly agitating that portion of the positive iron electrode which is immersed in the liquid, but without breaking the circuit.

Passing a variety of other interesting observations in Schœnbein's memoir, we come to that part of it which is most intimately connected with our subject. At p. 278 he states that when a communication is established between the mercury-cups by means of a wire of a certain length, there succeed each other, at certain intervals, a lively disengagement of gas on the negative electrode, and a time of cessation of the electrolysis in the cell of decomposition. He observes also that after some time the alternations cease, and the positive iron electrode takes a permanent inactivity.

My own experiments on the intermittent states of a positive electrode of iron differ from those of the physicist of Bâle with regard to the intensity of the voltaic apparatus employed. His was a powerful single pair (Grove's), mine was a series of from two to five cells of Daniell. Hence in Schœnbein's experiments, when the passive state was assumed by the iron the current was entirely cut off, because the battery used by him had not sufficient intensity to produce the electrolysis of water, except where the oxygen liberated could enter into combination with the positive metal; but in my own experiments, in consequence of the superior intensity of the battery employed, the passive state of the iron was accompanied with the regular decomposition of water into its gaseous elements.

It may also be remarked that Schœnbein did not observe the intermitting effects until the intensity of the current was much reduced by the opening of a new channel for it by connecting the poles of the cell by a wire of a certain length, whilst I have succeeded in obtaining the alternations of state when using the whole force of five large cells of Daniell in series.

List of Voltaic Intensities.

No.	Negative Elements.		Positive Elements.		Intensity.
1.	Platinum.	Nitric Acid.	Solution of Potash.	Amalgam of Potassium.	302
2.	Passive Iron.	”	” ”	Amalgamated Zinc.	220
3.	Coke.	”	” ”	” ”	225
4.	Gold.	”	” ”	” ”	234
5.	Platinum.	”	” ”	” ”	234
6.	”	”	” ”	Iron.	169
7.	”	”	” ”	Copper.	120
8.	”	”	” ”	Silver.	66
9.	”	”	” ”	Platinum.	31
10.	”	”	Solution of Common Salt.	Amalgamated Zinc.	198
11.	”	”	” ” ”	Iron.	146
12.	”	”	” ” ”	Copper.	116
13.	”	”	” ” ”	Silver.	95
14.	”	”	” ” ”	Platinum.	55
15.	”	”	Solution of Sulphate of Soda.	Amalgamated Zinc.	187
16.	”	”	” ” ”	Iron.	147
17.	”	”	” ” ”	Copper.	92
18.	”	”	” ” ”	Silver.	78
19.	”	”	” ” ”	Platinum.	17

20.	"	"	Dilute Sulphuric Acid.	Amalgamated Zinc.	187
21.	"	"	" "	Iron.	140
22.	"	"	" "	Copper.	91
23.	"	"	" "	Silver.	53
24.	"	"	" "	Platinum.	37
25.	"	Peroxide of Lead and Sulphuric Acid.	Solution of Potash.	Amalgamated Zinc.	277
26.	"	" " " "	" "	Iron.	177
27.	"	Peroxide of Manganese and Sulphuric Acid.	" "	Amalgamated Zinc.	237
28.	"	Peroxide of Manganese and Hydrochloric Acid.	" "	" "	237
29.	"	Bichromate of Potash.	" "	" "	161
30.	"	" "	Dilute Sulphuric Acid.	" "	102
31.	"	Bichromate of Potash and Sulphuric Acid.	Solution of Potash.	" "	207
32.	"	" " " "	Dilute Sulphuric Acid.	" "	161
33.	Copper.	" " " "	" "	" "	116
34.	"	Bichromate of Potash.	" "	" "	79
35.	"	Sulphate of Copper.	Solution of Potash.	" "	138
36.	"	" "	" "	Iron.	66
37.	"	" "	" "	Copper.	33
38.	"	" "	Solution of Common Salt.	Amalgamated Zinc.	106
39.	"	" "	" " "	Iron.	55
40.	"	" "	" " "	Copper.	28
41.	"	" "	Solution of Sulphate of Soda.	Amalgamated Zinc.	104
42.	"	" "	" " "	Iron.	59
43.	"	" "	" " "	Copper.	8
44.	"	" "	Dilute Sulphuric Acid.	Amalgamated Zinc.	100
45.	"	" "	" "	Iron.	49
46.	"	" "	" "	Copper.	4
47.	Platinized Silver.	Dilute Sulphuric Acid.	" "	Amalgamated Zinc.	65
48.	" "	" "	" "	Iron.	17
49.	" "	" "	Solution of Common Salt.	Amalgamated Zinc.	68
50.	" "	" "	Solution of Potash.	" "	98

It is evident, therefore, that a powerful intensity of current is not always able to retain a positive iron electrode in the passive state.

On the Intensities of various Voltaic Arrangements.

We know that the important law which has been established by the labours of Ohm, Fechner, and De la Rive is expressed by the formula $E=\frac{A}{R}$, where A is the electromotive force, R the resistance to conduction of the whole circuit, and E is the quantity of electricity circulating in a given time. Therefore, if the resistances of different voltaic circles are made equal to one another, the quantities of current will be proportional to the electromotive forces; and hence we derive the following simple method of determining the intensity or electromotive force of a battery. We take an accurate galvanometer, furnished with a coil of very great resistance, and connecting the arrangements under examination successively with it, we take the currents indicated by the instrument as the measure of their respective intensities. I have in this way obtained the annexed list of voltaic intensities (pp. 168, 169), using a galvanometer which, with the wires attached to it, had a resistance at least 300 times as great as that of most of the cells under examination. I have made the ordinary cell of Daniell the standard of comparison, calling its intensity 100.

The use of the peroxides of lead and manganese as negative elements of the voltaic pile has been recently pointed out by De la Rive*. By using the peroxide of lead with either dilute sulphuric acid or a saline solution, he has produced a battery of greater intensity than the pile of Grove †. It will be seen, on reference to the Table, that an arrangement consisting of peroxide of lead moistened with sulphuric

* Archives de l'Electricité, No. 8, p. 166.

† The use of peroxide of lead as the negative element was originally proposed by Prof. Grove himself, in Phil. Mag. ser. 3. vol. xv. p. 290.—*Edit. Phil. Mag.*

acid in contact with platinum, and solution of potassa in contact with amalgamated zinc, is half as intense again as the ordinary cell of Grove.

I may observe that a single cell of any of the arrangements given in the Table, the intensity of which is above 200, is able to decompose water into its elements with facility.

On the Changes of Temperature produced by the Rarefaction and Condensation of Air. By JAMES PRESCOTT JOULE*.

['Proceedings of the Royal Society,' June 20, 1844.]

IN order to estimate with greater accuracy than has hitherto been done the quantities of heat evolved or absorbed during the condensation or rarefaction of atmospheric air, the author contrived an apparatus where both the condensing-pump and the receiver were immersed in a large quantity of water, the changes in the temperature of which were ascertained by a thermometer of extreme sensibility. By comparing the amount of force expended in condensing air in the receiver with the quantity of heat evolved, after deducting that which was the effect of friction, it was found that a mechanical force capable of raising 823 pounds to the height of one foot must be applied in the condensation of air, in order to raise the temperature of 1 lb. of water 1° of Fahrenheit's scale. In another experiment, when air condensed in one vessel was allowed to pass into another vessel from which the air had been exhausted, both vessels being immersed in a large receiver full of water, no change of temperature took place, no mechanical power having been developed. The author considers these results as strongly corroborating the dynamical theory of the nature of heat, in opposition to that

* This abstract was made, I believe, by Dr. Roget, who took a kind interest in my early papers.—*Note*, 1881.

which ascribes to it materiality; but he reserves the further discussion of this question to a future communication, which he hopes soon to present to the Royal Society.

On the Changes of Temperature produced by the Rarefaction and Condensation of Air. *By* J. P. Joule, *Esq.**

['Philosophical Magazine,' ser. 3, May 1845.]

In a paper† which was read before the Chemical Section of the British Association at Cork, I applied Dr. Faraday's fine discovery of magneto-electricity in order to establish definite relations between heat and the ordinary forms of mechanical power. In that paper it was demonstrated experimentally that the mechanical power exerted in turning a magneto-electrical machine is *converted into the heat* evolved by the passage of the currents of induction through its coils; and, on the other hand, that the motive power of the electro-magnetic engine is obtained at the expense of the heat due to the chemical reactions of the battery by which it is worked. I hope, at a future period, to be able to communicate some new and very delicate experiments, in order to ascertain the mechanical equivalent of heat with the accuracy which its importance to physical science demands. My present object is to relate an investigation in which I believe I have succeeded in successfully applying the principles before maintained to the changes of temperature arising from the alteration of the density of gaseous bodies—an inquiry of great interest in a practical as well as theoretical point of view, owing to its bearing upon the theory of the steam-engine.

Dr. Cullen and Dr. Darwin appear to have been the first

* The experiments were made at Oak Field, Whalley Range, near Manchester.

† Phil. Mag. ser. 3. vol. xxiii. pp. 263, 347, 435.

who observed that the temperature of air is decreased by rarefaction and increased by condensation. Other philosophers have subsequently directed their attention to the subject. Dalton was, however, the first who succeeded in measuring the change of temperature with some degree of accuracy. By the employment of an exceedingly ingenious contrivance, that illustrious philosopher ascertained that about 50° of heat are evolved when air is compressed to one half of its original bulk, and that, on the other hand, 50° are absorbed by a corresponding rarefaction*.

Fig. 52. Scale $\frac{1}{12}$.

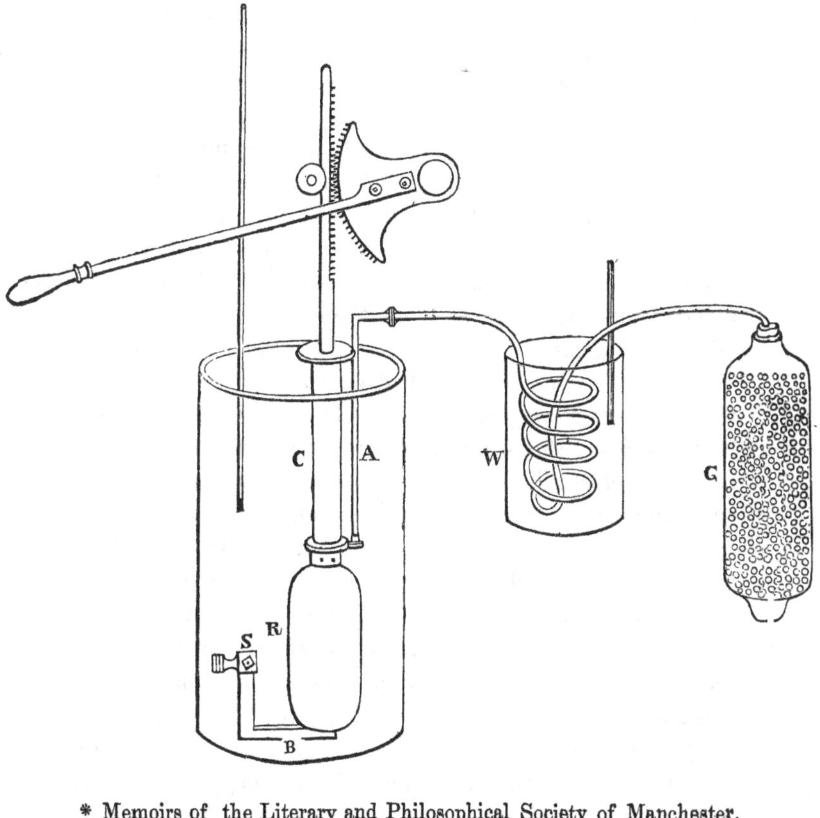

* Memoirs of the Literary and Philosophical Society of Manchester, vol. v. part 2, pp. 251–525.

There is every reason for believing that Dalton's results are very near the truth, especially as they have been exactly confirmed by the experiments of Dr. Ure with the thermometer of Breguet. But our knowledge of the specific heat of elastic fluids is of such an uncertain character, that we should not be justified in attempting to deduce from them the absolute quantity of heat evolved or absorbed. I have succeeded in removing this difficulty by immersing my condensing-pump and receiver into a large quantity of water, so as to transfer the calorific effect to a body which is universally received as the standard of capacity.

My apparatus will be understood on inspecting fig. 52. C represents the condensing-pump, consisting of a cylinder of gun-metal, and of a piston fitted with a plug of oiled leather, which works easily, yet tightly, through a stroke of 8 inches. The cylinder is $10\frac{1}{2}$ inches long, $1\frac{3}{8}$ inch in interior diameter, and $\frac{1}{4}$ of an inch in thickness of metal. The pipe A, for the admission of air, is fitted to the lower part of the cylinder; at the bottom of this pipe there is a conical valve, constructed of horn, opening downwards. A copper receiver, R, which is 12 inches long, $4\frac{1}{2}$ inches in exterior diameter, $\frac{1}{4}$ of an inch thick, and has a capacity of $136\frac{1}{2}$ cubic inches, may be screwed upon the pump at pleasure. This receiver is furnished with a conical valve of horn opening downwards, and, at the bottom, with a piece of brass, B, along the centre of which there is a bore of $\frac{1}{8}$ of an inch diameter. There is a stop-cock at S which I shall describe more particularly in the sequel.

Anticipating that the changes of temperature of the large quantity of water which was necessary in order to surround the receiver and pump would be very minute, I was at great pains in providing a thermometer of extreme sensibility and very great accuracy. A glass tube of narrow bore having been selected, a column of mercury, 1 inch long, was introduced, and gradually advanced in such a manner that the end of the column in one position coincided with the beginning of the column in the next. In each position the length of the column was ascertained to the $\frac{1}{4000}$ part of an inch, by means of an instrument invented for the purpose by Mr. Dancer*. Afterwards the tube was covered with a film of bees'-wax, and each of the previously measured spaces was divided into twenty equal parts by means of a steel point carried by the dividing instrument; it was then etched by exposure to the vapour of fluoric acid. The scale thus formed was entirely arbitrary; and as it only extended between 30° and 90°, it was necessary to compare the thermometer with another, constructed in the same manner, but furnished with a scale including the boiling- as well as the freezing-point. When this was done, it was found that ten divisions of the sensible thermometer (occupying about $\frac{1}{2}$ an inch) were nearly equal to the degree of Fahrenheit; therefore, since by practice I can easily estimate with the naked eye $\frac{1}{20}$ of each of these divisions, I could with this instrument determine temperatures to the $\frac{1}{200}$ part of a degree. The scale being arbitrary, the indications of the thermometer had to be reduced in every instance, a circumstance which accounts for my having given the temperatures in the tables to three places of decimals.

It was important to employ, for the purpose of containing the water, a vessel as impermeable to heat as possible. With this view, two jars of tinned iron, one of them every way an inch smaller than the other, having been provided, the smaller jar was placed within the larger one, and the interstice between the two was closed hermetically. By this means a stratum of air of nearly the same temperature as the water was kept in contact with the sides and bottom of the inner jar. The jars used in the other experiments which I shall bring forward were constructed in a similar manner. Among other precautions to ensure accuracy, proper screens were placed between the vessels of water and the experimenter.

My first experiments were conducted in the following

* Of the firm of Abraham and Dancer, Cross Street, Manchester. I have great pleasure in acknowledging here the skill displayed by this gentleman in the construction of the different parts of my apparatus; to it I must, in a great measure, attribute whatever success has attended the experiments detailed in this paper.

manner:—The pump and copper receiver were immersed in 45 lb. 3 oz. of water, into which the very sensible thermometer above described was then placed; whilst two other thermometers were employed in order to ascertain the temperature of the room and that of the water contained by the vessel W. Having stirred the water thoroughly, its temperature was carefully read off. The pump was then worked at a moderate degree of speed until about twenty-two atmospheres of air, dried by being passed through the vessel G full of small pieces of chloride of calcium, were compressed into the copper receiver. After this operation (which occupied from fifteen to twenty minutes) the water was stirred for five minutes so as to diffuse the heat equally through every part, and then its temperature was again read off.

The increase of temperature thus observed was owing partly to the condensation of the air, and partly also to the friction of the pump and the motion of the water during the process of stirring. To estimate the value of the latter sources of heat, the air-pipe A was closed, and the pump was worked at the same velocity and for the same time as before, and the water was afterwards stirred precisely as in the first instance. The consequent increase of temperature indicated heat due to friction, &c.

The jar was now removed, and the receiver having been immersed into a pneumatic trough, the quantity of air which had been compressed into it was measured in the usual manner, and then corrected for the force of vapour, &c. The result, added to 136·5 cubic inches, the quantity contained by the receiver at first, gave the whole quantity of compressed air.

The result given in Table I. is the difference between the effects of condensation and friction alone, corrected for the slight superiority of the cooling influence of the atmosphere in the experiments on friction. We must now, however, proceed to apply a further correction, on account of the circumstance that the friction of the piston was considerably greater during the condensing experiments than during the

TABLE I.

Source of heat.	Number of strokes of pump.	Barometrical pressure.	Quantity of air compressed in cubic inches.	Temp. of the air admitted.	Mean temp. of the room.	Difference.	Temp. of water.		Heat gained.
							Before expt.	After expt.	
Condensation, &c.	300	30·06	3047	56°·2	57°·5	2°·224−	54°·930	55°·622	0°·692
Friction, &c.	300	....	....	....	57·5	1·685−	55·652	55·979	0·327
Condensation, &c.	300	30·07	2924	54·8	53·5	0·817+	53·970	54·664	0·694
Friction, &c.	300	....	....	....	54·5	0·358+	54·675	55·042	0·367
Condensation, &c.	300	30·24	2870	53·7	52·5	0·380+	52·562	53·197	0·635
Friction, &c.	300	....	....	....	52·6	0·760+	53·197	53·524	0·327
Condensation, &c.	300	30·07	2939	58·8	57·5	1·794−	55·359	56·053	0·694
Friction, &c.	300	....	....	....	57·75	1·536−	56·053	56·375	0·322
Condensation, &c.	300	30·34	2924	55·7	53·5	2·184+	55·409	55·959	0·550
Friction, &c.	300	....	....	....	53·75	2·316+	55·962	56·170	0·208
Condensation, &c.	300	30·40	3033	58·1	60·0	0·174+	59·876	60·472	0·596
Friction, &c.	300	....	....	....	60·4	0·196+	60·478	60·713	0·235
Condensation mean	300	30·20	2956	56·2	....	0·078−	....	....	0·643
Friction, &c. mean	300	....	....	....	....	0·068+	....	....	0·297
Corrected result..	....	30·20	2956	....	....	......	....	....	0·344

experiments to ascertain the effect of friction. In the latter case the piston worked with a vacuum beneath it, whilst in the former the leather was pressed to the sides of the pump by a force of condensed air, averaging 32 lb. per square inch. I endeavoured to estimate the difference between the friction in the two cases, by removing the valve of the receiver and working the pump with about 32 lb. per square inch pressure below it. These experiments, alternated with others in which a vacuum was beneath the piston, showed that the heat given out in the two cases was, as nearly as possible, in the ratio of six to five. When the correction indicated in this manner has been applied to 0°·297 (see Table) and the result subtracted from 0°·643, we obtain 0°·285 as the effect of compressing 2956 cubic inches of dry air at a pressure of 30·2 inches of mercury, into the space of 136·5 cubic inches.

This heat was distributed through 45 lb. 3 oz. of water, 20½ lb. of brass and copper, and 6 lb. of tinned iron. It was therefore equivalent to 13°·628 per lb. avoirdupois of water.

The force necessary to effect the above condensation may

be easily deduced from the law of Boyle and Mariotte, which has been proved by the French academicians to hold good as far as the twenty-fifth atmosphere of pressure. Let fig. 53

Fig. 53.

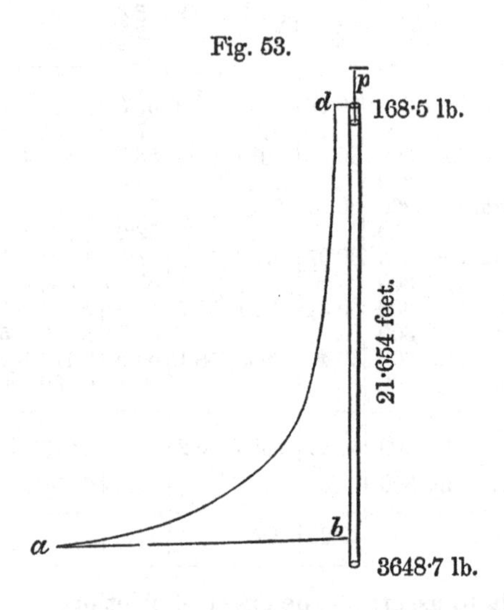

represent a cylinder closed at one end, the length of which is 21·654 feet, and the sectional area 11·376 square inches. Then one foot of it will have exactly the same capacity as the copper receiver used in the experiments, and its whole capacity will be 2956 cubic inches. It is evident, therefore, that the force used in pumping (considered to be without friction) was exactly equal to that which would push the piston p to the distance of a foot from the bottom of the cylinder. Excluding exterior atmospheric pressure, the force upon the piston, when at the top of the cylinder, will be 168·5 lb., the weight of a column of mercury 30·2 inches long and of 11·376 square inches section; and at a foot from the bottom it will be 21·654 times as much, or 3648·7 lb. The hyperbolic area, *a b c d*, will therefore represent the force employed in the condensation, including the assistance of the atmospheric pressure. Applying the formula for hyperbolic spaces, we have

$$s = 3648{\cdot}7 \times 2{\cdot}302585 \times \log 21{\cdot}654 = 11220{\cdot}2.$$

The force expended in condensation was therefore equivalent to that which can raise 11220·2 lb. to the perpendicular height of one foot.

Comparing this with the quantity of heat evolved, we have $\frac{11220{\cdot}2}{13^{\circ}{\cdot}628}=\frac{823}{1^{\circ}}$. So that a mechanical force capable of raising 823 lb. to the height of one foot must be applied in the condensation of air, in order to increase the temperature of a pound of water by one degree of Fahrenheit's scale.

The following Table contains the results of experiments similar to the last, except in the extent to which the compression of the air was carried.

TABLE II.

Source of heat.	Number of strokes of pump.	Barometrical pressure.	Quantity of air compressed in cubic inches.	Temp. of the air admitted.	Mean temp. of the room.	Difference.	Temp. of water. Before expt.	Temp. of water. After expt.	Heat gained.
Condensation, &c.	120	30·40	1410	54°·0	54°·2	0°·010+	54°·099	54°·322	0°·223
Friction, &c.	120	....	....	....	54·6	0·224−	54·332	54·421	0·089
Condensation, &c.	120	30·50	1467	56·6	56·5	0·308+	56·693	56·923	0·230
Friction, &c.	120	....	....	....	56·7	0·281+	56·926	57·036	0·110
Condensation, &c.	120	30·50	1440	62·6	63·6	1·763−	61·703	61·971	0·268
Friction, &c.	120	....	....	....	64·0	1·960−	61·976	62·105	0·129
Condensation, &c.	120	30·57	1442	59·0	58·4	0·400+	58·680	58·921	0·241
Friction, &c.	120	....	....	....	58·5	0·477+	58·921	59·033	0·112
Condensation, &c.	120	29·94	1405	55·2	57·0	1·566−	55·310	55·558	0·248
Friction, &c.	120	....	....	....	57·2	1·573−	55·563	55·692	0·129
Condensation mean	120	30·38	1433	57·5	....	0·522−	....	....	0·242
Friction mean ..	120	....	....	....	....	0·600−	....	....	0·114
Corrected result..	....	30·38	1433	....	....	......	....	....	0·128

After applying the proper correction for the increase of friction during condensation, and reducing the result, as before, to the capacity of a lb. of water, I find 5°·26 to be the mean quantity of heat evolved by compression of air in the above series of experiments.

The mechanical force spent in the condensation is represented in this instance by

$$s=1779{\cdot}3\times 2{\cdot}302585\times \log 10{\cdot}498=4183{\cdot}46.$$

Hence the equivalent of a degree of heat per lb. of water, as determined by the above series, is 795 lb. raised to the height of one foot.

The mechanical equivalents of heat derived from the foregoing experiments were so near 838 lb.*, the result of magnetical experiments in which "latent heat" could not be suspected to interfere in any way, as to convince me that the heat evolved was simply the manifestation, in another form, of the mechanical power expended in the act of condensation: I was still further confirmed in this view of the subject by the following experiments.

I provided another copper receiver (E, fig. 54) which had a capacity of 134 cubic inches. Like the former receiver, to

Fig. 54.

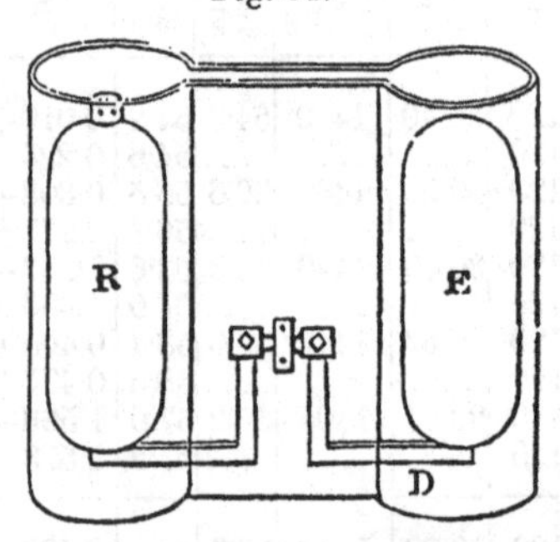

which it could be connected by a coupling nut, it had a piece D attached, in the centre of which there was a bore $\frac{1}{8}$ of an inch diameter, which could be closed perfectly by means of a proper stop-cock.

I must here be permitted to make a short digression, in order to explain the construction of the stop-cocks, as it may save those who shall in future attempt similar experiments the useless trouble of trying to make the ordinary stop-cock perfectly air-tight under high pressures. The one I have

* Phil. Mag. ser. 3. vol. xxiii. p. 441.

used is the invention of Mr. Ash, of this town, a gentleman well known for his great mechanical genius; and he has in the most obliging manner allowed me to give a full description of it. Fig. 55 is a full-sized sectional view of the stop-

Fig. 55.

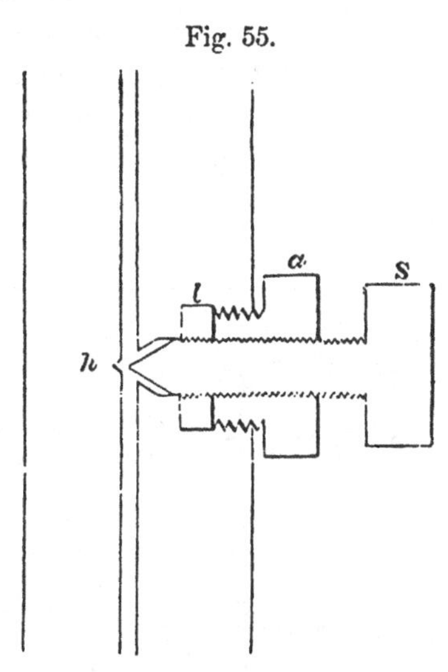

cock. *a* is a brass screw, by means of which a thick collar of leather, *l*, is very tightly compressed. The centre of *a* is perforated with a female screw, in which a steel screw, *s*, works, the threads of which press so tightly against the leather collar as effectually to prevent any escape of air in that direction. The end of the steel screw is smooth and conical, and the conical hole *h* is plugged with tin. When the stop-cock is shut, the smooth end of the steel screw presses against the soft metal, so as to prevent the escape of the least particle of air; but when opened, as represented in the figure, it leaves a passage for the air round the conical point. I have tested this stop-cock in the most severe manner, and have found it to answer perfectly.

Having filled the receiver R (fig. 54) with about 22 atmospheres of dry air, and having exhausted the receiver E by means of an air-pump, I screwed them together, and then put them into a tin can containing 16½ lb. of water. The water was first thoroughly stirred, and its temperature taken by the same delicate thermometer which was made use of in the former experiments. The stop-cocks were then opened by means of a proper key, and the air allowed to pass from the full into the empty receiver until equilibrium was established between the two. Lastly, the water was again stirred and its temperature carefully noted. The following Table contains the results of a series of experiments conducted in this way, alternated with others to eliminate the effects of stirring, evaporation, &c.

TABLE III.

Nature of experiment.	Barometrical pressure.	Quantity of air compressed in receiver R in cubic inches.	Mean temp. of the room.	Difference.	Temp. of water.		Gain or loss of heat.
					Before expt.	After expt.	
Expansion	30·20	2910	57°·4	0°·118+	57°·520	57°·517	0°·003 loss.
Alternation	....	....	57·0	0·906−	56·085	56·103	0·018 gain.
Expansion	30·44	2920	57·0	0·885−	56·103	56·128	0·025 gain.
Alternation	....	....	62·0	0·783−	61·217	61·217	0
Expansion	30·44	2910	62·1	0·873−	61·222	61·232	0·010 gain.
Alternation	....	....	58·5	0·233+	58·732	58·735	0·003 gain.
Expansion	30·44	2915	58·6	0·132+	58·732	58·732	0
Alternation	....	....	61·3	0·787−	60·508	60·518	0·010 gain.
Expansion	30·46	3200	61·3	0·780−	60·518	60·523	0·005 gain.
Alternation	....	....	58·0	0·186+	58·184	58·187	0·003 gain.
Expansion	30·50	2880	58·3	0·110−	58·190	58·190	0
Mean of expts. of expansion	30·41	2956	....	0·400−	....	....	0·0062 gain.
Mean of alternations	....	....	....	0·411−	....	....	0·0068 gain.
Corrected result..	30·41	2956	....	......	....	....	0

The difference between the means of the expansions and alternations being exactly such as was found to be due to the increased effect of the temperature of the room in the latter case, we arrive at the conclusion, that *no change of temperature occurs when air is allowed to expand in such a manner as not to develop mechanical power*.

In order to analyze the above experiments, I inverted the receivers, as shown in fig. 56, and immersed them, as well as

Fig. 56.

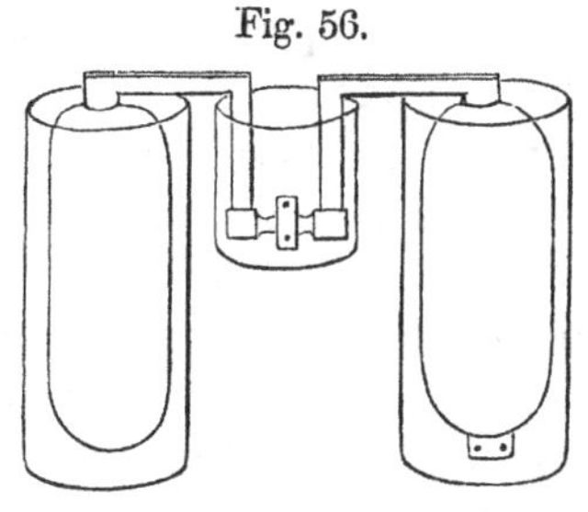

the connecting-piece, into separate cans of water. One of the receivers had 2828 cubic inches of dry air condensed into it, while the other was vacuous. After equilibrium was restored by opening the cocks, I found that 2°·36 of cold per lb. of water had been produced in the receiver from which the air had expanded, while 2°·38 of heat had been produced in the other receiver, and 0°·31 of heat also in the can in which the connecting-piece was immersed, the sum of the whole amounting nearly to zero. The slight redundance of heat was owing to the loss of cold during the passage of the air from the charged receiver to the stop-cocks, through a part of the pipe which could not be immersed in water.

A series of experiments was now made in the following manner:—The receiver was filled with dry compressed air, and a coiled leaden pipe, $\frac{1}{4}$ of an inch in internal diameter and 12 yards long, was screwed tightly upon the nozzle, as represented in fig. 57. The whole was then immersed into

Fig. 57.

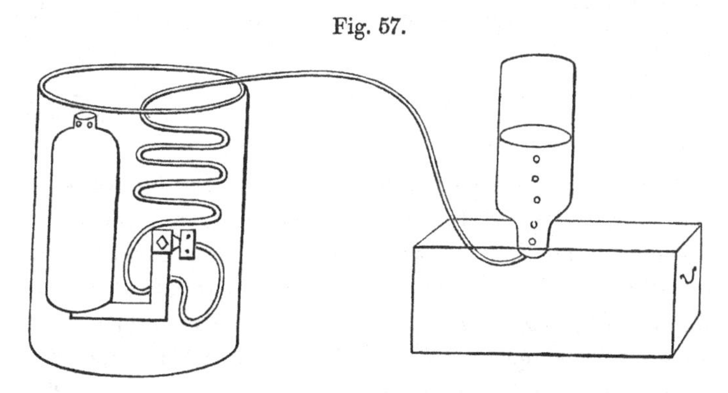

an oval can, which was constructed as before described, and

was also covered at top as perfectly as possible. Having ascertained the temperature of the water by means of the sensible thermometer before used, the stop-cock was opened and the air made to pass from the receiver through a pneumatic trough into a jar, by which it was carefully measured. After the air in the receiver had been reduced to the atmospherical pressure, the water was again well stirred and its temperature noted. An alternation was made after each of these experiments, in order to eliminate the effects of stirring, &c.

TABLE IV.

Nature of experiment.	Barometrical pressure.	Quantity of air compressed.	Quantity of air let out.	Mean temp. of the room.	Difference.	Temp. of water. Before expt.	Temp. of water. After expt.	Gain or loss of heat.
Expansion	30·04	2862	2726	55°·7	0°·405+	56°·207	56°·004	0°·203 loss.
Alternation	....	....	....	55·4	0·579+	56·004	55·954	0·050 loss.
Expansion	30·10	2807	2670	54·6	0·022+	54·714	54·530	0·184 loss.
Alternation	....	....	....	54·25	0·276+	54·536	54·516	0·020 loss.
Expansion	30·10	2723	2587	53·6	0·760+	54·460	54·259	0·201 loss.
Alternation	....	....	....	53·4	0·839+	54·259	54·219	0·040 loss.
Expansion	30·10	2807	2670	49·05	0·307+	49·456	49·258	0·198 loss.
Alternation	....	....	....	49·1	0·158+	49·258	49·258	0
Expansion	30·23	3039	2903	50·6	0·508−	50·176	50·008	0·168 loss.
Alternation	....	....	....	51·1	1·063−	50·017	50·057	0·040 gain.
Expansion	30·20	2919	2782	49·0	0·355−	48·728	48·563	0·165 loss.
Alternation	....	....	....	48·85	0·277−	48·573	48·573	0
Mean of expts. of expansion	30·13	2859	2723	....	0·105+	......	......	0·1865 loss.
Mean of alternations	....	....	....	....	0·085+	......	......	0·0117 loss.
Corrected result.	30·13	2859	2723	....	......	......	......	0·1738 loss.

The cold produced was diffused through 21·17 lb. of water, 14 lb. of copper, 8 lb. of lead, and 7 lb. of tinned iron. Hence we find that a quantity of cold was produced in the experiments sufficient to cause the temperature of a lb. of water to decrease by 4°·085. At the same time a mechanical force was developed, which could raise a column of the atmosphere, of an inch square at the base, to the altitude of 2723 inches; or, in other words, could raise 3352 lb. to the height of one foot. For each degree of heat lost there was therefore generated a force sufficient to raise 820 lb. to the height of one foot.

In the two following series the experiments were varied by compressing and measuring out different volumes of air.

TABLE V.

Nature of experiment.	Barometrical pressure.	Quantity of air compressed.	Quantity of air let out.	Mean temp. of the room.	Difference.	Temp. of water.		Gain or loss of heat.
						Before expt.	After expt.	
Expansion	30·06	1336	1200	52°·5	1°·441 –	51°·074	51°·044	0°·030 loss.
Alternation	....	....	....	52·55	1·460 –	51·069	51·110	0·041 gain.
Expansion	30·20	1343	1206	53·5	1·385 –	52·125	52·105	0·020 loss.
Alternation	....	....	....	53·6	1·457 –	52·115	52·171	0·056 gain.
Expansion	30·28	1386	1250	52·4	0·419 –	52·021	51·941	0·080 loss.
Alternation	....	....	....	52·55	0·588 –	51·951	51·974	0·023 gain.
Expansion	30·28	1387	1250	52·95	0·778 –	52·195	52·148	0·047 loss.
Alternation	....	....	....	53·2	1·017 –	52·171	52·195	0·024 gain.
Expansion	30·30	1434	1298	59·0	0·610 +	59·665	59·556	0·109 loss.
Alternation	....	....	....	58·65	0·888 +	59·551	59·526	0·025 loss.
Expansion	30·27	1405	1268	55·35	0·227 +	55·622	55·532	0·090 loss.
Alternation	....	....	....	55·1	0·534 +	55·647	55·622	0·025 loss.
Expansion	30·14	1400	1264	55·2	0·313 +	55·565	55·461	0·104 loss.
Alternation	....	....	....	55·3	0·158 +	55·461	55·456	0·005 loss.
Mean of expts. of expansion	30·22	1384	1248	....	0·410 –	......	......	0·0686 loss.
Mean of alternations	....	....	....	....	0·420 –	......	......	0·0127 gain.
Corrected result.	30·22	1384	1248	....	......	......	......	0·081 loss.

On reducing the results of these experiments in the manner before indicated, we find that in the experiments of Table V. 814 lb, and in those of Table VI. 760 lb. were raised to the height of a foot for every degree of heat per lb. of water lost.

These results are inexplicable if heat be a substance. If that were the case, the same quantity of heat would have been absorbed by the rarefaction which took place in the experiments of Table IV. as was evolved by the corresponding

TABLE VI.

Nature of experiment.	Barometrical pressure.	Quantity of air compressed.	Quantity of air let out.	Mean temp. of the room.	Difference.	Temp. of water.		Gain or loss of heat.
						Before expt.	After expt.	
Expansion	30·24	3116	1238	60·1	0·418−	59·724	59·641	0·083 loss.
Alternation	....	....	....	60·2	0·552−	59·641	59·655	0·014 gain.
Expansion	30·20	3198	1238	56·1	0·041+	56·185	56·098	0·087 loss.
Alternation	....	....	....	56·2	0·090−	56·103	56·108	0·005 gain.
Expansion	30·15	3192	1238	61·15	1·117+	62·328	62·207	0·121 loss.
Alternation	....	....	....	60·9	1·275+	62·195	62·155	0·040 loss.
Expansion	30·15	3143	1238	60·15	0·863+	61·063	60·964	0·099 loss.
Alternation	....	....	....	60·05	0·896+	60·959	60·934	0·025 loss.
Expansion	30·20	2966	1238	55·45	0·343+	55·835	55·751	0·084 loss.
Alternation	....	....	....	55·45	0·298+	55·751	55·746	0·005 loss.
Expansion	30·30	3160	1238	58·1	0·285+	58·432	58·337	0·095 loss.
Alternation	....	....	....	57·95	0·384+	58·337	58·332	0·005 loss.
Expansion	30·14	3188	1238	55·0	0·678+	55·733	55·624	0·109 loss.
Alternation	....	....	....	55·1	0·515+	55·624	55·607	0·017 loss.
Mean of expts. of expansion	30·20	3137	1238	....	0·416+	......	......	0·0968 loss.
Mean of alternations	....	....	....	....	0·389+	......	......	0·0104 loss.
Corrected result.	30·20	3137	1238	....	......	......	......	0·0855 loss.

condensation in the experiments of Table I.: also a certain quantity of cold would have been produced in the experiments given in Table III. The results are, however, such as might have been deduced *à priori* from any theory in which heat is regarded as a state of motion among the constituent particles of bodies. It is easy to understand how the mechanical force expended in the condensation of air may be communicated to these particles so as to increase the rapidity of their motion, and thus may produce the phenomenon of increase of temperature. In the experiments of Table III. no cold was produced, because the momentum of these particles was not permanently converted into mechanical power; but had the motion of the air from one vessel to the other been opposed in such a manner as to develop power at the outside of the jar, which might have been accomplished by means of a cylinder and piston, *then* loss of heat would have occurred, just as in

Tables IV., V., and VI., where the force was applied in lifting the atmosphere of the earth.

It is quite evident that the reason why the cold in the experiments of Table IV. was so much inferior in quantity to the heat evolved in those of Table I., is that all the force of the air, over and above that employed in lifting the atmosphere, was applied in overcoming the resistance of the stop-cock, and was there converted back again into its equivalent of heat.

The discovery of Dulong*, that equal volumes of all elastic fluids, taken at the same temperature and under the same pressure, when suddenly compressed or dilated to the same fraction of their volume, disengage or absorb the same absolute quantity of heat, accords perfectly with these principles.

The mechanical equivalents of heat determined by the various series of experiments given in this paper are 823, 795, 820, 814, and 760. The mean of the last three, which I take as least liable to error, is 798 lb., a result so near 838 lb., the equivalent which I deduced from my magnetical experiments, as to confirm, in a remarkable manner, the above explanation of the phenomena described in this paper, and to afford a new and, to my mind, powerful argument in favour of the dynamical theory of heat which originated with Bacon, Newton, and Boyle, and has been at a later period so well supported by the experiments of Rumford, Davy, and Forbes. With regard to the detail of the theory, much uncertainty at present exists. The beautiful idea of Davy, that the heat of elastic fluids depends partly upon a motion of particles round their axes†, has not, I think, hitherto received the attention it deserves. I believe that most phenomena may be explained by adapting it to the great electro-chemical discovery of Faraday, by which we know that each atomic element is associated with the same absolute quantity of electricity. Let us suppose that these atmospheres of electricity, endowed to a certain extent with the ordinary properties of matter, revolve

* Annales de Chimie, vol. xli. p. 156.

† Elements of Chemical Philosophy, vol. i. p. 94.

with great velocity round their respective atoms, and that the velocity of rotation determines what we call temperature. In an aeriform fluid we may suppose that the attraction of the atmospheres by their respective atoms, and that of the atoms towards one another, are inappreciable for all pressures to which the law of Boyle and Mariotte applies, and that, consequently, the centrifugal force of the revolving atmospheres is the sole cause of expansion on the removal of pressure. By this mode of reasoning, the law of Boyle and Mariotte receives an easy explanation, without recourse to the improbable hypothesis of a repulsion varying in a ratio different from that of the inverse square. The phenomena described in the present paper, as well as most of the facts of thermo-chemistry, agree with this theory; and in order to apply it to radiation, we have only to admit that the revolving atmospheres of electricity possess, in a greater or less degree, according to circumstances, the power of exciting isochronal undulations in the ether which is supposed to pervade space.

The principles I have adopted lead to a theory of the steam-engine very different from the one generally received, but at the same time much more accordant with facts. It is the opinion of many philosophers that the mechanical power of the steam-engine arises simply from the passage of heat from a hot to a cold body, no heat being necessarily lost during the transfer. This view has been adopted by Mr. E. Clapeyron in a very able theoretical paper, of which there is a translation in the 3rd part of Taylor's Scientific Memoirs. This philosopher agrees with Mr. Carnot in referring the power to *vis viva* developed by the caloric contained by the vapour in its passage from the temperature of the boiler to that of the condenser. I conceive that this theory, however ingenious, is opposed to the recognized principles of philosophy, because it leads to the conclusion that *vis viva* may be destroyed by an improper disposition of the apparatus: thus Mr. Clapeyron draws the inference that "the temperature of the fire being from 1000° (C.) to 2000° (C.) higher than that of the boiler, there is an enormous loss of *vis viva* in the passage of the heat from the furnace into the boiler."

Believing that the power to destroy belongs to the Creator alone, I entirely coincide with Roget and Faraday in the opinion that any theory which, when carried out, demands the annihilation of force, is necessarily erroneous. The principles, however, which I have advanced in this paper are free from this difficulty. From them we may infer that the steam, while expanding in the cylinder, loses heat in quantity exactly proportional to the mechanical force which it communicates by means of the piston, and that on the condensation of the steam the heat thus converted into power is *not* given back. Supposing no loss of heat by radiation &c., the theory here advanced demands that the heat given out in the condenser shall be less than that communicated to the boiler from the furnace, in exact proportion to the equivalent of mechanical power developed.

It would lengthen this paper to an undue extent were I now to introduce any direct proofs of these views, had I even leisure at present to make the experiments requisite for the purpose; I shall therefore reserve the further discussion of this interesting subject for a future communication, which I hope to have the honour of presenting to the Royal Society at no distant period.

Oak Field, near Manchester,
June 1844.

On Specific Heat. *By* J. P. Joule.

[Phil. Mag. ser. 3. vol. xxv. p. 334. Read before the British Association at York, September 27th, 1844.]

The well-known law which applies to the specific heat of many simple bodies, while interesting in itself as one of the doctrines of physical science, is of great importance to theoretical chemistry as a criterion for the determination of atomic weights. Dulong and Petit, the philosophers who first announced that the specific heat of simple bodies is inversely proportional to their atomic weights, proved their proposition by experiment in the cases of several solid (chiefly metallic) bodies. Subsequently several attempts have been made to discover the law of the specific heat of gases, liquids, and compound bodies. With regard to gases, Haycraft*, and subsequently De la Rive and Marcet†, have attempted to prove that under the same pressure and volume all gases have the same specific heat. Unfortunately, the practical difficulty of ascertaining the specific heat of aeriform fluids is so great that considerable uncertainty exists in the results obtained by the most skilful experimenters; and hence we find that the law of Haycraft, De la Rive and Marcet has not been confirmed by the researches of De la Roche and Berard, Dr. Apjohn‡, and Dulong. With compound solid bodies, however, experiments have met with better success. Neumann§, in the cases of the carbonates and sulphates of protoxides, has pointed out that in each of these classes the specific heat is inversely proportional to the atomic weight. His researches on some of the oxides and sulphurets also conducted him to a similar law for each of these species of compounds.

Of late years no philosopher has made more numerous or more accurate experiments on specific heat than Mr. V. Regnault. The investigations of this eminent chemist were, in the first instance, directed to the specific heat of simple solid bodies ||; subsequently they have been directed to a great variety of compound bodies¶. By these researches Regnault has given the law of Dulong and Petit all the confirmation that could be desired, and has also proved the correctness of Neumann's extension of that law to classes of chemical compounds. He has stated a general law in the following terms:—"In all compound bodies of the same atomic composition and of similar chemical constitution,

* Trans. R. S. Edin. vol. x. pt. 1, p. 215.

† Annales de Chimie, 1827, tome xxxv. p. 27.

‡ Trans. Royal Irish Acad. vol. xviii. pt. 1, p. 1C.

§ Poggendorff's 'Annalen,' vol. xxiii.

|| Annales de Chimie, vol. lxxv. p. 1 (1840).

¶ Ibid. vol. i. p. 129 (1841).

the specific heats are inversely proportional to the atomic weights."

Regnault remarks that the above law holds good only within certain limits, and that the slight differences which are observed between the results of theory and observation are not wholly to be attributed to mere errors of experiment. He says that if the specific heat were taken for the temperature at which the bodies present the greatest analogy in their

TABLE OF SPECIFIC HEATS.

Name of substance.	Formula adopted.	Number of atoms divided by equivalent.	Theoretical specific heat.	Actual specific heat.	Experimenter.
Water	H_2O	$\frac{3}{9}$	1000	1000	[Berard.
Hydrogen	H_2	$\frac{2}{1}$	6000	3294	De la Roche &
Oxygen	O	$\frac{1}{8}$	375	236	„
Sulphur	S	$\frac{1}{16}$	188	188	Dulong & Petit.
Iodine	I	$\frac{1}{63}$	48	54	V. Regnault.
Carbon	C	$\frac{1}{12}$	250	241	„
Lead	Pb	$\frac{1}{103}$	29	29	Dulong & Petit.
Zinc	Zn	$\frac{1}{33}$	91	93	„
Copper	Cu	$\frac{1}{32}$	94	95	„
Mercury	Hg	$\frac{1}{101}$	30	33	„
Protoxide of lead	PbO	$\frac{2}{111}$	54	51	V. Regnault.
Protoxide of copper	CuO	$\frac{2}{40}$	150	142	„
Magnesia	MgO	$\frac{2}{21}$	286	244	„
Peroxide of iron	Fe_2O_3	$\frac{5}{78}$	192	167	„
Subsulphuret of copper	Cu_2S	$\frac{3}{80}$	113	121	„
Protosulphuret of lead	PbS	$\frac{2}{119}$	50	50	„
Protosulphuret of iron	FeS	$\frac{2}{43}$	140	136	„
Bisulphuret of iron	FeS_2	$\frac{3}{59}$	153	130	„
Chloride of sodium	Na_2Cl_2	$\frac{4}{58}$	·207	214	„
Protochloride of copper	Cu_2Cl_2	$\frac{4}{100}$	120	138	„
Chloride of lead	$PbCl_2$	$\frac{3}{139}$	65	66	„
Iodide of silver	Ag_2I_2	$\frac{4}{234}$	51	62	„
Iodide of lead	PbI_2	$\frac{3}{230}$	39	43	„
Sulph. acid, sp. gr. 1·85	SO_3H_2O	$\frac{7}{49}$	429	350	Dalton.
Carbonate of potash	$2KO+CO_4$	$\frac{9}{139}$	194	216	V. Regnault.
Carbonate of lead	$2PbO+CO_4$	$\frac{9}{268}$	101	86	„
Sulphate of lead	$PbOSO_3$	$\frac{6}{152}$	118	87	„
Sulphate of potash	$KOSO_3$	$\frac{6}{88}$	204	190	„
Chlorate of potash	$KOCl_2O_5$	$\frac{9}{123}$	220	210	„
Nitrate of potash	KON_2O_5	$\frac{9}{102}$	265	238	„
Alcohol	$C_2H_{10}O+H_2O$	$\frac{16}{46}$	1043	622	Despretz.

chemical and physical properties, the most complete isomorphism, the law would probably hold good rigorously.

Now, without denying altogether the influence of a change of state on the specific heat of a body, I think it may be fairly doubted whether it is so great as quite to mask a general law. Impressed with this idea, I have drawn up the above Table, in which the theoretical specific heats of a variety of bodies impartially taken are calculated on the hypothesis that the capacity for heat of any simple atom remains the same in whatever chemical combination it enters. The law implied by this hypothesis is, that the specific heats are directly as the number of atoms in combination, and inversely as the equivalent.

The substances chosen for the construction of the above Table have been, as I have said, impartially selected. I have omitted some in which theory was found to agree perfectly with experiment, whilst I have inserted others (as, for instance, alcohol) which appear to disagree very considerably. On the whole, the coincidence between the theoretical and experimental results is such that I think chemists will agree with me in believing that the law of Dulong and Petit, with regard to simple atoms, is capable of a greater degree of generalization than we have hitherto been inclined to admit.

On a new Method for ascertaining the Specific Heat of Bodies. By JAMES P. JOULE, *Esq., Sec. Lit. & Phil. Soc., Mem. Chem. Society, &c.**

[Memoirs Manch. Lit. & Phil. Soc., 2nd ser.vol. vii. p. 559. Read before the Manchester Lit. and Phil. Soc., Dec. 2nd, 1845.]

THREE methods have been employed by natural philosophers in order to determine the specific heat of bodies. These are:—1st, the method of the calorimeter; 2nd, the method of mixtures; and, 3rd, the method of cooling.

* The experiments were made at Oak Field, Whalley Range, near Manchester.

The first of these methods consists, as is well known, in plunging a heated body, the temperature of which has been carefully ascertained, into ice, and observing the quantity of water produced by the cooling of the body to the freezing temperature. Lavoisier and Laplace employed this method in their researches on specific heat; and, by the invention of the *calorimeter*, carried it to the greatest degree of perfection of which it is in all probability capable. In practice, however, this method has been found exceedingly clumsy and tedious, and liable to a variety of errors, which have been pointed out by Wedgewood. It has in consequence been long ago abandoned by accurate experimenters.

The method of *mixtures* has been employed by Boerhaave, Black, Irvine, Wilcke, Crawford, and others. In this method, the heat lost by one body in cooling is received by another body which is heated. By knowing the weights of the bodies and the temperatures gained and lost by them, it is easy to deduce the relation of their specific heats.

The method of *cooling* was employed by Meyer, Leslie, and Dalton, and has been much improved by Dulong and Petit. It consists in ascertaining the rapidity with which different bodies cool when placed in similar circumstances with respect to the radiation of heat.

The methods of cooling, and especially of mixtures, appear to be founded upon correct theoretical principles. Both of them are, however, exposed to several sources of error which the utmost skill and ingenuity have not hitherto been able to remove. In proof of this I may mention that V. Regnault, one of the most expert experimenters in this department of physics, has found that the two methods give different results when applied to the determination of the capacity of the same body. It is sufficiently obvious therefore that although the processes hitherto used are, in skilful hands, capable of giving rough approximations to the truth, they are at the same time wholly unfitted for the determination of specific heat with that extreme degree of accuracy which is so desirable in the present state of science. It will not therefore, I hope, be deemed superfluous to point out a new and, as I confidently believe, far better method, founded upon those laws of the evolution of heat by voltaic electricity which I have developed in previous memoirs.

I propose in the first place to point out briefly the principles of the new method, and then to show how it may be applied in order to determine the specific heat of solids, liquids, and gases.

When any body, capable of conducting the voltaic electricity, is placed in the circuit of a battery, the quantity of heat evolved by it in a given time is proportional to its resistance to conduction and the square of the quantity of transmitted electricity. Consequently, if a wire traversed by a voltaic current be made to communicate to any body the heat which it evolves, the capacity for heat of that body and the wire taken together will be directly proportional to the square of the quantity of electricity transmitted in a given time, to the resistance of the wire, and to the time, and inversely proportional to the increase of temperature of the body. Hence we derive the general equation,

$$y=\frac{Cc^2rt}{h}, \quad . \quad . \quad . \quad . \quad . \quad . \quad (1)$$

where y is put for the capacity, c for the voltaic current, r for the resistance of the wire, t for the time, and h for the increase of temperature.

If we make the time and the resistance of the wire constant, the above equation becomes simplified to

$$y=\frac{Cc^2}{h}. \quad . \quad . \quad . \quad . \quad . \quad . \quad . \quad (2)$$

If the current and time are constant, we have

$$y=\frac{Cr}{h}. \quad . \quad . \quad . \quad . \quad . \quad . \quad . \quad (3)$$

I have made several series of experiments, using the same conducting-wire as a source of heat, a constant interval of time, and a variable current of electricity. The method adopted in these experiments was, to try the effect of a wire traversed by a current of electricity, first, when it was immersed in water, and, afterwards, when it was immersed in another liquid. Hence I obtained two determinations of y in equation (2)—one of them for water, the other for the liquid. The relation between these two quantities gave, of course, the capacity of the liquid compared with that of the water.

The results at which I thus arrived showed, by their agreement with one another, that the method just described was susceptible of great accuracy. But it will at once be seen that it requires very exact determinations of the intensity of the voltaic currents, and thus renders the use of a correct and delicate galvanometer indispensable. Such instruments are in the possession of very few, and, moreover, require many troublesome precautions in using them. It is therefore more advantageous to dispense with the measurement of the voltaic currents by employing equation (3), in which the current is constant and the resistance variable.

I take two pieces of thin wire, the conducting-powers of which have been previously ascertained, and immerse one of them in a quantity of water, and the other in the substance whose specific heat is to be determined. I then place the wires in different parts of the same voltaic circuit, so as to cause the same current to traverse both of them consecutively. After the current has passed for a proper length of time, I note the increase of temperature which has occurred in the water and in the substance under examination. Thus I obtain two determinations of y in equation (3)—one of them for water and the other for the substance under trial, from which the specific heat of the latter may be readily deduced.

The method I have just described requires an accurate knowledge of the resistance of the wires by which the heat is evolved. The principal ways of ascertaining the resistances of wires to the passage of electric currents are those of Ohm and Wheatstone. These methods, however, require the use of galvanometers and other delicate instruments, and are not therefore so well adapted for my purpose as the method of determining the resistances from the calorific powers, to which they are always directly proportional.

Having now given a rough outline of the new method, I

may proceed to point out in greater detail the course of experiments which may be advantageously pursued, in order to apply it to the determination of the specific heat of solids, liquids, and gases.

The first business will be to obtain two similar platinum wires, each about four inches long and $\frac{1}{100}$ of an inch thick*. It will be desirable to employ wires of exactly the same length and diameter; for if perfect equality could be attained in the resistances, equation 3 would become

$$y=\frac{C}{h}, \qquad \ldots \ldots \ldots \qquad (4)$$

and the numerical calculations would be much simplified. As it is, however, extremely difficult, if not impossible, to obtain exactly similar portions, even of the same length of wire, it will be always proper to examine the conducting-powers of the wires. For this purpose they must be immersed in similar vessels, containing the same quantity (say a pound) of water, and then placed in different parts of the same voltaic circuit. If the voltaic battery employed consists of six large Daniell's cells, the copper element of each of which exposes an active surface of two square feet, the heat evolved in the two jars will be at the rate of about one degree per minute. Connexion with the battery must be broken after the current has passed for five or ten minutes, when the increase of temperature must be ascertained to the hundredth part of a degree by the help of accurate thermometers. The influence of the surrounding atmosphere must be obviated as far as possible by arranging matters so that the temperature of the water shall be as much above that of the air at the conclusion as it was below it at the commencement of an experiment. A series of ten experiments will be sufficient to determine the relative heating-powers and resistances of the two wires to one thousandth at least; and still greater accuracy may be attained by taking the mean of this and another series, in which the wires shall have changed places

* The resistance of the wires ought to be equal to that of the battery, in order to produce the maximum calorific effect.

in the vessels of water. In this way the effect of any slight difference in the capacity for heat of the vessels will be eliminated.

The next step will be to determine the capacity for heat of the vessels which are to be employed in the experiments. These vessels ought to be as thin as possible, in order that their capacity for heat may be very small in comparison with that of their contents. They may be constructed of any material incapable of being acted upon by the liquids. In order to ascertain the capacity for heat of these vessels, one of them must be filled with water, and the other partly with water and partly with the material of which the vessels are made. A platinum wire, whose resistance or heating-power is already known, must be placed in each vessel, and then made to form part of the circuit of a voltaic battery, so that the same current shall traverse both wires. The resistances of the wires and the increase of the temperatures will then determine the relative capacities of the vessels and their contents, from which the specific heat of the material of which the vessels are made may be readily deduced. For, calling the quantity of water in the first vessel* a, the weight of the vessel when empty b, the resistance of its wire r, and the increment of temperature h, the quantity of water in the other vessel a', the weight of that vessel along with the quantity of the same material put into it b', the resistance r', and the increment of temperature h', also calling the specific heat of the material in question x, we obtain from equation (3)

$$a+bx : \frac{r}{h} :: a'+b'x : \frac{r'}{h'},$$

whence

$$x=\frac{ahr'-a'h'r}{b'h'r-bhr'}. \qquad (5)$$

Having thus ascertained the specific heat of the material of which the vessels are made, they may be employed as

* The quantities a and a' of course include the weight of water equal in capacity to the thermometers, conducting-wires, stirrers, &c.

follows, for determining the specific heat of various kinds of matter.

1. *Method with Liquids.*—One of the vessels must be filled with pure water, and the other with the liquid under examination. We may call the former the standard, and the latter the experimental vessel. The same current of electricity must then be made to traverse two platinum wires immersed in the vessels; and the consequent increments of temperature must be carefully noted. As the capacity for heat of the standard vessel is in this case known, our equation (5) becomes

$$x=\frac{ahr'-a'h'r}{b'h'r}, \quad . \quad . \quad . \quad . \quad . \quad . \quad (6)$$

where a is the weight of water in the standard vessel, including the water which is equivalent in capacity for heat to the vessel itself, r is the resistance of its platinum wire, and h the increase of its temperature, a' is the weight of water equivalent in capacity to the experimental vessel, b' the weight of the liquid under examination in the experimental vessel, r' the resistance of its platinum wire, and h' the increase of its temperature.

When the specific heat of a fluid, capable of conducting the voltaic current, is to be determined, it will be proper to protect the wire from immediate contact with it. This object may be accomplished by enclosing the platinum wire in a small platinum or glass vessel filled with water; or, instead of the platinum wire, a column of mercury, enclosed in a glass tube of narrow bore, may be chosen as the conducting medium.

2. *Method with Solids.*—If the solid body be insoluble in water, it must be broken into small pieces and placed in the experimental vessel, the remainder of which must then be filled with pure water. In this case, as well as in the last, equation (6) applies, the only difference being that the quantity a' includes the water in the experimental vessel as well as the vessel itself, considered as water.

Soluble bodies must be enclosed in small tubes or bottles, made of tinned iron, platinum, or glass, whose capacity for heat has been previously determined. The bottles must be completely immersed in the water contained in the experimental vessel. It will be necessary, in this case, to take care to let sufficient time elapse between the stoppage of the current and the determination of the temperature, in order to allow the salt to acquire the temperature of the water. Equation (6) applies in this case also, the quantity a' including, as in every other case, all the bodies in the experimental vessel whose capacities are known, considered as water.

3. *Method with Gases.*—The gases must be enclosed in proper vessels, whose capacity for heat has been previously determined, and then the experiments with them may be conducted as in the previous cases, using equation (6). Owing to the great space occupied by the gases, it will be important to employ as little water as possible in the standard vessel, and only as much water in the experimental vessel as will be sufficient to surround the flask containing the gas.

Of the three kinds of matter to which I have thus shown the applicability of the new method, the solids seem to me to present fewer difficulties to the old methods than the rest. Consequently they are the most fitting to be employed in order to compare the results of the old and new methods together. I have made the following experiments for this purpose.

Five pounds of water were poured into the standard vessel, into which a platinum wire, having a resistance called 100, was then immersed. A bundle of small sticks of lead weighing $11\frac{1}{2}$ lb. was placed in the experimental vessel, along with 4 lb. of water and a platinum conducting-wire whose resistance was 106. Each vessel was properly furnished with stirrers and thermometers of known capacity. A current from a battery of six large Daniell's cells was then passed through the platinum wires for 20 minutes, at the end of which time the increase of the temperature of the standard vessel was $3^{\circ}{\cdot}575$, and of the experimental vessel $4^{\circ}{\cdot}35$.

Applying equation (6), and turning all the weights into grains*, I had

$$x = \frac{35280 \times 3{\cdot}575 \times 106 - 28260 \times 4{\cdot}35 \times 100}{80500 \times 4{\cdot}35 \times 100} = 0{\cdot}03073.$$

Another experiment was conducted in the same manner; but in this instance the thermometers and platinum conducting-wires were made to change places in the vessels, so as to detect any small error in the values of the scales of the thermometers or in the resistances of the wires. The result of this second experiment was 0·0299. The mean of the two results is 0·0303. The result of Dulong and Petit is 0·0293; that of Regnault 0·0314.

I made two experiments also on the specific heat of wrought iron. The mean result of them was 0·10993. The result of Dulong and Petit is 0·11, and that of Regnault 0·114.

Although, in the above experiments, the quantities of water and metal were much larger than desirable, the coincidence of the results, both with themselves and with those which have been arrived at by the expert use of the best of the old methods, is such as must, I think, inspire confidence in the correctness of the new method. Hitherto I have not made many experiments with it besides the foregoing and those which I have incidentally found necessary in the course of other investigations. I have not, however, thought it right to delay the publication of a method which appears to possess eminent advantages over the old processes for the determination of specific heat, and which, in the hands of any painstaking experimenter, would, I doubt not, lead to the complete elucidation of the laws of a most important branch of physics.

* The capacities of the vessels, thermometers, &c. were found to be equivalent to those of 280 and 260 grms. of water. Hence in the equation $a = 35000 + 280$, and $a' = 28000 + 260$.

Note on the Employment of Electrical Currents for ascertaining the Specific Heat of Bodies. By J. P. JOULE, *Esq.*

['Memoirs Manch. Lit. & Phil. Soc.' 2nd ser. vol. viii. p. 375. Read before the Manchester Literary and Philosophical Society, July 13th, 1847.]

HAVING recently had occasion to ascertain the specific heat of sperm-oil, I employed for the purpose the new method described in the seventh volume, new series, of the Memoirs of this Society. Two platinum wires, each 4 inches long and $\frac{1}{100}$ of an inch in diameter, were immersed, one in a known quantity of water, and the other in the sperm-oil. A powerful current of electricity from six large constant cells was then transmitted through the wires for half an hour, and the increase of the temperature of the water and oil noted. The specific heat of the sperm-oil arrived at was 0·3757, a result so much lower than that of Dalton that I was led to examine whether I had fallen into any error. For this purpose I repeated the experiment, taking, however, the precaution to keep the liquid constantly agitated. The specific heat now came out 0·406. The cause of the smallness of the result became thus apparent. The oil could not carry off the heat from the wire as quickly as the water, and hence the wire which was immersed in the oil became highly heated, occasioning an increase of its resistance and a proportional increase in the quantity of heat evolved by it. This was easily proved by placing the finger in contact with the wire, which could not be retained in that position longer than one or two seconds.

The object of this communication is therefore to guard the experimenter against employing wires of so small a surface as those recommended in my paper on specific heat, whenever powerful currents are employed, especially when, at the same time, the specific heat of a viscous liquid of bad conductive power and small capacity for heat is sought. In such cases a large strip of platinum foil would be preferable to a wire, on account of the extensive surface which would thus be presented to the liquid.

On the Mechanical Equivalent of Heat.
By JAMES P. JOULE*.

[Brit. Assoc. Rep. 1845, Trans. Chemical Sect. p. 31. Read before the British Association at Cambridge, June 1845.]

THE author gave the results of some experiments, in order to confirm the views he had already derived from experiments on the heat evolved by magneto-electricity, and from experiments on the changes of temperature produced by the condensation and rarefaction of elastic fluids. He exhibited to the Section an apparatus consisting of a can of peculiar construction filled with water. A sort of paddle-wheel was placed in the can, to which motion could be communicated by means of weights thrown over two pulleys working in contrary directions. He stated that the force spent in revolving the paddle-wheel produced a certain increment in the temperature of the water; and hence he drew the conclusion that when the temperature of a pound of water is increased by one degree of Fahrenheit's scale, an amount of *vis viva* is communicated to it equal to that acquired by a weight of 890 pounds after falling from the altitude of one foot.

On the Existence of an Equivalent Relation between Heat and the ordinary Forms of Mechanical Power. By JAMES P. JOULE, *Esq.* [In a letter to the Editors of the 'Philosophical Magazine.']

['Philosophical Magazine,' ser. 3. vol. xxvii. p. 205.]

GENTLEMEN,

The principal part of this letter was brought under the notice of the British Association at its last meeting at Cambridge. I have hitherto hesitated to give it further publication, not because I was in any degree doubtful of the conclusions at which I had arrived, but because I intended

* The experiments were made at Oak Field, Whalley Range.

to make a slight alteration in the apparatus calculated to give still greater precision to the experiments. Being unable, however, just at present to spare the time necessary to fulfil this design, and being at the same time most anxious to convince the scientific world of the truth of the positions I have maintained, I hope you will do me the favour of publishing this letter in your excellent Magazine.

The apparatus exhibited before the Association consisted of a brass paddle-wheel working *horizontally* in a can of water. Motion could be communicated to this paddle by means of weights, pulleys, &c., exactly in the manner described in a previous paper*.

The paddle moved with great resistance in the can of water, so that the weights (each of four pounds) descended at the slow rate of about one foot per second. The height of the pulleys from the ground was twelve yards, and consequently, when the weights had descended through that distance, they had to be wound up again in order to renew the motion of the paddle. After this operation had been repeated sixteen times, the increase of the temperature of the water was ascertained by means of a very sensible and accurate thermometer.

A series of nine experiments was performed in the above manner, and nine experiments were made in order to eliminate the cooling or heating effects of the atmosphere. After reducing the result to the capacity for heat of a pound of water, it appeared that for each degree of heat evolved by the friction of water a mechanical power equal to that which can raise a weight of 890 lb. to the height of one foot had been expended.

The equivalents I have already obtained are:—1st, 823 lb., derived from magneto-electrical experiments†; 2nd, 795 lb., deduced from the cold produced by the rarefaction of air‡;

* Phil. Mag. ser. 3. vol. xxiii. p. 436. The paddle-wheel used by Rennie in his experiments on the friction of water (Phil. Trans. 1831, plate xi. fig. 1) was somewhat similar to mine. I employed, however, a greater number of "floats," and also a corresponding number of stationary floats, in order to prevent the rotatory motion of the water in the can.

† Phil. Mag. ser. 3. vol. xxiii. pp. 263, 347. ‡ Ibid. May 1845, p. 369.

and 3rd, 774 lb. from experiments (hitherto unpublished) on the motion of water through narrow tubes. This last class of experiments being similar to that with the paddle-wheel, we may take the mean of 774 and 890, or 832 lb., as the equivalent derived from the friction of water. In such delicate experiments, where one hardly ever collects more than half a degree of heat, greater accordance of the results with one another than that above exhibited could hardly have been expected. I may therefore conclude that the existence of an equivalent relation between heat and the ordinary forms of mechanical power is proved; and assume 817 lb., the mean of the results of three distinct classes of experiments, as the equivalent, until more accurate experiments shall have been made.

Any of your readers who are so fortunate as to reside amid the romantic scenery of Wales or Scotland could, I doubt not, confirm my experiments by trying the temperature of the water at the top and at the bottom of a cascade. If my views be correct, a fall of 817 feet will of course generate one degree of heat, and the temperature of the river Niagara will be raised about one fifth of a degree by its fall of 160 feet.

Admitting the correctness of the equivalent I have named, it is obvious that the *vis viva* of the particles of a pound of water at (say) 51° is equal to the *vis viva* possessed by a pound of water at 50° plus the *vis viva* which would be acquired by a weight of 817 lb. after falling through the perpendicular height of one foot.

Assuming that the expansion of elastic fluids on the removal of pressure is owing to the centrifugal force of revolving atmospheres of electricity, we can easily estimate the absolute quantity of heat in matter. For in an elastic fluid the pressure will be proportional to the square of the velocity of the revolving atmospheres, and the *vis viva* of the atmospheres will also be proportional to the square of their velocity; consequently the pressure will be proportional to the *vis viva*. Now the ratio of the pressures of elastic fluids at the temperatures 32° and 33° is 480 : 481; consequently the zero of temperature must be 480° below the freezing-point of water.

Fig. 59. scale 1/12

Fig 58. scale 1/12

Fig 60. full size.

Fig: 66. Scale 1/12

Fig. 64. Scale 1/12

Fig. 62. scale 1/12

Fig. 63. scale 1/12

Fig. 61. scale 1/12

Fig. 65. scale 1/12

We see then what an enormous quantity of *vis viva* exists in matter. A single pound of water at 60° must possess 480°+28°=508° of heat; in other words, it must possess a *vis viva* equal to that acquired by a weight of 415036 lb. after falling through the perpendicular height of one foot. The velocity with which the atmospheres of electricity must revolve in order to present this enormous amount of *vis viva* must of course be prodigious, and equal probably to the velocity of light in the planetary space, or to that of an electric discharge as determined by the experiments of Wheatstone.

I remain, Gentlemen,

Yours respectfully,

JAMES P. JOULE.

Oak Field, near Manchester,
August 6, 1845.

On the Heat disengaged in Chemical Combinations. By JAMES PRESCOTT JOULE, *F.R.S.** [In a letter to the Editors of the 'Philosophical Magazine.']

['Philosophical Magazine,' ser. 4. vol. iii. p. 481.]

(PLATE I.)

GENTLEMEN,

The following memoir was communicated to the French Academy several years ago†, in order to compete for the prize offered for the best essay on the heat of chemical combinations. Owing to my not having been aware of the regulations to be observed by the competitors, and to the delay in procuring a good translation, my paper was deemed ineligible, but was referred to the Commission appointed to examine the memoirs presented on the subject. The time has now arrived when I feel that I ought not to delay communicating

* The experiments were made at Oak Field, Whalley Range, near Manchester.

† See the 'Comptes Rendus' for February 9, 1846.

the results of a laborious investigation to the scientific world, and I therefore transmit to you an exact copy of the paper from which the French translation, now in the possession of the Academy of Sciences, was made. The necessity of making no alterations whatever in the paper has prevented me from citing additional authorities and making sundry small corrections which have occurred to me since it was written. I will therefore avail myself of the opportunity of making a few prefatory observations.

I would remark, in the first place, that the laws of the disengagement of heat by voltaic electricity originally propounded by me received, nearly at the same time that the experiments contained in the present paper were made, a new confirmation and important developments by the researches of Professor J. D. Botto of Turin. This valuable contribution to physical science will be found in the Memoirs of the Academy of Sciences of Turin, 2nd series, vol. viii. In the 'Bulletin des Sciences' of the Berlin Academy for the 25th of November, 1848, will also be found an able memoir on the same subject by Professor Poggendorff.

I am assured by Professor Thomson that the length of the needle of my galvanometer is so small in comparison with the diameter of the coil, that no sensible error could possibly arise from taking the tangents of the deflections as the measure of the currents traversing the coil. The amount of the correction that I have applied is, however, too trifling to affect materially the numerical results arrived at. Professor Thomson has kindly allowed me to describe the arrangement by means of which he has recently effected a very valuable improvement in the tangent galvanometer. The coil he employs is represented by Pl. I. fig. 66. It consists of two concentric circles of flat copper wire, the outer one being 12 inches, the inner 9 inches in diameter. It is furnished with three terminals, *a*, *b*, and *c*, which can be readily connected with a battery, or other apparatus, by means of the usual clamps. It will be seen that if the current pass from *a* to *c* it will traverse the outer circle, that if it pass from *b* to *c* it will traverse the inner circle, and that if it pass from *a* to *b* it will traverse both circles in opposite directions. The diameters of the circles being in the proportion of 4 to 3, it is obvious that the effect of a constant current on the needle will in the three arrangements be respectively proportional to 3, 4, and 1. So that by making two circles of unequal force oppose one another, Professor Thomson obtains the means of measuring powerful currents accurately, without needlessly increasing the size of the galvanometer.

I observe with pleasure that Dr. Woods has recently arrived at one of the results of the paper, viz. "that the decomposition of a compound body occasions as much cold as the combination of its elements originally produced heat," by the use of an elegant experimental process described in this Magazine for October 1851. I ought, however, to remark that previous to the year 1843 I had demonstrated "that the heat rendered latent in the electrolysis of water is at the expense of the heat which would otherwise have been evolved in a free state by the circuit"*—a proposition which Professor Thomson has laid down as rigorously true and the demonstrable consequence of the dynamical theory of heat †.

I have the honour to remain, Gentlemen,

Yours very respectfully,

JAMES P. JOULE.

Acton Square, Salford,
March 30, 1852.

Actioni contraria semper et æqualis est reactio.—*Newton.*

1. My object in the present memoir is to communicate to the Academy of Sciences an inverse method for ascertaining the quantity of heat evolved by combustion, together with the results at which I have arrived by its use. Being convinced of the accuracy of the experiments by Dulong, and knowing that distinguished philosophers were engaged in confirming and extending his results, I thought that by examining the electrical reactions I should best fulfil the wishes of the Academy.

* Philosophical Magazine, ser. 3. vol. xxiii. p. 263.

† Memoir on the Dynamical Theory of Heat, § 18.

2. Davy drew from his electro-chemical experiments the conclusion that the heat and light evolved in chemical combinations are caused by the reunion of the two electricities. Subsequently Berzelius has taken up Davy's theory, and, giving it new developments, has made it the foundation of modern chemistry. Ampère has still further modified the theory, in order to explain the permanency of chemical combinations. The views of these and other philosophers have been ably discussed by Becquerel in his 'Traité de l'Electricité'*; and therefore I need not attempt any criticism of them, were it, indeed, necessary to my design to do so. It will be sufficient for my purpose to admit, 1st, that when two atoms combine by combustion a current of electricity passes from the oxygen to the combustible; 2nd, that the quantity of this current of electricity is fixed and definite; and, 3rd, that it is the means of the evolution of light and heat, precisely as is any other current of electricity whatever.

The first of these propositions I consider to have been proved by the experiments of Pouillet and Becquerel†; and the second is naturally derived from the discoveries of Faraday. Therefore it is only necessary to obtain one element more, viz. the intensity or electromotive force of the electric currents passing between the atoms, in order to be able to estimate the heat due to chemical combinations.

3. But it is important to decide first by what laws the evolution of heat by electricity is governed. Brooke and Cuthbertson found that the length of wire melted by an electrical battery varied nearly with the square of its charge; and Children and Harris showed that more or less heat is evolved by frictional electricity in proportion to the goodness or badness of the conductors. P. Riess, however, appears to have made the most extensive and accurate experiments on the calorific effects of frictional electricity. This philosopher has shown that the heat evolved by an electrical discharge is proportional to the square of the quantity of fluid divided by

* 'Traité de l'Electricité,' vol. iii. p. 366.

† Becquerel, 'Traité de l'Electricité,' vol. ii. p. 85.

the extent of coated glass surface upon which it was induced; in other words, proportional to the quantity and density of the fluid*.

In the year 1840† I commenced experiments on the calorific effects of voltaic electricity, having at that time no knowledge of what Reiss had previously done in frictional electricity. By these experiments it was proved that, when a current of voltaic electricity is propagated along a metallic conductor, the heat evolved thereby in a given time is proportional to the resistance of the wire and the square of the quantity of electricity transmitted.

Pursuing the inquiry, I found that the law applied very well to liquid conductors; and hence I inferred that *the heat evolved by any voltaic pile is proportional to its intensity or electromotive force and the number of chemical equivalents electrolyzed in each cell of the circuit*; or, in other words, *proportional to the intensity of the pile and the quantity of transmitted electricity*‡.

The above law must be understood to hold good only when the pile is free from local or secondary action; for it is obvious that the heat evolved by any action not directly engaged in propelling the current ought to be eliminated. Those parts of the pile where these secondary and local actions are carried on may be regarded as minute voltaic circles respectively evolving heat in quantities determined by the law; but this heat is not to be confounded with that due to the direct and useful action of the pile.

In applying the law, the intensity or electromotive force of the pile must be taken at its maximum, and not when under the influence of the polarization of Ritter—a phenomenon which, as is well known, is occasioned by the deposit of electro-positive substances on the negative plates of the pile. When a pile is under the influence of this polarization its intensity is diminished; but I have shown that the dimi-

* Annales de Chimie et de Physique, vol. lxix. p. 113.

† Proceedings of the Royal Society, Dec. 17, 1840.

‡ Philosophical Magazine, ser. 3. vol. xix. p. 275.

nution of heat due to this diminution of the intensity of the pile is exactly counterbalanced by the evolution of an additional quantity of heat at the polarized plates; and hence it appears that the heat evolved is at all times proportional to the intensity of the pile as it exists when its plates are in the proper condition and free from the polarization of Ritter, multiplied by the quantity of transmitted electricity.

In a memoir* "On the Heat evolved during the Electrolysis of Water," I proved the three following propositions:—

1st. That the *resistance to conduction*, whether it exists in solid or in liquid conductors, occasions the evolution of a quantity of heat, which, for a given time, is proportional to the magnitude of the resistance to conduction and the square of the quantity of transmitted electricity.

2nd. That the *resistance to electrolysis* presented by water does not occasion the evolution of heat in the decomposing cell. At the same time, the heat evolved by the whole circuit, for a given quantity of transmitted electricity, is diminished on account of the decreased electromotive force of the current, owing to the resistance to electrolysis. It is reasonable to infer that this diminution of the heat evolved by the circuit is occasioned by the absorption of heat in the decomposing cell.

And 3rd. That the *resistance occasioned by the polarization of Ritter* occasions the evolution of heat at the surfaces on which this phenomenon takes place; and thus it happens that the diminution of the heat evolved by the circuit, in consequence of the diminished intensity of the pile, is exactly compensated for; so that the heat evolved by the whole circuit may be estimated by the chemical changes occurring in the pile, just as if no such polarization existed.

I have already given theoretical results† for the heat of combustion agreeing so well with the experiments of Dulong as to convince me of the accuracy of the principles above

* Memoirs of the Literary and Philosophical Society of Manchester, 2nd series, vol. vii. part 2.

† Phil. Mag. ser. 3. vol. xix. p. 276, and vol. xxii. p. 205, &c.

advanced. My method was to ascertain how much of the intensity of a pile is spent in overcoming the affinity of the combustible for oxygen, and then to calculate the heat due to such an intensity, which heat ought, on our principles, to be equal to the heat occasioned by the recombination of the elements in combustion. I hope in the present paper to be able to show that still more accurate results may be attained by the use of a method founded upon the same principles, but of greater simplicity. Previously, however, to the description of the new experiments, I intend to bring forward some new proofs of the correctness of the law upon which their accuracy entirely depends.

I am aware that M. Ed. Becquerel* and M. Lenz† have separately, and by numerous and skilfully-performed experiments, confirmed my discovery that the heat evolved by voltaic electricity is proportional to the resistance to conduction and the square of the current. Nevertheless I have made new experiments upon the subject, thinking it impossible to demonstrate too completely the accuracy of a law upon which all thermo-chemical phenomena depend. I have endeavoured to make the results of these experiments worthy of confidence by the employment of a galvanometer and thermometers of great delicacy and accuracy.

4. The galvanometer was, in all its essential parts, constructed similarly to Pouillet's compass of tangents. A stout circular mahogany board (*a a*, figs. 58, 59, Pl. I.), supported upon three levelling-screws, forms the base of the instrument. To the diameter of this a stout vertical board is fixed, having a semicircular hole in its centre for the reception of the graduated circle *b b*. This circle, constructed of brass entirely free from magnetic influence, is 6 English inches in diameter and divided to half-degrees; it is also furnished with levelling-screws, &c. At *c c c* are three screws furnished with nuts, for the purpose of affixing to the vertical board any coil that may be required.

* Annales de Chimie et de Physique, 1843, vol. ix. p. 21.
† Annalen der Physik und Chemie, vol. lxi. p. 18.

The magnetic needle, of which I have given a full-size representation in fig. 60, Pl. I., consists of two pieces of hard steel, each half an inch long, kept about a millimetre asunder by the intervention of a small piece of brass to which they are cemented. An exceedingly delicate glass pointer*, weighing only 7 or 8 milligrammes, is affixed to the top of the needle. The ends of the pointer are painted black, in order to be the more distinctly seen as they traverse the silvered edge of the circle. A piece of fine copper wire is affixed to the needle in such a manner as to afford the means of suspension by a single filament of silk, and also of a rest when the instrument is not in use. The filament of silk is suspended from a graduated circle, *d d*, fig. 59, Pl. I., by means of which torsion can be given to the filament. There is also a piece of apparatus, *e e*, by means of which the needle can be elevated, depressed, or adjusted to the centre of the divided circle *b b*.

One can, as I have already hinted, affix to the vertical board of the galvanometer a coil consisting of a fewer or larger number of turns of copper wire, according to the nature of the experiment intended to be made. In the present research I employed one consisting simply of a thick copper wire bent into a circle of a foot (English) diameter. The connexion between the coil of the galvanometer and the voltaic battery (which were placed at a distance from each other of about six metres) was established by means of clamps and screws at *f f*.

I need hardly remark that it was found necessary to protect the galvanometer against any vibrations of the floor of the laboratory in which the experiments were made. This was effected by boring three holes in the floor, and driving strong wooden stakes through them into the ground. The feet of the stool upon which the galvanometer was placed rested upon the tops of these stakes, so that the floor of the room formed a platform entirely independent of the instrument.

* The use of a glass pointer was suggested to me by Mr. Dancer, the maker of my galvanometer. I have since found that it has been employed by Professor Bunsen of Marburg in his galvanometer of tangents.

Although the glass pointer weighs, as I have already observed, only 7 or 8 milligrammes, the resistance to its motion presented by the air was so great as to bring the magnetic needle to which it was affixed to a state of perfect tranquillity in ten seconds after the circuit was shut. Therefore I found it quite unnecessary to employ the oil recommended by Lenz as an additional resisting medium. Nor did I find it necessary to make use of any verniers or micrometers, the only assistance to the naked eye being a glass prism, to look through horizontally when the pointer could not be well viewed from a vertical position. I found no difficulty in reading off the angles of deflection to 2 or 3 minutes of a degree. Those who are accustomed to work with galvanometers will admit that it would be useless to attempt to arrive at greater accuracy by the employment of means which must necessarily increase the time occupied in the observations.

The galvanometer was adjusted in the plane of the magnetic meridian, by changing its position until a current passing in one direction could produce a deflection of the needle exactly equal in extent to the deflection on the other side of the meridian occasioned by a current of the same intensity, but passing in the opposite direction. After the galvanometer had been thus adjusted with very great care to the magnetic meridian, it was found that the glass pointer stood 30′ from zero. This error arose from the difficulty of cementing the pointer to the needle so as to be exactly in the same plane with the magnetic axis of the latter; but it did not give rise to any serious inconvenience, as it only made it necessary to affect the observed deflections with an increase or diminution of 30′, according to the direction of the current. I may mention in this place, that in every observation the position of *both* ends of the pointer was noted.

Since the force of torsion of the filament of silk is so trifling that six complete twists of it only produce a deflection of the magnetic needle amounting to 12′, and since the length of the magnetic needle is only $\frac{1}{24}$ of the diameter of the coil, it could hardly be doubted that the tangents of the angles of deflection represent pretty accurately the intensity of the transmitted electricity. I have nevertheless made experiments in order to prove my galvanometer. Sixteen large cells of Daniell's pile having been arranged in a series of four, the deflection of the needle under this voltaic force was ascertained for both sides of the meridian. The mean of the deflections produced by each separate cell was also noted. It is evident that the resistance of sixteen cells in a series of four is equal to that of a single cell; consequently the deflections produced by the above arrangements ought to indicate currents whose ratio is as 4 to 1. Experiments of this kind having been repeated, with proper precautions against any alteration in the intensity of the battery-cells, it was found that $\frac{1}{200}$ had to be added to the tangent at a deflection of 55°, $\frac{1}{120}$ at 82°, &c.

My thermometers were constructed by a method very similar to that employed by Regnault and Pierre*. The calibre of the tube was first measured in every part by passing a short column of mercury along it. The surface of the glass having then been covered with a thin film of bees'-wax, the portions of tube previously measured were each divided into the same number of parts by a machine constructed for the purpose. The divisions were then etched by means of the vapour of fluoric acid. Two thermometers were employed in the present research, in one of which the value of each space was $\frac{1}{18 \cdot 14}$, in the other $\frac{1}{23 \cdot 38}$ of a degree Centigrade. A practised eye can easily estimate the tenth part of each of these spaces; consequently I could by these thermometers observe a difference of temperature not greater than 0°·005.

The voltaic pile that I made use of was one of very large dimensions, each cell being 2 feet high and 5 inches in diameter. The internal arrangements of the cells were similar to those of the ordinary pile of Daniell.

5. I shall now proceed to describe my experiments on the heat evolved by currents traversing metallic wires. The apparatus used consisted of a wire of pure silver, 8 metres long

* Annales de Chimie et de Physique, 1842, vol. v. p. 428, note.

and about 0·6 of a millimetre in thickness, coiled upon a thin chimney-glass, the several coils being prevented from touching one another by means of silken threads. The ends of the silver wire were connected metallically with two thick copper wires, the ends of which dipped into cups of mercury. The coil, thus mounted, was immersed in a jar of tinned iron, capable of containing two pounds and a half of water. In order to prevent, as well as possible, the influence of the surrounding atmosphere in raising or depressing the temperature of the water, the sides and bottom of this jar were made hollow by soldering two jars of unequal magnitude within each other. Pl. I. fig. 61 represents a section of this double can, *a a* being the hollow part between the internal and external cans. The positions of the coil, thermometer, and stirrer are also shown in the same figure.

At 7 o'clock A.M., Sept. 4, 1844*, having filled the jar with $2\frac{1}{2}$ lb. of distilled water, I immersed the coil of silver wire into it, and caused it to form part of a circuit in which a pile consisting of sixteen of the large Daniell's cells in series and the galvanometer were placed. The circuit remained closed for exactly five minutes, during which time the deflections of both ends of the pointer of the galvanometer were observed three times. The mean of all the observations, no two of which differed from each other more than a few minutes of a degree, when properly corrected for the error in the position of the pointer, was 72° 17′. The increase of the temperature of the water, ascertained with all proper precautions in stirring, &c., was indicated by 81·8 divisions of the scale of the thermometer, each division corresponding to $\frac{1}{18 \cdot 14}$ of a degree Cent. The temperature of the room was 2°·07 Cent. lower than the mean temperature of the water.

* My object in being so particular as to the dates of the experiments was to eliminate the effects of any variation in the intensity of the earth's magnetism. In the subsequent series of experiments I have not always thought it necessary to mention these dates, but I have nevertheless used the same precaution in all of them.

As soon as the experiment just described was finished, another was performed in exactly the same manner; only the direction of the current was reversed, in order that the deflections might be observed on the other side of the meridian.

At 7 o'clock P.M. of the same day, two experiments were again made in the manner above described; but in these the quantity of electricity passed through the silver wire was only about half as much as before, five cells of the pile being now employed, instead of sixteen as before.

On the morning of September 5, two experiments of the same kind were made with a pile consisting of two cells in series; and on the evening of the same day, two experiments were made using only one of the constant cells.

On the two succeeding days all the above experiments were gone over again in the reverse order, beginning with one cell and ending with sixteen. In this way I sought to get rid of the mischievous effects of any change in the intensity of the earth's magnetism during the experiments.

The Table of these results which I subjoin will easily be understood by means of the headings of the columns; and the only thing, therefore, which it will be necessary for me to say in explanation of it is, that the last column contains the results of observation corrected for the cooling or heating effect of the surrounding air. The amount of this correction was estimated by simple and decisive experiments, and was in no one instance found to exceed one tenth of the quantity of heat evolved, even in the experiments with one cell, in which the heat evolved was least.

TABLE I.

Date of the experiments.	Number of cells in the pile.	Deflections of the needle of the galvanometer.		Corrected tangents of the mean deflections.	Square of the corrected tangents reduced to facilitate comparison with column 6.	Heat evolved in 5 minutes, in divisions of the thermometer.	
			mean				mean
Sept. 4, 7 A.M.	16	72° 17′	72° 46′	3·2428	87·24	81·94	87·24
Sept. 4, 7½ A.M.	16	73 1½				89·13	
Sept. 7, 7 P.M.	16	72 47				87·97	
Sept. 7, 7½ P.M.	16	72 58				89·90	
Sept. 4, 7 P.M.	5	61 8½	60 50¼	1·8015	26·92	27·08	26·56
Sept. 4, 7½ P.M.	5	61 26				27·98	
Sept. 7, 7 A.M.	5	60 19				25·39	
Sept. 7, 7½ A.M.	5	60 28				25·79	
Sept. 5, 7 A.M.	2	41 8	40 41	0·8634	6·18	6·23	6·03
Sept. 5, 7½ A.M.	2	41 7				6·27	
Sept. 6, 7 P.M.	2	40 5				5·76	
Sept. 6, 7½ P.M.	2	40 23				5·87	
Sept. 5, 7 P.M.	1	25 57	25 11½	0·4721	1·85	2·01	1·71
Sept. 5, 7½ P.M.	1	25 50½				1·83	
Sept. 6, 7 A.M.	1	24 40½				1·49	
Sept. 6, 7½ A.M.	1	24 18				1·53	
1	2	3		4	5	6	

In order to carry on the experiments with electric currents of feebler tension, I now introduced into the circuit an electrolytic cell, consisting of two plates of zinc immersed in a solution of sulphate of zinc. The results thus obtained are arranged in the following table. In order to collect an appreciable quantity of heat, the experiments were carried on for an hour with the lower intensities, and for half an hour with the highest intensity of current; I have, however, reduced all the results to five minutes, in order that they might be more readily compared with those of Table I. Each of the results given in Table II. is the mean of four experiments tried at different times, according to the principles which guided me in the former experiments.

TABLE II.

Number of cells in the pile.	Quantity of zinc deposited on the negative electrode per 5 minutes, in milligrammes.	Deflections of the needle of the galvanometer.	Corrected tangents of the deflections.	Squares of the corrected tangents reduced to facilitate comparison with column 6.	Heat evolved in 5 minutes, in divisions of the thermometer.
4	143	19° 22′	0·3527	1·03	0·96
2	82	11 14½	0·1991	0·33	0·29
1	44	6 2½	0·1059	0·09	0·09
1	2	3	4	5	6

By comparing the last two columns of the foregoing tables with each other, we see that throughout a very extensive range of electric intensities the heat evolved in a given time remains proportional to the square of the quantity of transmitted electricity.

6. Having thus succeeded in giving another proof of the law of voltaic heat as far as regards a change in the intensity of the current, we may now proceed to consider the effects produced by a change in the resistance of the wire. It will not be necessary for me to enter very largely upon this part of the subject, inasmuch as it has long been admitted by philosophers that the heat evolved by a current of given intensity is proportional to the resistance of the wire. I will, however, give one series of experiments, in which I have compared a wire of mercury with the coil of silver wire used in the previous experiments. The comparison of a fluid with a solid metal was, I thought, eminently calculated to test the accuracy of the law.

A glass tube, 157 centimetres long and about 2·3 millimetres in internal diameter, was fashioned into a spiral, as represented in Pl. I. fig. 62. The tube was filled with mercury as high as the bulbs *a a*. Connexion could be established between the pile and the spiral by means of the copper wires *b b*, which dipped as far as the centre of the bulbs.

The coil of mercury, thus prepared, was immersed in 2 lb. 11 oz. of water contained in a double-cased can, similar to the one I have already described, and a current from a pile of five cells was transmitted through it for ten minutes. The heat evolved, the temperature of the room, and the deflections of the galvanometer during the experiment were carefully noted. Eight of these experiments were made, in four of which the deflections were on one side, and in the other four on the other side of the meridian.

Four experiments were made in a similar way with the coil of silver wire. In order to avoid the effects of any change in the intensity of the earth's magnetism, these four experiments were alternated with those made with the mercury coil. The thermometer used in all the experiments was one of great accuracy; and each division of its scale corresponded to $\frac{1}{23 \cdot 38}$ of a degree of the Centigrade scale.

TABLE III.

	Deflections of the galvanometer.	Corrected tangents of the deflections.	Squares of the corrected tangents.	Heat evolved in 10 minutes, in divisions of the thermometer.	Difference between the mean temperature of the water and of the room.
Experiments with the mercury spiral.	60° 24′	1·7696	3·1315	45·5	0°·61 C. +
	60 56	1·8086	3·2710	45·0	1·97 +
	57 20½	1·5683	2·4596	34·8	0·80 +
	58 17	1·6266	2·6458	38·4	0·80 +
	58 30½	1·6409	2·6926	40·1	0·87 −
	59 48	1·7271	2·9829	43·8	0·84 −
	57 11	1·5584	2·4286	35·2	1·11 −
	56 30	1·5184	2·3055	34·7	0·84 −
		Mean....	2·7397	39·69	0·05 +
Experiments with the coil of silver wire.	55 54½	1·4847	2·2043	42·6	0·04 −
	56 19½	1·5083	2·2750	44·2	0·75 +
	54 38	1·4159	2·0048	39·4	0·40 −
	54 52	1·4282	2·0398	39·5	0·19 −
		Mean....	2·1310	41·425	0·03 +

The resistance of the mercury wire in comparison with that of the silver wire was found by ascertaining the intensity of the current produced by a pile of five cells,—1st, when the pile was in direct communication with the galvanometer; 2nd, when the resistance of the circuit was increased by the addition to it of the coil of silver wire; and 3rd, when the mercury wire was substituted in the circuit for the silver wire. Calling the intensity of the current in the first instance A, and the resistance y; in the second instance B, and the resistance $1+y$; and in the third instance C, and the resistance $x+y$, we have, by the laws of Ohm and Pouillet,

$$x=\frac{B(A-C)}{C(A-B)}.$$

The observations from which I have deduced the constant quantities of the above formula are arranged in the following table. In these experiments the precaution was taken that the temperature of the water in which the coils of mercury and silver were immersed should be as nearly as possible the same as in the experiments of Table III., in order to obviate the possibility of an alteration of the resistance arising from an alteration of the temperature of the metals. I may mention also that each of the recorded deflections is the mean of two observations, one on one side, and the other (by reversing the direction of the current) on the other side of the magnetic meridian. The effect of any change in the intensity of the pile during any of the experiments was carefully guarded against by a repetition of each experiment in the reverse order, *i. e.* beginning with current C and ending with current A, and then taking the mean of the two sets of observations.

TABLE IV.

No. of experiment.	Deflection with resistance y.	Corrected tangent of deflection, or A.	Deflection with resistance $1+y$.	Corrected tangent of deflection, or B.	Deflection with resistance $x+y$.	Corrected tangent of deflection, or C.	Resistance of mercury spiral, or $\frac{B(A-C)}{C(A-B)}$.
1	73° 48	3·4635	57° 10	1·5574	60° 56	1·8085	0·74771
2	71 10	2·9492	55 2	1·4370	58 39	1·6501	0·74814
3	72 25	3·1753	56 21	1·5098	59 54½	1·7346	0·75292
4	75 24	3·8630	58 37	1·6475	62 25	1·9242	0·74927
5	76 11	4·0916	58 47	1·6583	62 42	1·9477	0·75015
						Mean...........	0·74964

On multiplying 0·74964 by 2·7397, the square of the

intensity of the current to which the mercury wire was exposed (see Table III.), we obtain 2·0538, a quantity which ought to be proportional to the heat evolved, if our law be correct. From Table III. we see also that, in the case of silver wire, the square of the current multiplied by its resistance (which we called unity) is 2·131, while the heat evolved was 41·425. Hence we have for the heat which ought to have been evolved by the mercury spiral,

$$\frac{2·0538}{2·1310} \times 41·425 = 39·924.$$

Referring again to Table III., we find that the heat actually evolved was 39·69. The difference between this number and the result of theory, trifling as it is, is almost entirely accounted for by the circumstance that the capacity for heat of the mercury spiral exceeded that of the coil of silver wire by a quantity equal to the capacity of 5·64 grms. of water. Hence we must apply a correction of $\frac{1}{222}$ to the observations with the mercurial apparatus. This brings the heat actually evolved up to 39·868—a quantity differing from 39·924, the theoretical result, only by 0·056 of a division of the thermometer, or 0°·0024 Centigrade.

7. Having thus given fresh proofs of the accuracy of the law of the evolution of heat by voltaic electricity, we may now proceed to apply it in order to determine the quantity of heat evolved in chemical combinations. The following is an outline of my process:—I take a glass vessel filled with the solution of an electrolyte, and properly furnished with electrodes. I place this electrolytic cell in the voltaic circuit for a given length of time, and carefully observe the quantity of decomposition and the heat evolved. By the law of Ohm I then ascertain the resistance of a wire capable of obstructing the current equally with the electrolytic cell. Then, by the law we have proved, I determine the quantity of heat which would have been evolved had a wire of such resistance been placed in the circuit instead of the electrolytic cell. This theoretical quantity, being compared with the heat actually evolved in the electrolytic cell, is always found to

exceed the latter considerably. The difference between the two results evidently gives the quantity of heat absorbed during the electrolysis, and is therefore equivalent to the heat which is due to the reverse chemical combination by combustion or other means.

Having thus given a short outline of the process, I shall at once proceed to describe the experiments in detail.

1st. *Heat evolved by the Combustion of Copper.*

I took a glass jar (Pl. I. fig. 63) filled with 3 lb. of a solution consisting of 24 parts of water, 7 parts of crystallized sulphate of copper, and 1 part of strong sulphuric acid. In this solution, two plates—one of platinum, the other of copper —were immersed, each being connected by means of a proper clamp with a thick copper wire passing through a cork in the mouth of the vessel, and terminating in a mercury-cup *a*. A very delicate thermometer, each of whose divisions was equal to $\frac{1}{23 \cdot 38}$ of a degree Cent., was also fixed in the cork so as to have its bulb nearly in the centre of the liquid. Lastly, a glass stirrer *b* was introduced.

The experiments were conducted in the following manner:— A pile consisting of four large cells of Daniell (*a*, fig. 64, Pl. I.) was connected with the galvanometer *b* by means of two thick copper wires, one of which was continuous, while the other was divided at the mercury-cups *c c*. The connexion between these mercury-cups was first established by means of a short thick copper wire, and the deflection of the needle noted. The quantity of current indicated by this deflection I shall call A. The thick copper wire was now removed from the cups at *c c*, and the standard coil of silver wire (immersed in water to keep it cool) was put there instead, and the deflection again noted. The current observed in this second instance I shall call B. The coil of silver wire was now removed, and the electrolytic cell above described being put in its stead, electrolysis was carried on for exactly 10 minutes, during which the deflections of the needle were noted at equal intervals of time. The current indicated by the mean of these observations I shall call C. Currents B and A were then again observed in the reverse order; and the mean of these and the former observations taken, so as to obviate the effects of any change that might be occurring in the intensity of the pile.

The temperature of the solution was observed, with the usual precautions, immediately before and after the electrolysis was carried on. The amount of electrolysis was obtained by weighing the negative copper electrode before and after each experiment.

Putting x for the resistance of a metallic wire capable of retarding the passage of the current equally with the electrolytic cell, and calling the resistance of the coil of silver wire unity, we have, as in the case of the coil of mercury,

$$x=\frac{(A-C)B}{(A-B)C}.$$

This value, multiplied by C^2, gives $\frac{(A-C)BC}{A-B}$ for the calorific effect of the current C passing along a wire whose resistance $=x$.

The calorific effects of the standard of silver wire were ascertained by experiments made on the day before and on the day after the experiments on electrolysis were performed. In this way I sought, as before, to avoid the injurious influence of a change, either in the intensity of the earth's magnetism, or in the resistance of the standard coil. The standard coil was immersed in a light tin can containing 2 lb. 12 oz. of distilled water. The thermometer employed was that used in the experiments on electrolysis.

TABLE V.—Experiments on the Electrolysis of the Solution of Sulphate of Copper, with a Pile of 4 Daniell's cells.

Current A.		Current B.		Current C.		Difference between the mean temperature of the solution and that of the room.	$\frac{A-C}{A-B} \times BC$.	Heat evolved in 10 minutes, in divisions of the thermometer.	Copper deposited on the negative electrode, in grammes.
Mean deflection.	Corrected tangent.	Mean deflection.	Corrected tangent.	Mean deflection.	Corrected tangent.				
73° 31′	3·4006	53° 52′	1·3764	35° 55½′	0·7275	1°·24 C. –	1·3223	20·4	0·5686
74 59	3·7510	54 40	1·4176	36 53	0·7535	0·43 –	1·3722	19·4	0·5777
75 21	3·8538	54 45	1·4220	37 18	0·7650	0·23 –	1·3817	19·45	0·5881
75 12	3·8084	54 47	1·4237	38 26	0·7968	1·78 +	1·4326	17·4	0·6153
				Mean		0·3 –	1·3772	19·162	0·5874
						* Corrected for difference 0°·03 –		19·113	

TABLE VI.—Experiments on the Heat evolved by the Standard Coil. Pile of 4 cells.

Mean deflection of the needle of the galvanometer.	Corrected tangent.	Difference between the mean temperature of the water and that of the room.	Square of the corrected tangent.	Heat evolved in 10 minutes, in divisions of the thermometer.
51° 7′	1·2462	1°·08 C. –	1·5530	30·0
51 18	1·2544	0·03 +	1·5735	31·0
53 38	1·3648	0·19 +	1·8627	37·1
54 47	1·4239	0·90 +	2·0275	36·9
Mean		0·01 +	1·7542	33·75
Corrected for the difference 0°·01+				33·76

In order to compare the results of the above tables, it now became necessary to ascertain the capacity for heat of the jars of liquid employed in the experiments. This was done,

* The corrections I have applied to the quantities of heat evolved were derived from experiments on the cooling of the liquids reduced by the law of Leslie to the difference between the mean temperature of the liquids and that of the room. The signs + or – signify that the temperature of the liquid is greater or less than that of the room.

in one or two instances, by the method of mixtures. The jar along with its contents was heated to a certain point, and then, having been immersed in a large can of cold water, the capacity was determined by the decrease of temperature in the former and the increase in the latter. I felt, however, that this plan was on several accounts incapable of giving results of extreme accuracy, and had therefore recourse to a method founded upon the law of the development of heat by electricity. The spiral glass tube (Pl. I. fig. 62), filled with mercury, was immersed up to the bulbs *a a* in the jar whose capacity for heat was to be determined. A current of electricity was then passed through the mercury for a given time, and the heat thereby evolved was observed with the usual precautions. The capacity of the jar and its contents was of course directly proportional to the square of the intensity of the current, and inversely to the increase of temperature.

Table VII.—Experiments on the Heat evolved by the Mercury Spiral in the jar of Solution of Copper used in the experiments of Table V. Pile of 4 cells.

Mean deflection of the galvanometer.	Corrected tangent.	Difference between the mean temperature of the solution and that of the room.	Square of the corrected tangent.	Heat evolved in 10^m, in divisions of the thermometer.
57° 34′	1·5815	0°·05 C. +	2·5011	39·1
58 10½	1·6193	1·59 +	2·6221	37·0
58 17	1·6261	1·23 −	2·6442	43·4
59 0	1·6725	0·37 −	2·7973	41·3
Mean......		0·01 +	2·6412	40·2
Corrected for difference 0°·01+				40·216
Corrected for capacity				40·622

Table VIII.—Experiments on the Heat evolved by the Mercury Spiral in the can of water used in the experiments of Table VI. Pile of 4 cells. 2 lb. 11 oz. of water in the can.

Mean deflection of the galvanometer.	Corrected tangent.	Difference between the mean temperature of the water and that of the room.	Square of the corrected tangent.	Heat evolved in 10^m, in divisions of the thermometer.
57° 26	1·5734	1°·16 C. −	2·4756	35·6
57 27	1·5744	0·25 +	2·4787	36·6
59 11½	1·6853	1·01 −	2·8402	40·3
58 27	1·6367	1·85 +	2·6788	38·2
Mean		0·02 −	2·6183	37·675
Corrected for difference 0°·02 −				37·65
Corrected for capacity........				36·99

Besides the correction on account of the difference between the mean temperature of the liquid and that of the room in which the experiments were made, it was necessary to supply the second correction given in the above tables, on account of the capacity for heat of the jars being necessarily somewhat different from what it was in the experiments of Tables V. and VI. In Table VI. the can contained 2 lb. 12 oz. of water and the coil of silver wire; whereas it contained 2 lb. 11 oz. and the coil of mercury in Table VIII. Again, in the experiments of Table V. there were 3 lb. of solution of copper along with the platinum and copper electrodes; whereas in those of Table VII. the mercury coil was substituted for the electrodes, whilst the weight of the solution was two grammes less than before. It would be tedious and unnecessary to give in detail the various reductions demanded by these circumstances; suffice it to say, that the calculations were founded upon the best tables of specific heat, and were made with the most scrupulous care.

The tin can containing (as in the experiments of Table VI.) the coil of silver wire and 2 lb. 12 oz. of water was found by careful calculations to be equivalent in its capacity for heat to 1283·7 grms. of water; consequently from Tables VII. and VIII. we obtain for the capacity of the jar of solution used in the experiments of Table V.,

$$\frac{2{\cdot}6412}{2{\cdot}6183}\times\frac{36{\cdot}99}{40{\cdot}622}\times 1283{\cdot}7=1179{\cdot}2.$$

Referring now to Tables V. and VI., and remembering that 23·38 divisions of the scale of the thermometer employed are equal to one degree of the Centigrade scale, we obtain for the quantity of heat due to $\frac{A-C}{A-B}\times BC$,

$$\frac{33{\cdot}76}{23{\cdot}38}\times\frac{1{\cdot}3772}{1{\cdot}7542}\times 1283{\cdot}7=1455^{\circ}{\cdot}3.$$

The quantity of heat actually evolved will be

$$\frac{19{\cdot}113}{23{\cdot}38}\times 1179{\cdot}2=963^{\circ}{\cdot}99.$$

Subtracting the latter from the former result, we obtain 491°·3 as the quantity of heat absorbed in the electrolysis of a quantity of sulphate of copper corresponding to 0·5874 of a gramme of copper. The quantity of heat absorbed per gramme of copper deposited will therefore be 836°·4.

Two other series of experiments conducted in precisely the same manner, excepting that in the former of the two the specific heat of the solution was obtained by the method of mixtures, gave, for the absorption of heat per gramme of copper deposited, respectively 856° and 796°·5. The mean of the three results is 829°·6.

The above quantity of heat is that absorbed in separating the copper and oxygen gas from a solution of sulphate of oxide of copper. It is therefore necessary to subtract the absorption due to the transfer of the sulphuric acid from the oxide of copper to water, in order to obtain the heat absorbed in the decomposition of oxide of copper into metal and oxygen gas. For this purpose 8 grammes of oxide of copper, prepared by adding potash to a solution of the sulphate of copper and then carefully washing and igniting the precipitate, were thrown into an acidulated solution of copper similar to that used in the above experiments, the capacity for heat of which had been previously ascertained. The mean of four experiments, tried in this way with every possible precaution, gave 236° as the heat due to the solution of 1·252 gramme, the quantity of oxide corresponding to a gramme of copper.

829°·6 − 236° = 593°·6 = the quantity of heat absorbed in the decomposition of oxide of copper into copper and oxygen gas, and which ought therefore to be the quantity of heat evolved by the combustion of a gramme of copper.

2nd. *Combustion of Zinc.*

My experiments on this metal were similar to those on copper; they will not therefore require a very detailed description. The solution employed was one consisting of 3 parts of crystallized sulphate of zinc and 8 parts of water, weighing 3 lb. 2 oz. The electrodes were plates of platinum and zinc, each plate exposing an active surface of about 8 square inches. At the conclusion of each experiment, oxide of zinc was thrown into the solution to replace that removed by electrolysis, in order to prevent the zinc electrode from being acted upon by free acid.

TABLE IX.—Experiments on the Heat evolved by the Electrolysis of Sulphate of Zinc. Pile of 7 cells.

Current A.		Current B.		Current C.		Difference between the mean temperature of the solution and that of the room.	$\frac{A-C}{A-B}\times BC.$	Heat evolved in 10^m, in divisions of the thermometer.	Zinc deposited on the negative electrode, in grammes.
Mean deflection.	Corrected tangent.	Mean deflection.	Corrected tangent.	Mean deflection.	Corrected tangent.				
74° 10½′	3·5500	61° 46′	1·8722	35° 31′	0·7167	0°·83 C. −	2·2659	32·2	0·5797
75 55½	4·0133	63 9	1·9858	35 14	0·7092	0·35 +	2·2951	30·0	0·5647
75 18	3·8356	62 30	1·9311	37 1½	0·7573	0·40 −	2·3638	31·3	0·6010
75 32½	3·9025	62 47	1·9546	36 56	0·7548	0·84 +	2·3841	30·5	0·5991
					Mean	0·01 −	2·3272	31·0	0·5861
							Corrected for difference 0°·01 −	30·984	

TABLE X.—Experiments on the Heat evolved by the Standard Silver Coil. Pile of 5 cells.

Mean deflection.	Corrected tangent.	Difference between the mean temperature of the water and that of the room.	Square of the corrected tangent.	Heat evolved in 10^m, in divisions of the thermometer.
53° 58½′	1·3820	1°·15 C. −	1·9099	36·3
53 48	1·3731	0·25 +	1·8854	37·0
56 26½	1·5150	0·44 −	2·2952	45·9
57 4½	1·5520	1·24 +	2·4087	44·4
Mean ……		0·025 −	2·1248	40·9
Corrected for difference 0°·025 −				40·869

The following tables give the results of the experiments for ascertaining the capacity for heat of the jar of solution.

TABLE XI.—Experiments on the Heat evolved by the Mercury Spiral in the jar of Solution of Sulphate of Zinc used in the experiments of Table IX. Pile of 5 cells.

Mean deflection.	Corrected tangent.	Difference between the mean temperature of the solution and that of the room.	Square of the corrected tangent.	Heat evolved in 10^m, in divisions of the thermometer.
61° 17½′	1·8355	0°·48 C. −	3·3691	51·1
60 30	1·7768	1·61 +	3·1570	45·6
58 19	1·6287	0·47 −	2·6527	42·7
59 18½	1·6935	1·11 +	2·8679	40·5
Mean ……		0·442 +	3·0117	44·975
Corrected for difference 0°·442 +				45·700
Corrected for capacity ……..				46·269

TABLE XII.—Experiments on the Heat evolved by the Mercury Spiral in the can of water used in the experiments of Table X. Pile of 5 cells. 2 lb. 11 oz. of water in the can.

Mean deflection.	Corrected tangent.	Difference between the mean temperature of the water and that of the room.	Square of the corrected tangent.	Heat evolved in 10m, in divisions of the thermometer.
60° 21′	1·7659	0°·66 C. −	3·1184	44·4
60 24	1·7696	1·25 +	3·1315	45·1
58 44½	1·6561	1·17 −	2·7427	40·7
59 16½	1·6913	0·51 +	2·8605	40·4
	Mean	0·018 −	2·9633	42·65
		Corrected for difference 0°·018 C. −		42·628
		Corrected for capacity		41·881

From the last two tables we obtain for the capacity of the jar of solution used in the experiments of Table IX.,

$$\frac{3{\cdot}0117}{2{\cdot}9633} \times \frac{41{\cdot}881}{46{\cdot}269} \times 1283{\cdot}7 = 1180{\cdot}9.$$

From Tables IX. and X. we obtain for the quantity of heat due to $\frac{A-C}{A-B} \times BC$,

$$\frac{40{\cdot}869}{23{\cdot}38} \times \frac{2{\cdot}3272}{2{\cdot}1248} \times 1283{\cdot}7 = 2457^{\circ}{\cdot}7\,;$$

and for the actual quantity of heat evolved during electrolysis,

$$\frac{30{\cdot}984}{23{\cdot}38} \times 1180{\cdot}9 = 1565^{\circ}.$$

Hence $2457^{\circ}{\cdot}7 - 1565^{\circ} = 892^{\circ}{\cdot}7$ = the quantity of heat absorbed in the electrolysis of a quantity of sulphate of zinc corresponding to 0·5861 of a gramme of zinc.

The quantity of heat absorbed by the electrolysis of a quantity of sulphate of zinc corresponding to a gramme of zinc will therefore be 1523°·1.

The results of two other series of experiments, conducted

in precisely the same manner as that I have just given, were 1547° and 1619° respectively. The mean of the three results is 1563°.

The heat absorbed by the transfer of the sulphuric acid from the oxide of zinc to the water was ascertained in the following manner. A solution of zinc similar to that employed in the experiments was acidulated with about 10 grammes of sulphuric acid. 7·9 grms. of oxide of zinc (prepared by igniting the carbonate) were thrown into this solution; and the heat evolved by its union with the free sulphuric acid was carefully ascertained, and properly corrected for the influence of the atmosphere. The capacity for heat of the jar of solution was then ascertained by the method of electrical currents. This being done, a fresh quantity of oxide of zinc was thrown into the solution, and the heat evolved again observed. The mean of the two experiments gave 378° for the quantity of heat evolved by the solution of 1·242 grm., the quantity of oxide of zinc corresponding to a gramme of zinc.

1563°−378°=1185°=the quantity of heat absorbed in the decomposition of oxide of zinc into zinc and oxygen gas, and which ought therefore to be the quantity of heat evolved by the combustion of a gramme of zinc.

3rd. *Combustion of Hydrogen Gas.*

The apparatus employed in the experiments on hydrogen is shown in Plate I. fig. 65. *a* represents a glass jar nearly full of a solution consisting of six parts of water and one of strong sulphuric acid, and containing platinum electrodes; *b* represents a glass tube for conveying the mixed gases to the pneumatic trough *c*. The glass stirrer *d*, being inserted in the small cork *e*, can, when not in use, be made perfectly tight by inserting the latter into the large cork which stops up the mouth of the jar. A coating of a viscid solution of rosin in turpentine was applied wherever it appeared necessary, in order to ensure perfect tightness. The quantity of mixed gases evolved was ascertained by the weight of water displaced in the bottle *f*; and hence the weight of liberated hydrogen was computed with the assistance of the best tables, regard being paid to the temperature of the gas, its hygrometric state, the barometric pressure, &c.

TABLE XIII.—Experiments on the Electrolysis of Dilute Sulphuric Acid, spec. grav. 1·103. Pile of 6 cells.

Current A.		Current B.		Current C.		Difference between the mean temperature of the solution and that of the room.	$\frac{A-C}{A-B}\times BC$.	Heat evolved in 10^m, in divisions of the thermometer.	Hydrogen liberated from the negative electrode, in grammes.
Mean deflection.	Corrected tangent.	Mean deflection.	Corrected tangent.	Mean deflection.	Corrected tangent.				
73° 35½′	3·4170	59° 26′	1·7020	58° 14′	1·6234	1°·30 C. −	2·8897	40·6	0·03978
75 41	3·9429	60 57	1·8098	60 31	1·7780	0·34 +	3·2658	37·8	0·04212
76 57	4·3412	62 0	1·8906	61 19	1·8374	0·71 +	3·5492	41·7	0·04372
76 3	4·0508	61 13	1·8298	60 50	1·8011	1·85 +	3·3382	38·8	0·04411
				Mean		0·40 +	3·2607	39·725	0·04243
				Corrected for difference 0°·40+				40·381	

TABLE XIV.—Experiments on the Heat evolved by the Standard Silver Coil. Pile of 5 cells.

Mean deflection.	Corrected tangent.	Difference between the mean temperature of the water and that of the room.	Square of the corrected tangent.	Heat evolved in 10^m, in divisions of the thermometer.
56° 12′	1·5012	0°·25 C. −	2·2536	42·1
55 40	1·4714	1·53 +	2·1650	41·1
55 37	1·4687	0·53 −	2·1571	43·4
56 27	1·5155	1·05 +	2·2967	41·7
Mean		0·45 +	2·2181	42·075
Corrected for difference 0°·45+				42·642

TABLE XV.—Experiments on the Heat evolved by the Mercury Spiral in the jar of Dilute Sulphuric Acid used in the experiments of Table XIII. Pile of 5 cells.

Mean deflection.	Corrected tangent.	Difference between the mean temperature of the solution and that of the room.	Square of the corrected tangent.	Heat evolved in 10^m, in divisions of the thermometer.
60° 15½′	1·7594	0°·46 C. –	3·0955	49·9
60 39½	1·7883	1·59 +	3·1980	46·0
59 34½	1·7117	1·55 –	2·9299	46·6
58 52½	1·6648	0·38 +	2·7716	43·9
Mean		0·01 –	2·9987	46·6
Corrected for difference 0°·01 –				46·584
Corrected for capacity..........				47·164

TABLE XVI.—Experiments on the Heat evolved by the Mercury Spiral in the can of water used in the experiments of Table XIV. Pile of 5 cells. 2 lb. 11 oz. of water in the can.

Mean deflection.	Corrected tangent.	Difference between the mean temperature of the water and that of the room.	Square of the corrected tangent.	Heat evolved in 10^m, in divisions of the thermometer.
59° 39′	1·7168	0°·47 C. –	2·9474	41·9
59 46½	1·7255	1·35 +	2·9774	42·6
59 10	1·6841	1·01 –	2·8362	42·8
59 37½	1·7151	0·74 +	2·9416	40·5
Mean		0·152 +	2·9256	41·95
Corrected for difference 0°·152 C.+				42·141
Corrected for capacity				41·402

From the above tables we obtain for the capacity for heat of the jar of dilute sulphuric acid used in the experiments of Table XIII.,

$$\frac{2·9987}{2·9256}\times\frac{41·402}{47·164}\times 1283·7=1155\,;$$

for the quantity of heat due to $\frac{A-C}{A-B} \times BC$,

$$\frac{42{\cdot}642}{23{\cdot}38} \times \frac{3{\cdot}2607}{2{\cdot}2181} \times 1283{\cdot}7 = 3441^{\circ}{\cdot}8\,;$$

and for the actual quantity of heat evolved in the electrolysis,

$$\frac{40{\cdot}381}{23{\cdot}38} \times 1155 = 1994^{\circ}{\cdot}9.$$

Hence $3441^{\circ}{\cdot}8 - 1994^{\circ}{\cdot}9 = 1446^{\circ}{\cdot}9$, the quantity of heat absorbed during the electrolysis of a quantity of sulphate of water corresponding to 0·04243 of a gramme of hydrogen.

The quantity of heat absorbed by the electrolysis of a quantity of sulphate of water corresponding to a gramme of hydrogen will therefore be 34101°.

Two other series of experiments conducted in precisely the same manner, excepting that in the former of the two the capacity for heat of the jar of dilute acid was obtained by the method of mixtures, gave 34212° and 32358° respectively as the heat absorbed per gramme of hydrogen liberated. The mean of the three results is 33557°.

A small portion of this quantity of heat absorbed is that due to the removal of water from the dilute acid; but the correction on this account is so exceedingly small as to be hardly worth applying. Subtracting 4°, however, on this account, we obtain 33553° as the quantity of heat absorbed during the electrolysis of water, which ought therefore to be equal to the quantity of heat evolved by the combustion of a gramme of hydrogen gas.

---

8. By the inverse method of electrical currents, then, we have found that the quantities of heat evolved by the combustion of copper, zinc, and hydrogen are respectively 594°, 1185°, and 33553°. These quantities agree so well with the results obtained by Dulong, that I think I may assume that the principles admitted in this paper are demonstrated suffi-

ciently to justify me in making them the basis of a few concluding observations.

The fact that the heat evolved in a given time by a metallic wire is proportional to the square of the quantity of transmitted electricity, proves that the action of the current is of a strictly mechanical character; for the force exerted by a fluid impinging against a solid body obeys the same law. Now I have shown in previous papers*, that when the temperature of a gramme of water is increased by 1° Centigrade, a quantity of *vis viva* is communicated to its particles equal to that acquired by a weight of 448 grammes after falling from the perpendicular height of one metre. Hence the mechanical force of a voltaic pile may be calculated from the heat which it evolves.

Hence also may the absolute force with which bodies enter into chemical combination be estimated by the quantity of heat evolved. Thus, from the data already given, the *vis viva* developed by the combustion of a gramme of copper, a gramme of zinc, and a gramme of hydrogen will be respectively equivalent to the *vis viva* possessed by weights of 266112, 530880, and 15031744 grammes, after falling from the perpendicular height of one metre.

On the Effects of Magnetism upon the Dimensions of Iron and Steel Bars. By James P. Joule†.

['Philosophical Magazine,' ser. 3. vol. xxx. pp. 76, 225.]

About the close of the year 1841 Mr. F. D. Arstall, an ingenious mechanist of Manchester, suggested to me a new form of electro-magnetic engine. He was of opinion that a bar of iron experienced an increase of bulk by receiving the

* Philosophical Magazine, ser. 3. vol. xxviii. p. 206.

† The experiments were made at Oak Field, Whalley Range, near Manchester.

magnetic condition, and that, by reversing its polarity rapidly by means of alternating currents of electricity, an available and useful motive power might be realized. At Mr. Arstall's request I undertook some experiments in order to decide how far his notions were well founded.

The results of my inquiries were brought before the public on the occasion of a *conversazione* held at the Royal Victoria Gallery of Manchester on the 16th of February, 1842, and are printed in the 8th vol. of Sturgeon's 'Annals of Electricity,' p. 219. In this lecture I made evident the fact that an increase of length of a bar of iron was produced by magnetizing it. I also stated my reasons for believing that whilst the bar was increased in length by the magnetic influence, it experienced a *contraction* at right angles to the magnetic axis, so as to prevent any change taking place in the bulk of the bar. I intended as soon as possible to bring this conjecture to the test of experiment, and I prepared some apparatus for the purpose; but, owing to other occupations, I was obliged to relinquish the experiments until the beginning of the summer of 1846. In the meantime the inquiry has been taken up by De la Rive, Matteucci, Wertheim, Wartmann, Marrian, Beatson, and others, whose ingenious experiments have invested the subject with additional interest. The researches of Beatson have taken a similar direction to my own; and he appears also to have employed a somewhat similar apparatus to that which I shall presently describe. I have confirmed several of the results at which this gentleman has arrived, and have added new facts, which I hope will throw further light on this rather obscure department of physics.

In order to ascertain how far my opinion as to the invariability of the *bulk* of a bar of iron under magnetic influence was well founded, I devised the following apparatus: —Ten copper wires, each 110 yards long and $\frac{1}{20}$ of an inch in diameter, were bound together by tape, so as to form a good and at the same time very flexible conductor. The bundle of wires thus formed was coiled upon a glass tube, 40 inches long and an inch and a half in diameter. One end of the tube was hermetically sealed; and the other end was furnished with an accurately ground glass stopper, which was itself perforated with a round orifice, into which a graduated capillary tube could be inserted. In making the experiments a bar of annealed iron, one yard long and half an inch square section, was placed in the tube, which was then filled up with water. The stopper was then adjusted, and the capillary tube inserted so as to force the water to a convenient height within it.

The bulk of the iron bar was 4,500,000 times that of each division of the graduated tube; consequently a very minute expansion of the former would have produced a perceptible motion of the water in the capillary tube; but on connecting the coil with a Daniell's battery of five to six large jars, which was quite equal to saturate the iron, no perceptible effect was produced either by making or breaking contact, whether the water was stationary in the capillary tube, or gradually rising or falling from a change of temperature. Now, had the usual increase of length been unaccompanied by a corresponding diminution of the diameter of the bar, the water would have been forced through twenty divisions of the capillary tube every time that contact was made with the battery.

Having thus ascertained that the bulk of the iron bar was invariable, I proceeded to repeat my first experiments with a more delicate apparatus, in order, by a more careful investigation of the laws of the increment of length, to ascend to the probable cause of the phenomenon.

A glass tube coiled with wire similar to that already described was fixed vertically in a wooden frame. When a bar of iron, one yard long, was introduced so as to rest on the sealed end, the coil at either end extended an inch beyond the bar. The apparatus for observing the increment of length consisted of two levers of the first order, and a powerful microscope situated at the extremity of the second lever. These levers were furnished with brass knife-edges resting upon glass. The connexion between the upper extremity of the bar under examination and the first lever, and that between the two levers, were established by means of exceedingly fine platinum wires.

The first lever multiplied the motion of the extremity of the bar 7·8 times, the second multiplied the motion of the first 8 times, and the microscope was furnished with a micrometer divided into parts, each corresponding to $\frac{1}{2220}$ of an inch. Consequently each division of the micrometer passed over by the index indicated an increment of the length of the bar amounting to $\frac{1}{138,528}$ of an inch.

The quantities of electricity passing through the coil were measured by a tangent galvanometer consisting of a circle of thick copper wire one foot in diameter, and a needle half an inch long, furnished with a suitable index.

The quantities of magnetic polarity communicated to the iron bar were measured by a magnet 18 inches long, suspended as a balance-beam at the distance of one foot from the centre of the coil. This magnetic bar was furnished with scales in the manner of an ordinary balance, and the weight required to bring it to a horizontal position indicated the intensity of the magnetism of the iron bar under examination.

After a few preliminary trials, a great advantage was found to result from filling the tube with water. The effect of the water was, as De la Rive had already remarked, to prevent the sound. It also checked the oscillations of the index, and had an important advantage in preventing any considerable irregularities in the temperature of the bar.

The first experiment which I shall record was made with a bar consisting of two pieces of very well-annealed rectangular iron wire, each one yard long, a quarter of an inch broad, and about one eighth of an inch thick.

The pieces were fastened together so as to form a bar of nearly a quarter of an inch square. The coil was placed in connexion with a single constant cell, the resistance being further increased by the addition of a few feet of fine wire. The instant that the circuit was closed, the index passed over one division of the micrometer. The needle of the galvanometer was then observed to stand at 7° 20′, while the magnetic balance required 0·52 of a grain to bring it to an equilibrium. It had been found by experiments made for the purpose that the coil alone, when traversed by a current of the same intensity, was capable of exerting a force of 0·03 grain on the balance; consequently the magnetic intensity of the bar was represented by 0·49 grain. On breaking the circuit the index was observed to retire 0·3 of a division, leaving a permanent elongation of 0·7 and a permanent polarity of 0·42 of a grain. More powerful currents were then used, and the observations repeated as before, with the results tabulated below. The numbers in the 2nd and last columns are multiplied by 1000.

EXPERIMENT 1.

Deflection of galvanometer.	Tangent of deflection.	Elongation or shortening of the bar.	Total elongation.	Magnetic intensity of the bar.	Square of magnetic intensity divided by total elongation.
$-7°\ 20'$	128	1·0 E	1·0	−0·49	240
0	0	0·3 S	0·7	−0·42	252
− 9 30	167	2·9 E	3·6	−0·93	240
0	0	1·2 S	2·4	−0·74	228
−14 48	264	5·9 E	8·3	−1·42	243
0	0	3·8 S	4·5	−1·00	222
−23 10	428	10·3 E	14·8	−1·87	236
0	0	7·6 S	7·2	−1·26	220
−47 25	1088	16·1 E	23·3	−2·22	211
0	0	13·9 S	9·4	−1·35	194
−58 50	1653	14·8 E	24·2	−2·21	202
0	0	13·3 S	10·9	−1·35	168

The electric current was now, in continuation of Experiment 1, transmitted in the + or contrary direction, so as first to remove the polarity acquired by the − current, and then to induce the opposite polarity. The total elongation in the fourth column is that of the bar in its original condition at the commencement of the first Table.

The next series of results, recorded under the head Experiment 2, was obtained with a fresh bar, of exactly the same size and temper as the preceding. To avoid an unnecessary occupation of space, I omit a fresh heading to the table when the direction of current is changed, simply designating the commencement of the fresh condition by ruling a line.

EXPERIMENT 1 (*continued*).

Deflection of galvanometer.	Tangent of deflection.	Elongation or shortening of the bar.	Total elongation.	Magnetic intensity of the bar.	Square of magnetic intensity divided by total elongation.
+ 6° 15′	109	3·4 S	7·5	−0·12	2
0	0	0	7·5	−0·17	4
+ 9 55	175	0·1 E	7·6	+0·57	43
0	0	1·0 S	6·6	+0·25	9
+15 40	280	3·7 E	10·3	+1·30	164
0	0	4·0 S	6·3	+0·78	97
+38 45	802	16·8 E	23·1	+2·30	229
0	0	14·5 S	8·6	+1·12	148
+51 30	1257	16·7 E	25·3	+2·35	218
0	0	16·3 S	9·0	+1·05	122

EXPERIMENT 2.

Deflection of galvanometer.	Tangent of deflection.	Elongation or shortening of the bar.	Total elongation.	Magnetic intensity of the bar.	Square of magnetic intensity divided by total elongation.
+ 5° 0′	87	0·1 E	0·1	+0·08	64
0	0	0	0·1	+0·03	9
+ 8 27	148	1·9 E	2·0	+0·50	125
0	0	1·0 S	1·0	+0·30	90
+13 27	239	5·8 E	6·8	+1·16	198
0	0	3·1 S	3·7	+0·69	129
+33 50	670	18·8 E	22·5	+2·20	215
0	0	14·3 S	8·2	+1·01	124
+53 50	1368	19·0 E	27·2	+2·32	198
0	0	17·1 S	10·1	+1·03	105
− 7 5	124	2·0 S	8·1	−0·15	3
0	0	0·1 S	8·0	−0·07	0
−55 10	1437	20·0 E	28·0	−2·20	173
0	0	14·6 S	13·4	−1·39	144

The next experiment was with a bar of well-annealed iron, one yard long and about half an inch square. Its weight was 45½ oz. I have introduced an additional column into the Table, in which the magnetic intensity is reduced to the section of the former bars, the weight of each of which was 8 oz.

Experiment 3.

Deflection of galvanometer.	Tangent of deflection.	Elongation or shortening of the bar.	Total elongation.	Magnetic intensity.	Reduced magnetic intensity.	Square of reduced magnetic intensity divided by elongation.
+5° 10′	90	0·4 E	0·4	+ 1·18	+0·21	110
0	0	0·1 S	0·3	+ 0·45	+0·08	21
+ 8 2	141	0·7 E	1·0	+ 1·82	+0·32	102
0	0	0·2 S	0·8	+ 0·67	+0·12	18
+14 43	262	2·0 E	2·8	+ 4·10	+0·72	185
0	0	1·0 S	1·8	+ 0·90	+0·16	14
+40 3	840	12·0 E	13·8	+11·08	+1·95	275
0	0	8·4 S	5·4	+ 1·20	+0·21	8
+54 0	1376	13·8 E	19·2	+13·53	+2·38	295
0	0	12·0 S	7·2	+ 1·20	+0·21	6
+62 5	1887	14·4 E	21·6	+14·13	+2·48	285
0	0	13·5 S	8·1	+ 1·20	+0·21	5
− 6 30	114	1·2 S	6·9	− 0·7	−0·12	2
0	0	0	6·9	− 0·3	−0·05	0
−14 25	257	0·7 E	7·6	− 3·8	−0·67	59
0	0	1·3 S	6·3	− 1·15	−0·2	7
−41 15	877	11·0 E	17·3	−11·33	−1·99	229
0	0	8·8 S	8·5	− 1·5	−0·26	8
−62 45	1941	16·0 E	24·5	−13·71	−2·41	237
0	0	13·0 S	11·5	− 1·55	−0·27	6
+ 5 35	98	0·8 S	10·7	+ 0·16	+0·03	0
0	0	0	10·7	− 0·40	−0·07	0
+ 9 0	158	0·2 S	10·5	+ 1·17	+0·21	4
0	0	0·2 S	10·3	+ 0·15	+0·03	0
+14 20	255	0·3 E	10·6	+ 3·30	+0·58	32
0	0	1·2 S	9·4	+ 0·50	+0·09	1
+24 45	461	3·3 E	12·7	+ 7·16	+1·26	125
0	0	3·4 S	9·3	+ 0·82	+0·14	2
+39 50	834	9·6 E	18·9	+11·43	+2·01	214
0	0	8·0 S	10·9	+ 0·95	+0·17	2
+54 15	1389	12·6 E	23·5	+13·47	+2·37	239
0	0	11·6 S	11·9	+ 1·0	+0·18	3
+60 45	1785	13·2 E	25·1	+13·84	+2·43	235
0	0	12·4 S	12·7	+ 1·0	+0·18	3
− 7 13	127	1·0 S	11·7	− 1·13	−0·2	3
0	0	0·1 S	11·6	− 0·50	−0·09	1
−10 25	184	0	11·6	− 2·16	−0·38	12
0	0	0·2 S	11·4	− 1·0	−0·18	3
−15 57	286	0·5 E	11·9	− 4·14	−0·73	45
0	0	1·1 S	10·8	− 1·25	−0·22	4
−26 0	488	3·5 E	14·3	− 7·45	−1·31	120
0	0	3·2 S	11·1	− 1·50	−0·26	6
−40 55	867	9·6 E	20·7	−11·48	−2·02	197
0	0	8·0 S	12·7	− 1·70	−0·3	7
−62 48	1946	14·6 E	27·3	−13·76	−2·42	214
0	0	13·0 S	14·3	− 1·73	−0·3	6

From the last column of each of the preceding tables we may, I think, safely infer that *the elongation is in the duplicate ratio of the magnetic intensity of the bar*, both when the magnetism is maintained by the influence of the coil, and, in the case of the permanent magnetism, after the current has been cut off. The discrepancies observable will, I think, be satisfactorily accounted for when we consider the nature of the magnetic actions taking place. When a bar experiences the inductive influence of a coil traversed by an electrical current, the particles near its axis do not receive as much polarity as those near its surface, because the former have to withstand the opposing inductive influence of a greater number of magnetic particles than the latter. This effect will be diminished in the extent of its manifestation with an increase of the electrical force, and will finally disappear when the current is sufficiently powerful to saturate the iron. Again, when the iron, after having been magnetized by the coil, is abandoned to its own retentive powers by cutting off the electrical current, the magnetism of the interior particles will suffer a greater amount of deterioration than that of the exterior particles. The polarity of the former may indeed be sometimes actually reversed, as Dr. Scoresby found it to be in some extensive combinations of steel bars. Now whenever such influences as these occur, so as to make the different parts of the bar magnetic in various degrees, the elongation will necessarily bear a greater proportion to the square of the magnetic intensity measured by the balance than would otherwise be the case.

For similar causes the interior of the bar will in general receive the neutralization and reversal of its polarity before the exterior; and hence we see in the tables that there is a considerable elongation of the bar after the reversal of the current, even when the effect upon the balance has become imperceptible owing to the opposite effects of the interior and exterior magnetic particles.

The bars employed in the preceding experiments were annealed as perfectly as possible. The next series was made with a bar of the same dimensions and quality as the bars employed in Experiments 1 and 2, excepting that it was not annealed.

EXPERIMENT 4.

Deflection of galvanometer.	Tangent of deflection.	Elongation or shortening of the bar.	Total elongation.	Magnetic intensity of the bar.	Square of magnetic intensity divided by total elongation.
+ 9° 20	164	0·2 E	0·2	+0·15	112
0	0	0	0·2	+0·08	32
+15 20	274	0·9 E	1·1	+0·50	227
0	0	0·7 S	0·4	+0·33	272
+38 32	796	7·1 E	7·5	+1·36	247
0	0	5·2 S	2·3	+0·8	278
+50 30	1213	10·2 E	12·5	+1·76	247
0	0	9·6 S	2·9	+0·97	324
+57 40	1580	13·0 E	15·9	+1·94	236
0	0	11·8 S	4·1	+1·0	244
+62 20	1907	14·0 E	18·1	+2·10	243
0	0	14·0 S	4·1	+1·01	249
− 6 50	120	1·2 S	2·9	+0·58	116
0	0	0	2·9	+0·65	145
−10 35	168	0·4 S	2·5	+0·21	17
0	0	0	2·5	+0·35	49
−14 57	267	0	2·5	−0·3	36
0	0	0·2 S	2·3	−0·13	7
−40 10	844	5·7 E	8·0	−1·36	231
0	0	4·6 S	3·4	−0·88	228
−53 30	1351	10·0 E	13·4	−1·7	215
0	0	9·5 S	3·9	−0·95	231
+ 9 27	166	1·3 S	2·6	−0·36	50
0	0	0·1 E	2·7	−0·4	59
+22 30	414	0·1 S	2·6	+0·38	55
0	0	0	2·6	+0·22	18
+38 27	794	4·9 E	7·5	+1·5	300
0	0	4·6 S	2·9	+0·97	324

In the foregoing series the discrepancies before adverted to do not make their appearance to any considerable extent except in the case of the reversal of the magnetic polarity. Taken altogether, the series is strikingly confirmatory of the law of elongation already announced.

The next series of observations was obtained with a piece of soft steel wire one yard long and a quarter of an inch in diameter. Its weight was exactly 8 oz.

Experiment 5.

Deflection of galvanometer.	Tangent of deflection.	Elongation or shortening of the bar.	Total elongation.	Magnetic intensity of the bar.	Square of magnetic intensity divided by total elongation.
+38° 10′	786	1·4 E	1·4	+0·94	631
0	0	0·6 S	0·8	+0·65	528
+50 45	1224	2·8 E	3·6	+1·43	568
0	0	1·8 S	1·8	+0·98	533
+60 25	1761	3·8 E	5·6	+1·71	521
0	0	3·1 S	2·5	+1·12	502
+67 50	2454	5·0 E	7·5	+1·88	471
0	0	4·5 S	3·0	+1·23	504
+69 20	2651	5·5 E	8·5	+1·97	456
0	0	4·5 S	4·0	+1·28	409
−41 40	890	1·3 S	2·7	−0·76	214
0	0	1·0 S	1·7	−0·35	72

Experiments 6 and 7 were made with fresh steel bars, similar in every respect to that used in Experiment 5.

Experiment 6.

Deflection of galvanometer.	Tangent of deflection.	Elongation or shortening of the bar.	Total elongation.	Magnetic intensity of the bar.	Square of magnetic intensity divided by total elongation.
+38° 25′	793	0·8 E	0·8	+0·78	760
0	0	0·5 S	0·3	+0·46	705
+60 50	1792	5·2 E	5·5	+1·6	466
0	0	3·4 S	2·1	+0·99	467
+70 30	2824	7·0 E	9·1	+1·88	388
0	0	5·8 S	3·3	+1·16	408
−16 28	295	1·8 S	1·5	+0·82	448
0	0	0·2 S	1·3	+0·94	680
−38 50	805	0	1·3	−0·64	315
0	0	0·3 S	1·0	−0·33	108

EXPERIMENT 7.

Deflection of galvanometer.	Tangent of deflection.	Elongation or shortening of the bar.	Total elongation.	Magnetic intensity of the bar.	Square of magnetic intensity divided by total elongation.
+58° 20′	790	1·4 E	1·4	+0·74	391
0	0	0·7 S	0·7	+0·46	302
+61 5	1810	5·3 E	6·0	+1·64	448
0	0	3·2 S	2·8	+1·07	409
+69 55	2735	4·7 E	7·5	+1·9	481
0	0	4·5 S	3·0	+1·2	480
−26 40	502	3·0 S	0	+0·2	
0	0	0·2 S	−0·2	+0·32	

The uniformity of the numbers in the last columns of the preceding tables shows that, where the metal possesses a considerable retentive power, the anomalies occasioned by the reaction of the magnetic particles upon one another, which have been already adverted to, do not exist to any considerable extent. Hence we have an additional confirmation of the law above stated, viz. that *the elongation is proportional, in a given bar, to the square of the magnetic intensity.*

I now made trial of a bar of steel one yard long, half an

EXPERIMENT 8.

Deflection of galvanometer.	Tangent of deflection.	Elongation or shortening of the bar.	Total elongation.	Magnetic intensity of the bar.	Reduced magnetic intensity.	Square of reduced magnetic intensity divided by total elongation.
+39° 0′	810	0	0	+1·11	+0·38	
0	0	0·2 E	0·2	+1·36	+0·47	1104
+52 35	1307	0·8 E	1·0	+4·09	+1·42	2016
0	0	0·3 E	1·3	+2·85	+0·99	754
+60 15	1750	0·5 E	1·8	+5·10	+1·77	1740
0	0	0·1 E	1·9	+3·52	+1·22	783
+69 45	2710	0·6 E	2·5	+5·91	+2·06	1697
0	0	0·2 E	2·7	+4·2	+1·46	790
−41 15	877	1·6 S	1·1	−0·43	−0·15	20
0	0	0·1 E	1·2	+0·35	+0·12	12
−56 5	1487	1·4 E	2·6	−3·9	−1·36	711
0	0	0·1 E	2·7	−2·63	−0·91	307

inch broad, and a quarter of an inch thick, weighing 23 oz. It was hardened considerably throughout its entire length, but not to such a degree as to enable it completely to resist the action of the file.

In the above table it will be observed that the steel bar was slightly increased in length every time that contact with the battery was broken, although a considerable diminution of the magnetism of the bar took place at the same time. I am disposed to attribute this effect to the state of tension in the hardened steel, for I find that soft iron wire presents a similar phenomenon when stretched tightly.

On inspecting the tables, it will be remarked that the elongation is, for the same magnetic intensity, greater in proportion to the softness of the metal. It is greatest of all in the well-annealed iron bars, and least in the hardened. This circumstance appears to me to favour the hypothesis that the phenomena are produced by the attractions taking place between the magnetic particles of the bars; an hypothesis in perfect accordance with the law of elongation which I have pointed out.

Part II.

With a view to ascertain whether the lengthening effects observed in the experiments detailed in the First Part of this paper were entirely independent of the diameter of the bars, I made an extensive series of experiments, in which fine wires, both of iron and steel, bundles of very fine iron wires, chains composed of copper and iron links, &c. were employed. In order to keep these flexible articles exactly in the axis of the coil, a weight was placed upon the system of levers so as to stretch them with a force of about eight ounces.

The results of the experiments in which wire $\frac{1}{20}$ of an inch thickness was employed accorded very well with the previous experiments made with thicker bars; but on employing iron wire which was only $\frac{1}{120}$ of an inch thick, the phenomena assumed quite a different character, for on transmitting the current through the coil, the length of the wire became suddenly *diminished* instead of being increased. This phenomenon appeared to me exceedingly anomalous, and it was some time before I found out its cause. At last, thinking that the wire was attracted by the coil, I varied its position from the centre to either side, and increased the amount of its tension. The former of these operations produced no sensible effect; but the increase of tension caused the shortening effect to be considerably augmented. It became manifest, therefore, that the weight of eight ounces, acting upon a very fine wire, produced the anomalies in question. This was further shown to be the case by diminishing the tension as far as possible. I then found that the phenomenon of elongation took place as in the case of the iron bars, only to a smaller extent, which was obviously owing to the degree of tension necessarily left in order to keep the wire in the axis of the coil.

Fig. 67. Scale $\frac{1}{12}$.

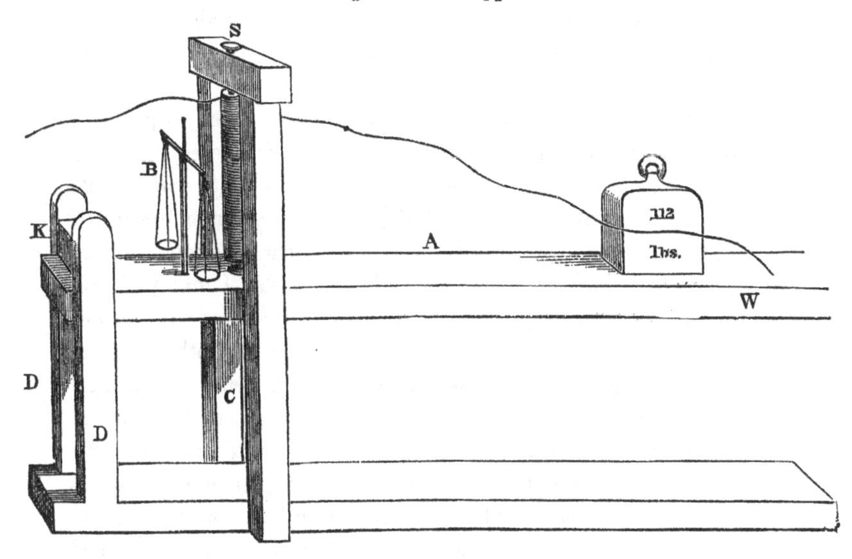

The new field of inquiry thus opened appeared to me to be one of great importance, and calculated eventually to become the means of throwing a great deal of light upon some of the most interesting questions connected with molecular actions and the real character of magnetism. I therefore constructed an apparatus whereby the effects of tension and pressure also might be studied. This apparatus, which is represented by the foregoing fig. 67, consists of a strong wooden lever A, furnished with a hard steel knife-edge at K, which rests against a hard steel plate held between the supports D, D. At the distance of one foot from the knife-edge a brass plate is let into the lever, into which a piece of iron or steel wire, one quarter of an inch in diameter, can be screwed; the upper end of the wire can also be screwed into a brass bolt, the head of which rests upon the cross-piece S. Weights can be placed on the lever at W to increase the tension. The magnetic balance, consisting of a bar magnet eight inches long, properly furnished with scales, is situated at B. The further extremity of the lever is connected with a fine lever multiplying eight times, the index of which is examined by the microscope employed in the first-described experiments. Each division of the micrometer passed over by the index indicates an elongation of the wire under examination equal to $\frac{1}{150960}$ of an inch.

When pressure instead of tension is employed, a pillar of iron or steel wire, one quarter of an inch in diameter, is placed at C, so as to support the weight of the lever; its ends, made perfectly square, abut upon flat plates of copper or brass.

Every precaution that I could think of was taken in order to give accuracy to the results. In particular I may mention that the coil was not permitted to be in contact with either the wire under examination, the lever, or any other part of the apparatus to which it might communicate motion through the change of its own molecular condition in consequence of the passage of the electrical current. In spite of these precautions the experiments were very troublesome, owing to the almost incessant vibrations of the index. Although my laboratory is situated quite out of the town, and detached from every dwelling, these vibrations were so extensive in the daytime that the experiments had in general to be carried on after 8 o'clock P.M., when the greater part of distant traffic had ceased. It was at all times impossible to make an observation when a cart was passing along a road at the distance of one or two hundred yards; nor could any thing be done as long as much wind was blowing. Owing to these circumstances it was frequently very difficult to estimate an effect equal to one tenth of a division of the micrometer. I believe, however, that the results recorded in the following tables are in no case more than two tenths of a division wide of the truth.

The experiment which I first made was with a piece of annealed iron wire, one foot long and one quarter of an inch in diameter. It was made as straight as possible, and its ends were ground perfectly true and flat. It was placed as a pillar under the lever, so as to support its pressure, which was equal to 82 lb. The coil by which it was magnetized was formed out of a copper wire 33 yards long and one tenth of an inch in diameter, well covered with cotton-thread. Its length was $11\frac{1}{2}$ inches, and its interior diameter 1 inch. The same coil was employed in all the experiments with wires of a foot long. The temperature of the wire was in every case about 45° Fahr.

The method of experimenting was the same that I employed in the First Part of this paper. A current was passed through the coil, the effects of which both on the length of the iron pillar and on its magnetism were noted. The current was then cut off; and the effect of so doing on the length of the pillar noted, as well as the quantity of magnetism remaining in it. A more powerful current was then passed, and the observations repeated as before, and so on, with increasingly powerful currents. The fifth column of the tables gives the effect upon the magnetic balance in grains, abstraction being made of the effect due to the coil itself, which had been previously determined by experiment. The last column is multiplied by 100.

Experiment 9.—Pressure 82 lb. One-foot bar.

Deflection of galvanometer.	Tangent of deflection.	Elongation or shortening of the pillar.	Total elongation.	Magnetic intensity of pillar.	Square of magnetic intensity divided by total elongation.
+ 6° 5′	106	0	0	+0·6	..
0	0	0	0	+0·3	..
+11 0	194	0·1 E	0·1	+1·4	1960
0	0	0·1 S	0	+0·4	..
+21 0	383	1·0 E	1·0	+2·8	784
0	0	1·0 S	0	+0·5	..
+35 5	702	3·3 E	3·3	+4·2	534
0	0	3·2 S	0·1	+0·6	..
+47 0	1072	5·4 E	5·5	+5·0	455
0	0	5·1 S	0·4	+0·7	..
+53 45	1364	6·4 E	6·8	+5·4	429
0	0	6·0 S	0·8	+0·7	..
− 6 45	118	0	0	−0·3	..
0	0	0	0	+0·1	..
−11 18	200	0·2 E	0·2	−1·2	720
0	0	0·2 S	0	−0·3	..
−21 25	392	1·0 E	1·0	−2·6	676
0	0	1·0 S	0	−0·4	..
−35 33	715	3·5 E	3·5	−4·1	480
0	0	3·2 S	0·3	−0·4	..
−45 40	1023	4·9 E	5·2	−4·9	461
0	0	5·0 S	0·2	−0·5	..
−54 5	1380	6·3 E	6·5	−5·4	449
0	0	6·3 S	0·2	−0·5	..

In the next experiment the same bar was subjected to a pressure of 480 lb. It possessed −0·5 magnetic intensity to begin with.

The numbers in the last columns of tables 9 and 10 show that the elongation follows a rather higher ratio than the square of the magnetic intensity. In the former section, in which all the bars employed were a yard long, the ratio was somewhat lower than that of the square of the intensity in the case of well-annealed iron. I am inclined therefore to think that the anomalies referred to at p. 242 were occasioned rather by the too great length of the iron bars, which prevented them from being magnetized as much at the ends as at the middle part, than by their different magnetic conditions at the centre and the surface.

EXPERIMENT 10.—Pressure 480 lb. One-foot bar.

Deflection of galvanometer.	Tangent of deflection.	Elongation or shortening of the pillar.	Total elongation.	Magnetic intensity of pillar.	Square of magnetic intensity divided by total elongation.
+ 6° 10′	108	0·15 S	0	+0·6	..
0	0	0	0	+0·2	..
+10 45	190	0·2 E	0·2	+1·2	720
0	0	0·2 S	0	+0·5	..
+20 0	364	0·8 E	0·8	+2·3	661
0	0	0·8 S	0	+0·7	..
+32 53	646	2·5 E	2·5	+3·6	518
0	0	2·5 S	0	+0·8	..
+42 35	919	4·6 E	4·6	+4·5	440
0	0	4·6 S	0	+0·8	..
+49 55	1188	6·8 E	6·8	+5·2	398
0	0	6·8 S	0	+0·8	..

From the above tables it appears evident that the augmentation of *pressure* does not make much difference in the amount of elongation for the same quantity of polarity. However, I thought it desirable to try the effect of a greater pressure. For this purpose I employed a piece of soft iron wire, six inches long and quarter of an inch in diameter. This, as an iron pillar, stood upon a small piece of flattened brass, resting upon a block of hard wood six inches high, in order that it might be sufficiently raised to abut against the lever. It was placed in the axis of a suitable coil, 5½ inches long and 1 inch in interior diameter, constructed of a covered copper wire 20 yards long and one tenth of an inch in diameter.

EXPERIMENT 11.—Pressure 82 lb. Six-inch pillar.

Deflection of galvanometer.	Tangent of deflection.	Elongation or shortening of the pillar.	Total elongation.	Magnetic intensity of pillar.	Square of magnetic intensity divided by total elongation.
+ 6° 30′	114	0	0	+0·8	..
0	0	0	0	+0·2	..
+11 15	199	0	0	+1·3	..
0	0	0	0	+0·4	..
+23 15	430	0	0	+2·9	..
0	0	0	0	+0·5	..
+41 55	898	0·6 E	0·6	+5·5	5040
0	0	0·6 S	0	+0·5	..
+51 50	1272	1·5 E	1·5	+7·2	3460
0	0	1·5 S	0	+0·5	..
+62 20	1907	2·6 E	2·6	+9·4	3400
0	0	2·6 S	0	+0·5	..

EXPERIMENT 12.—Six-inch pillar. Pressure 1380 lb. Permanent magnetism of the pillar to begin with −0·6.

Deflection of galvanometer.	Tangent of deflection.	Elongation or shortening of the pillar.	Total elongation.	Magnetic intensity of pillar.	Square of magnetic intensity divided by total elongation.
+ 6° 40′	117	0	0	+0·1	..
0	0	0	0	−0·3	..
+11 30	203	0	0	+0·6	..
0	0	0	0	−0·1	..
+23 25	433	0	0	+2·1	..
0	0	0	0	+0·3	..
+42 35	919	0·4 E	0·4	+4·5	5060
0	0	0·4 S	0	+0·5	..
+51 30	1257	0·9 E	0·9	+5·9	3870
0	0	0·9 S	0	+0·7	..
+63 30	2005	2·1 E	2·1	+8·6	3520
0	0	2·1 S	0	+0·7	..

A comparison between the last columns of the two preceding tables will show that pressure has no sensible effect upon the extent of the elongation. I had not sufficient voltaic force to saturate the short bars; but there appears no reason to doubt that their elongation, if saturated, would be one half that of the pillars one foot long, just as the latter were found to experience one third of the elongation observed in the bars of a yard long employed in the first section. I may remark in this place that the greater proximity of the magnetic balance to the coil increased the numbers in columns 5, representing the magnetic intensity of the six-inch pillars. The two preceding tables are therefore only comparable with themselves. In all the other experiments with one-foot bars the magnetic balance was at the uniform distance of $4\frac{1}{2}$ inches from the centre of the bar, so that they are strictly comparable with one another.

I now proceed to give an account of some experiments on the effects of the force of tension. The bar employed was a rod of soft iron wire $12\frac{2}{3}$ inches long and a quarter of an inch in diameter. Its extremities were formed into very fine threaded screws, extending one third of an inch, for the purpose of screwing into the brass plate and bolt, as shown in the figure already described. The effectual length of the bar, when screwed into its place, was exactly one foot. In the first experiment of this kind, the tension employed, being that occasioned by the weight of the lever alone, amounted to 80 lb. In the subsequent experiments the tension was further increased by placing weights on the lever.

EXPERIMENT 13.—Iron wire one foot long and a quarter of an inch in diameter. Tension 80 lb.

Deflection of galvanometer.	Tangent of deflection.	Elongation or shortening of the wire.	Total elongation.	Magnetic intensity of wire.	Square of magnetic intensity divided by total elongation.
$+6^\circ 30'$	114	0·1 E	0·1	+0·6	360
0	0	0·1 S	0	+0·2	. .
+11 30	203	0·4 E	0·4	+1·5	562
0	0	0·4 S	0	+0·3	. .
+21 25	392	1·0 E	1·0	+3·0	900
0	0	1·0 S	0	+0·4	. .
+35 55	724	3·2 E	3·2	+4·5	632
0	0	3·2 S	0	+0·4	. .
+46 38	1058	4·0 E	4·0	+5·1	650
0	0	4·0 S	0	+0·4	. .
+53 10	1335	4·6 E	4·6	+5·4	634
0	0	4·6 S	0	+0·4	. .
+61 25	1835	4·5 E	4·5	+5·6	697
0	0	4·5 S	0	+0·4	. .

EXPERIMENT 14.—Same wire with +0·4 permanent magnetism to begin with. Tension 408 lb.

Deflection of galvanometer.	Tangent of deflection.	Elongation or shortening of the wire.	Total elongation.	Magnetic intensity of wire.	Square of magnetic intensity divided by total elongation.
+ 6° 30′	114	0	0	+0·6	..
0	0	0	0	+0·2	..
+11 10	197	0	0	+1·4	..
0	0	0	0	+0·3	..
+20 55	382	0·2 E	0·2	+3·1	4805
0	0	0·2 S	0	+0·4	..
+35 5	702	0·6 E	0·6	+4·7	3682
0	0	0·6 S	0	+0·4	..
+45 20	1012	0·8 E	0·8	+5·1	3251
0	0	0·8 S	0	+0·4	..

EXPERIMENT 15.—Same wire. Tension 740 lb.

Deflection of galvanometer.	Tangent of deflection.	Elongation or shortening of the wire.	Total shortening.	Magnetic intensity of wire.	Current multiplied by magnetic intensity and divided by shortening effect.
+ 6° 20′	111	0	0	+0·9	..
0	0	0	0	+0·3	..
+20 0	364	0·1 S	0·1	+3·4	1237
0	0	0·1 E	0	+0·5	..
+35 0	700	0·4 S	0·4	+4·9	858
0	0	0·4 E	0	+0·6	..
+45 30	1017	0·6 S	0·6	+5·4	915
0	0	0·6 E	0	+0·6	..
+58 0	1600	1·3 S	1·3	+5·7	701
0	0	1·3 E	0	+0·7	..

In experiments 13 and 14 we notice the rapid decline of the effect of elongation, until at last, in experiment 15, with a tension of 740 lb. it ceases altogether, and the new condition of shortening commences. With a tension of about 600 lb., the effect on the dimensions of the wire would cease altogether in the limits of the electric currents employed in the above experiments. From the last column of table 15, which consists of the columns 2 and 5 multiplied together and divided by column 4, we gather that the shortening effect is very nearly proportional to the magnetism of the wire into the current traversing the coil. The law of the square of the magnetism will still indeed hold good where the iron is sufficiently below the point of saturation, on account of the magnetism being in that case nearly proportional to the intensity of the current. For the same reason, on examination of the previous tables, it will be found that the elongation is, below the point of saturation, very nearly proportional to the magnetism multiplied by the current. The necessity for changing the law arises from the fact that the elongation ceases to increase after the iron is fully saturated; whereas the shortening effect still continues to be augmented with the increase of the intensity of the current.

Experiment 16.—Same wire. +0·5 magnetism to begin with. Tension 1040 lb.

Deflection of galvanometer.	Tangent of deflection.	Elongation or shortening of the wire.	Total shortening.	Magnetic intensity of wire.	Current multiplied by magnetic intensity divided by shortening effect.
− 6° 20′	111	0	0	−0·4	..
0	0	0	0	0	..
−11 5	196	0·1 S	0·1	−1·3	255
0	0	0·1 E	0	−0·2	..
−20 55	382	0·2 S	0·2	−2·7	515
0	0	0·2 E	0	−0·4	..
−34 45	694	0·8 S	0·8	−4·0	347
0	0	0·8 E	0	−0·4	..
−46 45	1063	1·5 S	1·5	−4·7	333
0	0	1·5 E	0	−0·5	..
−61 0	1804	2·8 S	2·8	−5·0	322
0	0	2·8 E	0	−0·5	..
+20 45	379	0·3 S	0·3	+3·0	379
0	0	0·3 E	0	+0·5	..
+35 10	704	0·8 S	0·8	+4·4	387
0	0	0·8 E	0	+0·5	..
+48 5	1114	1·8 S	1·8	+4·9	303
0	0	1·8 E	0	+0·5	..
+61 20	1829	2·7 S	2·7	+5·4	366
0	0	2·7 E	0	+0·5	..

EXPERIMENT 17.—Same wire. Tension 1680 lb.

Deflection of galvanometer.	Tangent of deflection.	Elongation or shortening of the wire.	Total shortening.	Magnetic intensity of wire.	Current multiplied by magnetic intensity divided by shortening effect.
− 6° 25′	112	0	0	−0·8	..
0	0	0	0	−0·2	..
−11 10	197	0	0	−1·5	..
0	0	0	0	−0·4	..
−20 45	379	0·5 S	0·5	−2·7	205
0	0	0·5 E	0	−0·4	..
−34 50	696	1·5 S	1·5	−3·8	17
0	0	1·5 E	0	−0·4	..
−45 5	1003	2·4 S	2·4	−4·3	180
0	0	2·4 E	0	−0·4	..
−52 25	1299	3·3 S	3·3	−4·4	173
0	0	3·3 E	0	−0·4	..
−61 15	1823	4·5 S	4·5	−4·7	190
0	0	4·5 E	0	−0·4	..
+61 45	1861	4·4 S	4·4	+5·0	211
0	0	4·4 E	0	+0·5	..

The uniformity of the numbers contained in the last column of each of the two preceding tables affords conclusive evidence of the correctness of the law I have stated, viz., that *in the case of tension a shortening effect takes place, which, in a given instance, is proportional to the current traversing the coil multiplied by the magnetic intensity of the bar.*

In order to discover, if possible, what proportion the shortening effect bore to the force of tension, I have constructed the following table from the results, observed with currents of about 700 and 1000 intensities in the preceding experiments.

With the exception of the number 80, the results of the last column of the succeeding table agree sufficiently well together to render it extremely probable that the shortening effects are proportional *cæteris paribus* to the square root of the force of tension.

Number of Experiment.	Elongation or Shortening.	Departure from Elongation 4·4.	Tension.	Square of Departure divided by Tension.
9.	4·4 E	0	0	..
13.	3·6 E	0·8	80	80
14.	0·7 E	3·7	408	335
15.	0·5 S	4·9	740	324
16.	1·2 S	5·6	1040	301
17.	2·0 S	6·4	1680	244

Experiments with Cast Iron.

The following experiments were made with a bar of round cast iron one foot long and a quarter of an inch in diameter.

EXPERIMENT 18.—Cast iron. Tension 80 lb.

Deflection of galvanometer.	Tangent of deflection.	Elongation or shortening of the bar.	Total elongation.	Magnetic intensity of bar.	Square of magnetic intensity divided by total elongation.
– 6° 28′	113	0	0	–0·2	..
0	0	0	0	–0·1	..
–11 5	196	0	0	–0·4	..
0	0	0	0	–0·2	..
–21 18	390	0·1 E	0·1	–0·9	810
0	0	0·1 S	0	–0·5	..
–36 0	726	0·8 E	0·8	–2·0	500
0	0	0·5 S	0·3	–1·3	563
–46 10	1041	1·0 E	1·3	–2·5	481
0	0	1·0 S	0·3	–1·5	750
–58 35	1637	1·7 E	2·0	–3·4	450
0	0	1·7 S	0·3	–1·7	963
+19 45	359	0·5 S	–0·2	0	..
0	0	0	–0·2	–0·4	..
+59 25	1692	2·3 E	2·1	+3·2	488
0	0	1·7 S	0·4	+1·7	722

EXPERIMENT 19.—Same bar. Tension 654 lb. Permanent magnetism to begin with, −1·7.

Deflection of galvanometer.	Tangent of deflection.	Elongation or shortening of the bar.	Total elongation.	Magnetic intensity of bar.	Square of magnetic intensity divided by total elongation.
+19° 15′	349	0·2 S	0	−0·1	..
0	0	0	0	−0·4	..
+57 50	1590	1·4 E	1·4	+3·0	643
0	0	1·0 S	0·4	+1·7	722

On comparing Experiment 18 with Experiment 13 it will be observed that the elongation of the cast iron was equal, if not superior, to that of the soft iron when magnetized to the same extent. It will also be remarked that increase of tension does not produce half the diminution of elongation which it does in the case of soft iron.

Experiments with Soft Steel Wire.

The following experiments with soft steel wire were made in precisely the same manner as those with soft iron wire, already described.

EXPERIMENT 20.—Soft steel pillar, one foot long, a quarter of an inch in diameter. Pressure 82 lb.

Deflection of galvanometer.	Tangent of deflection.	Elongation or shortening of pillar.	Total elongation.	Magnetic intensity of pillar.	Square of magnetic intensity divided by total elongation.
−34° 0′	674	0·8 E	0·8	−2·1	551
0	0	0·4 S	0·4	−1·1	302
−47 0	1072	1·0 E	1·4	−3·6	926
0	0	0·6 S	0·8	−1·7	361
−59 30	1697	1·2 E	2·0	−4·5	1012
0	0	0·6 S	1·4	−2·4	411

EXPERIMENT 21.—Same soft steel pillar. Pressure 480 lb. Permanent magnetism to begin with, −2·0.

Deflection of galvanometer.	Tangent of deflection.	Elongation or shortening of pillar.	Total elongation.	Magnetic intensity of pillar.	Square of magnetic intensity divided by total elongation.
+18° 15′	330	0·2 S	0	−0·2	..
0	0	0	0	−0·4	..
+30 30	589	0·1 E	0·1	+1·3	1690
0	0	0	0·1	+0·6	360
+40 0	839	0·4 E	0·5	+2·6	1352
0	0	0·2 S	0·3	+1·3	566
+45 0	1000	0·6 E	0·9	+3·1	1068
0	0	0·6 S	0·3	+1·6	853
+61 10	1816	1·8 E	2·1	+4·1	800
0	0	1·6 S	0·5	+2·0	800

EXPERIMENT 22.—Soft steel wire, one foot long, a quarter of an inch in diameter. Tension 80 lb.

Deflection of galvanometer.	Tangent of deflection.	Elongation or shortening of the wire.	Total elongation.	Magnetic intensity of wire.	Square of magnetic intensity divided by total elongation.
+19° 50′	360	0	0	+1·0	..
0	0	0	0	+0·5	..
+34 40	691	0·4 E	0·4	+2·6	1690
0	0	0·1 S	0·3	+1·5	750
+46 10	1041	0·5 E	0·8	+3·5	531
0	0	0·2 S	0·6	+1·9	601
+56 30	1511	0·5 E	1·1	+4·0	1455
0	0	0·3 S	0·8	+2·1	551
−20 50	380	0·8 S	0	−0·2	..
0	0	0	0	+0·3	..
−34 50	696	0·2 E	0·2	−2·3	2645
0	0	0·1 S	0·1	−1·1	1210
−47 45	1101	0·7 E	0·8	−3·6	1620
0	0	0·3 S	0·5	−2·0	800
−60 0	1732	0·7 E	1·2	−4·4	1613
0	0	0·2 S	1·0	−2·4	576

EXPERIMENT 23.—Same soft steel wire. Tension 462 lb.

Deflection of galvanometer.	Tangent of deflection.	Elongation or shortening of the wire.	Total elongation.	Magnetic intensity of wire.	Square of magnetic intensity divided by total elongation.
+34° 55′	698	0	0	+2·6	..
0	0	0·1 E	0·1	+1·4	1960
+45 5	1003	0	0·1	+3·4	..
0	0	0·2 E	0·3	+2·0	1333
+60 15	1750	0·4 S	−0·1	+4·2	..
0	0	0·6 E	0·5	+2·4	1152

EXPERIMENT 24.—Same wire. Tension 1680 lb.
Magnetism to begin with, +1·8.

Deflection of galvanometer.	Tangent of deflection.	Elongation or shortening of the wire.	Total shortening effect.	Magnetic intensity of wire.	Current multiplied by magnetic intensity and divided by shortening effect.
−21° 13′	388	0·2 S	0·2	−0·4	77
0	0	0·2 E	0	+0·2	..
−35 13	706	0·8 S	0·8	−2·2	194
0	0	0·8 E	0	−1·1	..
−44 35	985	1·3 S	1·3	−2·9	219
0	0	1·3 E	0	−1·6	..
−58 45	1648	2·5 S	2·5	−3·7	244
0	0	2·5 E	0	−2·2	..

From the above experiments it will be seen that the same remarks which were made with regard to the soft iron apply also to the soft steel. The superior retentive power of the latter metal enables us, however, to trace the elongating effects of the permanent magnetism, which, it appears, diminish with the increase of tension, until at last, as in experiment 24, they cease altogether.

Experiments with Hardened Steel Wire.

The following experiments were made with steel wire of the same kind as that employed in the previous experiments. It was, however, thoroughly hardened, so as to resist the action of the file in every part.

EXPERIMENT 25.—Hardened steel wire, one foot long and a quarter of an inch in diameter. Tension 80 lb.

Deflection of galvanometer	Tangent of deflection.	Elongation or shortening of the wire.	Total shortening.	Magnetic intensity of wire.	Current multiplied by magnetic intensity and divided by shortening effect.
+ 6° 15′	109	0	0	0	..
0	0	0	0	0	..
+11 0	194	0	0	+0·2	..
0	0	0	0	0	..
+20 50	380	0	0	+0·5	..
0	0	0	0	+0·2	..
+35 20	709	0	0	+1·1	..
0	0	0	0	+0·5	..
+45 40	1023	0·1 S	0·1	+1·8	1841
0	0	0·1 E	0	+0·9	..
+52 0	1280	0·2 S	0·2	+2·3	1472
0	0	0·2 E	0	+1·2	..
+62 20	1907	0·4 S	0·4	+3·3	1573
0	0	0·4 E	0	+1·9	

EXPERIMENT 26.—Same hardened steel wire. Permanent magnetism to begin with, −1·8. Tension 408 lb.

Deflection of galvanometer.	Tangent of deflection.	Elongation or shortening of the wire.	Total shortening.	Magnetic intensity of wire.	Current multiplied by magnetic intensity and divided by shortening effect.
+19° 30′	354	0	0	−1·1	..
0	0	0	0	−1·4	..
+35 40	717	0	0	−0·2	..
0	0	0	0	−0·6	..
+46 50	1066	0	0	+1·2	..
0	0	0	0	+0·2	..
+60 20	1755	0·4 S	0·4	+2·8	1228
0	0	0·4 E	0	+1·5	..

EXPERIMENT 27.—Same hardened steel wire. Permanent magnetism to begin with, +1·3. Tension 1030 lb.

Deflection of galvanometer.	Tangent of deflection.	Elongation or shortening of wire.	Total shortening.	Magnetic intensity of wire.	Current multiplied by magnetic intensity and divided by shortening effect.
−21° 15′	389	0	0	+0·7	....
0	0	0	0	+1·1	....
−35 55	724	0	0	−0·1	....
0	0	0	0	+0·4	....
−48 33	1132	0·1 S	0·1	−1·3	1470
0	0	0·1 E	0	−0·5	....
−52 10	1287	0·15 S	0·15	−1·6	1373
0	0	0·15 E	0	−0·6	....
−61 10	1816	0·4 S	0·4	−2·4	1089
0	0	0·4 E	0	−1·2	....
+35 45	720	0	0	+0·2	....
0	0	0	0	−0·4	....
+61 10	1816	0·4 S	0·4	+2·7	1226
0	0	0·4 E	0	+1·3	....

From the above experiments we find that the induction of permanent magnetism produces no sensible effect on the length of a bar of perfectly hardened steel, and that the temporary shortening effect of the coil is proportional to the magnetism multiplied by the current traversing the coil. The shortening effect does not in this case sensibly increase with the increase of tension. I have made an experiment, in which a hard steel pillar was subjected to a pressure of 80 lb., which I have not inserted, because the coil not being the same as that I had hitherto employed, the experiment was not strictly comparable with the rest. Its result, however, showed that the hard steel pillar suffered a diminution of length equal to 0·1 of a division of the micrometer, with a current capable of giving a magnetic polarity of 1·7. This accords very well with the results obtained with a tension of 1030 lb.

Copper is the only non-ferruginous metal which I have experimented on. In the trials made with wires of this metal, pressure and tension were successively employed, and very powerful currents were transmitted through the coil; but I could in no case detect the slightest alteration in their dimensions.

I shall not prolong this paper by a discussion of the probable cause of the remarkable facts I have pointed out. The law of *elongation* naturally suggests the joint operation of the attractive and repulsive forces of the constituent particles of the magnet as the cause of that phenomenon. On the other hand, the fact that the *shortening effect* is proportional to the magnetic intensity of the bar multiplied by the current traversing the coil, seems to indicate that in this case the effect is produced by the attraction of the magnetic particles by the coil. But then it will be asked why so remarkable an augmentation of this effect is produced by increase of tension in the case of the soft iron bars. When we are able to answer this question in a satisfactory manner, we shall probably have a much more complete acquaintance with the real nature of magnetism than we at present possess.

Fig. 68.
Scale $\frac{1}{12}$.

I have already, in the former part of this paper, described an experiment which indicated that no alteration in the *bulk* of a bar of soft iron could be produced by magnetizing it. I thought, however, that it would be interesting to confirm the fact by an observation of the alteration of the dimensions of the iron at right angles to the direction of its polarity. For this purpose I took a piece of drawn iron gas-piping one yard long, $\frac{3}{16}$ of an inch in bore, and $\frac{3}{16}$ of an inch in thickness of metal. A piece of stout, covered copper wire was inserted into this tube, and bent along the sides of it, as indicated by fig. 68. The lower extremity of the iron tube being fixed, and the upper end being attached to the micrometrical apparatus described in the first section of this paper, each division of which corresponded to $\frac{1}{138528}$ of an inch, I obtained the following results :—

EXPERIMENT 28.—Drawn iron tube.

Deflection of galvanometer.	Tangent of deflection.	Shortening or elongation.	Total shortening effect.
+62° 30′	1921	3·2 S	3·2
0	0	1·2 E	2·0
+68 30	2539	1·0 S	3·0
0	0	1·0 E	2·0
− 5 0	87	0	2·0
0	0	0	2·0
−15 30	277	0·2 E	1·8
0	0	0	1·8
−30 0	577	1·0 E	0·8
0	0	0	0·8
−61 30	1842	2·0 S	2·8
0	0	1·0 E	1·8
+15 30	277	0·1 E	1·7
0	0	0	1·7
+30 20	585	0·7 E	1·0
0	0	0	1·0
+45 30	1017	0·1 S	1·1
0	0	0·1 E	1·0
+60 0	1732	1·2 S	2·2
0	0	0·6 E	1·6
+72 30	3171	1·8 S	3·4
0	0	0·8 E	2·6

The results of the above Table show that the length of the tube was diminished in order to make up for the increase of its diameter, which in this instance was in the direction of the polarity. The amount of the shortening effect, viz. 3·4, is, however, only one third of that due to the maximum elongation of soft iron bars as observed in the first section. This is probably owing to the grain of the iron being in cross directions with respect to the polarity in the two cases, and partly probable to the iron tube not being fully saturated with magnetism. The experiment is worth repeating, especially as it affords a means of studying the magnetic condition of closed magnetic circuits.

On Matter, Living Force, and Heat. By J. P. Joule, *Secretary of the Manchester Literary and Philosophical Society.*

[A Lecture at St. Ann's Church Reading-Room; and published in the Manchester 'Courier' newspaper, May 5 and 12, 1847.]

In our notion of matter two ideas are generally included, namely those of *impenetrability* and *extension*. By the extension of matter we mean the space which it occupies; by its impenetrability we mean that two bodies cannot exist at the same time in the same place. Impenetrability and extension cannot with much propriety be reckoned among the *properties* of matter, but deserve rather to be called its *definitions*, because nothing that does not possess the two qualities bears the name of matter. If we conceive of impenetrability and extension we have the idea of matter, and of matter only.

Matter is endowed with an exceedingly great variety of wonderful properties, some of which are common to all matter, while others are present variously, so as to constitute a difference between one body and another. Of the first of these classes, the attraction of gravitation is one of the most important. We observe its presence readily in all solid bodies, the component parts of which are, in the opinion of Majocci, held together by this force. If we break the body in pieces, and remove the separate pieces to a distance from each other, they will still be found to attract each other, though in a very slight degree, owing to the force being one which diminishes very rapidly as the bodies are removed further from one another. The larger the bodies are the more powerful is the force of attraction subsisting between them. Hence, although the force of attraction between small bodies can only be appreciated by the most delicate apparatus except in the case of contact, that which is occasioned by a body of immense magnitude, such as the earth, becomes very considerable. This attraction of bodies towards the earth constitutes what is called their *weight* or *gravity*, and is always exactly proportional to the quantity of matter. Hence, if any body be found to weigh 2 lb , while another only weighs 1 lb., the former will contain exactly twice as much matter as the latter; and this is the case, whatever the bulk of the bodies may be: 2-lb. weight of air contains exactly twice the quantity of matter that 1 lb. of lead does.

Matter is sometimes endowed with other kinds of attraction besides the attraction of gravitation; sometimes also it possesses the faculty of *repulsion*, by which force the particles tend to separate further from each other. Wherever these forces exist, they do not supersede the attraction of gravitation. Thus the weight of a piece of iron or steel is in no way affected by imparting to it the magnetic virtue.

Besides the force of gravitation, there is another very remarkable property displayed in an equal degree by every kind of matter—its perseverance in any condition, whether of rest or motion, in which it may have been placed. This faculty has received the name of *inertia*, signifying passiveness, or the inability of any thing to change its own state. It is in consequence of this property that a body at rest cannot be set in motion without the application of a certain amount of force to it, and also that when once the body has been set in motion it will never stop of itself, but continue to move straight forwards with a uniform velocity until acted upon by another force, which, if applied contrary to the direction of motion, will retard it, if in the same direction will accelerate it, and if sideways will cause it to move in a curved direction. In the case in which the force is applied contrary in direction, but equal in degree to that which set the body first in motion, it will be entirely deprived of motion whatever time may have elapsed since the first impulse, and to whatever distance the body may have travelled.

From these facts it is obvious that the force expended in setting a body in motion is carried by the body itself, and exists with it and in it, throughout the whole course of its motion. This force possessed by moving bodies is termed by mechanical philosophers *vis viva*, or *living force*. The term may be deemed by some inappropriate, inasmuch as there is no life, properly speaking, in question; but it is *useful*, in order to distinguish the moving force from that which is stationary in its character, as the force of gravity. When, therefore, in the subsequent parts of this lecture I employ the term *living force*, you will understand that I simply mean the force of bodies in motion. The living force of bodies is regulated by their weight and by the velocity of their motion. You will readily understand that if a body of a certain weight possess a certain quantity of living force, twice as much living force will be possessed by a body of twice the weight, provided both bodies move with equal velocity. But the law by which the *velocity* of a body regulates its living force is not so obvious. At first sight one would imagine that the living force would be simply proportional to the velocity, so that if a body moved twice as fast as another, it would have twice the impetus or living force. Such, however, is not the case; for if three bodies of equal weight move with the respective velocities of 1, 2, and 3 miles per hour, their living forces will be found to be proportional to those numbers multiplied by themselves, viz. to 1×1, 2×2, 3×3, or 1, 4, and 9, the squares of 1, 2, and 3. This remarkable law may be proved in several ways. A bullet fired from a gun at a certain velocity will pierce a block of wood to only one quarter of the depth it would if propelled at twice the velocity. Again, if a cannon-ball were found to fly at a certain velocity when propelled by a given charge of gunpowder, and it were required to load the cannon so as to propel the ball with twice that velocity, it would be found necessary to employ four times the weight of powder previously used. Thus, also, it will be found that a railway-train going at 70 miles per hour possesses 100 times the impetus, or living force, that it does when travelling at 7 miles per hour.

A body may be endowed with living force in several ways. It may receive it by the impact of another body. Thus, if a perfectly elastic ball be made to strike another similar ball of equal weight at rest, the striking ball will communicate the whole of its living force to the ball struck, and, remaining at rest itself, will cause the other ball to move in the same direction and with the same velocity that it did itself before the collision. Here we see an instance of the facility with which living force may be transferred from one body to another. A body may also be endowed with living force by means of the action of gravitation upon it through a certain distance. If I hold a ball at a certain height and drop it, it will have acquired when it arrives at the ground a degree of living force proportional to its weight and the height from which it has fallen. We see, then, that living force may be produced by the action of gravity through a given distance or space. We may therefore say that the former is of equal value, or *equivalent*, to the latter. Hence, if I raise a weight of 1 lb. to the height of one foot, so that gravity may act on it through that distance, I shall communicate to it that which is of equal value or equivalent to a certain amount of living force; if I raise the weight to twice the height, I shall communicate to it the equivalent of twice the quantity of living force. Hence, also, when we compress a spring, we communicate to it the equivalent to a certain amount of living force; for in that case we produce molecular attraction between the particles of the spring through the distance they are forced asunder, which is strictly analogous to the production of the attraction of gravitation through a certain distance.

You will at once perceive that the living force of which we have been speaking is one of the most important qualities with which matter can be endowed, and, as such, that it would be absurd to suppose that it can be destroyed, or even lessened, without producing the equivalent of attraction through a given distance of which we have been speaking. You will therefore be surprised to hear that until very recently the universal opinion has been that living force could be absolutely and irrevocably destroyed at any one's option. Thus, when a weight falls to the ground, it has been generally supposed that its living force is absolutely annihilated, and that the labour which may have been expended in raising it to the elevation from which it fell has been entirely thrown away and wasted, without the production of any permanent effect whatever. We might reason, *à priori*, that such absolute destruction of living force cannot possibly take place, because it is manifestly absurd to suppose that the powers with which God has endowed matter can be destroyed any more than that they can be created by man's agency; but we are not left with this argument alone, decisive as it must be to every unprejudiced mind. The common experience of every one teaches him that living force is not *destroyed* by the friction or collision of bodies. We have reason to believe that the manifestations of living force on our globe are, at the present time, as extensive as those which have existed at any time since its creation, or, at any rate, since the deluge—that the winds blow as strongly, and the torrents flow with equal impetuosity now, as at the remote period of 4000 or even 6000 years ago; and yet we are certain that, through that vast interval of time, the motions of the air and of the water have been incessantly obstructed and hindered by friction. We may conclude, then, with certainty, that these motions of air and water, constituting living force, are not *annihilated* by friction. We lose sight of them, indeed, for a time; but we find them again reproduced. Were it not so, it is perfectly obvious that long ere this all nature would have come to a dead standstill. What, then, may we inquire, is the cause of this apparent anomaly? How comes it to pass that, though in almost all natural phenomena we witness the arrest of motion and the apparent destruction of living force, we find that no waste or loss of living force has actually occurred? Experiment has enabled us to answer these questions in a satisfactory manner; for it has shown that, wherever living force is *apparently* destroyed, an equivalent is produced which in process of time may be reconverted into living force. This equivalent is *heat*. Experiment has shown that wherever living force is apparently destroyed or absorbed, heat is produced. The most frequent way in which living force is thus converted into heat is by means of friction. Wood rubbed against wood or against any hard body, metal rubbed against metal or against any other body—in short, all bodies, solid or even liquid, rubbed against each other are invariably heated, sometimes even so far as to become red-hot. In all these instances the quantity of heat produced is invariably in proportion to the exertion employed in rubbing the bodies together—that is, to the living force absorbed. By fifteen or twenty smart and quick strokes of a hammer on the end of an iron rod of about a quarter of an inch in diameter placed upon an anvil an expert blacksmith will render that end of the iron visibly red-hot. Here heat is produced by the absorption of the living force of the descending hammer in the soft iron; which is proved to be the case from the fact that the iron cannot be heated if it be rendered hard and elastic, so as to transfer the living force of the hammer to the anvil.

The general rule, then, is, that wherever living force is *apparently* destroyed, whether by percussion, friction, or any similar means, an exact equivalent of heat is restored. The converse of this proposition is also true, namely, that heat cannot be lessened or absorbed without the production of living force, or its equivalent attraction through space. Thus, for instance, in the steam-engine it will be found that the power gained is at the expense of the heat of the fire,—that is, that the heat occasioned by the combustion of the coal would have been greater had a part of it not been absorbed in producing and maintaining the living force of the machinery. It is right, however, to observe that this has not as yet been demonstrated by experiment. But there is no room to doubt that experiment would prove the correctness of what I have said; for I have myself proved that a conversion of heat into living force takes place in the expansion of air, which is analogous to the expansion of steam in the cylinder of the steam-engine. But the most convincing proof of the conversion of heat into living force has been derived from my experiments with the electro-magnetic engine, a machine composed of magnets and bars of iron set in motion by an electrical battery. I have proved by actual experiment that, in exact proportion to the force with which this machine works, heat is abstracted from the electrical battery. You see, therefore, that living force may be converted into heat, and that heat may be converted into living force, or its equivalent attraction through space. All, three, therefore—namely, heat, living force, and attraction through space (to which I might also add *light*, were it consistent with the scope of the present lecture)—are mutually convertible into one another. In these conversions nothing is ever lost. The same quantity of heat will always be converted into the same quantity of living force. We can therefore express the equivalency in definite language applicable at all times and under all circumstances. Thus the attraction of 817 lb. through the space of one foot is equivalent to, and convertible into, the living force possessed by a body of the same weight of 817 lb. when moving with the velocity of eight feet per second, and this living force is again convertible into the quantity of heat which can increase the temperature of one pound of water by one degree Fahrenheit. The knowledge of the equivalency of heat to mechanical power is of great value in solving a great number of interesting and important questions. In the case of the steam-engine, by ascertaining the quantity of heat produced by the combustion of coal, we can find out how much of it is converted into mechanical power, and thus come to a conclusion how far the steam-engine is susceptible of further improvements. Calculations made upon this principle have shown that at least ten times as much power might be produced as is now obtained by the combustion of coal. Another interesting conclusion is, that the animal frame, though destined to fulfil so many other ends, is as a machine more perfect than the best contrived steam-engine—that is, is capable of more work with the same expenditure of fuel.

Behold, then, the wonderful arrangements of creation. The earth in its rapid motion round the sun possesses a degree of living force so vast that, if turned into the equivalent of heat, its temperature would be rendered at least 1000 times greater than that of red-hot iron, and the globe on which we tread would in all probability be rendered equal in brightness to the sun itself. And it cannot be doubted that if the course of the earth were changed so that it might fall into the sun, that body, so far from being cooled down by the contact of a comparatively cold body, would actually blaze more brightly than before in consequence of the living force with which the earth struck the sun being converted into its equivalent of heat. Here we see that our existence depends upon the *maintenance* of the living force of the earth. On the other hand, our safety equally depends in some instances upon the *conversion* of living force into heat. You have, no doubt, frequently observed what are called *shooting-stars,* as they appear to emerge from the dark sky of night, pursue a short and rapid course, burst, and are dissipated in shining fragments. From the velocity with which these bodies travel, there can be little doubt that they are small planets which, in the course of their revolution round the sun, are attracted and drawn to the earth. Reflect for a moment on the consequences which would ensue, if a hard meteoric stone were to strike the room in which we are assembled with a velocity sixty times as great as that of a cannon-ball. The dire effects of such a collision are effectually prevented by the atmosphere surrounding our globe, by which the velocity of the meteoric stone is checked and its living force converted into heat, which at last becomes so intense as to melt the body and dissipate it into fragments too small probably to be noticed in their fall to the ground. Hence it is that, although multitudes of shooting-stars appear every night, few meteoric stones have been found, those few corroborating the truth of our hypothesis by the marks of intense heat which they bear on their surfaces.

Descending from the planetary space and firmament to the surface of our earth, we find a vast variety of phenomena connected with the conversion of living force and heat into one another, which speak in language which cannot be misunderstood of the wisdom and beneficence of the Great Architect of nature. The motion of air which we call *wind* arises chiefly from the intense heat of the torrid zone compared with the temperature of the temperate and frigid zones. Here we have an instance of heat being converted into the living force of currents of air. These currents of air, in their progress across the sea, lift up its waves and propel the ships; whilst in passing across the land they shake the trees and disturb every blade of grass. The waves by their violent motion, the ships by their passage through a resisting medium, and the trees by the rubbing of their branches together and the friction of their leaves against themselves and the air, each and all of them generate heat equivalent to the diminution of the living force of the air which they occasion. The heat thus restored may again contribute to raise fresh currents of air; and thus the phenomena may be repeated in endless succession and variety.

When we consider our own animal frames, "fearfully and wonderfully made," we observe in the motion of our limbs a continual conversion of heat into living force, which may be either converted back again into heat or employed in producing an attraction through space, as when a man ascends a mountain. Indeed the phenomena of nature, whether mechanical, chemical, or vital, consist almost entirely in a continual conversion of attraction through space, living force, and heat into one another. Thus it is that order is maintained in the universe—nothing is deranged, nothing ever lost, but the entire machinery, complicated as it is, works smoothly and harmoniously. And though, as in the awful vision of Ezekiel, "wheel may be in the middle of wheel," and every thing may appear complicated and involved in the apparent confusion and intricacy of an almost endless variety of causes, effects, conversions, and arrangements, yet is the most perfect regularity preserved—the whole being governed by the sovereign will of God.

A few words may be said, in conclusion, with respect to the real nature of heat. The most prevalent opinion, until of late, has been that it is a *substance* possessing, like all other matter, impenetrability and extension. We have, however, shown that heat can be converted into living force and into attraction through space. It is perfectly clear, therefore, that unless matter can be converted into attraction through space, which is too absurd an idea to be entertained for a moment, the hypothesis of heat being a substance must fall to the ground. Heat must therefore consist of either living force or of attraction through space. In the former case we can conceive the constituent particles of heated bodies to be, either in whole or in part, in a state of motion. In the latter we may suppose the particles to be removed by the process of heating, so as to exert attraction through greater space. I am inclined to believe that both of these hypotheses will be found to hold good,—that in some instances, particularly in the case of *sensible* heat, or such as is indicated by the thermometer, heat will be found to consist in the living force of the particles of the bodies in which it is induced; whilst in others, particularly in the case of *latent* heat, the phenomena are produced by the separation of particle from particle, so as to cause them to attract one another through a greater space. We may conceive, then, that the communication of heat to a body consists, in fact, in the communication of impetus, or living force, to its particles. It will perhaps appear to some of you something strange that a body apparently quiescent should in reality be the seat of motions of great rapidity; but you will observe that the bodies themselves, considered as wholes, are not supposed to be in motion. The constituent particles, or atoms of the bodies, are supposed to be in motion, without producing a gross motion of the whole mass. These particles, or atoms, being far too small to be seen even by the help of the most powerful microscopes, it is no wonder that we cannot observe their motion. There is therefore reason to suppose that the particles of all bodies, their constituent atoms, are in a state of motion almost too rapid for us to conceive, for the phenomena cannot be otherwise explained. The velocity of the atoms of water, for instance, is at least equal to a mile per second of time. If, as there is reason to think, some particles are at rest while others are in motion, the velocity of the latter will be proportionally greater. An increase of the velocity of revolution of the particles will constitute an increase of temperature, which may be distributed among the neighbouring bodies by what is called *conduction*—that is, on the present hypothesis, by the communication of the increased motion from the particles of one body to those of another. The velocity of the particles being further increased, they will tend to fly from each other in consequence of the centrifugal force overcoming the attraction subsisting between them. This removal of the particles from each other will constitute a new condition of the body—it will enter into the state of fusion, or become melted. But, from what we have already stated, you will perceive that, in order to remove the particles violently attracting one another asunder, the expenditure of a certain amount of living force or heat will be required. Hence it is that heat is always absorbed when the state of a body is changed from solid to liquid, or from liquid to gas. Take, for example, a block of ice cooled down to zero; apply heat to it, and it will gradually arrive at 32°, which is the number conventionally employed to represent the temperature at which ice begins to melt. If, when the ice has arrived at this temperature, you continue to apply heat to it, it will become melted; but its temperature will not increase beyond 32° until the whole has been converted into water. The explanation of these facts is clear on our hypothesis. Until the ice has arrived at the temperature of 32° the application of heat increases the velocity of rotation of its constituent particles; but the instant it arrives at that point, the velocity produces such an increase of the centrifugal force of the particles that they are compelled to separate from each other. It is in effecting this separation of particles strongly attracting one another that the heat applied is *then* spent; not in increasing the velocity of the particles. As soon, however, as the separation has been effected, and the fluid water produced, a further application of heat will cause a further increase of the velocity of the particles, constituting an increase of temperature, on which the thermometer will immediately rise above 32°. When the water has been raised to the temperature of 212°, or the boiling-point, a similar phenomenon will be repeated; for it will be found impossible to increase the temperature beyond that point, because the heat then applied is employed in separating the particles of water so as to form steam, and not in increasing their velocity and living force. When, again, by the application of cold we condense the steam into water, and by a further abstraction of heat we bring the water to the solid condition of ice, we witness the repetition of similar phenomena in the reverse order. The particles of steam, in assuming the condition of water, fall together through a certain space. The living force thus produced becomes converted into heat, which must be removed before any more steam can be converted into water. Hence it is always necessary to abstract a great quantity of heat in order to convert steam into water, although the temperature will all the while remain exactly at 212°; but the instant that all the steam has been condensed, the further abstraction of heat will cause a diminution of temperature, since it can only be employed in diminishing the velocity of revolution of the atoms of water. What has been said with regard to the condensation of steam will apply equally well to the congelation of water.

I might proceed to apply the theory to the phenomena of combustion, the heat of which consists in the living force occasioned by the powerful attraction through space of the combustible for the oxygen, and to a variety of other thermo-chemical phenomena; but you will doubtless be able to pursue the subject further at your leisure.

I do assure you that the principles which I have very imperfectly advocated this evening may be applied very extensively in elucidating many of the abstruse as well as the simple points of science, and that patient inquiry on these grounds can hardly fail to be amply rewarded.

On the Mechanical Equivalent of Heat, as determined from the Heat evolved by the Agitation of Liquids. By J. P. Joule, *Sec. Manchester Lit. & Phil. Soc.**

[Rep. Brit. Assoc. 1847, Sections, p. 55. Read before the British Association at Oxford, June 1847.]

The author exhibited and described an apparatus, consisting of a brass paddle-wheel working in a vessel filled with liquid, with which he had repeated the experiments brought before

* The experiments were made at Oak Field, Whalley Range, near Manchester.

the Cambridge Meeting of the Association. By these experiments he had shown that heat is invariably produced by the friction of fluids in exact proportion to the force expended. Two series of experiments had been made—one on the friction of water, the other on the friction of sperm-oil. In the former of these series the heat capable of raising the temperature of a pound of water 1° was found to be equal to the mechanical force capable of raising a weight of 781·5 pounds to the height of one foot; whilst in the series of experiments on the friction of sperm-oil, the same quantity of heat was found to be equal to a mechanical force represented by 782·1 pounds through one foot.

The author also stated the result of some experiments made more than a year previously, by which he had found that when a spiral steel spring was compressed no heat was evolved. The steel spring, whose particles were thus forcibly disturbed without a change of temperature being produced, was illustrative of the condition of a body possessing what is commonly called "latent heat."

On the Mechanical Equivalent of Heat, as determined by the Heat evolved by the Friction of Fluids. By J. P. Joule, *Secretary to the Literary and Philosophical Society of Manchester.*

[Phil. Mag. ser. 3. vol. xxxi. p. 173. Read before the Mathematical and Physical Section of the British Association at Oxford, June 1847.]
(Abstracted above.)

In the 'Philosophical Magazine' for September 1845 I gave a concise account of some experiments brought before the Cambridge Meeting of the British Association, by which I had proved that heat was generated by the friction of water produced by the motion of a horizontal paddle-wheel. These experiments, though abundantly sufficient to establish the equivalency of heat to mechanical power, were not adapted to determine the equivalent with very great numerical accuracy, owing to the apparatus having been situated in the open air, and having been in consequence liable to great cooling or heating effects from the atmosphere. I have now repeated the experiments under more favourable circumstances, and with a more exact apparatus, and have moreover employed sperm-oil as well as water with equal success.

The brass paddle-wheel employed had, as described in my former paper, a brass framework attached, which presented sufficient resistance to the liquid to prevent the latter being whirled round. In this way the resistance presented by the liquid to the paddle was rendered very considerable, although no splashing was occasioned. The can employed was of copper, surrounded by a very thin casing of tin. It was covered with a tin lid, having a capacious hole in its centre for the axle of the paddle, and another for the insertion of a delicate thermometer. Motion was communicated to the paddle by means of a drum fitting to the axle, upon which a quantity of twine had been wound, so as by the intervention of delicate pulleys to raise two weights, each of 29 lb., to the height of about $5\frac{1}{4}$ feet. When the weights in moving the paddle had descended through that space, the drum was removed, the weights wound up again, and the operation repeated. After this had been done twenty times, the increase of the temperature of liquid was ascertained. In the second column of the following table the whole distance through which the weights descended during the several experiments is given in inches. I may observe also that both the experiments on the friction of water, and the alternations made in order to ascertain the effect of the surrounding atmosphere, were conducted under similar circumstances, each occupying forty minutes.

TABLE I.—Friction of Distilled Water.

Nature of experiment.	Total descent of each weight of 29 lb., in inches.	Mean temperature of the room.	Difference.	Temperature of the water.		Gain or loss of heat.
				Before expt.	After expt.	
Friction	1268·5	60°·839	0°·040–	60°·452	61°·145	0°·693 gain.
Alternation	0	61·282	0·120–	61·145	61·180	0·035 gain.
Friction	1266·1	61·007	0·408+	61·083	61·748	0·665 gain.
Alternation	0	61·170	0·570+	61·752	61·729	0·023 loss.
Friction	1265·8	57·921	0·809–	56·752	57·472	0·720 gain.
Alternation	0	58·119	0·628–	57·472	57·511	0·039 gain.
Friction	1265·4	58·152	0·293–	57·511	58·207	0·696 gain.
Alternation	0	58·210	0·003+	58·207	58·219	0·012 gain.
Friction	1265·1	57·860	0·215+	57·735	58·416	0·681 gain.
Alternation	0	58·162	0·256+	58·416	58·420	0·004 gain.
Friction	1265·3	57·163	0·220+	57·050	57·716	0·666 gain.
Alternation	0	57·602	0·121+	57·716	57·731	0·015 gain.
Friction	1265·2	57·703	0·359+	57·731	58·393	0·662 gain.
Alternation	0	58·091	0·304+	58·393	58·397	0·004 gain.
Friction	1262·4	56·256	0·015–	55·901	56·582	0·681 gain.
Alternation	0	56·888	0·285–	56·590	56·617	0·027 gain.
Friction	1262·3	57·041	0·078–	56·617	57·310	0·693 gain.
Alternation	0	57·612	0·285–	57·310	57·344	0·034 gain.
Mean friction experiments	1265·13	....	0·0037–	....	....	0·6841 gain.
Mean of the alternations ..	0	....	0·0071+	....	....	0·0163 gain.
Corrected result..	1265·13	....	....	....	....	0·6680 gain.

We see then that the weights of 29 lb., in descending through the altitude of 1265·13 inches, generated 0°·668 in the apparatus. But in order to reduce these quantities, it became necessary in the first place to ascertain the friction of the pulleys and that of the twine in unwinding from the drum. This was effected by causing the twine to go once round a roller of the same diameter as the drum, working upon very fine pivots, the two extremities of the twine being thrown over the pulleys. Then it was found that, by adding a weight of 3150 grains to either of the two weights, the friction was just overcome. The actual force employed in the experiments would therefore be 406000 grs. − 3150 grs. = 402850 grs. through 1265·13 inches, or 6067·3 lb. through a foot.

The weight of water being 77617 grs., that of the brass paddle-wheel 24800 grs., the copper of the can 11237 grs., and the tin casing and cover 19396 grs., the whole capacity of the vessel and its contents was estimated at 77617 + 2319 + 1056 + 363 = 81355 grs. of water. Therefore the quantity of heat evolved in the experiments, referred to a pound of water, was 7°·7636.

The equivalent of a degree of heat in a pound of water was therefore found to be 781·5 lb., raised to the height of one foot.

I now made a series of experiments in which sperm-oil was substituted for the water in the can. This liquid, being that employed by engineers as the best for diminishing the friction of their machinery, appeared to me well calculated to afford another and even more decisive proof of the principles contended for.

In this second series the force employed, corrected as before for the friction of the pulleys, was equal to raise 6080·4 lb. to the height of one foot.

In estimating the capacity for heat of the apparatus, it was necessary to obtain the specific heat of the sperm-oil employed. For this purpose I employed the *method of mixtures*. 43750 grs. of water were heated in a copper vessel weighing 10403 grs. to 82°·697. I added to this 28597 grs. of oil at 55°·593, and after stirring the two liquids together, found the temperature of the mixture to be 76°·583. Having applied to these data the requisite corrections for the cooling of the liquids during the experiment, and for the capacity of the copper vessel, the specific heat of the sperm-oil came out 0·45561. Another experiment of the same kind, but in which the water was poured into the heated oil, gave the specific heat 0·46116. The mean specific heat was therefore 0·45838.

The weight of oil employed was 70273 grains, and the

TABLE II.—Friction of Sperm-Oil.

Nature of experiment.	Total descent of each weight of 29 lb., in inches.	Mean temperature of the room.	Difference.	Temperature of the oil.		Gain or loss of heat.
				Before expt.	After expt.	
Friction	1263·8	56°·677	0°·453+	56°·354	57°·906	1°·552 gain.
Alternation	0	57·316	0·595+	57·906	57·917	0·011 gain.
Friction	1269·0	56·198	1·024+	56·516	57·929	1·413 gain.
Alternation	0	56·661	1·221+	57·929	57·836	0·093 loss.
Friction	1268·7	57·958	0·588+	57·813	59·280	1·467 gain.
Alternation	0	57·051	0·773+	57·836	57·813	0·023 loss.
Friction	1268·5	58·543	1·685−	55·951	57·766	1·815 gain.
Alternation	0	57·153	1·504−	55·568	55·731	0·163 gain.
Friction	1268·1	59·097	0·534−	57·766	59·361	1·595 gain.
Alternation	0	57·768	1·927−	55·731	55·951	0·220 gain.
Friction	1268·3	56·987	0·186−	56·029	57·573	1·544 gain.
Alternation	0	57·156	0·413+	57·573	57·565	0·008 loss.
Friction	1268·7	57·574	0·734+	57·581	59·036	1·455 gain.
Alternation	0	57·336	0·237+	57·565	57·581	0·016 gain.
Friction	1267·6	58·537	0·829−	56·884	58·532	1·648 gain.
Alternation	0	59·641	0·364+	60·026	59·984	0·042 loss.
Friction	1268·0	59·131	0·148+	58·532	60·026	1·494 gain.
Alternation	0	60·164	0·138−	59·984	60·069	0·085 gain.
Mean friction experiments	1267·85	....	0·034+	....	......	1·5537 gain.
Mean of the alternations ..	0	....	0·004+	....	......	0·0366 gain.
Corrected result..	1267·85	....	....	....	......	1·5138 gain.

paddle, can, &c. were the same as employed in the first series of experiments; consequently the entire capacity in this instance will be equivalent to that of 35951 grs. of water. The heat evolved was therefore 7°·7747 when reduced to the capacity of a pound of water.

Hence the equivalent deduced from the friction of sperm-oil was 782·1, a result almost identical with that obtained from the friction of water. The mean of the two results is 781·8, which is the equivalent I shall adopt until further and still more accurate experiments shall have been made.

On the Theoretical Velocity of Sound.
By J. P. Joule.

['Philosophical Magazine,' ser. 3. vol. xxxi. p. 114.]

The celebrated French mathematician De Laplace has, it is well known, pointed out that the heat evolved by the compression of air is the cause of the velocity of sound, according to the theory of Newton, being so much less than that actually observed. He has also given a formula by which the velocity may be determined when the ratio of the specific heat of air at constant pressure to that at constant volume is known. The determination of the elevation of temperature in air by compression has, however, been hitherto attended with difficulty, and hence the theorem of De Laplace has never yet been fairly compared with experiment. I was therefore anxious to ascertain how far the mechanical equivalent of heat, as determined by my recent experiments on the friction of fluids, might be able to contribute to clear up this question.

The capacity of air at constant pressure, according to the experiments of De la Roche and Berard, is 0·2669. Consequently a quantity of heat capable of increasing the temperature of a lb. of water by 1° will give 1° also to 3·747 lb. of air, while the air will be expanded $\frac{1}{491}$, an expansion in which a force equal to 200·7 lb. through a foot is expended in raising the atmosphere of the earth. The equivalent of a degree of heat per lb. of water, determined by careful experiments made since those brought before the British Association at Oxford, is 775 lb. through a foot. Hence 200·7 lb. through a foot is equal to 0°·259.

We see, therefore, that for every degree of heat employed by De la Roche and Berard in expanding and heating air, 0°·259 was occupied in producing the mechanical effect, leaving 0°·741 as that actually employed in raising the temperature of the air. Hence the actual specific heat (commonly called capacity at constant volume) is 0·2669 × 0·741 = 0·1977. Taking this as the specific heat of air and the equivalent 775, it follows that if a volume of air of 171·6 cubic inches be compressed to 170·6 cubic inches, it will be heated 1°, a quantity of heat which will occasion an increased pressure of $\frac{1}{491}$. So that the celerity of sound will be increased by this means in the subduplicate ratio of 491 to 661·6, or in the simple ratio of 2216 to 2572, which will bring it up from Newton's estimate of 943 to 1095 feet per second, which is as near 1130, the actual velocity at 32°, as could be expected from the nature of the experiments on the specific heat of air, and fully confirms the theory of Laplace.

Oak Field, near Manchester,
July 17, 1847.

Expériences sur l'Identité entre le Calorique et la Force mécanique. Détermination de l'équivalent par la Chaleur dégagée pendant la friction du Mercure. Par M. J. P. JOULE *.

['Comptes Rendus,' August 23, 1847.]

PENDANT les quatre dernières années j'ai fait diverses expériences, dans le but de m'assurer que la chaleur était l'équivalent de la force mécanique. De ces expériences, peut-être les plus intéressantes sont celles que j'ai faites sur la friction des liquides. Quand l'eau était agitée par l'action d'une roue à pannes agissant dans le liquide, la quantité de chaleur dégagée était en proportion exacte à la force mécanique dépensée. La force mécanique capable d'élever un poids de 428·8 grammes à la hauteur de 1 mètre fut ainsi trouvée être l'équivalent d'une quantité de chaleur nécessaire pour élever la température de 1 gramme d'eau par 1 degré centigrade.

* [The Commissioners were Biot, Pouillet, and Regnault. I had the honour to present the iron vessel with its revolving paddle-wheel to the last-named eminent physicist.—*Note*, 1881.]

J'ai aussi fait des expériences semblables sur la friction de l'*huile de* baleine. Dans ce cas, le dégagement de chaleur fut encore plus considérable, la chaleur spécifique de l'huile étant bien inférieure comparativement à celle de l'eau. Quoi qu'il en soit de cette différence, les résultats auxquels je suis arrivé étaient à peu près les mêmes; c'est-à-dire que le développement de 1 degré de chaleur par gramme d'eau était égal à 429·1 grammes soulevé de 1 mètre.

Poursuivant mes recherches, j'ai aussi employé du *mercure* comme liquide frotté, et j'ai, en l'employant, obtenu des résultats si confirmatifs des expériences ci-dessus, que j'ose les communiquer à l'Académie des Sciences.

L'appareil dont je me suis servi était composé d'un vase cylindrique en fonte, dans l'intérieur duquel était placée horizontalement une roue à pannes en tôle. À la partie supérieure était vissé un couvercle également en fonte, lequel avait deux ouvertures: l'une au centre, pour la passage d'un axe par lequel le mouvement était communiqué à la roue; l'autre était pour servir à l'introduction d'un thermomètre. Lorsque la température du mercure était exactement déterminée, la roue à pannes était mise en mouvement par des poids avec lesquels elle était en communication par l'entremise de poulies. Après que le mercure était agité ainsi pendant un certain temps, l'augmentation de la température était déterminée par une nouvelle observation thermométrique. La valeur de la force employée était évaluée par l'espace qu'avaient parcouru les poids en descendant: on tenait compte de $\frac{1}{100}$ comme valeur de la friction des poulies. L'influence de l'air environnant fut déterminée par des expériences qui consistaient à placer l'appareil dans des atmosphères dont les températures étaient variées.

Désignation des expériences.	Force en grammes tombant de 1 mètre.	Température du laboratoire.	Différence.	Température du Mercure		Gain ou Perte du Température.
				au commencement de l'expérience.	à la fin de l'expérience.	
Friction..............	716977	15°·300	+0°·426	14°·628	16°·825	2°·197 gain.
Influence de l'air	0	15·135	−0·569	14·504	14·628	0·124 gain.
Friction..............	715297	16·001	−0·042	14·800	17·118	2·318 gain.
Influence de l'air	0	15·452	+1·247	16·808	16·590	0·218 perte.
Friction..............	715832	15·792	+0·369	15·067	17·255	2·188 gain.
Influence de l'air	0	16·098	+0·930	17·114	16·942	0·172 perte.
Friction..............	713992	15·548	+0·329	14·774	16·980	2·206 gain.
Influence de l'air	0	15·387	−0·518	14·822	14·917	0·095 gain.
Friction..............	714463	14·684	+0·628	14·250	16·374	2·124 gain.
Influence de l'air	0	15·806	+1·056	16·959	16·765	0·194 perte.
Friction..............	714822	14·869	−0·007	13·751	15·974	2·223 gain.
Influence de l'air	0	14·529	−0·359	14·138	14·203	0·065 gain.
Moyen { Friction de mercure	715230	....	+0·2838	....	....	2·2093 gain.
Moyen { Influence de l'air	0	...	+0·2978	....	....	0·0500 perte.
Résultat corrigé	715230	....	....	....	....	2·2568 gain.

Dans le tableau ci-joint, qui renferme mes résultats, chaque expérience alternative tient compte de l'influence de l'atmosphère, en élevant ou abaissant la température de l'appareil. Le poids du mercure dans l'appareil était de 13269 grammes, et il s'ensuivrait, d'après les expériences de M. Regnault sur la chaleur spécifique du mercure, qu'il serait égal, en capacité, à 442·12 grammes d'eau. Le poids du fer étant égal à 2569 grammes, sa capacité pour la chaleur serait, d'après mes déterminations, égale à 291·31 grammes d'eau. Par conséquent, la capacité totale de l'appareil pour la chaleur était égale à 733·43 grammes d'eau. L'absorption d'une force mécanique estimée par un poids de 715,230 grammes tombant de 1 mètre, fut ainsi accompagnée par le dégagement de 2°·2568 dans 733·43 grammes d'eau. Par conséquent, la chaleur capable d'augmenter la température de 1 gramme d'eau de 1 degré centigrade est égale à une force mécanique capable d'élever un poids de 432·1 grammes à 1 mètre de hauteur.

On Shooting-Stars. *By* J. P. Joule, *Corresponding Member of the Royal Academy of Sciences, Turin, Secretary to the Literary and Philosophical Society, Manchester.*

['Philosophical Magazine,' ser. 3. vol. xxxii. p. 349.]

I have read with much interest the valuable papers on shooting-stars inserted by Sir J. W. Lubbock in the numbers of the 'Philosophical Magazine' for February and March. This philosopher seems to have placed the subject in a fair way for satisfactory solution. He has advanced three hypotheses to account for the sudden disappearance of these bodies, the last of which he has enabled us to prove or disprove by actual observation.

I have for a long time entertained an hypothesis with respect to shooting-stars similar to that advocated by Chladni to account for meteoric stones, and have reckoned the *ignition* of these miniature planetary bodies by their violent collision with our atmosphere to be a remarkable illustration of the doctrine of the equivalency of heat to mechanical power or *vis viva.* In a popular lecture delivered in Manchester on the 28th of April, 1847, I said:—"You have, no doubt, frequently observed what are called *shooting-stars,* as they appear to emerge from the dark sky of night, pursue a short and rapid course, burst, and are dissipated in shining fragments. From the velocity with which these bodies travel, there can be little doubt that they are small planets which, in the course of their revolution round the sun, are attracted and drawn to the earth. Reflect for a moment on the consequences which would ensue if a hard meteoric stone were to strike the room in which we are assembled with a velocity sixty times as great as that of a cannon-ball. The dire effects of such a collision are effectually prevented by the atmosphere surrounding our globe, by which the velocity of the meteoric stone is checked, and its living force converted into heat, which at last becomes so intense as to melt the body and dissipate it in fragments too small probably to be noticed in their fall to the ground. Hence it is that, although multitudes of shooting-stars appear every night, few meteoric stones have been found, those few corroborating the truth of our hypothesis by the marks of intense heat which they bear on their surfaces" *.

The likelihood of the above hypothesis will be rendered evident if we suppose a meteoric stone, of the size of a six-inch cube, to enter our atmosphere at the rate of eighteen miles per second of time, the atmosphere being $\frac{1}{100}$ of its density at the earth's surface. The resistance offered to the motion of the stone will in this case be at least 51,600 lb.; and if the stone traverse twenty miles with this amount of resistance, sufficient heat will thereby be developed to give 1° Fahrenheit to 6,967,980 lb. of water. Of course by far the largest portion of this heat will be given to the displaced air, every particle of which will sustain the shock, whilst only the surface of the stone will be in violent collision with the atmosphere. Hence the stone may be considered as placed in a blast of intensely heated air, the heat being communicated from the surface to the centre by conduction. Only a small portion of the heat evolved will therefore be received by the stone; but if we estimate it at only $\frac{1}{100}$, it will still be equal to 1° Fahrenheit per 69,679 lb. of water, a quantity quite equal to the melting and dissipation of any materials of which it may be composed.

The dissolution of the stone will also be accelerated in most cases by its breaking into pieces, in consequence of the unequal resistance experienced by different parts of its surface, especially after its cohesion has been partially overcome by heat.

* 'Manchester Courier' newspaper, May 12, 1847.

It appears to me that the varied phenomena of meteoric stones and shooting-stars may all be explained in the above manner, and that the different velocities of the aerolites, varying from four to forty miles per second according to the direction of their motions with respect to the earth, along with their various sizes, will suffice to show why some of these bodies are destroyed the instant they arrive in our atmosphere, and why others arrive at the earth's surface with diminished velocity.

I cannot but be filled with admiration and gratitude for the wonderful provision thus made by the Author of nature for the protection of his creatures. Were it not for the atmosphere which covers us with a shield, impenetrable in proportion to the violence which it is called upon to resist, we should be continually exposed to a bombardment of the most fatal and irresistible character. To say nothing of the larger stones, no ordinary buildings could afford shelter from very small particles striking at the velocity of eighteen miles per second. Even dust flying at such a velocity would kill any animal exposed to it.

On the Mechanical Equivalent of Heat, and on the Constitution of Elastic Fluids. By J. P. Joule.

[Rep. Brit. Assoc. 1848, Sections, p. 21. Read before the British Association at Swansea, August 1848.]

At the last Meeting of the Association the author exhibited an apparatus which, by the agitation of fluids, produced heat in exact proportion to the mechanical power expended. Experiments were made with this apparatus on the heat evolved by the friction of three totally dissimilar fluids—water, mercury, and oil; and in all three cases the remarkable result appeared, that the mechanical power represented by the force necessary to raise 782 lb. one foot high produced the quantity of heat equal to raise the temperature of a pound of water one degree.

Since the above experiments were communicated to the Association, a slight alteration in the form of the apparatus, calculated to give greater exactness to the results, occurred to the author; and he therefore commenced a new and extensive series of experiments in order to determine the equivalent of heat with all the accuracy which its importance to physical science demands. The result arrived at after a series of forty experiments was an alteration of the equivalent before stated to 771, which is believed to be within $\frac{1}{200}$ of the truth, and therefore may for the present be assumed as a tolerably good basis for calculations.

The author conceives the following points to be established:—1st, his experiments on the friction of fluids, confirming the views and experiments of Davy and Rumford on the friction of solids, afford another and decisive proof that heat is simply a mechanical effect, not a substance; 2nd, his experiments, showing that the thermal effects of the condensation and rarefaction of air are the equivalents of the mechanical force expended in the one case and gained in the other, prove that the heat of elastic fluids consists simply in the *vis viva* of their particles; and, 3rd, the zero of temperature, determined by the expansion of gases, is at 491° below the freezing-point of water.

We may, the author thinks, employ the above propositions in order to calculate the specific heat of the gases. For, whether we conceive the particles to be revolving round one another, according to the hypothesis of Davy, or flying about in every direction according to Herapath's view, the pressure of the gas will be proportional to the *vis viva* of its particles. Thus it may be shown that the particles of hydrogen gas at the barometrical pressure of 30 inches and temperature 60° must move with a velocity of 6225·54 feet per second in order to produce the observed pressure of 14·714 pounds on the square inch. Now a pound moving at that velocity is equivalent to 781°·45 in a pound of water, which will therefore represent the absolute heat of a pound of hydrogen at 60°. But 60° is, as already stated, 519° from absolute zero, whence

$\frac{781 \cdot 45}{519} = 1 \cdot 5157$ will be the heat required to raise the temperature of a pound of hydrogen 1°, taking that which can raise a pound of water 1° as unity; in other words, 1·5157 will be the specific heat of the gas.

Further, since oxygen is sixteen times as heavy in the same space as hydrogen, its particles must move at one quarter the velocity in order to produce the same amount of pressure. Its specific heat will be therefore 0·09473, being, as in the case of all elastic fluids, inversely as the specific gravity.

	According to Theory.	Experiment of De la Roche and Berard reduced to constant volume.
Hydrogen	1·5157	2·3520
Aqueous vapour . .	0·1684	0·6050
Nitrogen	0·1074	0·1953
Oxygen	0·0947	0·1686
Carbonic acid . . .	0·0685	0·1579

Some Remarks on Heat and the Constitution of Elastic Fluids. By J. P. JOULE.

[Memoirs Manchester Lit. & Phil. Soc. vol. ix. p. 107. Read Oct. 3, 1848. Also Phil. Mag. ser. 4. vol. xiv. p. 211.]

IN a paper, "On the Heat evolved during the Electrolysis of Water," published in the 7th volume of the Memoirs of this Society, I stated that the magneto-electrical machine enabled us to convert mechanical power into heat, and that I had little doubt that, by interposing an electro-magnetic engine in the circuit of a voltaic battery, a diminution of the quantity of heat evolved per equivalent of chemical reaction would be observed, and that this diminution would be proportional to the mechanical power obtained.

The results of experiments in proof of the above proposition were communicated to the British Association for the

Advancement of Science in 1843*. They showed that whenever a current of electricity was generated by a magneto-electrical machine, the quantity of heat evolved by that current had a constant relation to the power required to turn the machine; and, on the other hand, that whenever an engine was worked by a voltaic battery, the power developed was at the expense of the calorific power of the battery for a given consumption of zinc, the mechanical effect produced having a fixed relation to the heat lost in the voltaic circuit.

The obvious conclusion from these experiments was, that heat and mechanical power were convertible into one another; and it became therefore evident that heat is either the *vis viva* of ponderable particles, or a state of attraction or repulsion capable of generating *vis viva*.

It now became important to ascertain the mechanical equivalent of heat, with as much accuracy as lay in my power to give it. For this purpose the magnetic apparatus was not very well adapted; and therefore I sought in the heat generated by the friction of fluids for the means of obtaining exact results. I found, first, that the expenditure of a certain amount of mechanical power in the agitation of a fluid uniformly produced a certain fixed quantity of heat; and, second, that the quantity of heat evolved in the friction of fluids was entirely uninfluenced by the nature of the liquid employed, for water, oil, and mercury, fluids as diverse from one another as could have been well selected, gave sensibly the same result, viz. that the quantity of heat capable of raising the temperature of a lb. of water 1° is equal to the mechanical power developed by a weight of 770 lb. in falling through one perpendicular foot†.

Believing that the discovery of the equivalent of heat furnished the means of solving several interesting phenomena, I commenced, in the spring of 1844, some experi-

* Philosophical Magazine, vol. xxiii. pp. 263, 347, 435.

† The equivalent I have since arrived at is 772 foot-pounds. See Phil. Trans. 1850, Part I.—J. P. J., May 1851.

ments on the changes of temperature occasioned by the rarefaction and compression of atmospheric air*. It had long been known that air, when forcibly compressed, evolves heat, and that, on the contrary, when air is dilated, heat is absorbed. In order to account for these facts, it was assumed that a given weight of air has a smaller capacity for heat when compressed into a small compass than when occupying a larger space. A few experiments served to show the incorrectness of this hypothesis: thus I found that by forcing 2956 cubic inches of air, at the ordinary atmospheric pressure, into the space of 136½ cubic inches, 13°·63 of heat per lb. of water were produced; whereas by the reverse process, of allowing the compressed air to expand from a stopcock into the atmosphere, only 4°·09 were absorbed instead of 13°·63, which is the quantity of heat which ought to have been absorbed according to the generally received hypothesis. I found, also, that when strongly compressed air was allowed to escape into a vacuum, no cooling effect took place on the whole, a fact likewise at variance with the received hypothesis. On the contrary, the theory I ventured to advocate† was in perfect agreement with the phenomena; for the heat evolved by compressing the air was found to be the equivalent of the mechanical power employed, and, *vice versâ*, the heat absorbed in rarefaction was found to be the equivalent of the mechanical power developed, estimated by the weight of the column of atmospheric air displaced. In the case of compressed air expanding into a vacuum, since no mechanical power was produced, no absorption of heat was expected or found. M. Seguin has confirmed the above results in the case of steam.

The above principles lead, indeed, to a more intimate acquaintance with the true theory of the steam-engine; for

* Philosophical Magazine, vol. xxvi.

† I subsequently found that M. Mayer had previously advocated a similar hypothesis, without, however, attempting an experimental demonstration of its accuracy ('Annalen' of Wöhler and Liebig for 1842). —J. P. J., May 1851.

they enable us to estimate the calorific effect of the friction of the steam in passing through the various valves and pipes, as well as that of the piston in rubbing against the sides of the cylinder; and they also inform us that the steam, while expanding in the cylinder, loses heat in quantity exactly proportional to the mechanical force developed*.

The experiments on the changes of temperature produced by the rarefaction and condensation of air give likewise an insight into the constitution of elastic fluids; for they show that the heat of elastic fluids is the mechanical force possessed by them; and since it is known that the temperature of a gas determines its elastic force, it follows that the elastic force, or pressure, must be the effect of the motion of the constituent particles in any gas. This motion may exist in several ways, and still account for the phenomena presented by elastic fluids. Davy, to whom belongs the signal merit of having made the first experiment absolutely demonstrative of the immateriality of heat, enunciated the beautiful hypothesis of a rotary motion. He says:—"It seems possible to account for all the phenomena of heat, if it be supposed that in solids the particles are in a constant state of vibratory motion, the particles of the hottest bodies moving with the greatest velocity and through the greatest space; that in fluids and elastic fluids, besides the vibratory motion, which must be considered greatest in the last, the particles have a motion round their own axes with different velocities, the particles of elastic fluids moving with the greatest quickness; and that in ethereal substances the particles move round their own axes, and separate from each other, penetrating in right lines through space. Temperature may be conceived to depend upon the velocity of the vibrations; increase of capacity on the motion being performed in greater

* A complete theory of the motive power of heat has been recently communicated by Professor Thomson to the Royal Society of Edinburgh. In this paper the very important law is established, that the fraction of heat converted into power in any perfect engine is equal to the range of temperature divided by the highest temperature above absolute zero.—J. P. J., May 1851.

space; and the diminution of temperature during the conversion of solids into fluids or gases may be explained on the idea of the loss of vibratory motion, in consequence of the revolution of particles round their axes at the moment when the body becomes fluid or aeriform, or from the loss of rapidity of vibration in consequence of the motion of the particles through greater space"*. I have myself endeavoured to prove that a rotary motion, such as that described by Sir H. Davy, can account for the law of Boyle and Mariotte, and other phenomena presented by elastic fluids†; nevertheless, since the hypothesis of Herapath—in which it is assumed that the particles of a gas are constantly flying about in every direction with great velocity, the pressure of the gas being owing to the impact of the particles against any surface presented to them—is somewhat simpler, I shall employ it in the following remarks on the constitution of elastic fluids, premising, however, that the hypothesis of a rotary motion accords equally well with the phenomena.

Let us suppose an envelope of the size and shape of a cubic foot to be filled with hydrogen gas, which, at 60° temperature and 30 inches barometrical pressure, will weigh 36·927 grs. Further, let us suppose the above quantity to be divided into three equal and indefinitely small elastic particles, each weighing 12·309 grs.; and, further, that each of these particles vibrates between opposite sides of the cube, and maintains a uniform velocity except at the instant of impact; it is required to find the velocity at which each particle must move so as to produce the atmospherical pressure of 14,831,712 grs. on each of the square sides of the cube. In the first place, it is known that if a body moving with the velocity of $32\frac{1}{6}$ feet per second be opposed, during one second, by a pressure equal to its weight, its motion will be stopped, and that, if the pressure be continued

* 'Elements of Chemical Philosophy,' p. 95.

† Mr. Rankine has given a complete mathematical investigation of the action of vortices, in his paper on the Mechanical Action of Gases and Vapours, Trans. R. S. Edin. vol. xx. part 1.—J. P. J., May 1851.

one second longer, the particle will acquire the velocity of $32\frac{1}{6}$ feet per second in the contrary direction. At this velocity there will be $32\frac{1}{6}$ collisions of a particle of 12·309 grs. against each side of the cubical vessel in every two seconds of time; and the pressure occasioned thereby will be $12{\cdot}309 \times 32\frac{1}{6} = 395{\cdot}938$ grs. Therefore, since it is manifest that the pressure will be proportional to the square of the velocity of the particles, we shall have for the velocity of the particles requisite to produce the pressure of 14,831,712 grs. on each side of the cubical vessel,

$$v = \sqrt{\left(\frac{14{,}831{,}712}{395{\cdot}938}\right) 32\tfrac{1}{6}} = 6225 \text{ feet per second.}$$

The above velocity will be found equal to produce the atmospheric pressure, whether the particles strike each other before they arrive at the sides of the cubical vessel, whether they strike the sides obliquely, and, thirdly, into whatever number of particles the 36·927 grs. of hydrogen are divided.

If only one half the weight of hydrogen, or 18·4635 grs., be enclosed in the cubical vessel, and the velocity of the particles be, as before, 6225 feet per second, the pressure will manifestly be only one half of what it was previously; which shows that the law of Boyle and Mariotte flows naturally from the hypothesis.

The velocity above named is that of hydrogen at the temperature of 60°; but we know that the pressure of an elastic fluid at 60° is to that at 32° as 519 is to 491. Therefore the velocity of the particles at 60° will be to that at 32° as $\sqrt{519} : \sqrt{491}$; which shows that the velocity at the freezing temperature of water is 6055 feet per second.

In the above calculations it is supposed that the particles of hydrogen have no sensible magnitude, otherwise the velocity corresponding to the same pressure would be lessened.

Since the pressure of a gas increases with its temperature in arithmetical progression, and since the pressure is proportional to the square of the velocity of the particles, in other words to their *vis viva*, it follows that the absolute temperature, pressure, and *vis viva* are proportional to one another, and that the zero of temperature is 491° below the freezing-point of water. Further, the absolute heat of the gas, or, in other words, its capacity, will be represented by the whole amount of *vis viva* at a given temperature. The specific heat may therefore be determined in the following simple manner :—

The velocity of the particles of hydrogen, at the temperature of 60°, has been stated to be 6225 feet per second, a velocity equivalent to a fall from the perpendicular height of 602,342 feet. The velocity at 61° will be $6225\sqrt{\frac{520}{519}}$ $=6230{\cdot}93$ feet per second, which is equivalent to a fall of 603,502 feet. The difference between the above falls is 1160 feet, which is therefore the space through which 1 lb. of pressure must operate upon each lb. of hydrogen, in order to elevate its temperature one degree. But our mechanical equivalent of heat shows that 770 feet is the altitude representing the force required to raise the temperature of water one degree; consequently the specific heat of hydrogen will be $\frac{1160}{778}=1{\cdot}506$, calling that of water unity.

The specific heats of other gases will be easily deduced from that of hydrogen; for the whole *vis viva* and capacity of equal bulks of the various gases will be equal to one another; and the velocity of the particles will be inversely as the square root of the specific gravity. Hence the specific heat will be inversely proportional to the specific gravity, a law which has been arrived at experimentally by De la Rive and Marcet.

In the following table I have placed the specific heats of various gases, determined in the above manner, in juxtaposition with the experimental results of Delaroche and Berard reduced to constant volume.

	Experimental specific heat.	Theoretical specific heat.
Hydrogen	2·352	1·506
Oxygen	0·168	0·094
Nitrogen	0·195	0·107
Carbonic acid. . . .	0·158	0·068

The experimental results of Delaroche and Berard are invariably higher than those demanded by the hypothesis. But it must be observed that the experiments of Delaroche and Berard, though considered the best that have hitherto been made, differ considerably from those of other philosophers. I believe, however, that the investigation undertaken by M. V. Regnault, for the French Government, will embrace the important subject of the capacity of bodies for heat, and that we may shortly expect a new series of determinations of the specific heat of gases, characterized by all the accuracy for which that distinguished philosopher is so justly famous. Till then, perhaps, it will be better to delay any further modifications of the dynamical theory, by which its deductions may be made to correspond more closely with the results of experiment*.

* If we assume that the particles of a gas are resisted uniformly until their motion is stopped, and that then their motion is renewed in the opposite direction, by the continued operation of the same cause, as in the projection upwards and subsequent fall of a heavy body, the maximum velocity of the particles will be to the uniform velocity required by the theory assumed in the text as the square root of two is to one, and the comparison of the theoretical with the experimental specific heat will be as follows :—

	Experimental specific heat.	Theoretical specific heat.
Hydrogen	2·352	3·012
Oxygen	0·168	0·188
Nitrogen..................	0·195	0·214
Carbonic oxide	0·158	0·136

I have just learned that the experiments of Regnault on the specific heat of elastic fluids are on the eve of publication, and doubt not that their accuracy will enable us to arrive at a decisive conclusion as to the correctness of the above hypothesis.—J. P. J., June 1851.

On the Mechanical Equivalent of Heat. By JAMES PRESCOTT JOULE, *F.C.S., Sec. Lit. and Phil. Society, Manchester, Cor. Mem. R.A., Turin, &c.** (*Communicated by* MICHAEL FARADAY, *D.C.L., F.R.S., Foreign Associate of the Academy of Sciences, Paris, &c. &c. &c.*)

['Philosophical Transactions,' 1850, Part I. Read June 21, 1849.]

(PLATE II.)

"Heat is a very brisk agitation of the insensible parts of the object, which produces in us that sensation from whence we denominate the object hot; so what in our sensation is *heat*, in the object is nothing but *motion*."—LOCKE.

"The *force* of a moving body is proportional to the square of its velocity, or to the height to which it would rise against gravity."—LEIBNITZ.

IN accordance with the pledge I gave the Royal Society some years ago, I have now the honour to present it with the results of the experiments I have made in order to determine the mechanical equivalent of heat with exactness. I will commence with a slight sketch of the progress of the mechanical doctrine, endeavouring to confine myself, for the sake of conciseness, to the notice of such researches as are immediately connected with the subject. I shall not therefore be able to review the valuable labours of Mr. Forbes and other illustrious men, whose researches on radiant heat and other subjects do not come exactly within the scope of the present memoir.

For a long time it had been a favourite hypothesis that heat consists of "a force or power belonging to bodies"†, but it was reserved for Count Rumford to make the first experiments decidedly in favour of that view. That justly

* The experiments were made at Oak Field, Whalley Range, near Manchester.

† Crawford on Animal Heat, p. 15.

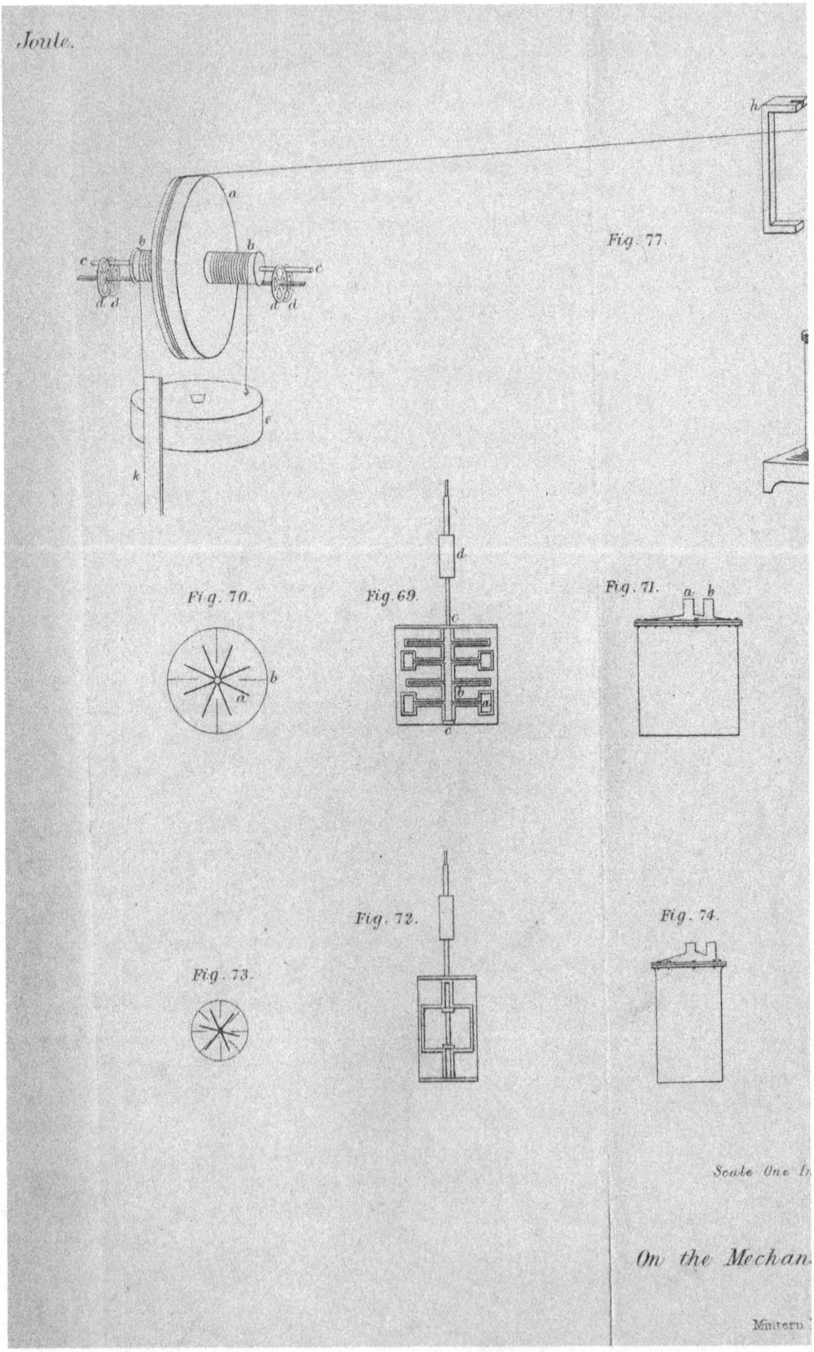

Joule.

Fig. 77.

Fig. 70.

Fig. 69.

Fig. 71.

Fig. 72.

Fig. 73.

Fig. 74.

Scale One In

On the Mechan

Mintern

Plate II.

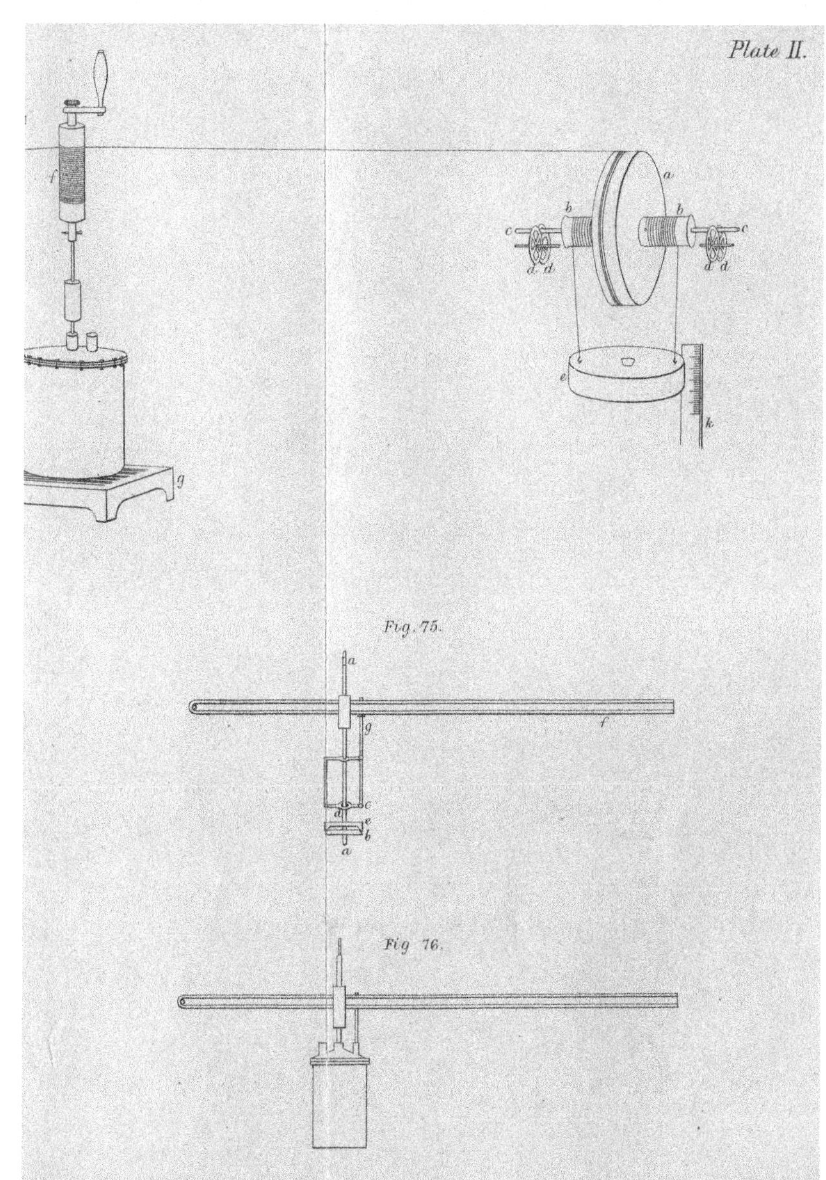

ich to a foot.

ical Equivalent of Heat.

Bros. lith.

celebrated natural philosopher demonstrated by his ingenious experiments that the very great quantity of heat excited by the boring of cannon could not be ascribed to a change taking place in the calorific capacity of the metal; and he therefore concluded that the motion of the borer was communicated to the particles of metal, thus producing the phenomena of heat. "It appears to me," he remarks, "extremely difficult, if not quite impossible, to form any distinct idea of anything capable of being excited and communicated in the manner the heat was excited and communicated in these experiments, except it be motion" *.

One of the most important parts of Count Rumford's paper, though one to which little attention has hitherto been paid, is that in which he makes an estimate of the quantity of mechanical force required to produce a certain amount of heat. Referring to his third experiment, he remarks that the "total quantity of ice-cold water which, with the heat actually generated by friction, and accumulated in $2^h\ 30^m$, might have been heated 180°, or made to boil, =26·58 lb."† In the next page he states that "the machinery used in the experiment could easily be carried round by the force of one horse (though, to render the work lighter, two horses were actually employed in doing it)." Now the power of a horse is estimated by Watt at 33,000 foot-pounds per minute, and therefore if continued for two hours and a half will amount to 4,950,000 foot-pounds, which, according to Count Rumford's experiment, will be equivalent to 26·58 lb. of water raised 180°. Hence the heat required to raise a lb. of water 1° will be equivalent to the force represented by 1034 foot-pounds. This result is not very widely different from that which I have deduced from my own experiments related in this paper, viz. 772 foot-pounds; and it must be observed that the excess of Count Rumford's equivalent is just such as might have been anticipated from the circumstance, which he himself mentions, that "no estimate was made of the heat

* "An Inquiry concerning the Source of Heat which is excited by Friction," Phil. Trans., Abridged, vol. xviii. p. 286.

† Ibid. vol. xviii. p. 283.

accumulated in the wooden box, nor of that dispersed during the experiment."

About the end of the last century Sir Humphry Davy communicated a paper to Dr. Beddoes's West Country Contributions, entitled "Researches on Heat, Light, and Respiration," in which he gave ample confirmation to the views of Count Rumford. By rubbing two pieces of ice against one another in the vacuum of an air-pump, part of them was melted, although the temperature of the receiver was kept below the freezing-point. This experiment was the more decisively in favour of the doctrine of the immateriality of heat, inasmuch as the capacity of ice for heat is much less than that of water. It was therefore with good reason that Davy drew the inference that "the immediate cause of the phenomena of heat is motion, and the laws of its communication are precisely the same as the laws of the communication of motion"*.

The researches of Dulong on the specific heat of elastic fluids were rewarded by the discovery of the remarkable fact that "equal volumes of all the elastic fluids, taken at the same temperature and under the same pressure, being compressed or dilated suddenly to the same fraction of their volume, disengage or absorb the same *absolute quantity of heat*"†. This law is of the utmost importance in the development of the theory of heat, inasmuch as it proves that the calorific effect is, under certain conditions, proportional to the force expended.

In 1834 Dr. Faraday demonstrated the "Identity of the Chemical and Electrical Forces." This law, along with others subsequently discovered by that great man, showing the relations which subsist between magnetism, electricity, and light, have enabled him to advance the idea that the so-called imponderable bodies are merely the exponents of different forms of Force. Mr. Grove and M. Mayer have also given their powerful advocacy to similar views.

My own experiments in reference to the subject were

* Elements of Chemical Philosophy, p. 94.

† Mémoires de l'Académie des Sciences, t. x. p. 188.

commenced in 1840, in which year I communicated to the Royal Society my discovery of the law of the heat evolved by voltaic electricity, a law from which the immediate deductions were drawn,—1st, that the heat evolved by any voltaic pair is proportional, *cæteris paribus*, to its intensity or electromotive force*; and 2nd, that the heat evolved by the combustion of a body is proportional to the intensity of its affinity for oxygen†. I thus succeeded in establishing relations between heat and chemical affinity. In 1843 I showed that the heat evolved by magneto-electricity is proportional to the force absorbed, and that the force of the electro-magnetic engine is derived from the force of chemical affinity in the battery, a force which otherwise would be evolved in the form of heat. From these facts I considered myself justified in announcing "that the quantity of heat capable of increasing the temperature of a lb. of water by one degree of Fahrenheit's scale is equal to, and may be converted into, a mechanical force capable of raising 838 lb. to the perpendicular height of one foot"‡.

In a subsequent paper, read before the Royal Society in 1844, I endeavoured to show that the heat absorbed and evolved by the rarefaction and condensation of air is proportional to the force evolved and absorbed in those operations§. The quantitative relation between force and heat deduced from these experiments is almost identical with that derived from the electro-magnetic experiments just referred to, and is confirmed by the experiments of M. Seguin on the dilatation of steam ||.

From the explanation given by Count Rumford of the heat arising from the friction of solids, one might have anticipated, as a matter of course, that the evolution of heat would also be detected in the friction of liquid and gaseous bodies. Moreover there were many facts, such as, for instance, the warmth of the sea after a few days of stormy weather, which had long been commonly attributed to fluid friction. Nevertheless the

* Phil. Mag. ser. 3. vol. xix. p. 275.
† Ibid. vol. xx. p. 111.
‡ Ibid. vol. xxiii. p. 441.
§ Ibid. vol. xxvi. pp. 375, 379.
|| Comptes Rendus, t. xxv. p. 421.

scientific world, preoccupied with the hypothesis that heat is a substance, and following the deductions drawn by Pictet from experiments not sufficiently delicate, have almost unanimously denied the possibility of generating heat in that way. The first mention, so far as I am aware, of experiments in which the evolution of heat from fluid friction is asserted was in 1842 by M. Mayer*, who states that he has raised the temperature of water from 12° C. to 13° C. by agitating it, without, however, indicating the quantity of force employed, or the precautions taken to secure a correct result. In 1843 I announced the fact that "heat is evolved by the passage of water through narrow tubes"†, and that each degree of heat per lb. of water required for its evolution in this way a mechanical force represented by 770 foot-pounds. Subsequently, in 1845‡ and 1847§, I employed a paddle-wheel to produce the fluid friction, and obtained the equivalents 781·5, 782·1, and 787·6 respectively from the agitation of water, sperm-oil, and mercury. Results so closely coinciding with one another, and with those previously derived from experiments with elastic fluids and the electro-magnetic machine, left no doubt on my mind as to the existence of an equivalent relation between force and heat; but still it appeared of the highest importance to obtain that relation with still greater accuracy. This I have attempted in the present paper.

Description of Apparatus.—The thermometers employed had their tubes calibrated and graduated according to the method first indicated by M. Regnault. Two of them, which I shall designate by A and B, were constructed by Mr. Dancer of Manchester; the third, designated by C, was made by M. Fastré of Paris. The graduation of these instruments was so correct, that when compared together their indications coincided to about $\frac{1}{100}$ of a degree Fahr. I also possessed

* 'Annalen' of Wöhler and Liebig, May 1842.

† Phil. Mag. ser. 3. vol. xxiii. p. 442. ‡ Ibid. vol. xxvii. p. 205.

§ Ibid. vol. xxxi. p. 173, and Comptes Rendus, tome xxv. p. 309.

another exact instrument made by Mr. Dancer, the scale of which embraced both the freezing- and boiling-points. The latter point in this standard thermometer was obtained, in the usual manner, by immersing the bulb and stem in the steam arising from a considerable quantity of pure water in rapid ebullition. During the trial the barometer stood at 29·94 inches, and the temperature of the air was 50°; so that the observed point required very little correction to reduce it to 0·760 metre and 0° C., the pressure used in France, and I believe the Continent generally, for determining the boiling-point, and which has been employed by me on account of the number of accurate thermometrical researches which have been constructed on that basis*. The values of the scales of thermometers A and B were ascertained by plunging them along with the standard in large volumes of water kept constantly at various temperatures. The value of the scale of thermometer C was determined by comparison with A. It was thus found that the number of divisions corresponding to 1° Fahr. in the thermometers A, B, and C were 12·951, 9·829, and 11·647 respectively. And since constant practice had enabled me to read off with the naked eye to $\frac{1}{20}$ of a division, it followed that $\frac{1}{200}$ of a degree Fahr. was an appreciable temperature.

Pl. II. fig. 69 represents a vertical and fig. 70 a horizontal plan of the apparatus employed for producing the friction of water, consisting of a brass paddle-wheel furnished with eight sets of revolving arms, *a, a,* &c., working between four sets of stationary vanes, *b, b,* &c., affixed to a framework also in sheet brass. The brass axis of the paddle-wheel worked freely, but without shaking, on its bearings at *c c,* and at *d*

* A barometrical pressure of 30 inches of mercury at 60° is very generally employed in this country, and fortunately agrees almost exactly with the continental standard. In the "Report of the Committee appointed by the Royal Society to consider the best method of adjusting the Fixed Points of Thermometers" (Philosophical Transactions, Abridged, xiv. p. 258) the barometrical pressure 29·8 is recommended, but the temperature is not named—a remarkable omission in a work so exact in other respects.

was divided into two parts by a piece of boxwood intervening, so as to prevent the conduction of heat in that direction.

Pl. II. fig. 71 represents the copper vessel into which the revolving apparatus was firmly fitted: it had a copper lid, the flange of which, furnished with a very thin washer of leather saturated with white-lead, could be screwed perfectly water-tight to the flange of the copper vessel. In the lid there were two necks, *a*, *b*, the former for the axis to revolve in without touching, the latter for the insertion of the thermometer.

Besides the above I had a similar apparatus for experiments on the friction of mercury, which is represented by Pl. II. figs. 72, 73, and 74. It differed from the apparatus already described in its size, number of vanes (of which six were rotary and eight sets stationary), and material, which was wrought iron in the paddle-wheel, and cast iron in the vessel and lid.

Being anxious to extend my experiments to the friction of solids, I also procured the apparatus represented by Pl. II. fig. 75, in which *a a* is the axis revolving along with the bevelled cast-iron wheel *b*, the rim of which was turned true. By means of the lever *c*, which had a ring in its centre for the axis to pass through, and two short arms *d*, the bevel-turned cast-iron wheel *e* could be pressed against the revolving wheel, the degree of force applied being regulated by hand by means of the wooden lever *f* attached to the perpendicular iron rod *g*. Fig. 76 represents the apparatus in its cast-iron vessel.

Pl. II. fig. 77 is a perspective view of the machinery employed to set the frictional apparatus just described in motion. *a a* are wooden pulleys, 1 foot in diameter and 2 inches thick, having wooden rollers *b b*, *b b*, 2 inches in diameter, and steel axles *c c*, *c c*, one quarter of an inch in diameter. The pulleys were turned perfectly true and equal to one another. Their axles were supported by brass friction-wheels *d d d d*, *d d d d*, the steel axles of which worked in holes drilled into brass plates attached to a very strong wooden framework firmly fixed into the walls of the apartment*.

The leaden weights *e, e,* which in some of the ensuing experiments weighed about 29 lb., and in others about 10 lb. a piece, were suspended by string from the rollers *b b, b b*; and fine twine attached to the pulleys *a a* connected them with the central roller *f*, which, by means of a pin, could with facility be attached to, or removed from, the axis of the frictional apparatus.

The wooden stool *g*, upon which the frictional apparatus stood, was perforated by a number of transverse slits, so cut out that only a very few points of wood came in contact with the metal, whilst the air had free access to almost every part of it. In this way the conduction of heat to the substance of the stool was avoided.

A large wooden screen (not represented in the figure) completely obviated the effects of radiant heat from the person of the experimenter.

The method of experimenting was simply as follows:— The temperature of the frictional apparatus having been ascertained and the weights wound up with the assistance of the stand *h*, the roller was refixed to the axis. The precise height of the weights above the ground having then been determined by means of the graduated slips of wood *k, k*, the roller was set at liberty and allowed to revolve until the weights reached the flagged floor of the laboratory, after accomplishing a fall of about 63 inches. The roller was then removed to the stand, the weights wound up again, and the friction renewed. After this had been repeated twenty times, the experiment was concluded with another observation of the temperature of the apparatus. The mean temperature of the laboratory was determined by observations made at the commencement, middle, and termination of each experiment.

Previously to, or immediately after, each of the experiments

* This was a spacious cellar, which had the advantage of possessing a uniformity of temperature far superior to that of any other laboratory I could have used.

I made trial of the effect of radiation and conduction of heat to or from the atmosphere in depressing or raising the temperature of the frictional apparatus. In these trials the position of the apparatus, the quantity of water contained by it, the time occupied, the method of observing the thermometers, the position of the experimenter, in short every thing, with the exception of the apparatus being at rest, was the same as in the experiments in which the effect of friction was observed.

1*st Series of Experiments.*—Friction of Water. Weight of the leaden weights along with as much of the string in connexion with them as served to increase the pressure, 203066 grs. and 203086 grs. Velocity of the weights in descending, 2·42 inches per second. Time occupied by each experiment, 35 minutes. Thermometer employed for ascertaining the temperature of the water, A. Thermometer for registering the temperature of the air, B.

TABLE I.

No. of experiment and cause of change of temperature.	Total fall of weights in inches.	Mean temperature of air.	Difference between mean of columns 5 and 6 and column 3.	Temperature of apparatus.		Gain or loss of heat during experiment.
				Commencement of experiment.	Termination of experiment.	
1 Friction ...	1256·96	57°·698	2°·252—	55°·118	55°·774	0°·656 gain
1 Radiation...	0	57·868	2·040—	55·774	55·882	0·108 gain
2 Friction ...	1255·16	58·085	1·875—	55·882	56·539	0·657 gain
2 Radiation...	0	58·370	1·789—	56·539	56·624	0·085 gain
3 Friction ...	1253·66	60·788	1·596—	58·870	59·515	0·645 gain
3 Radiation...	0	60·926	1·373—	59·515	59·592	0·077 gain
4 Friction ...	1252·74	61·001	1·110—	59·592	60·191	0·599 gain
4 Radiation...	0	60·890	0·684—	60·191	60·222	0·031 gain
5 Friction ...	1251·81	60·940	0·431—	60·222	60·797	0·575 gain
5 Radiation...	0	61·035	0·237—	60·797	60·799	0·002 gain
6 Radiation...	0	59·675	0·125+	59·805	59·795	0·010 loss
6 Friction ...	1254·71	59·919	0·157+	59·795	60·357	0·562 gain
1	2	3	4	5	6	7

TABLE I. (*continued*).

No. of experiment and cause of change of temperature.	Total fall of weights in inches.	Mean temperature of air.	Difference between mean of columns 5 and 6 and column 3.	Temperature of apparatus.		Gain or loss of heat during experiment.
				Commencement of experiment.	Termination of experiment.	
7 Radiation...	0	59°·888	0°·209–	59°·677	59°·681	0°·004 gain
7 Friction ...	1254·02	60·076	0·111–	59·681	60·249	0·568 gain
8 Radiation...	0	58·240	0·609+	58·871	58·828	0·043 loss
8 Friction ...	1251·22	58·237	0·842+	58·828	59·330	0·502 gain
9 Friction ...	1253·92	55·328	0·070+	55·118	55·678	0·560 gain
9 Radiation...	0	55·528	0·148+	55·678	55·674	0·004 loss
10 Radiation...	0	54·941	0·324–	54·614	54·620	0·006 gain
10 Friction ...	1257·96	54·985	0·085–	54·620	55·180	0·560 gain
11 Radiation...	0	55·111	0·069+	55·180	55·180	0·000
11 Friction ...	1258·59	55·229	0·227+	55·180	55·733	0·553 gain
12 Friction ...	1258·71	55·433	0·238+	55·388	55·954	0·566 gain
12 Radiation...	0	55·687	0·265+	55·954	55·950	0·004 loss
13 Friction ...	1257·91	55·677	0·542+	55·950	56·488	0·538 gain
13 Radiation...	0	55·674	0·800+	56·488	56·461	0·027 loss
14 Radiation...	0	55·579	0·583–	54·987	55·006	0·019 gain
14 Friction ...	1259·69	55·864	0·568–	55·006	55·587	0·581 gain
15 Radiation...	0	56·047	0·448–	55·587	55·612	0·025 gain
15 Friction ...	1259·89	56·182	0·279–	55·612	56·195	0·583 gain
16 Friction ...	1259·64	55·368	0·099+	55·195	55·739	0·544 gain
16 Radiation...	0	55·483	0·250+	55·739	55·728	0·011 loss
17 Friction ...	1259·64	55·498	0·499+	55·728	56·266	0·538 gain
17 Radiation...	0	55·541	0·709+	56·266	56·235	0·031 loss
18 Radiation...	0	56·769	1·512–	55·230	55·284	0·054 gain
18 Friction ...	1260·17	56·966	1·372–	55·284	55·905	0·621 gain
19 Radiation...	0	60·058	1·763–	58·257	58·334	0·077 gain
19 Friction ...	1262·24	60·112	1·450–	58·334	58·990	0·656 gain
20 Radiation...	0	60·567	1·542–	58·990	59·060	0·070 gain
20 Friction ...	1261·94	60·611	1·239–	59·060	59·685	0·625 gain
21 Friction ...	1264·07	58·654	0·321–	58·050	58·616	0·566 gain
21 Radiation...	0	58·627	0·018–	58·616	58·603	0·013 loss
22 Friction ...	1262·97	58·631	0·243+	58·603	59·145	0·542 gain
22 Radiation...	0	58·624	0·505+	59·145	59·114	0·031 loss
1	2	3	4	5	6	7

TABLE I. (*continued*).

No. of experiment and cause of change of temperature.	Total fall of weights in inches.	Mean temperature of air.	Difference between mean of columns 5 and 6 and column 3.	Temperature of apparatus.		Gain or loss of heat during experiment.
				Commencement of experiment.	Termination of experiment.	
23 Friction ...	1264·72	59°·689	1°·100 −	58°·284	58°·894	0°·610 gain
23 Radiation...	0	59·943	1·027 −	58·894	58·938	0·044 gain
24 Radiation...	0	60·157	1·160 −	58·977	59·017	0·040 gain
24 Friction ...	1263·94	59·811	0·505 −	59·017	59·595	0·578 gain
25 Radiation...	0	59·654	0·061 −	59·595	59·591	0·004 loss
25 Friction ...	1263·49	59·675	0·185 +	59·591	60·129	0·538 gain
26 Radiation...	0	59·156	0·609 −	58·541	58·554	0·013 gain
26 Friction ...	1263·49	59·333	0·488 −	58·554	59·137	0·583 gain
27 Friction ...	1263·99	59·536	0·198 −	59·054	59·623	0·569 gain
27 Radiation ..	0	59·726	0·101 −	59·623	59·627	0·004 gain
28 Friction ...	1263·99	59·750	0·155 +	59·627	60·183	0·556 gain
28 Radiation...	0	59·475	0·102 +	59·585	59·569	0·016 loss
29 Friction ...	1263·31	58·695	0·182 −	58·230	58·796	0·566 gain
29 Radiation...	0	58·906	0·108 −	58·796	58·801	0·005 gain
30 Radiation...	0	59·770	1·286 −	58·454	58·515	0·061 gain
30 Friction ...	1263·99	60·048	1·223 −	58·515	59·135	0·620 gain
31 Friction ...	1263·49	59·343	0·022 +	59·091	59·639	0·548 gain
31 Radiation...	0	59·435	0·198 +	59·639	59·627	0·012 loss
32 Radiation...	0	59·374	0·357 −	59·015	59·020	0·005 gain
32 Friction ...	1263·49	59·407	0·105 −	59·020	59·585	0·565 gain
33 Radiation...	0	59·069	0·201 −	58·867	58·870	0·003 gain
33 Friction ...	1263·49	59·234	0·081 −	58·870	59·436	0·566 gain
34 Friction ...	1262·99	56·328	0·331 +	56·387	56·932	0·545 gain
34 Radiation...	0	56·643	0·287 +	56·932	56·929	0·003 loss
35 Friction ...	1262·99	56·790	0·413 +	56·929	57·477	0·548 gain
35 Radiation...	0	56·772	0·687 +	57·477	57·442	0·035 loss
36 Radiation...	0	55·839	0·304 −	55·527	55·543	0·016 gain
36 Friction ...	1262·99	56·114	0·281 −	55·543	56·124	0·581 gain
37 Radiation...	0	56·257	0·127 −	56·124	56·137	0·013 gain
37 Friction ...	1262·99	56·399	0·024 +	56·137	56·709	0·572 gain
38 Radiation...	0	55·826	0·065 −	55·759	55·764	0·005 gain
38 Friction ...	1262·99	55·951	0·093 +	55·764	56·325	0·561 gain
1	2	3	4	5	6	7

TABLE I. (*continued*).

No. of experiment and cause of change of temperature.	Total fall of weights in inches.	Mean temperature of air.	Difference between mean of columns 5 and 6 and column 3.	Temperature of apparatus.		Gain or loss of heat during experiment.
				Commencement of experiment.	Termination of experiment.	
39 Radiation...	0	56°·101	0°·220 +	56°·325	56°·317	0°·008 loss
39 Friction ...	1262·99	56·182	0·409 +	56·317	56·865	0·548 gain
40 Friction ...	1262·99	56·108	0·100 +	55·929	56·488	0·559 gain
40 Radiation...	0	56·454	0·036 +	56·488	56·492	0·004 gain
Mean Friction .	1260·248	......	0·305075 −	......	......	0·575250 gain
Mean Radiation	0	......	0·322950 −	......	......	0·012975 gain
1	2	3	4	5	6	7

From the various experiments in the above Table in which the effect of radiation was observed, it may be readily gathered that the effect of the temperature of the surrounding air upon the apparatus was, for each degree of difference between the mean temperature of the air and that of the apparatus, 0°·04654. Therefore, since the excess of the temperature of the atmosphere over that of the apparatus was 0°·32295 in the mean of the radiation experiments, but only 0°·305075 in the mean of the friction experiments, it follows that 0°·000832 must be added to the difference between 0°·57525 and 0°·012975, and the result, 0°·563107, will be the proximate heating effect of the friction. But to this quantity a small correction must be applied on account of the mean of the temperatures of the apparatus at the commencement and termination of each friction experiment having been taken for the true mean temperature, which was not strictly the case, owing to the somewhat less rapid increase of temperature towards the termination of the experiment when the water had become warmer. The mean temperature of the apparatus in the friction experiments ought therefore to be estimated 0°·002184 higher, which will diminish the heating effect of the atmosphere by 0°·000102. This, added to

0°·563107, gives 0°·563209 as the true mean increase of temperature due to the friction of water*.

In order to ascertain the absolute quantity of heat evolved, it was necessary to find the capacity for heat of the copper vessel and brass paddle-wheel. That of the former was easily deduced from the specific heat of copper according to M. Regnault. Thus, capacity of 25541 grs.† of copper ×0·09515=capacity of 2430·2 grs. of water. A series of seven very careful experiments with the brass paddle-wheel gave me 1783 grs. of water as its capacity, after making all the requisite corrections for the heat occasioned by the contact of the water with the surface of the metal, &c. But on account of the magnitude of these corrections, amounting to one thirtieth of the whole capacity, I prefer to avail myself of M. Regnault's law, viz. *that the capacity in metallic alloys is equal to the sum of the capacities of their constituent metals*‡. Analysis of a part of the wheel proved it to consist of a very pure brass containing 3933 grs. of zinc to 14968 grs. of copper. Hence

* This increase of temperature was, it is necessary to observe, a mixed quantity, depending partly upon the friction of the water, and partly upon the friction of the vertical axis of the apparatus upon its pivot and bearing, *c c*, Pl. II. fig. 69. The latter source of heat was, however, only equal to about $\frac{1}{80}$ of the former. Similarly also, in the experiments on the friction of solids hereafter detailed, the cast-iron disks revolving in mercury rendered it impossible to avoid a very small degree of friction among the particles of that fluid. But since it was found that the quantity of heat evolved was the same, for the same quantity of force expended, in both cases, *i. e.* whether a minute quantity of heat arising from friction of solids was mixed with the heat arising from the friction of a fluid, or whether, on the other hand, a minute quantity of heat arising from the friction of a fluid was mingled with the heat developed by the friction of solids, I thought there could be no impropriety in regarding the heat as if developed from a simple source—in the one case entirely from the friction of a fluid, and in the other entirely from the friction of a solid body.

† The washer, weighing only 38 grs., was reckoned as copper in this estimate.

‡ Ann. de Ch. 1841, t. i.

Cap. 14968 grs. copper $\times$ 0·09515 = cap. 1424·2 grs. water.
Cap. 3933 grs. zinc $\times$ 0·09555 = cap. 375·8 grs. water.

Total cap. brass wheel = cap. 1800 grs. water.

The capacity of a brass stopper which was placed in the neck *b*, Pl. II. fig. 71, for the purpose of preventing the contact of air with the water as much as possible, was equal to that of 10·3 grs. of water; the capacity of the thermometer had not to be estimated, because it was always brought to the expected temperature before immersion. The entire capacity of the apparatus was therefore as follows:—

Water	93229·7
Copper as water . . .	2430·2
Brass as water . . .	1810·3
Total . .	97470·2

So that the total quantity of heat evolved was 0°·563209 in 97470·2 grs. of water, or, in other words, 1° Fahr. in 7·842299 lb. of water.

The estimate of the force applied in generating this heat may be made as follows:—The weights amounted to 406152 grs., from which must be subtracted the friction arising from the pulleys and the rigidity of the string; which was found by connecting the two pulleys with twine passing round a roller of equal diameter to that employed in the experiments. Under these circumstances, the weight required to be added to one of the leaden weights in order to maintain them in equable motion was found to be 2955 grs. The same result, in the opposite direction, was obtained by adding 3035 grs. to the other leaden weight. Deducting 168 grs., the friction of the roller on its pivots, from 3005, the mean of the above numbers, we have 2837 grs. as the amount of friction in the experiments, which, subtracted from the leaden weights, leaves 403315 grs. as the actual pressure applied.

The velocity with which the leaden weights came to the ground, viz. 2·42 inches per second, is equivalent to an altitude of 0·0076 inch. This, multiplied by 20, the number of times the weights were wound up in each experiment, produces 0·152 inch, which, subtracted from 1260·248, leaves 1260·096 as the corrected mean height from which the weights fell.

This fall, accompanied by the above-mentioned pressure, represents a force equivalent to 6050·186 lb. through one foot; and $0{\cdot}8464 \times 20 = 16{\cdot}928$ foot-lb. added to it, for the force developed by the elasticity of the string after the weights had touched the ground, gives 6067·114 foot-pounds as the mean corrected force.

Hence $\frac{6067{\cdot}114}{7{\cdot}842299} = 773{\cdot}64$ foot-pounds will be the force which, according to the above experiments on the friction of water, is equivalent to 1° Fahr. in a lb. of water.

2nd Series of Experiments.—Friction of Mercury. Weight of the leaden weights and string, 203026 grs. and 203073 grs. Velocity of the weights in descending, 2·43 inches per second. Time occupied by each experiment, 30 minutes. Thermometer for ascertaining the temperature of the mercury, C. Thermometer for registering the temperature of the air, B. Weight of cast-iron apparatus, 68446 grs. Weight of mercury contained by it, 428292 grs.

TABLE II.

No. of experiment and cause of change of temperature.	Total fall of weights in inches.	Mean temperature of air.	Difference between mean of columns 5 and 6 and column 3.	Temperature of apparatus.		Gain or loss of heat during experiment.
				Commencement of experiment.	Termination of experiment.	
1 Friction ...	1265·42	58°·491	1°·452+	58°·780	61°·107	2°·327 gain
1 Radiation...	0	58·939	2·056+	61·107	60·884	0·223 loss
2 Radiation...	0	58·390	0·237−	58·119	58·188	0·069 gain
2 Friction ...	1265·77	58·949	0·467+	58·188	60·644	2·456 gain
3 Friction ...	1265·73	57·322	1·203+	57·325	59·725	2·400 gain
3 Radiation...	0	57·942	1·678+	59·725	59·515	0·210 loss
4 Radiation...	0	57·545	0·010−	57·518	57·553	0·035 gain
4 Friction ...	1264·72	58·135	0·624+	57·553	59·965	2·412 gain
5 Friction ...	1265·73	57·021	0·907+	56·715	59·141	2·426 gain
5 Radiation...	0	57·596	1·474+	59·141	58·999	0·142 loss
1	2	3	4	5	6	7

TABLE II. (*continued*).

No. of experiment and cause of change of temperature.	Total fall of weights in inches.	Mean temperature of air.	Difference between mean of columns 5 and 6 and column 3.	Temperature of apparatus.		Gain or loss of heat during experiment.
				Commencement of experiment.	Termination of experiment.	
6 Radiation...	0	56°·406	0°·174+	56°·565	56°·595	0°·030 gain
6 Friction ...	1265·65	57·057	0·749+	56·595	59·017	2·422 gain
7 Friction ...	1269·55	58·319	0·049+	57·115	59·622	2·507 gain
7 Radiation...	0	58·771	0·831+	59·622	59·583	0·039 loss
8 Radiation...	0	60·363	0·612−	59·691	59·811	0·120 gain
8 Friction ...	1257·70	60·842	0·209+	59·811	62·292	2·481 gain
9 Friction ...	1255·77	60·282	1·044+	60·129	62·524	2·395 gain
9 Radiation...	0	60·862	1·576+	62·524	62·352	0·172 loss
10 Friction ...	1255·33	60·725	0·764+	60·266	62·713	2·447 gain
10 Radiation...	0	61·340	1·313+	62·713	62·593	0·120 loss
11 Radiation...	0	58·654	0·109+	58·755	58·772	0·017 gain
11 Friction ...	1266·47	59·234	0·746+	58·772	61·189	2·417 gain
12 Radiation...	0	56·436	0·247+	56·673	56·694	0·021 gain
12 Friction ...	1265·80	57·240	0·673+	56·694	59·133	2·439 gain
13 Friction ...	1264·70	55·002	1·808+	55·638	57·982	2·344 gain
13 Radiation...	0	55·633	2·213+	57·982	57·711	0·271 loss
14 Friction ...	1265·20	54·219	1·273+	54·290	56·694	2·404 gain
14 Radiation...	0	54·595	1·972+	56·694	56·441	0·253 loss
15 Radiation...	0	53·476	0·174+	53·633	53·667	0·034 gain
15 Friction ...	1265·63	53·995	0·872+	53·667	56·067	2·400 gain
16 Radiation...	0	52·082	0·254+	52·332	52·341	0·009 gain
16 Friction ...	1265·45	52·479	1·047+	52·341	54·711	2·370 gain
17 Friction ...	1257·50	50·485	1·453+	50·772	53·105	2·333 gain
17 Radiation...	0	50·821	2·164+	53·105	52·865	0·240 loss
18 Radiation...	0	48·944	0·450−	48·434	48·554	0·120 gain
18 Friction ...	1257·50	49·330	0·462+	48·554	51·031	2·477 gain
19 Friction ...	1257·50	48·135	1·273+	48·219	50·598	2·379 gain
19 Radiation...	0	48·725	1·780+	50·598	50·413	0·185 loss
20 Radiation...	0	48·878	0·148−	48·687	48·773	0·086 gain
20 Friction ...	1257·50	49·397	0·597+	48·773	51·216	2·443 gain
Mean Friction .	1262·731	......	0·8836+	......	......	2·41395 gain
Mean Radiation	0	......	0·8279+	......	......	0·06570 loss
1	2	3	4	5	6	7

From the above Table it appears that the effect of each degree of difference between the temperature of the laboratory and that of the apparatus was 0°·13742. Hence 2°·41395 + 0°·0657 + 0°·007654 = 2°·487304 will be the proximate value of the increase of temperature in the experiments. The further correction on account of the mean temperature of the apparatus in the friction experiments having been in reality 0°·028484 higher than is indicated by the table, will be 0°·003914, which, added to the proximate result, gives 2°·491218 as the true thermometrical effect of the friction of the mercury.

In order to obtain the absolute quantity of heat evolved, it was requisite to ascertain the capacity for heat of the apparatus. I therefore caused it to be suspended by iron wire from a lever so contrived that the apparatus could be moved with rapidity and ease to any required position. The temperature of the apparatus having then been raised about 20°, it was placed in a warm air-bath, in order to keep its temperature uniform for a quarter of an hour, during which time the thermometer C, immersed in the mercury, was from time to time observed. The apparatus was then rapidly immersed in a thin copper vessel containing 141826 grs. of distilled water, the temperature of which was repeatedly observed by thermometer A. During the experiment the water was repeatedly agitated by a copper stirrer; and every precaution was taken to keep the surrounding atmosphere in a uniform state, and also to prevent the disturbing effects of radiation from the person of the experimenter. In this way I obtained the following results :—

	Time of observation.	Temperature of water.	Temperature of apparatus.
	min.	°	°
Apparatus in air-bath	0	47·705	70·518
	5	47·705	70·492
	10	47·713	70·518
Instant of immersion...	11		
Apparatus immersed in water	$13\frac{1}{2}$	49·836	57·673
	16	50·493	52·641
	21	50·694	50·941
	26	50·690	50·778
	31	50·667	50·744
	36	50·636	50·709

By applying the correction to the temperature of the water due to its observed increase during the first ten minutes of the experiment, and the still smaller correction due to the rise of the water in the can covering 60 square inches of copper at the temperature of the atmosphere, 47°·714 was found to be the temperature of the water at the instant of immersion. To remove the apparatus from the warm air-bath and to immerse it into the water occupied only 10 seconds, during which it must (according to preliminary experiments) have cooled 0°·027. The heating effect of the air-bath during the remaining 50 seconds (estimated from the rate of increase of temperature between the observations at 5 min. and 10 min.) will be 0°·004. These corrections, applied to 70°·518, leave 70°·495 as the temperature of the apparatus at the moment of immersion.

The temperature of the apparatus at 26 min. was 50°·778, indicating a loss of 19°·717. That of the water at the same time of observation, being corrected for the effect of the atmosphere (deduced from the observations of the cooling from 26 min. to 36 min. and of the heating from 0 min. to 10 min.), will be 50°·777, indicating a gain of 3°·063. Twenty such results, obtained in exactly the same manner, are collected in the following Table:—

TABLE III.

No.	Corrected temperature of water.		Gain of heat by the water.	Corrected temperature of apparatus.		Loss of heat by the apparatus.
	Commencement of experiment.	Termination of experiment.		Commencement of experiment.	Termination of experiment.	
1.	47°·714	50°·777	3°·063	70°·495	50°·778	19°·717
2.	48·127	51·113	2·986	70·518	51·147	19·371
3.	48·453	51·430	2·977	70·642	51·452	19·190
4.	47·543	50·598	3·055	70·674	50·684	19·990
5.	44·981	48·449	3·468	70·901	48·468	22·433
6.	45·289	48·701	3·412	70·769	48·657	22·112
7.	45·087	48·497	3·410	70·504	48·494	22·010
8.	46·375	49·614	3·239	70·678	49·662	21·016
9.	47·671	50·832	3·161	71·500	50·873	20·627
10.	47·693	50·801	3·108	70·878	50·821	20·057
11.	48·728	51·714	2·986	70·947	51·714	19·233
12.	47·240	50·414	3·174	71·006	50·392	20·614
13.	48·324	51·345	3·021	70·939	51·362	19·577
14.	49·079	51·905	2·826	70·332	51·937	18·395
15.	49·635	52·490	2·855	71·012	52·504	18·508
16.	47·207	50·282	3·075	70·265	50·263	20·002
17.	46·227	49·402	3·175	69·877	49·314	20·563
18.	46·053	49·296	3·243	70·367	49·258	21·109
19.	45·733	48·981	3·248	70·068	49·001	21·067
20.	47·170	50·317	3·147	70·741	50·332	20·409
Mean	......	......	3·13145	.....	......	20·300

I did not consider these experiments on the capacity of the apparatus sufficiently complete until I had ascertained the heat produced by the wetting of the surface of the iron vessel. For this purpose the following trials were made in a similar manner to the above, with the exception that the observations did not require to be extended beyond 26 min.

TABLE IV.

No.	Corrected temperature of water.		Gain or loss of heat by water.	Corrected temperature of apparatus.		Gain or loss of heat by apparatus.
	Commencement of experiment.	Termination of experiment.		Commencement of experiment.	Termination of experiment.	
1.	50°·558	50°·556	0°·002 loss	50°·565	50°·589	0°·024 gain
2.	49·228	49·232	0·004 gain	49·239	49·254	0·015 gain
3.	48·095	48·106	0·011 gain	48·034	48·099	0·065 gain
4.	47·416	47·425	0·009 gain	47·384	47·429	0·045 gain
5.	47·484	47·532	0·048 gain	48·103	47·782	0·321 loss
6.	47·429	47·439	0·010 gain	47·703	47·610	0·093 loss
7.	47·624	47·637	0·013 gain	47·870	47·790	0·080 loss
8.	47·705	47·712	0·007 gain	47·915	47·859	0·056 loss
9.	47·685	47·702	0·017 gain	47·891	47·837	0·054 loss
10.	48·733	48·793	0·060 gain	49·498	49·112	0·386 loss
11.	49·689	49·694	0·005 gain	49·946	49·842	0·104 loss
12.	48·191	48·168	0·023 loss	47·972	48·134	0·162 gain
13.	48·101	48·119	0·018 gain	48·310	48·254	0·056 loss
14.	49·413	49·390	0·023 loss	49·249	49·413	0·164 gain
15.	49·243	49·241	0·002 loss	49·343	49·318	0·025 loss
16.	49·103	49·103	0	49·172	49·172	0
17.	46·991	46·902	0·089 loss	46·204	46·923	0·719 gain
18.	46·801	46·814	0·013 gain	47·139	46·953	0·186 loss
19.	46·624	46·624	0	46·652	46·652	0
20.	46·266	46·158	0·108 loss	45·369	46·167	0·798 gain
Mean	......	......	0·0016 loss	......	......	0·03155 gain

By adding these results to those of the former table, we have a gain of temperature in the water of 3°·13305, and a loss in the apparatus of 20°·33155. Now the capacity of the can of water was estimated as follows:—

Water	141826 grs.
15622 grs. copper as water	1486 grs.
Thermometer and stirrer as water......	118 grs.
Total	143430 grs.

Hence $\frac{3 \cdot 13305}{20 \cdot 33155} \times 143430 = 22102 \cdot 27$, the capacity of the apparatus as tried. The addition of 21·41 (the capacity of 643 grs. of mercury which had been removed in order to admit of the expansion of 70°) to, and the subtraction of 52 grs. (the capacity of the bulb of thermometer C and of the iron wire employed in suspending the apparatus) from this result leaves 22071·68 grs. of water as the capacity of the apparatus employed in the friction of mercury.

The temperature 2°·491218 in the above capacity, equivalent to 1° in 7·85505 lb. of water, was therefore the absolute mean quantity of heat evolved by the friction of mercury.

The leaden weights amounted to 406099 grs., from which 2857 grs., subtracted for the friction of the pulleys, leaves 403242 grs. The mean height from which they fell, as given in Table II., was 1262·731 inches, from which 0·152 inch, subtracted for the velocity of fall, leaves 1262·579 inches. This height, combined with the above weight, is equivalent to 6061·01 foot-lb., which, increased by 16·929 foot-lb. on account of the elasticity of the string, gives 6077·939 foot-lb. as the mean force employed in the experiments.

$\frac{6077 \cdot 939}{7 \cdot 85505} = 773 \cdot 762$; which is therefore the equivalent derived from the above experiments on the friction of mercury. The next series of experiments were made with the same apparatus, using lighter weights.

3rd Series of Experiments.—Friction of Mercury. Weight of the leaden weights and string, 68442 grs. and 68884 grs. Velocity of the weights in descending, 1·4 inch per second. Time occupied by each experiment, 35 minutes. Thermometer for ascertaining the temperature of the mercury, C. Thermometer for registering the temperature of the air, B.

TABLE V.

No. of experiment and cause of change of temperature.	Total fall of weights in inches.	Mean temperature of air.	Difference between mean of columns 5 and 6 and column 3.	Temperature of apparatus.		Gain or loss of heat during experiment.
				Commencement of experiment.	Termination of experiment.	
1 Friction ...	1292·12	49°·539	0°·399+	49°·507	50°·370	0°·863 gain
1 Radiation...	0	50·165	0·226+	50·370	50·413	0·043 gain
2 Friction ...	1292·00	49·865	0·189+	49·606	50·503	0·897 gain
2 Radiation...	0	50·363	0·159+	50·503	50·542	0·039 gain
3 Friction ...	1293·18	50·139	0·460+	50·168	51·030	0·862 gain
3 Radiation...	0	50·617	0·408+	51·030	51·021	0·009 loss
4 Radiation...	0	50·750	0·146+	50·873	50·920	0·047 gain
4 Friction ...	1293·25	51·401	0·013–	50·920	51·856	0·936 gain
5 Radiation...	0	49·936	0·121+	50·031	50·083	0·052 gain
5 Friction ...	1294·92	50·551	0·020–	50·083	50·980	0·897 gain
6 Radiation...	0	50·638	0·135+	50·752	50·795	0·043 gain
6 Friction ...	1294·43	51·172	0·065+	50·795	51·680	0·885 gain
7 Radiation...	0	51·553	0·260–	51·237	51·349	0·112 gain
7 Friction ...	1294·07	52·194	0·371–	51·349	52·298	0·949 gain
8 Friction ...	1293·30	52·774	0·019–	52·298	53·212	0·914 gain
8 Radiation...	0	53·029	0·204+	53·212	53·255	0·043 gain
9 Friction ...	1294·05	51·513	0·306+	51·379	52·259	0·880 gain
9 Radiation...	0	52·093	0·177+	52·259	52·281	0·022 gain
10 Friction ...	1293·95	51·197	0·180+	50·907	51·847	0·940 gain
10 Radiation...	0	51·960	0·079–	51·847	51·916	0·069 gain
11 Friction ...	1292·80	50·577	0·652+	50·804	51·654	0·850 gain
11 Radiation...	0	51·055	0·577+	51·654	51·611	0·043 loss
12 Radiation...	0	51·416	0·483–	50·860	51·006	0·146 gain
12 Friction ...	1293·25	52·057	0·551–	51·006	52·006	1·000 gain
13 Radiation...	0	51·747	0·246–	51·456	51·547	0·091 gain
13 Friction ...	1293·25	52·403	0·389–	51·547	52·482	0·935 gain
14 Friction ...	1293·45	52·703	0·054+	52·294	53·221	0·927 gain
14 Radiation...	0	53·201	0·050+	53·221	53·281	0·060 gain
15 Friction ...	1293·93	53·644	0·088+	53·281	54·183	0·902 gain
15 Radiation...	0	54·061	0·145+	54·183	54·230	0·047 gain
16 Radiation...	0	51·492	0·318+	51·821	51·800	0·021 loss
16 Friction ...	1292·83	52·011	0·242+	51·800	52·706	0·906 gain
1	2	3	4	5	6	7

TABLE V. (*continued*).

No. of experiment and cause of change of temperature.	Total fall of weights in inches.	Mean temperature of air.	Difference between mean of columns 5 and 6 and column 3.	Temperature of apparatus.		Gain or loss of heat during experiment.
				Commencement of experiment.	Termination of experiment.	
17 Radiation...	0	51°·350	0°·055−	51°·272	51°·319	0°·047 gain
17 Friction ...	1292·83	52·057	0·264−	51·319	52·268	0·949 gain
18 Friction ...	1292·84	52·576	0·147+	52·268	53·178	0·910 gain
18 Radiation...	0	52·906	0·276+	53·178	53·187	0·009 gain
19 Radiation...	0	50·119	0·142−	49·928	50·027	0·099 gain
19 Friction ...	1292·33	50·760	0·272−	50·027	50·950	0·923 gain
20 Friction ...	1293·01	51·004	0·147−	50·370	51·345	0·975 gain
20 Radiation...	0	51·798	0·385−	51·345	51·482	0·137 gain
21 Radiation...	0	52·194	0·646−	51·482	51·615	0·133 gain
21 Friction ...	1292·83	52·383	0·298−	51·615	52·555	0·940 gain
22 Friction ...	1292·33	50·389	0·374+	50·332	51·195	0·863 gain
22 Radiation...	0	50·958	0·239+	51·195	51·199	0·004 gain
23 Radiation...	0	51·218	0·498−	50·636	50·804	0·168 gain
23 Friction ...	1294·69	51·848	0·546−	50·804	51·800	0·996 gain
24 Friction ...	1294·33	50·582	0·286+	50·435	51·302	0·867 gain
24 Radiation...	0	51·223	0·092+	51·302	51·328	0·026 gain
25 Radiation...	0	51·665	0·406−	51·190	51·328	0·138 gain
25 Friction ...	1294·33	52·281	0·464−	51·328	52·306	0·978 gain
26 Friction ...	1294·34	52·652	0·105+	52·306	53·208	0·902 gain
26 Radiation...	0	52·957	0·259+	53·208	53·225	0·017 gain
27 Friction ...	1293·83	49·463	0·277+	49·293	50·188	0·895 gain
27 Radiation...	0	50·068	0·142+	50·188	50·233	0·045 gain
28 Radiation...	0	48·420	0·145+	48·537	48·593	0·056 gain
28 Friction ...	1294·33	49·132	0·093−	48·593	49·486	0·893 gain
29 Friction ...	1294·84	49·142	0·092+	48·773	49·696	0·923 gain
29 Radiation...	0	49·783	0·053−	49·696	49·765	0·069 gain
30 Radiation...	0	50·251	0·422−	49·765	49·894	0·129 gain
30 Friction ...	1294·33	50·597	0·246−	49·894	50·808	0·914 gain
Mean Friction .	1293·532	......	0·00743⅓+	......	......	0·9157 gain
Mean Radiation	0	......	0·0048 +	......	......	0·0606 gain
1	2	3	4	5	6	7

The effect of each degree of difference between the temperature of the laboratory and that of the apparatus being 0°·18544, 0°·9157−0°·0606+0°·000488=0°·855588 will be the proximate mean increase of temperature in the above series of experiments. The correction, owing to the mean temperature of the mercury in the friction experiments being 0°·013222 higher than appears in the table, will be 0°·002452, which, being added to the proximate result, gives 0°·85804 as the true thermometrical effect. This, in the capacity of 22071·68 grs. of water, is equal to 1° in 2·70548 lb. of water.

The leaden weights amounted to 137326 grs., from which 1040 grs. must be subtracted for the friction of the pulleys, leaving 136286 grs. as the corrected weight. The mean height of fall was 1293·532 inches, from which 0·047 inch, subtracted on account of the velocity with which the weights came to the ground, leaves 1293·485 inches. This fall, combined with the above corrected weight, is equivalent to 2098·618 foot-lb., which, with 1·654 foot-lb., the force developed by the elasticity of the string, gives 2100·272 foot-lb. as the mean force employed in the experiments.

$\frac{2100{\cdot}272}{2{\cdot}70548}=776{\cdot}303$ will therefore be the equivalent from the above series of experiments, in which the amount of friction of the mercury was moderated by the use of lighter weights.

4th Series of Experiments.—Friction of Cast Iron. Weight of cast-iron apparatus, 44000 grs. Weight of mercury contained by it, 204355 grs. Weight of the leaden weights and string attached, 203026 grs. and 203073 grs. Average velocity with which the weights fell, 3·12 inches per second. Time occupied by each experiment, 38 minutes. Thermometer for ascertaining the temperature of the mercury, C. Thermometer for registering the temperature of the air, A.

TABLE VI.

No. of experiment and cause of change of temperature.	Total fall of weights in inches.	Mean temperature of air.	Difference between mean of columns 5 and 6 and column 3.	Temperature of apparatus.		Gain or loss of heat during experiment.
				Commencement of experiment.	Termination of experiment.	
1 Friction ...	1257·90	46°·362	2°·544 +	46°·837	50°·976	4°·139 gain
1 Radiation...	0	46·648	3·950 +	50·976	50·220	0·756 loss
2 Radiation...	0	47·296	0·455 −	46·730	46·953	0·223 gain
2 Friction ...	1258·97	47·891	1·247 +	46·953	51·323	4·370 gain
3 Friction ...	1261·80	47·705	1·830 +	47·352	51·718	4·366 gain
3 Radiation...	0	48·547	2·950 +	51·718	51·276	0·442 loss
4 Radiation...	0	47·825	0·044 −	47·756	47·807	0·051 gain
4 Friction ...	1260·35	48·385	1·598 +	47·807	52·160	4·353 gain
5 Radiation...	0	48·323	0·248 −	48·009	48·142	0·133 gain
5 Friction ...	1260·15	48·833	1·494 +	48·142	52·513	4·371 gain
6 Friction ...	1259·95	48·049	1·995 +	47·902	52·186	4·284 gain
6 Radiation...	0	48·632	3·283 +	52·186	51·645	0·541 loss
7 Radiation...	0	50·385	0·240 −	50·053	50·237	0·184 gain
7 Friction ...	1263·13	51·018	1·408 +	50·237	54·616	4·379 gain
8 Friction ...	1262·12	48·385	1·096 +	47·249	51·714	4·465 gain
8 Radiation...	0	49·199	2·343 +	51·714	51·371	0·343 loss
9 Friction ...	1257·20	49·721	2·495 +	50·160	54·273	4·113 gain
9 Radiation...	0	50·338	3·643 +	54·273	53·689	0·584 loss
10 Radiation...	0	48·439	0·821 +	49·271	49·250	0·021 loss
10 Friction ...	1258·70	49·690	2·282 +	49·877	54·067	4·190 gain
Mean Friction .	1260·027	...	1·7989+	......	......	4·303 gain
Mean Radiation	0	...	1·6003+	......	......	0·2096 loss
1	2	3	4	5	6	7

From the above Table it appears that there was a thermometrical effect of 0°·20101 for each degree of difference between the temperature of the laboratory and that of the apparatus. Hence 4°·303 + 0° 2096 + 0°·03992 = 4°·55252 will be the proximate mean increase of temperature. The correction, owing to the mean temperature of the mercury in the friction experiments appearing 0°·07625 too low in the Table, will be 0°·01533, which, added to the proximate result, gives 4°·56785 as the true mean increase of temperature.

The capacity of the apparatus was obtained by experiments made in precisely the same manner that I have already described in the case of the mercurial apparatus for fluid friction. Their results are collected into the following Table :—

TABLE VII.

No.	Corrected temperature of water.		Gain of heat by the water.	Corrected temperature of apparatus.		Loss of heat by the apparatus.
	Commencement of experiment.	Termination of experiment.		Commencement of experiment.	Termination of experiment.	
1.	45°·535	47°·305	1°·770	71°·112	47°·421	23°·691
2.	46·210	47·937	1·727	71·292	48·073	23·219
3.	47·334	49·023	1·689	71·454	49·151	22·303
4.	49·007	50·555	1·548	71·152	50·632	20·520
5.	47·895	49·498	1·603	71·249	49·636	21·613
6.	48·784	50·357	1·573	71·445	50·460	20·985
7.	50·323	51·757	1·434	70·793	51·808	18·985
8.	47·912	49·525	1·613	71·253	49·653	21·600
9.	48·449	50·013	1·564	70·798	50·083	20·715
10.	49·836	51·337	1·501	71·356	51·375	19·981
11.	46·870	48·559	1·689	71·026	48·657	22·369
12.	48·562	50·151	1·589	71·291	50·199	21·092
Mean	......	......	1·60833	......	......	21·42275

By adding 0°·00071 and 0°·0141, the loss and gain of Table IV. reduced to the surface of the solid-friction apparatus, to the above mean results, we have a gain of 1°·60904 by the water and a loss of 21°·43685 by the apparatus. The capacity of the can of water was in this instance as follows :—

Water	155824 grs.
Copper can as water	1486 grs.
Thermometer and stirrer as ditto	118 grs.
Total	157428 grs.

Hence $\frac{1{\cdot}60904}{21{\cdot}43685} \times 157428 = 11816{\cdot}47$ will be the capacity of the apparatus as tried. By applying the two corrections, one additive on account of the absence during the trials of 300 grs. of mercury, the other subtractive on account of the capacity of the thermometer C and suspending wire, we obtain 11796·07 grs. of water as the capacity of the apparatus during the experiments.

The temperature $4^{c}{\cdot}56785$ in the above capacity, equivalent to 1° in 7·69753 lb. of water, was therefore the mean absolute quantity of heat evolved by the friction of cast iron.

The leaden weights amounted to 406099 grs., from which 2857 grs., subtracted on account of the friction of the pulleys, leaves 403242 grs. as the pressure applied to the apparatus.

Owing to the friction being in the simple ratio of the velocity, it required a good deal of practice to hold the regulating lever so as to cause the weights to descend to the ground with any thing like a uniform and moderate velocity. Hence, although the mean velocity was 3·12 inches per second, the force with which the weights struck the ground could not be correctly estimated by that velocity as in the case of fluid friction. However, it was found that the noise produced by the impact was on the average equal to that produced by letting the weights fall from the height of one eighth of an inch. It generally happened also that in endeavouring to regulate the motion, the weights would stop suddenly before arriving at the ground. This would generally happen once, sometimes twice, during the descent of the weights, and I estimate the force thereby lost as equal to that lost by impact with the ground. Taking therefore the total loss at one fourth of an inch in each fall, we have twenty times that quantity, or 5 inches, as the entire loss, which, subtracted from 1260·027, leave 1255·027 inches as the corrected height through which the weight of 403242 grs. operated. These numbers are equivalent to 6024·757 foot-lb.; and adding 16·464 foot-lb. for the effect of the elasticity of the string, we have 6041·221 foot-lb. as the force employed in the experiments.

The above force was not, however, entirely employed in generating heat in the apparatus. It will be readily conceived that the friction of a solid body like cast iron must have produced a considerable vibration of the framework upon which the apparatus was placed, as well as a loud sound. The value of the force absorbed by the former was estimated by experiment at 10·266 foot-lb. The force required to vibrate the string of a violoncello, so as to produce a sound which could be heard at the same distance as that arising from the friction, was estimated by me, with the concurrence of another observer, at 50 foot-lb. These numbers, subtracted from the previous result, leave 5980·955 foot-lb. as the force actually converted into heat.

$\frac{5980 \cdot 955}{7 \cdot 69753} = 776 \cdot 997$ will therefore be the equivalent derived from the above experiments on the friction of cast iron. The next series of experiments was made with the same apparatus, using lighter weights.

5th Series of Experiments.—Friction of Cast Iron. Weight of leaden weights, 68442 grs. and 68884 grs. Average velocity of fall, 1·9 inch per second. Time occupied by each experiment, 30 minutes. Thermometer for ascertaining the temperature of the mercury, C. Thermometer for registering the temperature of the laboratory, A.

TABLE VIII.

No. of experiment and cause of change of temperature.	Total fall of weights in inches.	Mean temperature of air.	Difference between mean of columns 5 and 6 and column 3.	Temperature of apparatus.		Gain or loss of heat during experiment.
				Commencement of experiment.	Termination of experiment.	
1 Friction ...	1281·07	47°·404	0°·852 +	47°·494	49°·018	1°·524 gain
1 Radiation...	0	48·003	0·998 +	49·018	48·984	0·034 loss
2 Radiation...	0	48·269	0·702 +	48·984	48·958	0·026 loss
2 Friction ...	1280·74	48·516	1·189 +	48 958	50·452	1·494 gain
3 Radiation...	0	49·003	0·133 −	48·812	48·928	0·116 gain
3 Friction ...	1285·10	49·728	0·022 +	48·928	50·572	1·644 gain
4 Friction ...	1283·89	50·138	1·172 +	50·572	52·049	1·477 gain
4 Radiation...	0	50·408	1·581 +	52·049	51·929	0·120 loss
5 Friction ...	1282·45	46·798	0·558 +	46·554	48·159	1·605 gain
6 Friction ...	1281·29	47·296	1·571 +	48·159	49·576	1·417 gain
5 Radiation...	0	47·535	1·929 +	49·576	49·353	0·223 loss
6 Radiation...	0	47·651	1·607 +	49·353	49·164	0·189 loss
7 Radiation...	0	46·261	0·298 −	45·880	46·047	0·167 gain
8 Radiation...	0	46·748	0·617 −	46·047	46·215	0·168 gain
7 Friction ...	1276·07	46·810	0·978 +	47·022	48·554	1·532 gain
8 Friction ...	1275·17	47·366	1·883 +	48·554	49·945	1·391 gain
9 Radiation...	0	46·771	0·271 −	46·425	46·575	0·150 gain
9 Friction ...	1276·95	47·126	0·258 +	46·575	48·194	1·619 gain
10 Friction ...	1276·84	47·238	1·655 +	48·194	49·593	1·399 gain
10 Radiation...	0	47·335	2·142 +	49·593	49·361	0·232 loss
Mean Friction .	1279·957	......	1·0138+	......	......	1·5102 gain
Mean Radiation	0	......	0·764 +	......	......	0·0223 loss
1	2	3	4	5	6	7

From the above Table it appears that the effect of each degree of difference between the temperature of the laboratory and that of the apparatus was 0°·1591. Hence 1°·5102+0°·0223+0°·03974=1°·57224 will be the proximate heating effect. To this the addition of 0°·00331, on account of the mean temperature of the apparatus in the friction experiments having been in reality 0°·02084 higher than appears in the Table, gives the real increase of temperature in the experiments at 1°·57555, which, in the capacity of 11796·07 grs. of water, is equivalent to 1° in 2·65504 lb. of water.

The leaden weights amounted to 137326 grs., from which 1040 grs., subtracted for the friction of the pulleys, leaves 136286 grs. The velocity of descent, which was in this case much more easily regulated than when the heavier weights were used, was 1·9 inch per second: Twenty impacts with this velocity indicate a loss of fall of 0·094 inch, which, subtracted from 1279·957, leaves 1279·863 inches as the corrected height from which the weights fell.

The above height and weight are equivalent to 2076·517 foot-lb., to which the addition of 1·189 foot-lb. for the elasticity of the string gives 2077·706 foot-lb. as the total force applied. The corrections for vibration and sound (deduced from the data obtained in the last series, on the hypothesis that they were proportional to the friction by which they were produced) will be 3·47 and 16·9 foot-lb. These quantities, subtracted from the previous result, leave 2057·336 foot-lb. as the quantity of force converted into heat in the apparatus.

$\frac{2057 \cdot 336}{2 \cdot 65504} = 774 \cdot 88$ will therefore be the equivalent as derived from this last series of experiments.

The following Table contains a summary of the equivalents derived from the experiments above detailed. In its fourth column I have supplied the results with the correction necessary to reduce them to a vacuum.

TABLE IX.

No. of series.	Material employed.	Equivalent in air.	Equivalent *in vacuo*.	Mean.
1.	Water..............	773·640	772·692	772·692
2.	Mercury...........	773·762	772·814	774·083
3.	Mercury...........	776·303	775·352	
4.	Cast iron	776·997	776·045	774·987
5.	Cast iron	774·880	773·930	

It is highly probable that the equivalent from cast iron was somewhat increased by the abrasion of particles of the metal during friction, which could not occur without the absorption of a certain quantity of force in overcoming the attraction of cohesion. But since the quantity abraded was not considerable enough to be weighed after the experiments were completed, the error from this source cannot be of much moment. I consider that 772·692, the equivalent derived from the friction of water, is the most correct, both on account of the number of experiments tried and the great capacity of the apparatus for heat. And since, even in the friction of fluids, it was impossible entirely to avoid vibration and the production of a slight sound, it is probable that the above number is slightly in excess. I will therefore conclude by considering it as demonstrated by the experiments contained in this paper,—

1st. *That the quantity of heat produced by the friction of bodies, whether solid or liquid, is always proportional to the quantity of force expended.* And,

2nd. *That the quantity of heat capable of increasing the temperature of a pound of water (weighed* in vacuo, *and taken at between* 55° *and* 60°) *by* 1° *Fahr. requires for its evolution the expenditure of a mechanical force represented by the fall of 772 lb. through the space of one foot* *.

Oak Field, near Manchester,
June 4th, 1849.

* A third proposition, suppressed in accordance with the wish of the Committee to whom the paper was referred, stated that friction consisted in the conversion of mechanical power into heat.

On a remarkable Appearance of Lightning. By J. P. Joule, *F.R.S.** [In a letter to the Editors of the 'Philosophical Magazine.']

['Philosophical Magazine,' ser. 3. vol. xxxvii. p. 127.]

Gentlemen,

On the 16th inst., after a very sultry morning, this town was, in common with a large tract of country, visited at 4 p.m. by a thunderstorm, accompanied with heavy rain. In the evening of the same day, about 9 o'clock, we had an opportunity of witnessing a most magnificent display of electrical discharges, which continued almost uninterruptedly for the space of one hour, accompanied, however, by only a few drops of rain. I had never before seen lightning of such an extraordinary character. Each discharge appeared to emanate from a mass of clouds in the S.W., and travelled six to ten miles in the direction of the spectator, dividing into half a dozen or more sparks, or zigzag streams of light—in some instances the termination of each of these streams being, as represented in the adjoining sketch, again subdivided into a

Fig. 77 bis.

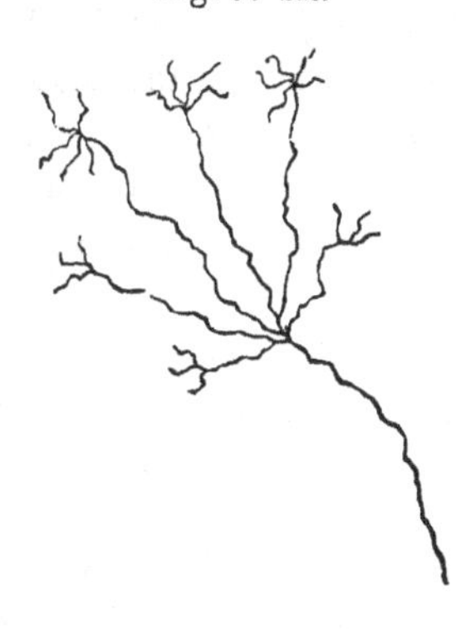

number of smaller sparks. I did not observe any of the discharges to strike the ground; and from the interval of time between the appearance of those which crossed the zenith and

* See also a paper by Mr. P. Clare on the same storm, Phil. Mag. ser. 3. vol. xxxvii. p. 329.

the thunder, I estimate their general elevation above the surface of the earth at $3\frac{1}{2}$ miles and upwards.

The diverging form arose no doubt from the extensive negative surface presented by the clouds overhead, and may be imitated on a small scale by filling a glass jar with water and using it as a Leyden phial. If such a jar be discharged by bringing one ball of the discharging-rod towards the exterior glass surface, the other ball being in connexion with the water, the spark will, in restoring the electrical equilibrium, be seen to diverge over the entire glass surface.

Another remarkable feature in the lightning was the *sensible time* occupied in travelling towards the spectator. The main streams of light were always formed before the diverging sparks; and, when formed, remained steady for an appreciable time, until the whole disappeared together. My brothers, Benjamin and John Joule, who observed the lightning two miles westward of my station, formed exactly the same impression of its character; and in addition they and several other parties were witnesses of a phenomenon which, if owing to the electrical state of the atmosphere, was, I believe, without a recorded precedent. At half-past 8 o'clock a bright red light appeared among the clouds, bearing nearly due south, and having an elevation of about 30° above the horizon. It appeared as if the sun were behind a cloud illuminating its edges, and throwing a brilliant light upon the neighbouring clouds. It lasted for about five minutes, and then gradually disappeared.

I ought to mention that, during the above-described phenomena, violent thunderstorms were taking place in different parts of the county and in Cheshire, but without any apparent connexion with them.

Yours, &c.,

JAMES P. JOULE.

Acton Square, Salford, Manchester,
July 19, 1850.

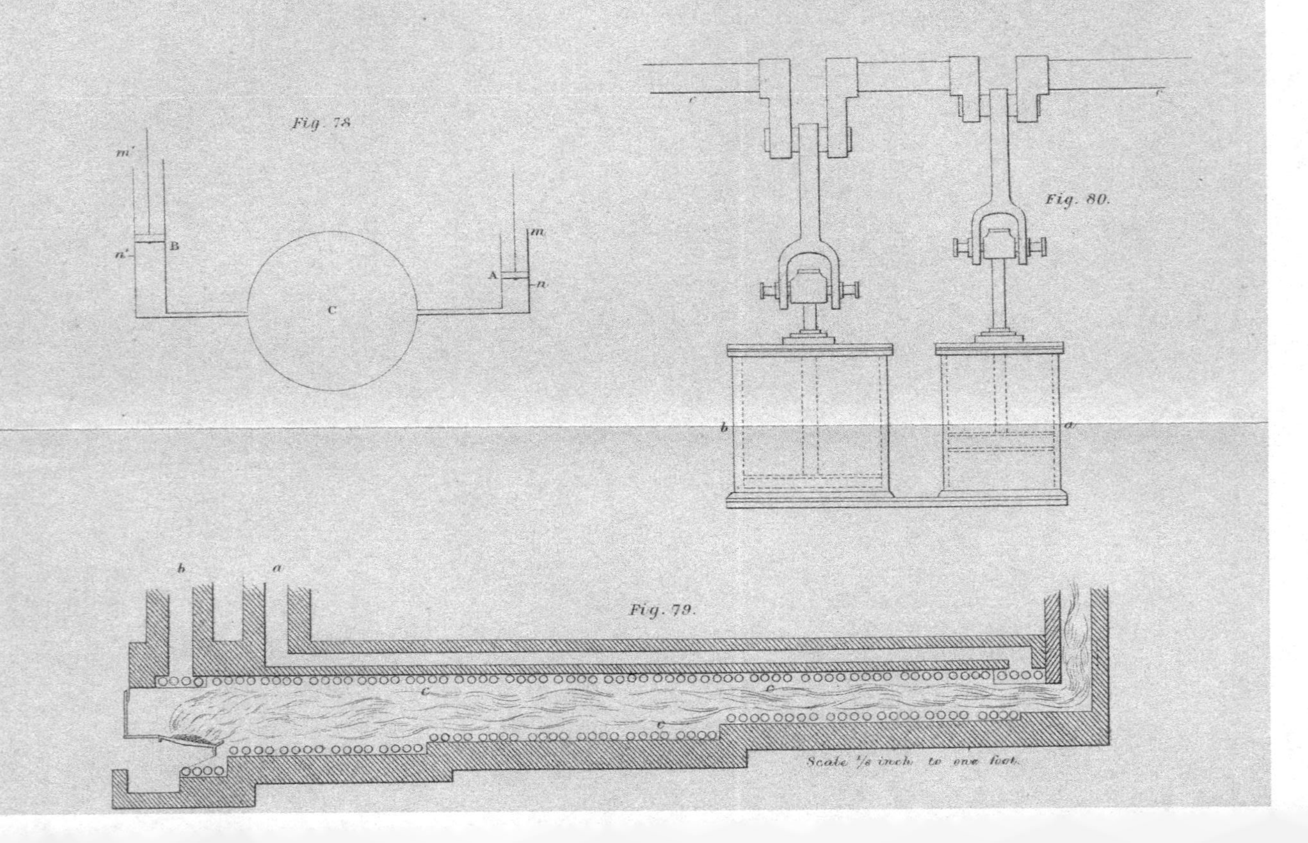
Fig. 78
m'
n'
B
C
A
m
n
Fig. 80.
c
c
b
a
Fig. 79.
b
a
c
c
c
Scale 1/8 inch to one foot.

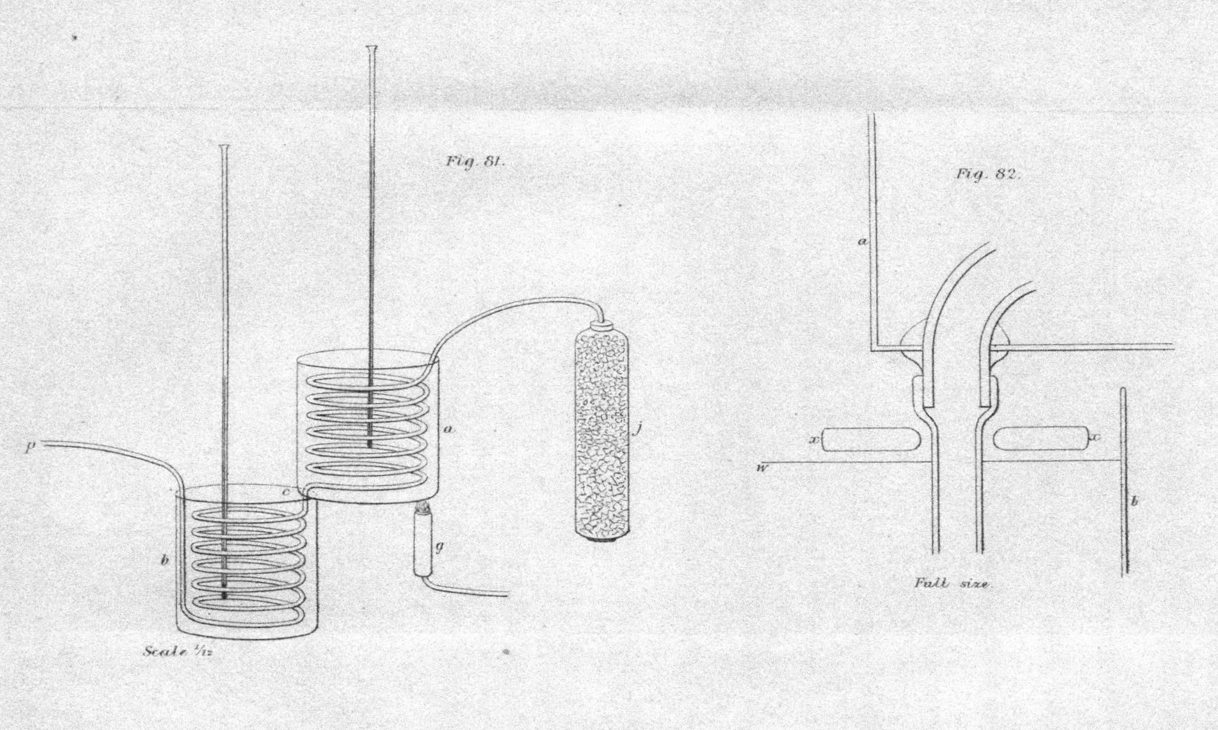

Fig. 81.

Scale 1/12

Fig. 82.

Full size.

Mintern Bros. lith.

On some Amalgams. By J. P. Joule, *F.R.S.*

[Rep. Brit. Assoc. 1850, Sections, p. 55. Read before the British Association at Edinburgh, August 1850.]

The author had procured an amalgam of iron by precipitating it on mercury by the electrotype process. He had subsequently pursued the research with a view to form definite amalgams by a simple chemical or mechanical process. When mercury was made negative under a solution of sulphate of copper, an amalgam of copper was formed, which, when fully saturated with copper, was found to be represented by the formula Cu + Hg.

The author also exhibited a small apparatus whereby amalgams could be made to endure a pressure of 60 tons per square inch of surface. The superfluous mercury was thus expelled through the openings in the sides of the press, leaving an amalgam of definite chemical composition. In this way he had procured the following compounds:—

$$Pt + 2Hg.$$
$$Ag + 2Hg.$$
$$Cu + Hg.$$
$$Fe + Hg.$$
$$2Zn + Hg.$$
$$2Pb + Hg.$$
$$7Sn + Hg.$$

On the Air-Engine. By James Prescott Joule, *F.R.S., F.C.S., Corr. Mem. R.A. Turin, Sec. Lit. and Phil. Soc. Manchester, &c.**

['Philosophical Transactions,' 1852, Part I. p. 65. Read June 19, 1851.]

(Plate III.)

It has long been suspected that important advantages might be derived from the substitution of air for steam as a prime mover of machinery. It has been alleged that the air-engine

* The experiments were made at Acton Square, Salford.

would be safer, lighter, and more economical in the expenditure of fuel than the steam-engine. Until comparatively recent times, however, experimental science was hardly in the state of advancement requisite to enable the physicist, in his investigation of this important subject, to arrive at conclusions sufficiently certain to give confidence to the practical machinist. Professor Thomson, Mr. Rankine, and M. Clausius have of late, however, published papers of great value on the mechanical action of gases, and particularly of steam, founded on tolerably correct experimental data. I hope that the following remarks founded on the same general principles, but applied to a particular kind of air-engine, may be interesting to the Royal Society.

The air-engine, the performance of which I propose to discuss, consists of two parts, in one of which the air is compressed into a receiver, where its elasticity is increased by the application of heat, and in the other it is allowed to escape again from the receiver into the atmosphere. By the former work is absorbed, by the latter it is evolved in a larger quantity, the excess constituting the work evolved by the engine on the whole. The simple question, therefore, is to determine the quantity of work so evolved, together with the heat applied to increase the elasticity of the air in the receiver.

In Plate III. fig. 78, let A be the pump by which air is forced into the receiver C, where heat may be communicated to it from an external source, and B the cylinder by which the same quantity is allowed to escape again into the atmosphere. Moreover, let the material of which the apparatus is made, with the exception of that part through which heat may be communicated to the air in C, be impervious to and destitute of capacity for heat. Such a machine may be conceived to work in the following manner.

The cylinder of the pump A being filled with air of the atmospheric temperature and pressure, the piston compresses the air until, at a point n, its pressure is rendered equal to that of the air in the receiver C, which has been previously filled with air of an elevated temperature and pressure. The work absorbed by this action will be that communicated to the air in the cylinder, minus the work due to the atmospheric pressure through *m n*. The moment the piston has passed the point *n* the valve will open, admitting the air into the receiver C; and as this receiver may be conceived to be of indefinite magnitude, the alteration of pressure in it, consequent upon the introduction of fresh air, may be neglected. Heat is then communicated to the air in the receiver, in order to restore its temperature to the intensity which existed before the admission of air at a lower temperature. The air is then allowed to escape from the receiver into the base of the cylinder B, evolving work until, on the arrival of the piston at *n'*, the same quantity has been removed from the receiver as was forced into it by the pump. The further supply of air from the receiver is then cut off, and that which has entered the cylinder expands, evolving work until, on the arrival of the piston at *m'*, its pressure is reduced to that of the atmosphere. By opening valves at the bases of A and B, the pistons are then brought to their first positions.

The problem which must be solved in order to estimate the power and consumption of fuel in an engine similar to that just described is as follows:—To determine the pressure and temperature for any point of the stroke of a piston which compresses a given volume of air, and the quantity of work absorbed in forcing the piston to that point. For the temperature and pressure Poisson has furnished the following formulæ—

$$\frac{T'}{T}=\left(\frac{V}{V'}\right)^{k-1}$$

and

$$\frac{P'}{P}=\left(\frac{V}{V'}\right)^{k},$$

where T, P, and V are the temperature from absolute zero (estimated at 491° Fahr. below the freezing-point of water), pressure, and volume of the air before compression; T', P', and V' the temperature from absolute zero, pressure, and volume of air after compression; and *k* is the ratio of the specific heat of air at constant pressure to that at constant volume. Professor W. Thomson has deduced, as a consequence of the above, the following formula for the work absorbed,

$$W = PV \frac{1}{k-1} \left\{ \left(\frac{V}{V'} \right)^{k-1} - 1 \right\} *$$

From the foregoing formulæ I have calculated the work absorbed by compressing air in a cylinder 1 foot long and of the capacity of 12 cubic inches, the absolute temperature of the air, and its pressure at each tenth of an inch of the piston's

* The above formula was kindly communicated to the author by Professor Thomson, in a letter dated January 15, 1851, from which the following is an extract:—"It is required to find the work necessary to compress a given mass of air to a given fraction of its volume, when no heat is permitted to leave the air. Let P, V, T be the primitive pressure, volume, and temperature, respectively; let p, v, and t be the pressure, volume, and temperature at any instant during the compression; and let P′, V′, and T′ be what they become when the compression is concluded. Then if k denote the ratio of the specific heat of air at constant pressure to the specific heat of air kept in a space of constant volume, and if, as appears to be nearly, if not rigorously true, k be constant for varying temperatures and pressures, we shall have by the investigation in Miller's 'Hydrostatics' (edit. 1835, p. 22)—

$$\frac{1+Et}{1+ET} = \left(\frac{V}{v} \right)^{k-1}$$

But

$$\frac{pv}{PV} = \frac{1+Et}{1+ET},$$

therefore

$$pv = PV \left(\frac{V}{v} \right)^{k-1}$$

Now the work done in compressing the mass from volume v to volume $v - dv$ will be pdv, or, by what precedes,

$$PV \, . \, V^{k-1} \frac{dv}{v^k}.$$

Hence by the integral calculus we readily find, for the work, W, necessary to compress from V to V′

$$W = PV \, . \, \frac{1}{k-1} \left\{ \left(\frac{V}{V'} \right)^{k-1} - 1 \right\}."$$

progress. The following data were employed in the computation :—Weight of 100 cubic inches of atmospheric air of 15 lb. pressure on the square inch, and 491° Fahr. from the absolute zero, 33·2237 grs.; specific heat of air at constant volume, 0·19742. Ratio of the specific heat of air at constant pressure to that at constant volume, as determined from the experiments of Delaroche and Berard, and the mechanical equivalent of heat, 1·3519325*. The results are shown in Table I.

I now proceed to give some estimates of the performance of an air-engine similar in principle to that already described, worked at various pressures and temperatures, those of the atmospheric air being 15 lb. on the square inch, and 32° Fahr. or 491° Fahr. from the absolute zero. In order to render the results easily available in calculating the duty of engines of greater size, I shall assume that the condensing-pump is 12 inches long and has a sectional area equal to 1 square inch, and that the cylinder, also of 1 inch section, has a length which may be made to vary according to the pressure and temperature employed.

I take as the first example a case in which the receiver C contains air of the atmospheric density, and of which the absolute temperature is 849°·464 Fahr. or 390°·464 of the scale of Fahrenheit's thermometer. The pressure in the receiver will then be 25·95104 lb. on the square inch, as given in the third column of Table II. The air in the pump A will be brought to the same pressure and to the absolute temperature 566°·3094 after the piston has traversed 4 inches. The work absorbed by the air will be 6·537154 foot-pounds, from which, by subtracting 5 foot-pounds, the work communicated by the pressure of the atmosphere following the piston, we obtain 1·537154 foot-pounds as the work of the engine absorbed by the first part of the stroke. This result is consigned to column 6. Immediately after the piston has passed the

* The experiments of Desormes and Clement give 1·354; those of Gay-Lussac and Welter 1·375; and those described under the article "Hygrometry" (Enc. Brit.) 1·333. See Art. "Sound," Enc. Brit., 7th edit.

fourth inch of the pump, the valve will be opened admitting the compressed air into the receiver C. The work of the engine absorbed by the remaining 8 inches of the piston's stroke will be $\frac{8}{12}(25{\cdot}95104-15)=7{\cdot}300693$ foot-pounds, as given in the seventh column. The air thus forced into the receiver at the absolute temperature $566^\circ{\cdot}3094$ Fahr. must then be raised to $849^\circ{\cdot}464$ Fahr., the constant absolute temperature of the receiver. The heat necessary for this purpose, being that due to the capacity for heat of air at constant pressure, will be that which is able to raise the temperature of 1 lb. of water $0^\circ{\cdot}04304312$ Fahr., as given in column 15. On leaving the receiver, the air enters the cylinder of expansion B, and having propelled the piston through 12 inches, the same quantity of air will have passed out of the receiver as was pumped into it by A. The further supply of air is then cut off, and the air, after expanding through the remaining 6 inches of the cylinder (which in this case must be 18 inches long), will be reduced to the pressure of 15 lb. on the square inch, and the absolute temperature $\frac{3}{2}(491^\circ)=736^\circ{\cdot}5$. The work evolved by the piston will also be to that absorbed in the condensing-pump as the volume of the cylinder B is to that of the pump A; from which we find $\frac{3}{2}(7{\cdot}300693)=10{\cdot}95104$ foot-pounds, and $\frac{3}{2}(1{\cdot}537154)=2{\cdot}305731$ foot-pounds, the work evolved by the first and second parts of the piston's stroke, as given in columns 11 and 12. The work evolved by the engine on the whole, being the difference between the work evolved by B and the work absorbed by A, will be equal to one third of the former, or one half of the latter, or $4{\cdot}418924$ foot-pounds, as given in column 14. Dividing this by $0^\circ{\cdot}04304312$, we obtain $102{\cdot}66276$ foot-pounds as the work evolved by the engine out of each 1° Fahr. per lb. of water communicated to the receiver. This result, which is consigned to the sixteenth column, informs us of the economical value of the engine, which is of course great in proportion to its approach to 772 foot-pounds, the theoretical maximum. The seventeenth column contains the theoretical duty according to Professor Thomson's law, viz. that the range of temperature divided by the maximum absolute temperature is equal to the fraction of heat converted into force by any perfect engine *.

It will be observed that the numbers in column 16, representing the work evolved out of each unit of heat, increase with the temperature and pressure of the air in the receiver. In every example given, with the exception of the first, the economical value of the air-engine is greater than that of the steam-engine calculated by Mr. Rankine in his paper on the Mechanical Action of Heat †. In considering the relative merits of the engines, we must not, however, lose sight of a most important fact discovered by Rankine and Clausius, viz. that a portion of the heat employed to evaporate water in the boiler is afterwards evolved in the form of work, in consequence of the liquefaction, in the cylinder, of a portion of the expanding vapour. This fact would induce the hope that a great portion of the latent heat of evaporation, which is at present almost entirely lost, might, by an increase of temperature and by extending the principle of expansion, be converted into mechanical effect.

If, as would appear from the experiments of De la Rive and Marcet, Haycraft and Dulong, the capacity for heat of a given volume is the same in all gases taken at the same pressure and temperature, the results of the Tables will be equally true whatever elastic fluid may be employed.

It now only remains to offer a few observations, with a view to facilitate the labours of those who may be desirous of constructing a good practical air-engine.

* See Professor Thomson's "Investigation of the Duty of a perfect Thermo-Dynamic Engine," at the end of this paper.

† Transactions of the Royal Society of Edinburgh, vol. xx. part 1. Professor Thomson, in a paper "On the Dynamical Theory of Heat," recently read before the Royal Society, Edinburgh, gives 209 foot-pounds as the duty of an absolutely perfect steam-engine, with a range of temperature between 30° and 140° Centigrade.

TABLE I.

Distance traversed by piston, in inches.	Work absorbed, in foot-pounds.	Temperature from absolute zero, in degrees Fahr.	Pressure on the piston, in lbs.	Distance traversed by piston, in inches.	Work absorbed, in foot-pounds.	Temperature from absolute zero, in degrees Fahr.	Pressure on the piston, in lbs.
0	0	491	15	6·0	11·77479	626·6480	38·28805
0·1	0·1257008	492·4481	15·17066	6·1	12·09828	630·3747	39·16857
0·2	0·2528426	493·9128	15·34473	6·2	12·42768	634·1694	40·08375
0·3	0·3814514	495·3944	15·52230	6·3	12·76566	638·0630	41·03738
0·4	0·5113882	496·8913	15·70343	6·4	13·11172	642·0498	42·03120
0·5	0·6432696	498·4106	15·88841	6·5	13·46626	646·1341	43·06763
0·6	0·7763749	499·9440	16·07709	6·6	13·82962	650·3201	44·14936
0·7	0·9111464	501·4966	16·26974	6·7	14·20221	654·6124	45·27926
0·8	1·047533	503·0678	16·46643	6·8	14·58441	659·0155	46·46043
0·9	1·185586	504·6582	16·66731	6·9	14·97668	663·5345	47·69625
1·0	1·325341	506·2682	16·87248	7·0	15·37948	668·1749	48·99042
1·1	1·466805	507·8979	17·08209	7·1	15·79334	672·9426	50·34692
1·2	1·610032	509·5479	17·29627	7·2	16·21879	677·8438	51·77015
1·3	1·755090	511·2190	17·51516	7·3	16·65633	682·8845	53·26481
1·4	1·901962	512·9110	17·73892	7·4	17·10672	688·0730	54·83624
1·5	2·050727	514·6248	17·96770	7·5	17·57047	693·4156	56·49008
1·6	2·201444	516·3611	18·20167	7·6	18·04842	698·9216	58·23268
1·7	2·354089	518·1196	18·44097	7·7	18·54128	704·5995	60·07098
1·8	2·508791	519·9018	18·68582	7·8	19·04988	710·4585	62·01267
1·9	2·665542	521·7076	18·93638	7·9	19·57510	716·5093	64·06620
2·0	2·824402	523·5377	19·19283	8·0	20·11797	722·7632	66·24102
2·1	2·985424	525·3927	19·45538	8·1	20·67947	729·2318	68·54755
2·2	3·148667	527·2733	19·72426	8·2	21·26078	735·9287	70·99750
2·3	3·314184	529·1801	19·99966	8·3	21·86317	742·8683	73·60395
2·4	3·481993	531·1133	20·28182	8·4	22·48795	750·0659	76·38145
2·5	3·652224	533·0744	20·57099	8·5	23·13667	757·5393	79·34657
2 6	3·824870	535·0633	20·86740	8·6	23·81096	765·3072	82·51786
2·7	4·000022	537·0811	21·17132	8·7	24·51263	773·3907	85·91642
2·8	4·177719	539·1282	21·48302	8 8	25·24353	781·8109	89·56590
2·9	4 358080	541·2060	21·80279	8·9	26·00607	790·5954	93·49398
3·0	4·541123	543·3147	22·13095	9·0	26·80260	799·7716	97·73172
3·1	4·726945	545·4554	22·46778	9·1	27·63584	809·3707	102·3153
3·2	4 915606	547·6288	22·81364	9·2	28·50889	819·4284	107·2862
3·3	5·107206	549·8361	23·16887	9·3	29·42512	829·9836	112·6929
3·4	5·301779	552·0776	23·53383	9·4	30·38842	841·0810	118·5920
3·5	5·499456	554·3549	23·90892	9·5	31·40316	852·7710	125·0499
3·6	5·700287	556·6685	24·29452	9·6	32·47432	865·1110	132·1452
3·7	5·904370	559·0196	24·69107	9·7	33·60755	878·1660	139·9716
3·8	6·111816	561·4094	25·09902	9·8	34·80946	892·0122	148·6412
3·9	6·322706	563·8389	25·51884	9·9	36·08755	906·7362	158·2897
4·0	6·537154	566·3094	25·95104	10·0	37·45073	922·4402	169·0827
4·1	6·755242	568·8218	26·39613	10·1	38·90927	939·2430	181·2239
4·2	6·977122	571·3779	26·85467	10·2	40·47547	957·2860	194·9666
4·3	7·202863	573·9785	27·32725	10·3	42·16396	976·7377	210·6300
4·4	7·432590	576·6250	27·81448	10·4	43·99234	997·8010	228·6204
4·5	7·666465	579·3193	28·31703	10·5	45·98215	1020·724	249·4640
4·6	7·904577	582·0624	28·83559	10·6	48·15980	1045·811	273·8522
4·7	8·147090	584·8562	29·37090	10·7	50·55854	1073·445	302·7106
4·8	8·394129	587·7021	29·92373	10·8	53·22073	1104·114	337·3058
4·9	8·645876	590·6023	30·49494	10·9	56·20106	1138·448	379·4122
5·0	8·902414	593·5577	31·08536	11·0	59·57200	1177·282	431·5900
5·1	9·164006	596·5713	31·69598	11·1	63·43234	1221·754	497·6597
5·2	9·430732	599·6440	32·32776	11·2	67·92089	1273·463	583·5623
5·3	9 702832	602·7787	32·98178	11·3	73·23965	1334·736	699·0180
5·4	9·980470	605·9771	33·65917	11·4	79·69880	1409·147	860·9854
5·5	10·26387	609·2420	34·36112	11·5	87·80466	1502·528	1101·650
5·6	10·55323	612·5754	35·08897	11·6	98·46010	1625·281	1489·565
5·7	10 84876	615·9800	35·84405	11·7	113·4919	1798·450	2197·699
5·8	11·15067	619·4581	36·62784	11·8	137·4363	2074·295	3802·170
5·9	11·45928	623·0133	37·44196	11·9	187·1806	2647·359	9705·187

TABLE II.

No. of Example.	Receiver C.			Pump of Compression A. Length 12 inches. Sectional Area=1 square inch.				Cylinder of Expansion B. Sectional Area=1 square inch.					Work evolved by the engine by each stroke of the piston, in foot-pounds.	Heat communicated to the air in receiver C, in degrees Fahr. per capacity of a lb. of water.	Work evolved out of each degree Fahr. in the capacity of a lb. of water, in foot-pounds.	Difference between the numbers in columns 4 and 13, divided by the numbers in column 4, and multiplied by the mechanical equivalent of heat.
	Density of the air, that of the atmosphere being called unity.	Pressure of the air in lb. on the square inch.	Absolute temperature of the air, in degrees Fahr. from the absolute zero.	Length of the first part of the stroke.	Work of the engine absorbed by the first part of the stroke of the piston, in foot-pounds.	Work of the engine absorbed by the second part of the stroke of the piston, in foot-pounds.	Absolute temperature of the air forced into receiver C, in degrees Fahr. from the absolute zero.	Length of cylinder B, in inches.	Length of the first part of the piston's stroke, in inches.	Work communicated to the engine by the first part of the stroke of the piston, in foot-pounds.	Work communicated to the engine by the second part of the stroke, in foot-pounds.	Absolute temperature of the air escaping into the atmosphere, in degrees Fahr. from absolute zero.				
1.	1	25·95104	849·464	4	1·537154	7·300693	566·3094	18	12	10·95104	2·305731	736·5	4·418924	0·04304312	102·6628	102·6626
2.	1	66·24102	2168·289	8	10·11797	17·08034	722·7632	36	12	51·24102	30·35391	1473·0	54·39662	0·2197384	247·5517	247·5515
3.	2	66·24102	1084·145	8	10·11797	17·08034	722·7632	18	6	25·62051	15·17695	736·5	13·59915	0·0549346	247·5517	247·5515
4.	2	169·0827	2767·321	10	24·95073	25·68045	922·4402	36	6	77·04135	74·85219	1473·0	101·2624	0·2804454	361·0769	361·0770
5.	4	169·0827	1383·660	10	24·95073	25·68045	922·4402	18	3	38·52067	37·42610	736·5	25·31569	0·07011137	361·0769	361·0770
6.	8	431·59	1765·923	11	45·82200	34·71583	1177·282	18	1·5	52·07375	68·73300	736·5	40·26892	0·08948094	450·0278	450·0279
7.	20	1101·65	1803·034	11·5	73·42966	45·27708	1502·528	14·4	0·6	54·33250	88·11559	589·2	23·74135	0·04568073	519·7236	519·7238
8.	100	9705·187	3176·831	11·9	172·3056	80·75156	2647·359	14·4	0·12	96·90187	206·7667	589·2	50·61143	0·0804865	628·8189	628·8189
1	2	3	4	5	6	7	8	9	10	11	12	13	14	15	16	17

It may be remarked in the first place that the receiver C need not be of much greater capacity than the cylinder B; for in the reciprocating engine the air could be introduced from the pump A at the same time that an equal amount would be expelled into the cylinder B. It would therefore be only requisite to pass the air through tubes heated by a proper furnace, as in Neilson's *hot blast*, the tubes themselves constituting the receiver C. For a temperature under the red heat, these tubes might be constructed of wrought or cast iron. They might be either straight, like the tubes of a locomotive boiler, or arranged in the form of a coil, as represented by fig. 79, in which *a* is the pipe which conveys the air from the pump, *c c c* &c. is the coil of wrought or cast-iron tubing, and *b* is the pipe which conveys the heated air to the cylinder. The coil is surrounded by a massive arch of brickwork, which serves at once to support the pipes and to prevent waste of heat. To prevent the temperature exceeding the proper limits, the pipe *b* might, as it expands by the heat of the inclosed air, move a piece of mechanism in connexion with the damper of the flue. I may remark that, on the scale adopted, fig. 79 represents the size of receiver which would be required for an engine the cylinder of which is 3 feet in diameter.

I would here venture to suggest whether the combustion of the fuel could not, by suitable mechanical arrangements, be carried on within the receiver C; if this could be accomplished, the heat, which in the form of receiver already described is lost up the chimney, would be economized, and a great saving of weight and space would be effected. An engine furnished with a receiver of this kind would be strikingly analogous to the electro-magnetic engine, and present a beautiful illustration of the evolution of mechanical effect from chemical forces.

In both of the above forms of receiver it would be desirable, as already hinted, that the introduction of the air into the receiver should be simultaneous with the expulsion of the same quantity into the cylinder. This is necessary in order both to keep the pressure in the receiver uniform and to promote the smooth action of the engine. For this purpose the piston-rods of the pump and cylinder, *a* and *b* (fig. 80), must be attached to cranks on different parts of the circumference of the revolving shaft *c c*, so contrived that the piston shall arrive at the top or bottom of the cylinder the moment that the pump-valve opens, admitting a fresh supply of air into the receiver. The cylinder should of course be provided with proper expansion-gear to cut off the air at the required part of the stroke, which must be a constant quantity for each engine. The valves of the pump would of course be self-acting.

In an engine similar to that described, it will be obvious that if the temperature of the receiver be kept constant, the pressure of air in it will also remain constant. For whilst the same quantity of air is always introduced into the receiver by each stroke of the pump, the quantity expelled out of it would increase with an augmentation and decrease with a diminution of pressure.

In conclusion, I would recommend the examples No. 3 and No. 5 of Table II. to the attention of those who may be willing to construct an air-engine. In both of these cases the capacity of the pump is two thirds of that of the cylinder. In the cylinder of No. 3 the air is to be cut off at one third of the stroke; and in that of No. 5 at one sixth of the stroke. The temperature of the air in the receiver (supposing that of the atmosphere to be 32° Fahr.) is 625°·145 Fahr. in No. 3, and 924°·66 Fahr. in No. 5. The consumption of fuel in No. 3 need not exceed one half, nor that in No. 5 one third of that in the most perfect steam-engines at present constructed.

Acton Square, Salford, Manchester,
May 6, 1851.

Note to the foregoing Paper, with a New Experimental Determination of the Specific Heat of Atmospheric Air. *

[Received March 23, 1852.]

SINCE the above was written, Professor W. H. Miller has directed my attention to the probable incorrectness of the value of k, as deduced from the experiments of Delaroche and Berard on the specific heat of air and my own determination of the mechanical equivalent of heat, in comparison with the value deduced from the numerous and excellent experiments on the velocity of sound. Mr. Rankine considers that the discrepancy between the two values arises from the incorrectness of Delaroche and Berard's result, an opinion which would seem to be justified by the entire want of accordance between the determination of these philosophers and those of Suermann, and Clement and Desormes. I have therefore been induced to make the following careful experiments in order to obtain a fresh and, if possible, more correct value of the specific heat of air at constant pressure.

The apparatus I employed is represented by fig. 81, in which a and b are two vessels, each of which contains a coil of leaden piping eight yards long and one quarter of an inch in internal diameter. The coil of the upper vessel passes three eighths of an inch through the bottom, to which it is soldered at c, and is thence connected with the coil of the lower vessel by a piece of vulcanized india-rubber tubing. This part of the apparatus will be better understood by a reference to fig. 82, in which a section of it is represented—a being the upper, b the lower vessel, and w the surface of the water in the latter. x x is a pair of wooden pincers by means of which the india-rubber tube could be compressed so as to prevent, when desired, any communication between the air in the two coils of piping. Referring again to fig. 81, g is a gas-lamp to maintain the water in the upper vessel at a constant high temperature, and j is a tall jar filled with

* The experiments were made at Acton Square, Salford.

coarsely pounded chloride of calcium, in passing through which the air was entirely deprived of aqueous vapour; a length of vulcanized india-rubber tubing, *p*, connects the coil of the lower vessel with a good air-pump, each barrel of which was found to have a capacity of 12·77 cubic inches. The temperature of the pump could be ascertained by means of a small thermometer, the bulb of which was kept in contact with one of the barrels.

The method of experimenting was as follows:—The lower vessel being filled with cold water, and the upper with water raised to about 190°, their exact temperatures were read off, with the usual precautions, from the scales of delicate and accurate thermometers. The pump was then worked at a uniform velocity for twenty-six minutes, the water in the lower vessel being agitated from time to time by a stirrer. The examination of the barometer and thermometers a second time occupied four minutes more; so that the whole time occupied by each experiment was exactly half an hour. The pincers were now applied so as to cut off all communication between the air in the two coils, and the effect of the various causes of a change of temperature in the lower vessel, unconnected with the current of heated air, was observed during another half-hour. Experiments of both the above kinds were repeated several times with the results tabulated below.

I may remark in this place that I have ascertained, by preliminary experiments, that the air passed from the coils of the vessels sensibly at the temperatures registered by the thermometers plunged into the surrounding water.

It will be observed that the excess of the temperature of the room above the mean temperature of the water in the lower vessel was, in the experiments with heated air 2°42, but in the experiments on the effect of radiation 2°·459. A comparison of the several experiments with one another furnished the means of determining the amount of the small correction due to this circumstance. Hence 0°·925 + 0°·002 − 0°·448 = 0°·479 will be the corrected mean increase of temperature due to the current of heated air. The material

SERIES I.—Pump worked 26 minutes, at the rate of twenty-four strokes per minute.

No. of experiment.	Source of calorific effect.	Height of barometer.	Temperature of barometer.	Temperature of air-pump.	Temperature of upper vessel.	Temperature of the room.	Temperature of the lower vessel.		Increase of temperature.
							Commencement of experiment.	Termination of experiment.	
1.	Radiation	....	°	°	°	° 46·081	° 41·270	° 41·814	° 0·544
1.	Heated air and radiation	30·195	46	49·3	189·28	46·188	41·814	42·802	0·988
2.	Radiation	....	....	....	....	46·497	42·802	43·304	0·502
2.	Heated air and radiation	30·205	46·75	50·3	189·43	46·785	43·304	44·246	0·942
3.	Radiation	....	....	....	....	46·948	44·246	44·694	0·448
3.	Heated air and radiation	30·22	47·5	51·1	189·89	47·068	44·694	45·590	0·896
4.	Radiation	....	....	....	....	47·197	45·590	45·983	0·393
4.	Heated air and radiation	30·235	48	51·7	194·85	47·283	45·983	46·856	0·873
5.	Radiation	....	....	....	....	47·455	46·856	47·211	0·355
Mean.	Heated air and radiation	30·214	47·06	50·6	190·862	46·831	43·949	44·874	0·925
Mean.	Radiation	....	....	....	....	46·836	44·153	44·601	0·448

SERIES II.—Pump worked 26 minutes, at the rate of forty strokes per minute.

No. of experiment.	Source of calorific effect.	Height of barometer.	Temperature of barometer.	Temperature of air-pump.	Temperature of upper vessel.	Temperature of the room.	Temperature of the lower vessel.		Increase of temperature.
							Commencemen of experiment.	Terminationt of experiment.	
1.	Radiation	....	....	....	....	47·223	44·200	44·648	0·448
1.	Heated air and radiation	30·6	47·75	52	197·71	47·558	44·648	45·902	1·254
2.	Radiation	....	....	....	....	47·841	45·902	46·319	0·417
2.	Heated air and radiation	30·602	48·25	53·5	198·63	48·099	46 319	47·516	1·197
3.	Radiation	....	....	....	....	48·339	47·516	47·860	0·344
3.	Heated air and radiation	30·61	49·5	55·4	202·42	49·107	49·327	50·443	1·116
4.	Radiation	....	....	....	....	49·524	50·443	50·728	0·285
4.	Heated air and radiation	30·607	50·25	56·4	203·13	49·850	50·728	51·809	1·081
5.	Radiation	....	....	....	....	50·030	51·809	52·037	0·228
Mean.	Heated air and radiation	30·605	48·94	54·32	200·472	48·653	47·755	48·917	1·162
Mean.	Radiation	....	....	....	....	48·591	47·974	48·318	0·344

in which this increase took place consisted of 175500 grs. of water, 15635 grs. of copper, and 53370 grs. of lead, the whole having a capacity for heat equivalent to that of 178535 grs. of water. The volume of air passed through the pump was $12{\cdot}77 \times 26 \times 24 = 7968{\cdot}48$ cubic inches, which, at the observed barometric pressure and the temperature $50^{\circ}{\cdot}6$, would weigh 2537·94 grs. We have therefore for the specific heat of atmospheric air at constant pressure,

$$\frac{178535 \times 0{\cdot}479}{2537{\cdot}94 \times 146{\cdot}45} = 0{\cdot}23008.$$

In Series II., $1^{\circ}{\cdot}162 + 0^{\circ}{\cdot}006 - 0^{\circ}{\cdot}344 = 0^{\circ}{\cdot}824$ will be the corrected mean increase of temperature due to the current of heated air. The material in which this increase took place consisted of 175000 grs. of water, 15635 grs. of copper, and 53370 grs. of lead, the whole having a capacity for heat equivalent to that of 178035 grs. of water. The volume of air passed through the pump was $12{\cdot}77 \times 26 \times 40 = 13280{\cdot}8$ cubic inches, which, at the observed barometric pressure and the temperature $54^{\circ}{\cdot}32$, would weigh 4252·7 grs. Hence we have for the specific heat,

$$\frac{178035 \times 0{\cdot}824}{4252{\cdot}7 \times 152{\cdot}136} = 0{\cdot}22674.$$

By another series of experiments, in which the air-pump was worked at the velocity of twenty strokes per minute for twenty minutes, I obtained the value 0·2325. The mean of the three results is 0·22977, or nearly 0·23, which we may take as the specific heat of air at constant pressure determined by the above experiments.

Professor W. H. Miller has remarked that Moll's experiments, when correctly reduced, give a velocity of sound equal to 332·475 metres per second in dry air at 32°. Hence he deduces 1·41029 as the value of k. Calling it in round numbers 1·41, and the mechanical equivalent of heat 772, we obtain 0·238944 as the value of the specific heat of air at constant pressure—a result sufficiently near the experimental determination to show that the value of k, as deduced by Professor Miller, is much nearer the truth than that upon which Tables I. & II. of the foregoing paper are founded.

The values of k, as determined by the experiments of Desormes and Clement, Gay-Lussac and Welter, and Mr. Meikle, referred to in the note to page 335, are respectively only 1·354, 1·375, and 1·333. In these experiments, a small portion of air having been drawn from a large receiver, the equilibrium was re-established by opening for an instant a large aperture communicating with the external air, and then, after the receiver and its contents had regained their original temperature, the alteration of pressure, indicating the sudden rise of temperature which had taken place on the admission of the air, was noted. But it is obvious that the sudden admission of the air would cause the development of *sound*, and that, a portion of the *vis viva* escaping in this form, the increase of temperature and the deduced ratio of the specific heats would be diminished accordingly.

I subjoin Tables, similar to Tables I. and II., calculated from the data $k=1{\cdot}41$, and the specific heat of air at constant volume $=0{\cdot}169464$, or at constant pressure $=0{\cdot}238944$.

In Table IV., the examples 9, 10, and 11 may be suggested to the notice of the practical engineer, the temperature of the receiver being in all those cases below that of redness. I may remind the reader that the table is founded on the supposition that the air which enters the pump has 491° of temperature from the absolute zero, and that its pressure is 15 lb. on the square inch. If this initial temperature be altered, the whole of the other temperatures in the table must be altered in the same proportion, but the pressure, work, and economical duty will remain unchanged. If the initial pressure be altered, all the other pressures and work will suffer a proportionate change, but the temperatures and economical duty will remain the same. The above are obvious deductions from the formulæ on which the Tables are founded.

Acton Square, Salford,
March 20, 1852.

Table III.

Distance traversed by piston, in inches.	Work absorbed, in foot-pounds.	Temperature from absolute zero, in degrees Fahr.	Pressure on the piston, in lbs.	Distance traversed by piston, in inches.	Work absorbed, in foot-pounds.	Temperature from absolute zero, in degrees Fahr.	Pressure on the piston, in lbs.
0	0	491	15	6·0	12·025096	652·3847	39·86055
0·1	0·1257463	492·6876	15·17803	6·1	12·36122	656·8958	40·81647
0·2	0·2529680	494·3950	15·35970	6·2	12·70547	661·5159	41·81223
0·3	0·3817395	496·1232	15·54513	6·3	13·05820	666·2498	42·85023
0·4	0·5120683	497·8723	15·73438	6·4	13·41977	671·1023	43·93309
0·5	0·6439993	499·6429	15·92769	6·5	13·79055	676·0785	45·06354
0·6	0·7775396	501·4351	16·12503	6·6	14·17096	681·1838	46·24464
0·7	0·9127567	503·2498	16·32661	6·7	14·56146	686·4245	47·47966
0·8	1·049665	505·0872	16·53253	6·8	14·96245	691·8060	48·77217
0·9	1·188301	506·9478	16·74292	6·9	15·37448	697·3358	50·12596
1·0	1·328719	508·8323	16·95793	7·0	15·79807	703·0207	51·54529
1·1	1·471827	510·7529	17·17773	7·1	16·23377	708·8680	53·03472
1·2	1·615031	512·6748	17·40240	7·2	16·68220	714·8862	54·59923
1·3	1·761007	514·6339	17·63216	7·3	17·14397	721·0835	56·24433
1·4	1·908922	516·6190	17·86716	7·4	17·61984	727·4700	57·97600
1·5	2·058809	518·6306	18·10755	7·5	18·11050	734·0550	59·80080
1·6	2·210724	520·6694	18·35353	7·6	18·61680	740·8498	61·72605
1·7	2·364719	522·7361	18·60529	7·7	19·13958	747·8659	63·75969
1·8	2·520828	524·8312	18·86299	7·8	19·67980	755·1160	65·91058
1·9	2·679114	526·9555	19·12686	7·9	20·23844	762·6133	68·18854
2·0	2·839636	529·1098	19·39710	8·0	20·81664	770·3732	70·60445
2·1	3·002421	531·2945	19·67392	8·1	21·41559	778·4115	73·17045
2·2	3·167547	533·5106	19·95757	8·2	22·03660	786·7459	75·90002
2·3	3·335073	535·7589	20·24830	8·3	22·68110	795·3954	78·80838
2·4	3·505041	538·0400	20·54632	8·4	23·35061	804·3808	81·91249
2·5	3·677529	540·3549	20·85193	8·5	24·04690	813·7254	85·23164
2·6	3·852596	542·7044	21·16540	8·6	24·77180	823·4542	88·78743
2·7	4·030240	545·0885	21·48700	8·7	25·52742	833·5950	92·60451
2·8	4·210744	547·5110	21·81705	8·8	26·31602	844·1786	96·71086
2·9	4·393947	549·9697	22·15585	8·9	27·14016	855·2390	101·1385
3·0	4·580211	552·4695	22·50375	9·0	28·00265	866·8144	105·9244
3·1	4·769047	555·0038	22·86105	9·1	28·90667	878·9468	111·1106
3·2	4·961064	557·5808	23·22823	9·2	29·85575	891·6840	116·7460
3·3	5·156197	560·1996	23·60557	9·3	30·85385	905·0792	122·8892
3·4	5·354524	562·8613	23·99352	9·4	31·90550	919·1930	129·6058
3·5	5·556132	565·5670	24·39246	9·5	33·01577	934·0936	136·9750
3·6	5·761092	568·3177	24·80292	9·6	34·19049	949·8591	145·0905
3·7	5·969523	571·1150	25·22533	9·7	35·43632	966·5791	154·0638
3·8	6·181561	573·9607	25·66015	9·8	36·76094	984·3564	164·0289
3·9	6·397243	576·8553	26·10795	9·9	38·17335	1003·312	175·1490
4·0	6·616735	579·8010	26·56929	10·0	39·68387	1023·584	187·6224
4·1	6·840106	582·7988	27·04473	10·1	41·30480	1045·338	201·6945
4·2	7·067466	585·8501	27·53490	10·2	43·05077	1068·770	217·6720
4·3	7·299071	588·9584	28·04045	10·3	44·93904	1094·112	235·9413
4·4	7·534902	592·1234	28·56207	10·4	46·99081	1121·648	256·9966
4·5	7·775119	595·3475	29·10049	10·5	49·23177	1151·723	281·4802
4·6	8·019950	598·6331	29·65651	10·6	51·69402	1184·768	310·2390
4·7	8·269468	601·9818	30·23092	10·7	54·41741	1221·318	344·4105
4·8	8·523854	605·3958	30·82462	10·8	57·45355	1262·065	385·5594
4·9	8·783274	608·8774	31·43856	10·9	60·86887	1307·901	435·8856
5·0	9·047890	612·4287	32·07366	11·0	64·75245	1360·021	498·5821
5·1	9·317890	616·0523	32·73102	11·1	69·22596	1420·059	578·4345
5·2	9·593477	619·7509	33·41175	11·2	74·46110	1490·318	682·9350
5·3	9·874827	623·5267	34·11703	11·3	80·71017	1574·184	824·4195
5·4	10·16216	627·3830	34·84815	11·4	88·36276	1676·887	1024·575
5·5	10·45581	631·3240	35·60646	11·5	98·06077	1807·041	1324·918
5·6	10·75570	635·3486	36·39342	11·6	110·9605	1980·164	1814·815
5·7	11·06235	639·4641	37·21060	11·7	129·4314	2228·056	2722·671
5·8	11·37594	643·6727	38·05962	11·8	159·4567	2631·016	4822·635
5·9	11·69678	647·9786	38·94231	11·9	223·8930	3495 794	12815·505

TABLE IV.

No. of Example.	Receiver C.			Pump of Compression A. Length 12 inches. Sectional Area=1 square inch.				Cylinder of Expansion B. Sectional Area=1 square inch.					Work evolved by the engine by each stroke of the piston, in foot-pounds.	Heat communicated to the air in receiver C, in degrees Fahr. per capacity of a lb. of water.	Work evolved out of each degree Fahr. in the capacity of a lb. of water, in foot-pounds.	Difference between the numbers in columns 4 and 13, divided by the numbers in column 4, and multiplied by the mechanical equivalent of heat.
	Density of the air, that of the atmosphere being called unity.	Pressure of the air in pounds on the square inch.	Absolute temperature of the air, in degrees Fahr. from the absolute zero.	Length of the first part of the stroke.	Work of the engine absorbed by the first part of the stroke of the piston, in foot-pounds.	Work of the engine absorbed by the second part of the stroke of the piston, in foot-pounds.	Absolute temperature of the air forced into the receiver C, in degrees Fahr. from the absolute zero.	Length of the cylinder B, in inches.	Length of the first part of the piston's stroke, in inches.	Work communicated to the engine by the first part of the stroke, in foot-pounds.	Work communicated to the engine by the second part of the stroke, in foot-pounds.	Absolute temperature of the air escaping into the atmosphere, in degrees Fahr. from absolute zero.				
1.	1	26·56929	869·7014	4	1·616735	7·71286	579·801	18	12	11·56929	2·425102	736·5	4·664797	0·03945268	118·2378	118·2377
2.	1	70·60445	2311·119	8	10·81664	18 53482	770·3732	36	12	55·60446	32·44992	1473·0	58·70292	0·2096809	279·9632	279·9630
3.	2	70·60445	1155·559	8	10·81664	18·53482	770·3732	18	6	27·80223	16·22496	736·5	14·67573	0·05242022	279·9632	279·9628
4.	2	187·6224	3070·753	10	27·18387	28·7704	1023·584	36	6	86·3112	81·55161	1473·0	111·90854	0·2786002	401·6815	401·6817
5.	4	187·6224	1535·3765	10	27·18387	28·7704	1023·584	18	3	43·1556	40·7758	736·5	27·97713	0·06965	401·6815	401·6815
6.	8	498·5821	2040·032	11	51·00245	40·29851	1360·021	18	1·5	60·44776	76·50367	736·5	45·65047	0·092543	493·2893	493·2897
7.	20	1324·918	2168·449	11·5	83·68577	54·57992	1807·041	14·4	0·6	65·4959	100·4229	589·2	27·65313	0·04918419	562·2364	562·2361
8.	100	12815·505	4194·953	11·9	209·018	106·6709	3495·794	14·4	0·12	128·0051	250·8216	589·2	63·1378	0·0951489	663·5685	663·5692
9.	3	105·92437	1155·7525	9	16·75265	22·73109	866·8144	16	4	30·30812	22·33687	654·666	13·16125	0·03932171	334·7069	334·7069
10.	2·5	105·92437	1386·903	9	16·75265	22·73109	866·8144	19·2	4·8	36·36974	26·80424	785·6	23·69024	0·0707791	334·7068	334·7068
11.	4·5	187·6224	1364·779	10	27·18387	28·7704	1023·584	16	$2\frac{2}{3}$	38·36053	36·24516	654·666	18·65142	0·0464334	401·6814	401·6816
1	2	3	4	5	6	7	8	9	10	11	12	13	14	15	16	17

Additional Note on the preceding Paper. By WILLIAM THOMSON, *M.A., F.R.S., F.R.S.E., Fellow of St. Peter's College, Cambridge, and Professor of Natural Philosophy in the University of Glasgow.*

1. *Synthetical Investigation of the Duty of a Perfect Thermo-Dynamic Engine, founded on the Expansions and Condensations of a Fluid for which the gaseous laws hold and the ratio* (k) *of the specific heat under constant pressure to the specific heat in constant volume is constant; and modification of the result by the assumption of* MAYER'S *hypothesis.*

LET the source from which the heat is supplied be at the temperature S, and let T denote the temperature of the coldest body that can be obtained as a refrigerator. A cycle of the following four operations, *being reversible in every respect*, gives, according to Carnot's principle, first demonstrated for the Dynamical Theory by Clausius, the greatest possible statical mechanical effect that can be obtained in these circumstances from a quantity of heat supplied from the source.

(1) Let a quantity of air contained in a cylinder and piston, at the temperature S, be allowed to expand to any extent, and let heat be supplied to it to keep its temperature constantly S.

(2) Let the air expand further, without being allowed to take heat from or to part with heat to surrounding matter, until its temperature sinks to T.

(3) Let the air be allowed to part with heat so as to keep its temperature constantly T, while it is compressed to such an extent that at the end of the fourth operation the temperature may be S.

(4) Let the air be further compressed, and prevented from either gaining or parting with heat, till the piston reaches its primitive position.

The amount of mechanical effect gained on the whole of this cycle of operations will be the excess of the mechanical effect obtained by the first and second above the work spent in the third and fourth. Now if P and V denote the primitive pressure and volume of the air, and P_1 and V_1, P_2 and V_2, P_3 and V_3, P_4 and V_4 denote the pressure and volume respectively, at the ends of the four successive operations, we have by the gaseous laws, and by Poisson's formula and a conclusion from it quoted above, the following expressions :—

Mechanical effect obtained by the first operation

$$=PV \log \frac{V_1}{V}.$$

Mechanical effect obtained by the second operation

$$=P_2V_2 \cdot \frac{1}{k-1} \cdot \left\{ \left(\frac{V_2}{V_1}\right)^{k-1} - 1 \right\}.$$

Work spent in the third operation

$$=P_3V_3 \log \frac{V_2}{V_3}.$$

Work spent in the fourth operation

$$=P_3V_3 \cdot \frac{1}{k-1} \left\{ \left(\frac{V_3}{V_4}\right)^{k-1} - 1 \right\}$$

Now, according to the gaseous laws, we have

$$P_1V_1 = PV; \quad P_2V_2 = P_1V_1 \frac{1+ET}{1+ES};$$

$$P_3V_3 = P_2V_2; \quad \text{and (since } V_4 = V) \; P_4 = P.$$

Also by Poisson's formula,

$$\left(\frac{V_2}{V_1}\right)^{k-1} = \left(\frac{V_3}{V}\right)^{k-1} = \frac{1+ES}{1+ET}.$$

By means of these we perceive that the work spent in the fourth operation is equal to the mechanical effect gained in the second; and we find, for the whole gain of mechanical effect (denoted by M), the expressions

$$M=(PV-P_3V_3)\log\frac{V_1}{V}=PV\log\frac{V_1}{V}.\frac{E(S-T)}{1+ES}.$$

All the preceding formulæ are founded on the assumption of the gaseous laws and the constancy of the ratio (k) of the specific heat under constant pressure to the specific heat in constant volume, for the air contained in the cylinder and piston, and involve no other hypothesis*. If now we add the assumption of Mayer's hypothesis, which for the actual circumstance is $PV\log\frac{V_1}{V}=JH$, where H denotes the heat abstracted by the air from the surrounding matter in the first operation, and J the mechanical equivalent of a thermal unit, we have

$$M=JH.\frac{E(S-T)}{1+ES}.$$

The investigation of this formula given in my paper on the Dynamical Theory of Heat shows that it would be true for every perfect thermo-dynamic engine, if Mayer's hypothesis were true for a fluid subject to the gaseous laws of pressure and density, whether, for such a fluid (did it exist), k were constant or not.

* From the sole hypothesis that k is constant for a single fluid fulfilling the gaseous laws, and having E for its coefficient of expansion, I find it follows, as a necessary consequence, that Carnot's function would have the form $\frac{JE}{1+Et+C}$; where C denotes an unknown absolute constant, and t the temperature measured by a thermometer founded on the equable expansions of that gas. From this it follows that for such a gas, subjected to the four operations described in the text, we must have

$$PV\log\frac{V_1}{V}=JH\frac{1+ES}{1+ES+C};$$

and consequently

$$M=JH\frac{E(S-T)}{1+ES+C},$$

which is Mr. Rankine's general formula.

It was first obtained by using, in the formula

$$M = JH\epsilon^{-\frac{1}{J}\int_{T}^{S}\mu dt},$$

which involves no hypothesis, the expression

$$\mu = \frac{J}{\frac{1}{E}+t}$$

for Carnot's function, which Mr. Joule had suggested to me, in a letter dated December 9, 1848, as the expression of Mayer's hypothesis, in terms of the notation of my "Account of Carnot's Theory" *. Mr. Rankine† has arrived at a formula agreeing with it (with the exception of a constant term in the denominator, which, as its value is unknown, but probably small, he neglects in the actual use of the formula), as a consequence of the fundamental principles assumed in his Theory of Molecular Vortices, when applied to any fluid whatever, experiencing a cycle of four operations satisfying Carnot's criterion of reversibility (being, in fact, precisely analogous to those described above, and originally invented by Carnot); and he thus establishes Carnot's law as a consequence of the equations of the mutual conversion of heat and expansive power, which had been given in the first section of his paper on the Mechanical Action of Heat‡.

2. *Note on the Specific Heats of Air.*

Let N be the specific heat of unity of weight of a fluid at the temperature t, kept within constant volume v; and let kN be the specific heat of the same fluid mass, under constant pressure p. Without any other assumption than that of

* Royal Society of Edinburgh, January 2, 1849, Transactions, vol. xvi. part 5.

† "On the Economy of Heat in Expansive Engines." Royal Society of Edinburgh, April 21, 1851, Transactions, vol. xx. part 2.

‡ Royal Society of Edinburgh, February 4, 1850, Transactions, vol. xx. part 1.

Carnot's principle, the following equation is demonstrated in my paper* on the Dynamical Theory of Heat, § 48,

$$kN-N=\frac{\left(\frac{dp}{dt}\right)^2}{\mu\times-\frac{dp}{dv}},$$

where μ denotes the value of Carnot's function for the temperature t, and the differentiations indicated are with reference to v and t considered as independent variables, of which p is a function. If the fluid be subject to Boyle's and Mariotte's law of compression, we have

$$\frac{dp}{dv}=-\frac{p}{v};$$

and if it be subject also to Gay-Lussac's law of expansion,

$$\frac{dp}{dt}=\frac{Ep}{1+Et}.$$

Hence, for such a fluid,

$$kN-N=\frac{E^2pv}{\mu(1+Et)^2}\dagger.$$

In the case of dry air these laws are fulfilled to a very high degree of approximation, and for it, according to Regnault's observations,

$$\frac{pv}{1+Et}=26215,\ E=\cdot00366$$

(a British foot being the unit of length, and the weight of a British pound at Paris the unit of force).

We have consequently, for dry air,

$$kN-N=\frac{26215E^2}{\mu(1+Et)}\quad . \ . \ . \ . \ . \ . \ . \quad (1)$$

* Royal Society of Edinburgh, March 17, 1851, Transactions, vol. xx. part 2.

† This equation expresses a proposition first demonstrated by Carnot. See "Account of Carnot's Theory," Appendix III. (Transactions Royal Society of Edinburgh, vol. xvi. part 5).

Now it is demonstrated, without any other assumption than that of Carnot's principle, in my "Account of Carnot's Theory" (Appendix III.), that

$$\frac{E}{\mu(1+Et)}=\frac{H}{W},$$

if W denote the quantity of work that must be spent in compressing a fluid subject to the gaseous laws, to produce H units of heat when its temperature is kept at t. Hence

$$kN-N=26215E\times\frac{H}{W}=95{\cdot}947\times\frac{H}{W}. \quad . \quad . \quad . \quad (2)$$

If we adopt the values of μ shown in Table I. of the "Account of Carnot's Theory," depending on no uncertain data except the densities of saturated steam at different temperatures, which, for want of accurate experimental data, were derived from the value $\frac{1}{1693{\cdot}5}$ for the density of saturated vapour at 100°, by the assumption of the "gaseous laws" of variation with temperature and pressure, we find 1357 and 1369 for the values of $\frac{E}{\mu(1+Et)}$ at the temperatures 0 and 10° respectively; and hence, for these temperatures,

$$\left.\begin{array}{ll}(t=0) & kN-N=\frac{95{\cdot}947}{1357}={\cdot}07071 \\ (t=10^\circ) & kN-N=\frac{95{\cdot}947}{1369}={\cdot}07008\end{array}\right\} \quad . \quad . \quad . \quad (a)$$

Or, if we adopt Mayer's hypothesis, according to which $\frac{W}{H}$ is equal to the mechanical equivalent of the thermal unit*,

* The number 1390, derived from Mr. Joule's experiments on the friction of fluids, cannot differ by $\frac{1}{100}$, and probably does not differ by $\frac{1}{300}$, of its own value from the true value of the mechanical equivalent of the thermal unit.

we have $\frac{W}{H}=1390$; and hence, for all temperatures,

$$kN-N=\frac{95{\cdot}947}{1390}={\cdot}06903. \quad . \quad . \quad . \quad . \quad (a')$$

The very accurate observations which have been made on the velocity of sound in air, taken in connexion with the results of Regnault's observations on its density &c., lead to the value 1·410 for k, which is probably true in three, if not in four, of its figures. Now k being known, the preceding equations enable us to determine the absolute values of the two specific heats (kN and N) according to the hypotheses used in (a) and in (a') respectively; and we thus find,

	Specific heat of air under constant pressure (kN).	Specific heat of air in constant volume (N).
for $t=0$, ...	·2431 . . .	·1724,
for $t=10$, ...	·2410 . . .	·1709,

according to the tabulated values of Carnot's function.

Or, for all temperatures, .	·2374 . . .	·1684,

according to Mayer's hypothesis.

By the adoption of hypotheses involving that of Mayer, and taking 1389·6 and 1·4 as the values of J and k respectively, Mr. Rankine finds ·2404 and ·1717 as the values of the two specific heats.

Hence it is probable that the values of the specific heat of air under constant pressure, found by Suermann (·3046) and by De la Roche and Berard (·2669), are both considerably too great; and the true value, to two significant figures, is probably ·24.

Glasgow College,
February 19, 1852.

Account of Experiments with a powerful Electro-Magnet. *By* J. P. JOULE, *F.R.S. &c.**

['Philosophical Magazine,' ser. 4. vol. iii. p. 32.]

SOME years ago I announced that if a particle of wire conducting a voltaic current be made to act upon a very large surface of iron, the intensity of the induced magnetism will not be much diminished by an increase in the distance of that particle from the surface of the iron. Guided by this principle, I constructed a very powerful electro-magnet in 1843†, and soon after prepared the iron of the electro-magnet employed in the experiments related in the present paper. This was a plate of the best wrought iron, 1 inch thick, 22 inches long, 12 inches broad at the centre, but tapered thence to the breadth of 3 inches, as represented in the adjoining sketch (fig. 83). The plate was then bent into a semicircular shape, so as to bring its ends within 12 inches of one another. Previously to fitting up this bar as an electro-magnet, I made a few experiments with a view to test the principle above named more completely than I had hitherto done.

Fig. 83.

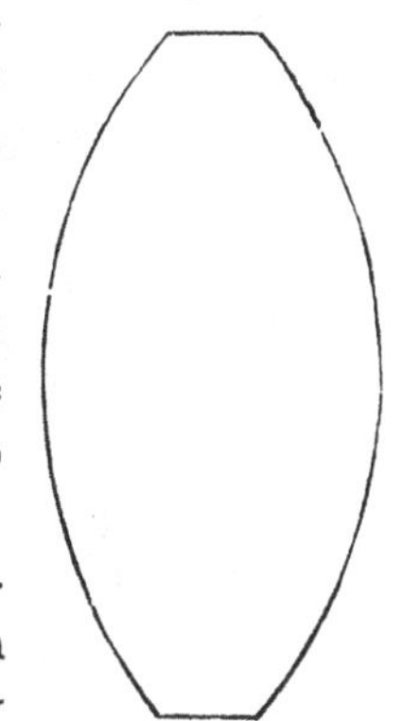

A length of about eight yards of insulated copper wire, $\frac{1}{20}$ of an inch in diameter, was divided into two exactly equal portions, one of which was wound four times round the broadest part of the iron, and close to its surface; the other was also wound four times round the broadest part of the iron, but was kept at the distance of one inch from its

* The experiments were made at Oak Field, Whalley Range, near Manchester.

† Philosophical Magazine, ser. 3. vol. xxiii. p. 268.

surface by means of interposed pieces of wood. A constant current of electricity was alternately passed through the wires; and the deflections of a magnetic needle half an inch long, placed at the distance of two feet from the iron bar, were observed to be as follows:—

6° 23′ with the wire close to the surface of the iron.
6° 9′ with the wire at the distance of one inch from the surface of the iron;

showing only a trifling diminution of effect in consequence of the removal of the wire to the distance of one inch from the surface.

Having been thus fortified in my previous conclusion as to the propriety of enveloping broad electro-magnets with a very large quantity of coils, even though the outer ones should be removed to a considerable distance from the surface of the iron, I proceeded to fit up the large bar already described with a coil consisting of a bundle of copper wires 68 yards long, and weighing 100 lb. The electro-magnet thus formed was placed in a wooden box, on the side of which two large brass clamps were screwed, the latter being soldered to the terminals of the coil. The accompanying sketch represents

Fig. 84. Scale $\frac{1}{12}$.

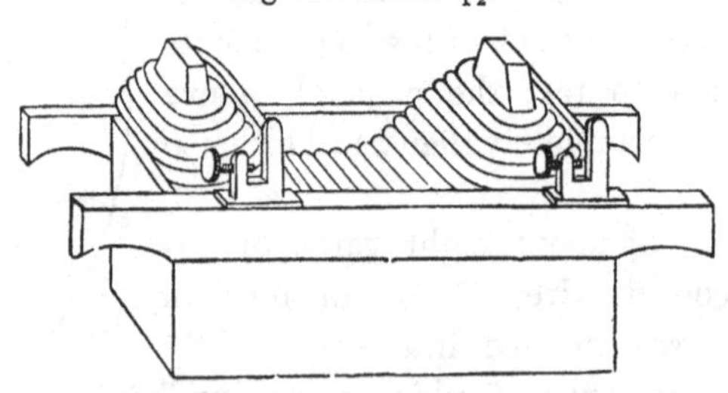

the apparatus in its completed state, excepting, however, two brass straps by means of which the coil is kept securely in its place, which are omitted for the sake of clearness.

In experimenting with the electro-magnet, I employed a battery consisting of sixteen Daniell's cells, the copper of each exposing an active surface of nearly two square feet.

They were arranged so that I could with facility use either one cell alone, four cells in a series of two, or sixteen in a series of four elements. The cells and the liquids in them being similar in every respect, it was evident that these arrangements must produce through the electro-magnetic coils currents represented by 1, 2, and 4. I therefore was enabled to dispense with the use of a galvanometer, which would have been acted upon by the powerful electro-magnet, even if it had been placed at the distance of many yards from it.

Experiment I.—A magnetic needle, 1½ inch long, was suspended at the distance of three feet from the electro-magnet measured on a line at right angles to that joining the poles. The northward tendency of the needle having been counteracted by means of a permanent magnet, I observed the following vibrations per minute resulting from the action of the electro-magnet :—

With	1 cell	in a series of	1	. .	48 vibrations.
„	4 cells	„ „	2	. .	63 „
„	16 „	„ „	4	. .	96 „

The vibrations are evidently in the ratio of the square roots of the quantities of current circulating around the electro-magnet; and consequently we may infer that the magnetism induced in the latter was simply in proportion to the current*.

Experiment II.—Having provided a pair of tapered poles terminating in vertical edges 1 inch long and ⅛ of an inch in breadth, I caused them to be slid on the poles of the electro-magnet until within 1¼ inch from each other. A cylindrical bar of bismuth 1⅜ inch long, ¼ of an inch in diameter, and weighing 174 grains, was suspended by a fila-

* In a paper lately contributed to the 'Philosophical Transactions,' I find that when the coercive force is great, as it is in a steel bar, an electro-magnet made of a very long bar, or an electro-magnet with the armature in contact, the magnetism permanently developed by a weak current is nearly in the proportion to the square of the current to which the iron was exposed.—*Note*, 1856.

ment of silk from a proper support, so as to vibrate between the tapered poles. The average numbers of vibrations in each minute of time through the quadrant of a circle were then found to be—

With 1 cell in a series of 1	. .	$4\frac{1}{4}$	vibrations.
„ 4 cells „ „ 2	. .	$9\frac{1}{2}$	„
„ 16 „ „ „ 4	. .	17	„

The currents being as 1, 2, and 4, and the vibrations $4\frac{1}{2}$, $9\frac{1}{2}$, and 17, or nearly in the same ratio, it follows that the repulsive action of the magnetic poles was as the square of the current, and consequently that the diamagnetism of the bismuth is a quality not self-inherent, but induced by the magnetic action to which it is exposed. I am happy to have been thus enabled to confirm the important fact, discovered by M. Ed. Becquerel and Dr. Tyndall, by experiments made without any knowledge* of the researches they were conducting almost simultaneously on the same subject.

Experiment III.—The tapered poles remaining at $1\frac{1}{4}$ inch asunder, I suspended a piece of soft iron, 3 inches long, 1 inch deep, and $\frac{1}{5}$ of an inch thick, at the distance of a quarter of an inch above the poles. Using one cell of the battery, this small piece of iron was attracted with a force of $6\frac{3}{4}$ oz.; but with 16 cells in a series of 4, with a force of no less than $71\frac{1}{2}$ oz. In this instance we notice a slight falling away from the theoretical attraction, owing no doubt to the gradual approach of the limit to magnetizability in the small bar of iron.

Experiment IV.—The tapered poles having been removed, a flat bar of soft iron, 14 inches long, 3 inches in breadth, and 1 inch thick, was placed at various distances from the poles of the electro-magnet, and the attractions measured as follows:—

* The electro-magnet with which the above experiments had been made was sent to the Exhibition of Industry in the middle of February. M. Becquerel's paper was published in the *Annales de Chimie* for May, Dr. Tyndall's in this Magazine for September, after having been previously communicated to the Ipswich Meeting of the British Association.

	¼ in. dist. oz.	½ in. dist. oz.	1 in. dist. oz.	2 in. dist. oz.
With 1 cell	102	38	13¾	3¾
With 16 cells in a series of 4	976	320	140	47

Here, again, we have evidences of an approach towards the limit of magnetizability; for the attractions with a current of 4 are only ten times, instead of sixteen times, as great as those observed with a current of 1.

The electro-magnet I described some years ago* consisted of a core of iron, half an inch thick, enveloped by a coil of wires weighing 60 lb. With a battery of ten cells, similar to those employed in the present experiments, a bar of iron, 3 inches broad and ½ an inch thick, was attracted at the distance of ¼ of an inch with a force of 480 oz., at ½ an inch with a force of 168 oz., and at 1 inch with a force of 77 oz. Both electro-magnets having been constructed on the same principle, their attractive powers ought to be proportional to the weight of coil and number of cells; and therefore to be represented by $60 \times 10 = 600$, and $100 \times 16 = 1600$. As this is tolerably well borne out on comparing the actual results of the above experiments, we may infer that little or no advantage was obtained by increasing the thickness of the core of iron from half an inch to one inch.

Experiment V.—A flat bar of iron, 1¾ inch deep and ⅕ of an inch thick, being placed with its thin edge in contact with the poles of the electro-magnet, the following weights had to be applied in order to overcome the attraction in contact:—

With 1 cell in a series of	1	. .	64 lb.
„ 4 cells „ „	2	. .	72 „
„ 16 „ „ „	4	. .	96 „

But when the bar of iron used in Experiment IV. was placed in contact with the poles, so as just to leave ¼ of an inch in breadth for the place of contact of the flat bar, the attraction of the latter with 16 cells was found to be only 82 lb. Thus it would appear that 14 lb. out of the 96 lb. in the previous ex-

* Philosophical Magazine, ser. 3. vol. xxiii. p. 268.

periment were owing to the distant attraction of that part of the poles not in contact with the bar. We may therefore conclude that while the attraction in contact, using one cell, was 64 lb., minus say 1 lb. for distant attraction, that produced by a current four times as great was only increased to 82 lb. And it must be remarked that the greater part of this small increase was doubtless owing to the action of the broader part of the iron core which still remained unneutralized. It would therefore appear that the greatest observed attraction in contact was, in this electro-magnet, about $70 \times 5 = 350$ lb. per square inch of the surface of each pole; or, otherwise, that the greatest magnetic attraction of one square inch of surface for another square inch was 175 lb. Several years ago I gave 140 lb. as the apparent limit of attraction in contact*. The force of current employed in obtaining this result was only one tenth of that which, in the present instance, did not produce a greater attraction than 175 lb. It is therefore improbable that any force of current could give an attraction equal to 200 lb. per square inch†.

Experiment VI.—The magnetic needle was suspended as in Experiment I., and its vibrations, with 4 cells in a series of 2, were found to be 63 per minute. I then placed the large bar used in Experiment IV. across the poles, so as to neutralize their action. The number of vibrations of the needle per minute was then found to be 62, or only 1 less than before; showing that the neutralization of the magnetic tension of the poles (which were only 3 inches in breadth, that of the core at its greatest being 12 inches) permitted the tension of the remaining unneutralized breadth of 9 inches to be increased so as to prevent almost any diminution in the action on the needle.

* Philosophical Magazine, ser. 4. vol. ii. p. 453.

† That is, the greatest weight which could be lifted by an electro-magnet formed of a bar of iron one inch square, bent into a semicircular shape, would not exceed 400 lb.—*Note*, 1852.

On the Economical Production of Mechanical Effect from Chemical Forces. By J. P. JOULE, *F.R.S., &c.*

['Manchester Memoirs,' vol. x. p. 173. Read April 6th, 1852.]

PERHAPS the most important applications of dynamical theory are those which refer to the production of motive power from chemical and other actions. To point out the rules for constructing an engine which shall approach perfection as nearly as possible, and to determine the quantity of work which ought to be evolved by a perfect engine of any given class, are objects of the greatest consequence in the present state of society, and which have in fact been to a great extent already accomplished by the labours of those who have taken a correct view of the nature of heat. I intend on the present occasion to submit to the Society some of the laws which have been recently arrived at by Professor Thomson and myself, and to offer some hints as to the means of carrying out into practice the deductions of theory.

Engines which derive their power from the operation of chemical forces may be divided into three classes. The first class comprises those exquisite machines in which chemical forces operate by the mysterious intervention of life, whether in the animal or vegetable creation. The second class includes machines in which the chemical forces act through the intervention of electrical currents, as in the ordinary revolving electro-magnetic apparatus. The third comprises those engines in which the chemical forces act through the intervention of the heat they produce; these, which may be termed thermo-dynamic engines, include steam-engines, air-engines, &c.

The process whereby muscular effort is developed in the living machine is, as might be expected, involved in great obscurity. Professor Magnus has endeavoured to prove that the oxygen inspired by an animal does not immediately enter into combination with the blood, but is mechanically conveyed by it to the capillary vessels within the muscles, where it combines with certain substances, converting them into carbonic acid and water. The carbonic acid, instead of oxygen, is then absorbed by the blood, and is discharged therefrom when it reaches the lungs. Taking this view, we may admit, with Liebig, that at each effort of an animal a portion of muscular fibre unites with oxygen, and that the whole force of combination is converted by some mysterious process into muscular power, without any waste in the form of heat. This conclusion, which is confirmed by the experiments related in a joint memoir by Dr. Scoresby and myself, shows that the animal frame, though destined to fulfil so many other ends, is as an engine more perfect in the economy of *vis viva* than any human contrivance.

The electro-magnetic engine presents some features of similarity to the living machine, and approaches it in the large proportion of the chemical action which it is able to evolve as mechanical force. If we denote the intensity of current electricity when the engine is at rest by a, and the intensity of current when the engine is at work by b, the proportion of chemical force converted into motive force will be $\frac{a-b}{a}$, and the quantity wasted in the form of heat will be $\frac{b}{a}$. Now, from my own experiments, I find that each grain of zinc consumed in a Daniell's battery will raise the temperature of a lb. of water 0°·1886, and that the heat which can increase the temperature of a pound of water by one degree is equal to the mechanical force which is able to raise a weight of 772 lb. to the height of one foot, or, according to the expression generally used, to 772 foot-pounds. Therefore the work developed by a grain of zinc consumed in a Daniell's battery is given by the equation

$$W = \frac{145{\cdot}6\ (a-b)}{a}.$$

We now come to the third class of engines, or those in which the chemical forces act through the intervention of heat. In the most important of these the immediate agent is the elasticity of vapour or permanently elastic fluids. In a very valuable paper on the dynamical theory of heat, Professor Wm. Thomson has demonstrated that if the heat evolved by compressing an elastic fluid be equivalent to the force absorbed in the compression, the proportion of heat converted into mechanical effect by any perfect thermo-dynamic engine will be equal to the range of temperature divided by the highest temperature from the absolute zero of temperature. Therefore, if in a perfect steam-engine a be the temperature of the boiler from the absolute zero, and b be the absolute temperature of the condenser, the fraction of the entire quantity of heat communicated to the boiler which will be converted into mechanical force will be represented by $\frac{a-b}{a}$, which is analogous to the fraction representing the proportion of chemical force converted into mechanical effect in the electro-magnetic engine*. The extreme simplicity of this very important deduction which Professor Thomson has drawn from the dynamical theory of heat is of itself a strong argument in favour of that theory, even if it were not already established by decisive experiments.

Now estimating the heat generated by the combustion of a grain of coal at 1°·634 per lb. of water, its absolute mechanical value will amount to 1261·45 foot-pounds. Hence, according to Professor Thomson's formula, the work performed by any perfect thermo-dynamic engine will, for each grain of coal consumed, be represented by the equation

$$W = \frac{1261{\cdot}45\,(a-b)}{a},$$

which applies, as before intimated, not only to air-engines, but also to those steam-engines in which the principle of expansion is carried to the utmost extent, providing always that no waste of power is allowed to take place in friction, and

* Referring to this analogy, Professor Thomson writes as follows:—"I am inclined to think that an electric current circulating in a closed conducter *is heat*, and becomes capable of producing thermometric effects by being frittered down into smaller local circuits or 'molecular vortices.'"—*Letter to the Author, dated March* 31*st*, 1852.

that the entire heat of combustion of the coal is conveyed to the boiler or air-receiver.

Professor Thomson was the first to point out the great advantages to be anticipated from the air-engine, in consequence of the extensive range of temperature which it may be made to possess; and in a paper communicated to the Royal Society soon afterwards, I described a very simple engine which fulfils the criterion of perfection according to Professor Thomson's formula. This engine consists of three parts, viz. a condensing air-pump, a receiver, and an expansion-cylinder. The pump forces atmospheric air into the receiver; in the receiver its elasticity is increased by the application of heat, and then the air enters the expansion-cylinder, of which the volume is to that of the pump as the absolute temperature of the air in the receiver is to that of the air entering it. The cylinder is furnished with expansion-gear to shut off the air, when the same quantity has been expelled from the receiver as was forced into it by one stroke of the pump. By this disposition the air is expelled from the expansion-cylinder at the atmospheric pressure and at the absolute temperature corresponding with b in Professor Thomson's formula.

As an example of the above kind of air-engine, I will take one working in atmospheric air of 15 lb. pressure on the square inch and 50° Fahr. I will suppose that the expansive action in the cylinder is to exist through three fourths of its length. Then, as the action of the compressing-pump is the reverse of that of the cylinder, the piston of the former must traverse three fourths of its length before the air is sufficiently compressed to enter the receiver by its own pressure. The temperature of the air entering the receiver, determined by Poisson's equation $\frac{t'}{t}=\left(\frac{V}{V'}\right)^{k-1}$, will be 439°·59 Fahr., and its pressure will be 105·92 lb. on the square inch. Supposing now that the volume of the cylinder is to that of the pump as 4 to 3, the density of the air in the receiver to that forced into it by the pump must be as 3 to 4 in order to keep the quantity of air in the receiver constant. The temperature of the air in the receiver will also require to be kept at 739°·12 Fahr. in order to maintain the pressure of 105·92 lb. on the square inch. The air entering the cylinder at the above pressure and temperature will escape from it at the end of the stroke at the atmospheric pressure and at the temperature 219⅔°.

It will be remarked that there are two ranges of temperatures in the engine I have described, viz. that of the pump and that of the cylinder. Owing, however, to the exact proportion which subsists between the two, the same result is arrived at by the application of Professor Thomson's formula to either of them. Taking, therefore, the range of the cylinder, and converting the temperatures of the air entering and discharged from the cylinder into the absolute temperatures from the real zero by adding to them 459°, we obtain for the work evolved by the consumption of a grain of coal,

$$W = \frac{1261{\cdot}45\ (1198{\cdot}12 - 678{\cdot}66)}{1198{\cdot}12} = 546{\cdot}92 \text{ foot-pounds.}$$

In order to compare the foregoing result with the duty of a steam-engine approaching perfection as nearly as possible, I will admit that steam may be safely worked at a pressure of 14 atmospheres. The temperature of the boiler corresponding to that pressure will, according to the experiments of the French Academicians, be 387° Fahr. The temperature of the condenser might be kept at 80°. Reducing the above to temperatures reckoned from the absolute zero, we obtain for the work evolved by the combustion of each grain of coal,

$$W = \frac{1261{\cdot}45\ (846 - 539)}{846} = 457{\cdot}76 \text{ foot-pounds.}$$

It would therefore appear, even in the extreme case which I have adduced, that the performance of the steam-engine is considerably inferior to that of the air-engine. The superiority of the latter would have been still more evident had I also taken an extreme case as an illustration of its economy. It must, moreover, be remarked that the heated air escaping from the engine at a temperature so high as 219⅔° might be made available in a variety of ways to increase still more the quantity of work evolved. A part of this heated air might also be employed in the furnaces instead of cold atmospheric air.

We may also hope eventually to realize the great advantage which would be secured to the air-engine by causing the air, in its passage from the pump to the cylinder, to come into contact with the fuel by the combustion of which its elasticity is to be increased. It appears to me that the air might pass through a number of air-tight chambers, each containing ignited fuel, and that whenever any one of the chambers required replenishing, its connexion with the engine might be cut off by means of proper valves, until, by removing an air-tight lid or door, the chamber could be filled again with fuel. By means of suitable valves it would be easy to regulate the quantity of air passing through each chamber, so as to keep its temperature uniform; and by a separate pipe, furnished also with valves, by which the air could be carried from the pump to the upper part of the chambers without traversing the fuel, the engine-man would be enabled to keep the temperatures of the chambers, as well as the velocity of the engine, under proper control.

An Account of some Experiments with a large Electro-Magnet. By J. P. Joule, *F.R.S.*

[Read by Prof. Thomson at the British Association, Glasgow, 1855.]

The relation of the exciting force to the sustaining power of a magnet was the subject which it was the author's desire to examine, the laws arrived at being very divergent from those usually received. The soft iron made use of in this magnet was of such a nature that, after magnetization by moderate currents, it always—probably on account of intense magnetization on some former occasion—retained a residual polarity which was always in the same direction. The magnet might be excited by a current which developed a polarity opposed to the residual one; but on the interruption of the current, the latter reappeared. With high power, the lifting-power fell short of being proportional to the square of the current; but with feeble excitation he found the sustaining force to vary nearly as the fourth power of the current-strength employed.

Introductory Research on the Induction of Magnetism by Electrical Currents. By J. P. JOULE, *F.R.S., Corr. Mem. R.A., Turin, Hon. Mem. of the Philosophical Society, Cambridge, &c.**

['Philosophical Transactions,' 1856, p. 287. Read June 21, 1855.]

THE researches of Jacobi and Lenz led them some years ago to the announcement as a law, that when two bars of iron of different diameters, but equal to one another in length and surrounded with coils of wire of the same length, carry equal streams of electricity, the magnetism developed in the bars is proportional to their respective diameters. Experiments which I made about the same time threw doubts on my mind as to the general accuracy of the above proposition; for I found that the magnetism induced in straight bars of a variety of dimensions varying from $\frac{1}{3}$ to 1 inch in diameter, and from 7 inches to one yard in length, was nearly proportional to the length of the wire and the intensity of the current it conveyed, irrespectively of the shape or magnitude of the bars. The valuable experimental researches which have recently been made by Weber, Robinson, Müller, Dub, and others, refer chiefly to the attraction of the keeper or submagnet, and are not calculated to confirm or disprove either of the above propositions; and the correct view is

* The experiments were made at Oak Field, Whalley Range, near Manchester.

probably that of Professor Thomson, who considers both of them as corollaries (applying to the particular conditions under which the experiments were made) of the general law, that "similar bars of different dimensions, similarly rolled with lengths of wire proportional to the squares of their linear dimensions and carrying equal currents, cause equal forces at points similarly situated with reference to them"*. I have been induced to undertake some further experiments with an endeavour to elucidate the subject, and also to open the way to the investigation of the molecular changes which occur during magnetization.

I procured four iron bars one yard long and of the respective diameters $\frac{1}{6}$, $\frac{1}{4}$, $\frac{1}{2}$ and 1 inch, their weights being 1736, 3802, 14,560, and 55,060 grs. Each bar was wound with 56 feet of copper wire $\frac{1}{40}$ of an inch in diameter covered with silk, the number of convolutions being 1020, 712, 388, and 207 respectively. The smallest bar was closely covered throughout its entire length; but, on account of the larger surface of the other bars, the coils had to be distributed upon them as evenly as possible. Four other bars were also procured, of the same diameters as the above. They were, however, twice as long, weighing 3500, 7624, 29,944, and 108,574 grs., and were wrapped with double the length of wire, forming 2060, 1435, 768, and 418 convolutions respectively.

To measure the electrical currents, I employed a galvanometer of tangents, the needle of which, half an inch long, carried a glass index over a divided circle 6 inches in diameter. This instrument was furnished with a coil of sixteen circumvolutions of 1 foot diameter, which could be exchanged for a single circle of 1 foot diameter when the intensities to be measured were very considerable. It was ascertained by experiment that the tangent of deflection by the former coil was exactly sixteen times that of the latter when the same intensity of current was employed. For convenience' sake I have reduced all the observations to the latter standard, the unit current being therefore that which, passing through

* Letter to the Author.

a circle 1 foot in diameter, is able to deflect the needle through 45°.

The amount of magnetism induced in a bar was ascertained by placing it vertically with its lower end at a distance of 6 or 12 inches from a magnetized needle $\frac{3}{16}$ of an inch long and $\frac{1}{40}$ of an inch in diameter, suspended by a filament of silk, and having a fine glass index traversing over a graduated circle 6 inches in diameter. The force of torsion of the filament was found to be so trifling, that the tangents of the deflections of the needle could be taken as representing, without sensible error, the magnetism of the bar. Observations with so small a needle were made with great facility, the pointer moving steadily up to and attaining a new angle of deflection in eight or ten seconds after the electrical circuit was completed, the resistance of the air to the motion of the pointer being such as to prevent the smallest degree of oscillation. This resistance, however, of the air, so useful in bringing the needle speedily to rest, renders it necessary to guard carefully against any irregularity of the temperature of the case in which it is enclosed. A ray of sunlight would speedily occasion a deflection of several degrees*; and I found that the heat of the hand held over a part of the thick glass case 45° in advance of the pointer was sufficient, after penetration through the glass, to produce a current of air causing a steady deflection of no less than 30°, a deflection which subsided with extreme regularity and great slowness after the hand was removed. I would suggest that this circumstance points to the means of constructing a new and exceedingly sensible thermometer which would be valuable in many researches, particularly those on the conduction of heat.

Previously to employing electric currents, I made some experiments simply with a view to ascertain the inductive power of the earth's magnetism on the bars, and in which the action on the suspended needle was observed both at the distance of 12 and 6 inches, in order to determine the

* Dr. Tyndall has drawn attention to the importance of guarding against these effects of heat on a delicately poised needle. 'Philosophical Magazine,' 4th series, vol. iii. p. 127.

influence of distance for the convenience of future reductions. Having noticed the deflection produced by any bar, it was reversed and the observation repeated, the sum of the tangents of deflection showing the total effect produced on the magnetism of the bar by its reversion. I may here remark that both ends of the pointer of the needle were invariably observed, though, to save unnecessary detail, the tangent of the mean is only given.

Effect of Reversal of Bars two yards long.

Diameter of bar.	Sum of tangents of deflection. At 6 inches distance.	At 12 inches distance.
$\frac{1}{6}$ inch	·0450	·0088
$\frac{1}{4}$ inch	·0850	·0300
$\frac{1}{2}$ inch	·5912	·1922
1 inch	1·3910	·4598

The magnetism induced in the smaller bars appears to be nearly proportional to the square of the diameter, as might have been anticipated. The ratio of the attraction at 6 inches to that at 12 inches is 2·98.

Effect of Reversal of Bars one yard long.

Diameter of bar.	Sum of tangents of deflection. At 6 inches distance.	At 12 inches distance.
$\frac{1}{6}$ inch	·0480	·0138
$\frac{1}{4}$ inch	·1260	·0384
$\frac{1}{2}$ inch	·4926	·1430
1 inch	1·0380	·3084

The magnetism induced in the smaller bars of the above set is nearly proportional to the square of the diameter, the greater amount of discrepancy arising in all probability from the inferior length of the bars compared with those of the last set. The ratio of the attractions at the two distances is as 3·39 to unity.

In the following experiments on the induction of magnetism in the above bars by electrical currents, the method employed was:—1st, to observe the magnetism of a bar under the influence of the current; 2nd, that left permanently developed; 3rd, to observe the magnetism when the current was reversed; and, 4th, the magnetism remaining after the current was the second time cut off. The difference between the first and third observations gives the entire change in the magnetism of the bar consequent on the reversal of the current; the difference between the second and fourth gives the entire permanent change, or, as I may term it for convenience, the *magnetic set.*

The results were obtained by using currents of four degrees of intensity, in the first two of which the needle was at 6 inches distance, in the last two at 12 inches. The latter results are reduced to the action at 6 inches distance by employing the data arrived at from the foregoing experiments.

TABLE I.

Attraction, at 6 inches, of bars one yard long wrapped with 56 feet of wire.

Diameter of bar.	Intensity of current.	Total change of magnetism by reversal of current.	Magnetic set.	Total change minus magnetic set.	Set divided by square of current.	Total change minus set, divided by current.
$\frac{1}{8}$ inch	·0044	·0164	·0014	·0150	72·31	3·409
	·0197	·1012	·0266	·0746	68·54	3·787
	·0417	·3020	·1085	·1935	62·40	4·640
	·1450	2·7747	1·7036	1·0711	81·03	7·387
$\frac{1}{4}$ inch	·0041	·0364	·0038	·0326	226·05	7·951
	·0197	·2336	·0628	·1708	161·82	8·670
	·0414	·8798	·4085	·4713	238·34	11·384
	·1446	8·2871	4·9179	3·3692	235·20	23·300
$\frac{1}{2}$ inch	·0045	·0857	·0113	·0744	558·02	16·533
	·0194	·4573	·0882	·3691	234·35	19·026
	·0419	1·2162	·3207	·8955	182·67	21·372
	·1460	8·6948	2·7628	5·9320	129·61	40·630
inch..	·0045	·1017	·0128	·0889	632·10	19·755
	·0195	·5089	·0817	·4272	214·86	21·908
	·0416	1·0935	·1377	·9558	79·57	22·976
	·1404	5·6858	1·0248	4·6610	51·99	33·198
1	2	3	4	5	6	7

Although the covered wire was fine and wound close to the iron, it could not be expected to act with exactly equal advantage in the bars of small as of large diameter, chiefly on account of the circuit taken by the wire being, relatively to the circumference of the bar, greater in the small than in the large bars. In comparing the results together, it should therefore be borne in mind that those obtained with the bar of $\frac{1}{6}$ of an inch in diameter are somewhat diminished from the above circumstance.

A very cursory inspection of the results convinced me that the *magnetic set* followed a very different law from that which regulated the magnetic action under the influence of the current. I have therefore subtracted the former from the latter in the 5th column of the Table. Even after this separation has been effected, it will be seen from column 7 that the magnetic action over and above the set increases with considerably greater rapidity than the intensity of the current—a result which is, I believe, owing to a portion of the set actually existing during the action of the current being destroyed on the breaking of the circuit. It will be remarked, on inspecting column 6, that the set of the bars of $\frac{1}{6}$ and $\frac{1}{4}$ inch diameter increases nearly in proportion to the square of the current, but that with the thicker bars the ratio is diminished; so that, although the set of the bars of small diameter is greater than that of the large bars when a current of powerful intensity is employed, the reverse takes place when a weak stream is used. From the 7th column it may be gathered that the magnetism induced by an equal current, increasing at first nearly with the section of the bars, becomes ultimately almost independent of their thickness, the attractions of the $\frac{1}{2}$ inch and inch bars being almost exactly equal to one another.

Table II.

Attraction, at 6 inches, of bars two yards long wrapped with 112 feet of wire.

Diameter of bar.	Intensity of current.	Total change of magnetism by reversal of current.	Magnetic set.	Total change minus magnetic set.	Set divided by square of current.	Total change minus set, divided by current.
⅛ inch	·0042	·0150	·0009	·0141	51·02	3·357
	·0160	·0826	·0190	·0636	74·22	3·975
	·0281	·1440	·0410	·1030	51·92	3·665
	·0988	1·6581	1·0030	·6501	102·75	6·580
¼ inch	·0042	·0451	·0037	·0414	209·75	9·857
	·0167	·2555	·0513	·2042	183·94	12·227
	·0297	·6227	·2392	·3835	271·17	12·912
	·1048	6·5007	4·3887	2·1120	399·59	20·152
½ inch	·0044	·0937	·0095	·0842	490·70	19·136
	·0192	·5275	·0870	·4405	236·00	22·943
	·0386	1·2243	·2597	·9646	174·30	24·990
	·1338	10·6557	4·9784	5·6773	278·08	42·429
inch..	·0043	·1280	·0128	·1152	692·27	26·791
	·0178	·6088	·0822	·5266	259·44	29·584
	·0316	1·0440	·1833	·8607	183·56	27·237
	·1154	6·1017	1·6200	4·4817	121·65	38·836
1	2	3	4	5	6	7

An inspection of the above results, obtained from bars of double length wrapped with twice the length of wire, leads to conclusions similar to those we drew from Table I.

It appeared to me a matter of very great importance to investigate more closely the laws which regulate the *magnetic set*, and to determine with certainty whether the proportionality between the set and the square of the current, leading as it inevitably would to the better understanding of the nature of the molecular changes which occur in a magnetized bar, existed, and to what modifications it was subject. Seeing, therefore, that the supposed law began to fail when the thicker bars were employed, in which the mutual action of the particles distributed over a large section would naturally tend to counteract the magnetic induction developed on the exterior surface, I constructed two straight electro-magnets—one of an iron wire one yard long and $\frac{1}{25 \cdot 6}$ of an inch in diameter, the other of an iron wire one yard long and $\frac{1}{17 \cdot 2}$ of an inch in diameter. The former was wrapped with a single layer of covered copper wire $\frac{1}{40}$ of an inch in diameter and 21 feet long, the latter similarly with wire 27 feet long. The attractions of these wire electro-magnets were ascertained at distances of 2 and 6 inches. They are all, however, reduced to the latter distance by means of the data derived from the comparison of the action of the wire electro-magnets at the respective distances.

In the next Table all the results, except the last six, were obtained at 2 inches distance, and the observations are divided by 8·96, the relative attraction at 2 inches to that at 6 inches, called unity: the first recorded magnetic set was deduced from the mean of thirty-six experiments on the attraction at 2 inches distance. The mean deflection amounted to no more than ·247 of a minute of a degree; and as the error incident to any single observation is from 1 to 2 minutes of a degree, it follows that no great reliance can be placed on this first result.

From the results of Table III. it appears that, through the range of electrical intensities from ·0065 to ·0841, the *set* of the wire electro-magnet is proportional to the square of the current; that from the latter intensity to ·1060 the set increases with much greater rapidity, varying at one point with the 6th or 7th power of the current; and that from the intensity ·1060 the rate of increase rapidly declines as the limit of magnetization is approached. From the last column of the table it will be seen that the magnetic effect of the current, separated from the set, increases very uniformly with the current, though a little more rapidly. Similar conclusions may be drawn from the results of experiments with the electro-magnet of thicker wire contained Table IV., in which all the observations but the last four were made at 2 inches distance, and are reduced to the standard of the rest by dividing by 6·668, the observed action on the needle at the distance of 2 inches compared with that at 6 inches.

TABLE III.

Attraction, at 6 inches, of wire electro-magnet $\frac{1}{25\cdot6}$ inch diameter, wrapped with 21 feet of wire.

Number of experiments forming the mean result.	Intensity of current.		Total change of magnetism by reversal of current.	Magnetic set.	Total change minus magnetic set.	Set divided by square of current.		Total change minus set, divided by current.	
36	·0044	·0065	·00072	·00001	·00071	·516	·934	·161	·159
32	·0086		·00145	·00010	·00135	1·352		·157	
18	·0195		·00377	·00029	·00348	·763		·178	
20	·0391		·00929	·00152	·00777	·994		·198	
9	·0568		·01528	·00330	·01198	1·023		·211	
8	·0787		·02657	·00782	·01875	1·263		·238	
8	·0806		·02798	·00855	·01943	1·316		·241	
8	·0848	·0841	·02998	·00939	·02059	1·306	1·315	·243	·246
8	·0870		·03220	·01001	·02219	1·323		·255	
8	·0908		·03529	·01228	·02301	1·489		·253	
8	·0961	·0963	·03976	·01488	·02488	1·611	1·765	·259	·256
8	·0992		·04570	·02090	·02480	2·124		·250	
8	·0992		·04413	·01809	·02604	1·838		·262	
8	·1019		·04573	·01904	·02669	1·834		·262	
8	·1046	·1060	·04838	·02047	·02791	1·871	2·742	·267	·287
8	·1085		·05338	·02355	·02983	2·000		·275	
8	·1089		·09969	·06240	·03729	5·262		·342	
8	·1134		·05972	·02835	·03137	2·205		·277	
8	·1151	·1156	·10190	·06580	·03610	4·967	3·168	·314	·291
8	·1184		·06622	·03269	·03353	2·332		·283	
8	·1653	·1703	·14570	·09900	·04670	3·623	4·125	·283	·287
8	·1753		·19320	·14220	·05100	4·628		·291	
8	·3041	·3043	·29710	·21420	·08290	2·316	2·393	·272	·298
8	·3045		·32810	·22900	·09910	2·470		·325	
8	·4372		·38750	·24760	·13990	1·295		·320	
6	1·2919		·52980	·26400	·26580	·158		·206	
1	2		3	4	5	6		7	

TABLE IV.

Attraction, at 6 inches, of wire electro-magnet $\frac{1}{17\cdot2}$ inch diameter, 1 yard long, wrapped with 27 feet of wire.

Number of experiments forming the mean result.	Intensity of current.	Total change of magnetism by reversal of current.	Magnetic set.	Total change minus magnetic set.	Set divided by square of current.	Total change minus set, divided by current.
44	·0043	·00213	·00007	·00206	3·786	·479
20	·0089	·00443	·00027	·00416	3·408	·469
20	·0248	·01498	·00180	·01318	2·927	·531
10	·0493	·03835	·00719	·03116	2·958	·632
10	·0900	·10720	·03611	·07109	4·458	·790
10	·1171 } ·1188	·18702	·08508	·10194	6·205 } 6·739	·871 } ·802
10	·1205 }	·19404	·10560	·08844	7·273 }	·734 }
10	·1998	·45360	·31840	·13520	7·976	·677
10	·3448	·68450	·43310	·25140	3·643	·729
6	1·1633	1·07320	·48640	·58680	·359	·504
1	2	3	4	5	6	7

My next experiments, recorded in the following Table, were made with a bar of hard steel $7\frac{3}{4}$ inches long, $\frac{1}{4}$ of an inch in diameter, wound with 34 feet of silked copper wire $\frac{1}{40}$ of an inch in diameter, distributed in two layers. The first five observations were obtained at the distance of 3 inches, and are reduced to the standard of the remaining observations at 9 inches by dividing by 22·762, the number of times that the attraction at 3 inches was observed to surpass that at 9 inches.

TABLE V.

Attraction, at 9 inches, of steel electro-magnet $7\frac{3}{4}$ inches long, $\frac{1}{4}$ inch diameter, wound with 34 feet of wire.

Number of experiments forming the mean result.	Intensity of current.	Total change of magnetism on reversing the current.	Magnetic set.	Total change minus magnetic set.	Set divided by square of current.	Total change minus set, divided by current.
40	·0045	·00281	·0000092	·00280	·454	·622
40	·0089	·00543	·0000448	·00539	·566	·606
20	·0263	·01663	·0002157	·01641	·312	·624
10	·0489	·03132	·0008769	·03044	·367	·622
8	·0921	·06046	·0032278	·05723	·381	·621
20	·1594	·22992	·02356	·20636	·927	1·294
8	·3201	·65241	·17791	·47450	1·736	1·482
6	·4582	1·09119	·39722	·69397	1·892	1·514
6	·5688	1·45540	·58421	·87119	1·806	1·531
6	·8381	2·22020	1·03410	1·18610	1·472	1·415
2	1·5108	2·96510	1·29880	1·66630	·569	1·103
1	2	3	4	5	6	7

From the preceding Table it appears that the *set* of the steel bar increases almost exactly with the square of the current from the intensity ·0045 to ·0921; that thence to 1594 it increases more rapidly than the cube of the current; and that from that point it increases in a gradually diminishing ratio as the point of saturation is approached. It will be remarked that the first five numbers of column 7 are nearly equal to one another, but that when the set begins to increase more rapidly than with the square of the current, the magnetism of the bar over and above the *set* increases more rapidly than the current.

There is a striking and instructive analogy between the phenomena above pointed out and those relating to the set and elasticity of materials. Professor Hodgkinson has pointed out that the set or permanent change of figure in any beam is proportional to the square of the force which has been applied, a law which of course is transgressed near the breaking-point. May we not with propriety term the point at which, in the foregoing experiments, the set increases so abruptly, the *magnetic breaking-point*? Mr. Thomson has propounded the view, that the elasticity of all bodies is perfect when abstraction is made of the effect of set. The foregoing Tables indicate approximately the same law respecting what might be termed the *magnetic elasticity*. The analogy thus established between magnetic and ordinary molecular actions, when viewed in connexion with those changes of dimension which take place in iron bars by magnetization, and which I propose to study more deeply, promises to afford a point of view whence a more perfect insight into the nature of magnetism than we at present possess may ultimately be attained.

Oak Field, Moss Side, Manchester,
June 20, 1855.

Postscript.

Since the above was written, I have made the subjoined experiments on the electro-magnetic attraction of contact. A cylinder of wrought iron, 9 inches long and 4 inches in diameter, had a hole 3 inches in diameter bored along its axis. The thickness of the metal of the hollow cylinder thus formed was exactly half an inch. This was cut longitudinally into two exactly equal pieces, the surfaces of which were then carefully finished. Each of these semicylinders was wound with 25 feet of covered copper wire $\frac{1}{10}$ inch in diameter, and making fifteen complete convolutions. One of the semicylindrical electro-magnets thus formed was firmly secured with its flat surfaces upwards; and the other, with its flat surfaces downwards, was suspended to the beam of a balance sensitive to 2 or 3 grains when several pounds were in each scale. A cup containing mercury was affixed to one of the terminals of the wire of the subelectro-magnet, into which a terminal of the wire of the suspended magnet dipped. And, similarly, a mercury-cup attached to the other terminal of the suspended electro-magnet was dipped into by a wire in connexion with the voltaic battery, so as to counteract any effect on the balance which might be produced by the other mercury-cup. Each semicylindrical electro-magnet was thus acted upon by the same current of electricity, and the resulting attractions are tabulated below, each recorded number being the mean of four experimental determinations—two with the current in one direction, and the other two with it in the reverse direction.

Table VI.

Attraction in contact of two semicylindrical electro-magnets.

Intensity of current.	Attraction in lbs. avoirdupois.	Attraction of magnetic set.	Attraction divided by the 4th power of the current.	Attraction of magnetic set divided by the 4th power of the current.
·0410	·0365	·0045	12917	1592
·0690	·242	·0185	10676	816
·1013	1·203	·0835	11424	793
·1388	5·595	·3280	15074	883
·2074	17·937	2·5095	9694	1356
·2364	32·812	4·9685	10506	1590
·3682	not observed	17·5	......	952
·7013	not observed	40·25	......	166
1	2	3	4	5

The numbers in column 5 show that the magnetic set obtained by a weak current is nearly proportional to the square of its intensity. On inspecting column 4, it will also appear that the magnetism existing during the flow of the current follows the same law so nearly, that we may infer that it possesses the character of the magnetic set. Experiments that I have recently made on the attraction of ordinary electro-magnets for their armatures lead to the same conclusions.

December 21, 1855.

On the Fusion of Metals by Voltaic Electricity. *By* J. P. Joule, *F.R.S. &c.*

['Manchester Memoirs,' vol. xiv. p. 49. Read March 4th, 1856.]

The attention of practical scientific men has of late been much occupied with the question, how far it is possible to forge large masses of iron without destroying the tenacity and other valuable qualities of the metal employed. In welding iron, the metal is raised to the high temperature at which it assumes a soft and incipient viscid consistency. Two pieces of iron in this condition will adhere together slightly, if merely placed in contact with one another. That a firm junction cannot be made in this way is simply owing to the fact that few particles are brought into contact, and that the metallic continuity is only established at those points. The hammer is therefore employed to cause the entire surfaces to meet together. The same end has also been attained by the employment of great pressure; and probably we shall ultimately see large masses of forged iron formed by simply subjecting a bundle of smaller pieces, raised to the welding temperature, to the operation of great pressure. To succeed in the latter process, it would, however, be requisite to press clean unoxidized surfaces together. Indeed, the importance of presenting clean surfaces together in ordinary welding cannot be too strongly insisted on; for, if oxide of iron be present, a portion of it will not fail to remain at, or in the neighbourhood of, the juncture, and seriously impair the quality of the iron at those points.

It occurred to me some months ago that it might be possible to employ the calorific agency of the electric current in the working of metals. By the use of a voltaic battery there appeared to be no doubt but that small pieces of metal could be fused into one lump. If so, it was obvious that by employing a battery of adequate size the largest masses of wrought iron could be produced, the question resolving itself simply into one of cost. It was not before the last month that I had an opportunity of witnessing an experiment on a small scale. It was performed at the laboratory of Professor Thomson. He surrounded a bundle of iron wires with charcoal, and, after transmitting a powerful current through it for some time, the wires were found in one part to be completely fused together.

More recently I have made several experiments in which the wires were placed in glass tubes, surrounded with charcoal, &c. With a battery of six Daniell's cells I have thus succeeded in fusing several steel wires into one, uniting steel wires with brass, platina with iron, &c. I doubt not but that in many instances the process would advantageously supersede that of soldering, especially when, for thermo-electric or other purposes, it is desirable to join metals of difficult fusibility without the intervention of another metal which melts at a lower temperature.

Having demonstrated the possibility of obtaining perfect junctions by means of the voltaic current, let us inquire what expenditure of battery-materials would be necessarily involved. In the outset it may be remarked that were not heat continually removed, by conduction, convection, and radiation, from a wire carrying a current of however low a degree of intensity, the wire would ultimately attain an excessively high temperature on account of the continuous augmentation of heat within it. Now the escape of heat may be largely prevented by means of non-conducting substances, and will be nearly proportional to the surfaces, so that by employing sufficiently large masses of metal, and surrounding them with non-conducting materials, it may be reduced to almost any extent. The quantity of zinc required to fuse a large mass of iron may therefore be estimated as follows:—

I have shown, in a paper already communicated to the Society, that the quantity of heat due to the intensity of a Daniell's cell is 6°·129 per pound of water for every 33 grains of zinc dissolved*. In working with a voltaic battery, it is generally an advantageous arrangement to make the resistance of the battery one half that of the entire circuit. Hence, as the quantity of heat evolved in any part of the circuit is proportional to its resistance, we may take half the above, or 3°·064 per pound of water, as the heat which may be advantageously produced outside a Daniell's battery by the dissolution of 33 grains of zinc. Calling the temperature of incipient fusion of iron 4000° above the ordinary temperature of the atmosphere and the specific heat of iron 0 11, we find 4740 grains of zinc to be the quantity consumed in the voltaic battery in order to raise one pound of iron to the temperature of fusion. But as a considerable quantity of heat will be rendered "latent," 5000 grains may be taken as the estimate of minimum consumption of zinc, in a Daniell's battery, in order to effect the fusion of one pound of iron.

The same effect is due to the heat evolved by the combustion of 500 grains of coal; but, on account of the large quantity of heat which must necessarily escape up the chimney of a hot furnace, we may estimate the minimum actual consumption at 1000 grains.

The quantity of zinc consumed in the voltaic process is therefore nearly equal to that of the iron to be melted; but it would be possible to effect the same object in a more economical manner, by availing ourselves of the use of the magneto-electrical machine. This machine enables us to obtain heat from ordinary mechanical force, which mechanical force may again be derived from the conversion of heat, as in the steam-engine. In a steam-engine it is practically possible to convert

* Memoirs of the Literary and Philosophical Society, vol. vii. p. 94.

at least one fifth of the heat due to the combustion of coal into force; and one half of this force applied to work a magneto-electrical machine may be evolved in the shape of heat. Hence, then, it is possible to arrange machinery so as to produce currents of electricity which shall evolve one tenth of the quantity of heat due to the combustion of the coal employed. So that 5000 grains of coal used in this way would suffice for the fusion of one pound of iron.

Note on Dalton's Determination of the Expansion of Air by Heat. *By* J. P. Joule, *LL.D., F.R.S., &c.*

['Manchester Memoirs,' vol. xv. p. 143. Read November 2nd, 1858.]

In the twenty-first volume of the Memoirs of the Academy of Sciences, p. 23, Mr. Regnault, in the course of a discussion of the various coefficients given for the expansion of air, by different experimenters, remarks that Rudberg had brought to recollection an observation made by Gilbert, in his 'Annals,' to the effect that the experiments of Dalton and Gay-Lussac, which had been considered as giving almost identical results, differed, on the contrary, very considerably from one another. Then, referring to Dalton's experiment, related in the 'Manchester Memoirs,' vol. v. part ii. p. 599, Regnault shows that if 1000 measures of air at 55° Fahrenheit expand to 1325 at 212°, 1000 measures taken at 32° will become 1391 at 212°. Upon this he goes on to remark that Dalton did not appear to have been aware of the error which had crept into his calculations, for he says, in his 'New System of Chemical Philosophy,' that the volume of air, according to Gay-Lussac's and his own experiments, being taken 1000 at 32° becomes 1376 at 212°.

On reading the remarks of the eminent French physicist, the extreme improbability that a man so notoriously exact and careful in his mathematical and arithmetical computations as Dalton should have made the gross error imputed to him at once occurred to me. I therefore, on consulting Dalton's works, was not surprised to find that his commentators had entirely misunderstood the facts of the case. These are as follow:—Dalton, in his "Experimental Essays," read before this Society in the month of October 1801, describes experiments on the expansion of air by heat, the results of which, referred to the freezing-point, are accurately stated by Regnault. But in the 'New System of Chemical Philosophy,' published in 1808, under the article "Temperature," Dalton, while explaining his New Table of Temperature, writes:—"The volume at 32° is taken 1000, and at 212° 1376, according to Gay-Lussac's and my own experiments. As for the expansion at intermediate degrees, General Roi makes the temperature at midway of total expansion, 116½° old scale; from the results of my former experiments (Manch. Mem. vol. v. part ii. p. 599) the temperature may be estimated at 119½°; but I had not then an opportunity of having air at 32°. By more recent experiments I am convinced that dry air of 32° will expand the same quantity from that to 117° or 118° of common scale, as from the last term to 212°" The first part of the above extract contains the passage quoted by Regnault, but its meaning is obviously not that which he infers. The experiments which Dalton states to agree with Gay-Lussac's are clearly some unpublished ones made subsequently to those described in the 'Manchester Memoirs.' He nowhere, that I can discover, advances the assertion, attributed to him by Gilbert and adopted by Rudberg and Regnault, that his former "experiments" agree exactly with those of Gay-Lussac. They were, however, highly important at the time when they were made, and justified the approximately correct conclusion he drew, that *all elastic fluids under the same pressure expand equally by heat.*

Dalton was at once aware of the immense importance of this law, and in a sentence prophetic of the advancement of the theory of heat in recent times, pointed to the force of heat as the *sole* and immediate source of expansion in elastic fluids, and predicted that a study of their phenomena would ultimately lead to general laws respecting the absolute quantity and the nature of heat.

On the Utilization of the Sewage of London and other large Towns. By J. P. Joule, *LL.D., F.R.S., &c.*

['Manchester Memoirs,' vol. xv. p. 146.]

I have learned with regret that a system of metropolitan drainage has been adopted, and is about to be attempted by the Metropolitan Board of Works, which I consider to be a stride in the wrong direction, and which, if persevered in and copied by other towns, must be fraught with disastrous consequences to the national prosperity. I cannot, however, say that I felt much surprised at the intelligence, as I knew that the advice and assistance of scientific men had not been sought, except in one or two solitary instances, and that the professional engineers consulted had been limited to only a few, however eminent, individuals, who differ among themselves as to the contemplated works. On a question of such vital importance as that which has been raised, I think all ought to contribute what experience, information, or common sense they possess, and so form a concentrated expression of opinion which cannot be disregarded. I therefore, though the subject is somewhat new to me, hesitate not to introduce it to the Society, in the hope of eliciting the opinions of those who may have studied it better than myself.

From the Report* of Messrs. Hawksley, Bidder, and Bazalgette, I find that the history of the present question of London sewage dates from the year 1847, when, instead of the eight separate commissions which had previously existed, a consolidated one was appointed. This was shortly followed by a second, which advertised for and obtained 116 plans. A third and fourth commission reported against those plans, and appointed Mr. Forster as their engineer, who, after preparing a plan for the drainage of the north district of the Thames, died in consequence of the anxieties of his position.

* Report presented to the Metropolitan Board of Works, ordered by the House of Commons to be printed 13th July, 1858.

A fifth commission was embarrassed by the plans of the "Great London Drainage Company," which, after occupying a great part of the session of 1853, were ultimately rejected by Parliament. In 1854 Messrs. Bazalgette and Haywood prepared a scheme; but another proposal by Mr. Ward having received the sanction of the Secretary of State, a sixth commission was appointed, which invited plans, but arrived at no conclusions. In 1856 the Metropolitan Board instructed their engineer to report and prepare plans. Sir B. Hall proposed modifications which were adopted. This final plan was submitted to three referees, viz., Captain Galton, R.E., James Simpson, C.E., and Thomas E. Blackwell, C.E., who reported thereon in July, 1857 *, and, on their report being objected to by the Board, suggested material modifications of the plans proposed in the report submitted to Parliament by Her Majesty's first Commissioner of Works. A further communication from Messrs. Galton and Simpson, involving a third plan with further modifications, was made in January and February, 1858. Finally, at the request of the Metropolitan Board, Messrs. Bidder, Hawksley, and Bazalgette reported, on the 6th of April, 1858, on the plans of the Government Referees as from time to time modified. This last report, which in general opposes the plans of the Referees, appears to be the one finally adopted by the Board of Works on June the 29th, 1858.

In justice to the eminent engineers I have named, it is needful to premise that the duty they were called on to perform was rather to carry out a system predetermined by the hasty voice of public opinion, than to devise a plan entirely agreeable to their own views.

It would be beyond my province as well as my ability to describe the vast works which are now being attempted in conformity with the last resolution of the Metropolitan Board. It will, however, be sufficient to describe the general principle, in order to enable us to decide how far the two great objects, which any reasonable person must place before

* Report on Metropolitan Drainage, ordered by the House of Commons to be printed 3rd August, 1857.

him, will be met. These are, first, and I say especially, as it to a great extent includes the second, the economical use of sewage; second, the beauty and healthfulness of the metropolis. And here it is most deeply to be regretted that the projectors generally, instead of applying themselves to the fair consideration of both the above objects, have hastily abandoned the first one, so that, even if the plans answer the intention of the designers, the first great object will be further than ever from its realization. In fact, to illustrate how steadily, and I may say determinedly, the opposition to economy has been carried on, I have only to quote the following language of the Government Referees in page 33 of their Report:—"We consider it very inexpedient for the Metropolitan Board of Works to adopt any plan which is based upon the deodorization or the utilization of sewage; that if an attempt is to be made to utilize London Sewage, it should made by private enterprize;" and in page 43:—"That the value of the fertilizing matter contained in London sewage is undoubtedly great; but that the large quantity of water with which it is diluted precludes the possibility of separating more than about one seventh part of this fertilizing matter by any known economical process; that a copious dilution of the sewage is necessary to the health of the inhabitants of the metropolis; and that therefore the sacrifice entailed by the dilution must be endured."—The plain meaning of all this I take to be: *We will take care to dilute and remove the sewage, and then when, as we have shown, private enterprize will be unremunerative, we will invite it.*

Sketch of the Scheme of the Metropolitan Board.—On the north side the scheme consists of a main high-level sewer, to intercept the fall from the higher parts, extending from Hampstead; a main middle-level sewer from Kensal Green; and a main low-level sewer from Vauxhall-Bridge Road. All these terminate near Bow, whence the united streams pass in a channel formed of a triple culvert of brickwork to Barking Creek. On the south side a similar system of high-level and low-level sewers is to extend from Clapham and Putney to Greenwich, and thence to be carried forward in one main through Woolwich to Crossness Point, a place midway between Woolwich and Erith. Pumps are to be employed to raise the sewage at certain points, and storm-overflows are to enable the mains to discharge themselves through the previous system of sewers into the river within the limits of the metrepolis, whenever in consequence of a sudden fall of rain the former are overcharged. Two large reservoirs are to be placed at the outfalls of the two great mains, with the object of retaining the sewage until after full tide, when it is to be discharged into the Thames.

The utilization of sewage is ignored in the scheme of which I have just given an outline. Will it answer the object for which it is solely designed—that of purifying the Thames, and increasing the healthfulness of the district? To reply to this question, we must consider, 1st, the operation of the principle of intercepting and diverting the sewage from it original course. The present sewers in their usual functions will have to be considered as taking their rise at the points where they are crossed by the mains. Hence their size will be larger than it ought to be for the diminished current, and accumulations will result, which latter will be carried in time of storm-overflow into the Thames. A striking proof that such accumulations are, even under the present system, liable to take place and be carried off during storms is adduced by Dr. Hofmann and Mr. Witt. These chemists state that when, after a sudden and heavy fall of rain, the flow of the Savoy-Street sewer had increased sixfold, they found that, instead of the sewage being thereby diluted, a given volume actually contained more than twice the quantity of solid constituents which it contained under normal circumstances.

2nd. I doubt whether mains built of brick, however well cemented, can be depended upon to convey sewage. Brick is usually porous, and in that state it cannot be doubted that sewage-water will filtrate through it and thus gradually contaminate the adjacent ground. The injurious effects of such infiltration ought not to be overlooked in a system of mains extending to a total length of sixty miles.

3rd. That portion of sewage which arrives at the outfalls will not be entirely prevented from returning to the metropolis. I arrive at this conclusion from the fact that the sea-water penetrates occasionally as far as London Bridge. The river is frequently brackish at Barking Creek and Woolwich. Experiments with floats may induce fallacious conclusions in this respect, since it is probable that the scour of the flood-tide at the bottom of the estuary is greater than that of the ebb-tide.

4th. The Thames will be rendered particularly noxious at the point where so vast a quantity of offensive matter is to be concentrated. By what justice a nuisance can be removed from ourselves to be placed under the noses of our neighbours I know not. Nor can I appreciate the wisdom of sacrificing the purity of the air inhaled by the inhabitants of Greenwich, Woolwich, Gravesend, &c., and the immense floating population *, in the doubtful attempt to make the air of the metropolis more wholesome.

5th. The air confined in the new drains will be a serious increase to the already enormous volume of putrid gases in the sewers. The Government Referees make just and forcible remarks upon this evil. They state that " the effect of trapping the street gully-drains, without providing other ventilation of the sewers, is, that the noxious gases generated in the sewers are forced into the houses when the flow of sewage increases, the syphon-traps of water-closets and sinks being the points at which the least resistance is presented to their escape from the sewers. To obviate these evils, the plan has been partially adopted of providing in the middle of the street untrapped openings in the sewers. These openings must be endured until a better mode of ventilation shall be adopted, although the foul smells they emit are frequently very great nuisances " †. In addition to the cause assigned

* " It is extremely undesirable in a sanitary point of view to cause sewage-water to be intermixed with sea-water."—*Messrs. Hawksley, Bidder, and Bazalgette's Report*, p. 52.

† Report of Government Referees, p. 40.

by the Referees for the expulsion of the poisonous gases, I will mention the changes of atmospheric pressure. The fall of the barometer of one inch will of course occasion the liberation of one thirtieth of the entire volume of gas. The smell so generally observed to arise in the neighbourhood of drains before rain may probably be referred to this cause. It is also worthy of remark, that in winter the comparatively warm air of the sewers will have a tendency to rise. May not the greater mortality during that part of the year be partly attributed to this circumstance?

6th. The proposed system must be considered a filthy one, as instead of removing sewage to the soil, which is the natural deodorizer, it will cause its accumulation in the bed of the river at a distance of only a few miles from the city. Even the liquid portion will remain for months near the spot where it is introduced, as is proved by the experiments of Mr. Forster, who found that a float put into the river at Barking advances only five miles in its course towards the sea in an entire fortnight *.

The above are some of my reasons for believing that the proposed plan of the Board of Works will fail in promoting the object to attain which the promoters have sacrificed what ought to have been their first consideration. I enter not now on various points, such as the destruction of fish in the river and in the wells of ships, and consequent interference with a useful trade; the formation of banks apprehended by some, and the consequent impediment to navigation; the expense; and other details which must be of minor importance in a question ultimately involving the life and subsistence of an entire population.

The Government Referees remark that the pollution of streams by sewage throughout the country is an evil which

* Report of Government Referees, p. 172. Walter Crum, Esq., F.R.S., has suggested to me as very possible that there may be times when a greater quantity of water enters the Thames by flood than goes out by ebb-tide, owing to the large quantity of water taken from the river by evaporation in dry and hot weather.—See Dalton on Rain, Evaporation, &c., 'Manch. Memoirs,' vol. v. p. 346.

is increasing with improved house-drainage *, and they sought to place the outfalls as low as Sea Reach. The present is a plan which supplements and perpetuates the evil of which they complain; a new patch is to be added to the old garment, and as a natural consequence the rent will be made worse. An evil ought to be honestly and fairly met, not merely slurred over and disguised.

I now pass from the consideration of works which when completed will, there is every reason to believe, result in total failure, and I will endeavour to show the practicability of realizing the first great object of removing sewage, viz. its utilization. Much has been said on this subject since Mr. John Martin in 1828 drew attention to the waste which was even then going on; but, judging from the acts of public bodies, it would seem to be still doubtful in the minds of a great portion of the community whether the saving of the manure of cities is a matter of any considerable importance. It is therefore desirable that the actual facts should be constantly brought under review. For this purpose I might bring the evidence of nearly every scientific chemist, but will content myself with quoting from Liebig, both on account of the great attention he has paid to the subject and the circumstance that, having been interested in artificial manures, he is not liable to be unduly biassed in favour of natural ones. At page 177 of his 'Chemistry of Agriculture and Physiology' this distinguished philosopher says:—"The mineral ingredients of food have been obtained from our fields, having been removed from them in the form of seeds, of roots, and of herbs. In the vital processes of animals the combustible elements of the food are converted into compounds of oxygen, while the urine and fæces contain the constituents of the soil abstracted from our fields; so that by incorporating these excrements with our land we restore it to its original state of fertility. If they are given to a field

* The pollution of springs is a still more serious evil, in many instances involving the necessity of conveying the rain which falls on the moors through a long series of pipes. It is to be doubted whether such water is as good for drinking-purposes as uncontaminated spring-water.

deficient in ingredients necessary for the growth of plants, it will be rendered fertile for all kinds of crops. A part of a crop taken from a field is used in feeding and fattening animals which are afterwards consumed by man. Another part is used directly in the form of potatoes, meal, or vegetables; while a third part, consisting of the remnants of plants, is employed as litter in the form of straw, &c. It is evident that all the constituents of the fields, removed from it in the form of animals, corn, and fruit, may again be obtained in the liquid and solid excrements of man, and in the bones and blood of slaughtered animals. It altogether depends upon us to keep our fields in a constant state of composition and fertility by the careful collection of these substances. We are able to calculate how much of the ingredients of the soil are removed by a sheep, by an ox, or in the milk of a cow, or how much we convey from it in a bushel of barley, wheat, or potatoes. From the known composition of the excrements of man, we are also able to calculate how much of them it is necessary to supply to a field to compensate for the loss that it has sustained." Again, in page 181, he says:—"In the solid and liquid excrements of man and of animals we restore to our field the ashes of the plants which served to nourish these animals. These ashes consist of certain soluble salt and insoluble earths which a fertile soil must yield, for they are indispensable to the growth of cultivated plants. It cannot admit of a doubt that, by introducing these excrements to the soil, we give to it the power of affording food to a new crop, or, in other words, we reinstate equilibrium which had been disturbed. Now that we know that the constituents of the food pass over into the urine and excrements of the animal fed upon it, we can with great ease determine the different value of various kinds of manure. *The solid and liquid excrements of an animal are of the highest value as manure for those plants which furnished food to the animal.*"

From the above incontestable principles we may easily calculate the magnitude of the loss which is sustained by the waste of sewage. If the excrements of an animal are not returned to the soil, the food of that animal cannot be reproduced. Hence the amount of barrenness communicated to the soil by the system now endeavoured to be enforced in our towns may, considering no food or manure to be imported from other countries, be directly estimated by the food consumed by the inhabitants. Dalton, in the fifth volume of the 'Manchester Memoirs,' 2nd series, has given the aggregate of the articles of food consumed by himself in fourteen days, his habits, daily occupations, and manner of living being exceedingly regular. They are :—

Bread	163 oz.	Milk	435½ oz.
Oatcake	79 ,,	Beer	230 ,,
Oatmeal	12 ,,	Tea	76 ,,
Butcher's meat ..	54½ ,,		
Potatoes........	130 ,,		
Pastry	55 ,,		
Cheese	32 ,,		
Total	525½ ,, solids.		741½ ,, fluids.

Much more than the above quantities are consumed by the luxurious, much less by the aged and invalid. I think on the whole, and for our present purpose, that we may take them as the food of every man, woman, and child in the metropolis. Hence we may infer that the 2,600,000 inhabitants of London consume every day provisions equivalent to—

	1,316 tons of bread.	505,000 gallons of milk.
	282 tons of butcher's meat.	267,000 gallons of beer.
	674 tons of potatoes.	88,000 gallons of tea.
	285 tons of pastry.	
	166 tons of cheese.	
Total ..	2,723 tons of solid, and	860,000 gallons of liquid food.

This is therefore the daily rate at which the productive power of the country suffers by the waste of one large town, and this is done in the face of a rapidly increasing population. Yet there are many who treat the subject entirely as a commercial one; and if the cost of transit is such as to prevent sewage competing with guano in the market, they argue that it ought to be thrown away as refuse. But this is a fallacious, narrow-minded, and selfish view of the subject. In order apparently to save ourselves a little money at the present moment it entails a heavy burden on the inhabitants of the country in subsequent years. Guano will not last for ever. According to the Peruvian Survey, the Chincha Islands can yield 18,200,000 tons. Of this quantity Great Britain alone consumed in 1857 no less than 288,362 tons, which, if we consider the entire waste of sewage in Great Britain to be double that of London, will almost exactly make up for it in money value. If the produce of the above islands, which afford the best guano, be reserved for the sole use of Great Britain, it will last only sixty-three years at the present rate of consumption. I am aware that other supplies have been found in various parts of the world, and that there is a trifling additional deposit each year. But when we consider the competition which will eventually take place on the part of other countries to secure so valuable a manure, and also the ever increasing difficulties of obtaining it, we cannot trust to our being able to import it in the quantities we now do for even so long a period as that above named.

In addition to the help derived from guano, the soil of Britain is relieved from the present effects of sewage waste by large importations of corn and cattle and of bone-manure. But such a supply can continue only so long as foreign governments remain in ignorance of the permanent injury sustained by their fields. Liebig, in his 'Agricultural Chemistry,' complains that, if the exportation of bones continued on the then scale, the German soil would become gradually exhausted.

Besides, we ought not to be satisfied with merely keeping the productive power of our agriculture from decline. With a rapidly increasing population the wisest course would be to reserve such supplies of guano as we may be able to obtain for the purpose for which nature appears to have designed it—that of forming a fertile soil where sterility at present exists.

In concluding this part of my subject I would urge the

importance of recollecting that, in the estimate of fertility, regard should be paid not only to the weight of a crop but also to its nutritive value, determined in each case by chemical analysis. Liebig states that an increase of animal manure gives rise not only to an increase of the number of seeds, but also to a most remarkable increase in the proportion of those nitrogenous substances which are the most important constituents of food.

Having endeavoured to show the imperative necessity, I will say a few words on the means, of putting a stop to the present waste. The first step I conceive should be to prohibit the introduction into the sewage of any organic matter which can be avoided. For instance, scavengers should be constantly employed in collecting and removing horse-dung from the streets. The present system of sending carts round at long intervals of time allows by far the larger portion of this manure to be washed by rain into the sewers, thus forming a very serious addition to their impurity.

Then why should slaughter-houses be tolerated? If only meat slaughtered in the country were admitted into the town, I submit that the meat would be cheaper in regard to its intrinsic nutritive value. The distress suffered by the animals in their passage from the field to the town slaughter-house destroys the richness and flavour of their meat, even if it do not render it positively unwholesome. By the present system a large quantity of offal and blood is removed from the country, where it would be a valuable manure, to the town, where it is a dangerous nuisance.

I may mention in this place the subject of intramural interments, which even at the present day have not been entirely discontinued. Tens of thousands of human bodies in a disintegrated and decomposed state have floated down the sewers of London into the Thames*. The drainage of burial grounds into sewers is, in fact, enjoined by act of Parliament. Now the body of any human being after death ought, in accordance with the Divine ordinance, to be permitted to return to the dust whence it came. For this purpose metallic coffins are

* See Walker, on intramural graveyards.

unsuitable; and the body should be placed at a moderate depth below a soil on which there is a vegetable growth. I cannot enter into details on this highly important subject; but I am satisfied that the object of rapid conversion into vegetable life may be attained without in the least degree hurting, but rather subserving, those feelings of affection and reverence with which we regard the dead.

After prohibiting the unnecessary introduction of organic matter, the next step will be to deal with the sewage proper. And here we find at the outset that the enormous quantity of water mixed with it in the drains prevents the possibility of using it in that state for agricultural purposes. Messrs. Bidder, Hawksley, and Bazalgette, among other objections, come to the following conclusions in their Report. *First*, "That the fertilizing properties of the organic matters contained in town refuse are for the most part destroyed by the long-continued action of water." *Second*, "That the cost and difficulties attending the application of liquid sewage in large quantities are absolutely prohibitory of its use." *Third*, "That liquid sewage cannot in general be used with advantage in this climate, except in particular states of the weather, and in certain stages of the growth of the crops to which it is applied"*. The precipitation processes by lime &c., even though at present commercially valueless, ought to be persisted in, if it is only in our power to deal with largely diluted sewage. But, according to Hofmann and Witt, not more than one third of the fertilizing constituents can be thus separated. It is obvious therefore that we should deal with the sewage in a more concentrated form, and before it is diluted with rain and other comparatively clear water. The separate system has been frequently advocated; but there is some doubt whether sewage in a concentrated form would flow through a long series of pipes of very moderate inclination. With these facts before me I see no alternative but a return to the cesspool system, to which I believe no inconvenience or nuisance attaches, except where it is attempted carelessly and with inefficient mechanical and other appliances. The following is a plan which I venture

* Report, p. 104.

to recommend in places where water-closets are generally used.

I would place in the centre of the streets cesspools having a capacity of about 1000 gallons. Each cesspool to be for the use of some 400 inhabitants, say 50 houses, and to collect water from urinals, cab-stands, &c. The present sewers to be solely employed in carrying off rain and other comparatively clear water.

A drain, or generally two drains of considerable inclination extending from the cesspool up and down the street, to receive the water-closet pipes from the houses on both sides of the street. The total length of drain would be about 200 yards.

A force-pump, permanently fixed in the cesspool, to be used every night for the purpose of pumping out the sewage collected in it during the last 24 hours.

The sewage thus pumped to be discharged into tanks, and then conveyed to a railway to be carried to reservoirs situated at convenient localities in the country.

Each tank might have a capacity of ten tons, and would then hold the contents of five cesspools. It might be drawn by a traction steam-engine, which also might be employed for the pumping. The discharge pipe of the force-pump, as well as its piston-rod, might rise to the level of the street, and the requisite connections be screwed or clamped on when required. Immediately after emptying the cesspool a portion of McDougall and Smith's disinfecting powder* might be thrown in. This, acting on the sewage at an early period, would, Dr. Smith states, have the best effect in deodorizing and in preserving the fertilizing property.

I believe that in the above way the sewage of London might be conveyed to the fields, and a very large annual profit realized, instead of the dead loss of three millions sterling which must be incurred if the plan of the Metropolitan Board is carried out. The other advantages would consist in:—1st, Easy construction and repair. 2nd, Total prevention of infiltration of sewage, and the effects of accumulations of noxious gas. 3rd, Rapid removal of sewage before

* Sulphite of magnesia and lime, and carbonate of lime.

decomposition has had time to take place. 4th, An unpolluted river.

[From the estimates of loss by our sewage system there ought to be set off the immense stores which return to us in the shape of mollusks and fish. It has been observed that the best and most nutritious fish are found in seas contiguous to fertile lands whose rivers bring down organic matter. On the whole, considering the enormous quantities of imported food, this country must be acquiring great fertilizing potentiality at the expense of America and other lands which are being impoverished to supply our present needs.—*Note*, 1882.]

Notice of Experiments on the Heat developed by Friction in Air. By J. P. Joule, *LL.D., F.R.S.**

[Report Brit. Assoc. 1859 (Aberdeen), Sections, p. 12.]

The research which Professor Thomson and myself have undertaken on the thermal effects of fluids in motion naturally led us to examine the thermal phenomena experienced by a body in rapid motion through the air. The experiments which we first made for this purpose were of a very simple kind. We attached a string to the stem of a sensible thermometer, and whirled it alternately slowly and rapidly. In this way we uniformly obtained a slight effect; there was a higher temperature observed immediately after rapid than after slow whirling. A thermo-electric junction rapidly whirled also gave us an appreciable thermal effect, indicated by the deflection of the needle of a galvanometer.

Afterwards a more accurate set of experiments was made by us—using a lathe, to the spindle of which an arm was attached carrying one of Professor Thomson's delicate ether- or chloroform-thermometers. The thermometers employed were so extremely sensitive that each division of their scales had a value of not more than $\frac{1}{300}$ of a degree Centigrade.

* The experiments were made at Oak Field, Whalley Range, near Manchester.

The great value of Professor Thomson's thermometers in the whirling experiments was further enhanced by the light specific gravity of ether comparatively with mercury; the pressure produced by centrifugal force operating on a long column of mercury would have probably broken a mercurial thermometer whirled at high velocity.

The results arrived at by Professor Thomson and myself were as follows:—

1st. The rise of temperature in the whirled thermometer was, except at very slow velocities, proportional to the square of the velocity.

2nd. The velocity with which the bulb had to travel in order that its temperature should be raised 1° Cent. was 182 feet per second.

3rd. At very low velocities the quantity of thermal effect appeared to be somewhat greater than that due to the square of the velocity calculated from the above datum; and we surmised that this was owing to a sort of fluid friction different from the source of resistance at high velocities. We therefore made several attempts to increase this particular fluid friction, the most successful result being obtained by wrapping fine wire over the bulbs. By this means we succeeded in obtaining $\frac{1}{6}$ of a degree Cent. with a velocity of only 30 feet per second—a quantity five or six times as great as that which took place when the naked bulb was revolved at the same velocity.

We resumed the whirling experiments last May; and it is owing to the circumstance that it has happened that I have myself been principally engaged in making those which I am about to communicate to the Section, that Professor Thomson has requested me to give an account of this part of our joint labours.

Our object was to repeat the former experiments under new circumstances, so as to verify and extend the results already obtained. A very brief outline can only be given in this place, as we intend shortly to incorporate them in a joint paper for the Royal Society, to whose assistance we owe the means of prosecuting the inquiry.

The lathe was again used as the whirling-apparatus, but instead of the ether-thermometer we whirled thermo-electric junctions of iron and copper wires. We obtained the following results :—

1st. The thermal effect was, as with the ether-thermometer, proportional to the square of the velocity.

2nd. The rise of temperature was independent of the thickness of the wire which formed the thermo-electric junction which was whirled. This was decided by experiments on wires of various diameters, ranging from $\frac{1}{100}$ to $\frac{1}{8}$ of an inch. The rise of temperature for the same velocity was in every case the same as that obtained with the ether-thermometer, the bulb of which was nearly half an inch in diameter.

3rd. The thermal effect appeared to be independent of the shape of the whirled body, little difference occurring in whatever position the wire was placed relatively to the direction of motion.

4th. The average result was that the wire was warmed 1° Cent. by moving at the velocity of 175 feet per second.

The highest velocity obtained was 372 feet per second, which gave a rise of 5°·3 Cent.; and there was no reason to doubt that the thermal effect would go on continually increasing with the square of the velocity. Thus at a mile per second the rise of temperature would be in round numbers 900° Cent.; and at 20 miles per second, which may be taken as the average velocity with which meteorites strike the atmosphere of the earth, 360,000°.

The temperature due to the stoppage of air at the velocity of 143 feet per second is one degree Centigrade. Hence we may infer that the rise observed in the experiments was that due to the stoppage of air, less a certain quantity, of which probably the greater part is owing to loss by radiation. It being also clear that the effect is independent of the density of the air, there remains no doubt as to the real nature of "shooting-stars." These are small bodies which come into the earth's atmosphere at velocities of 20 miles per second and upwards. The instant they touch the atmosphere their surfaces become heated far beyond the point of fusion or even of volatilization; and the consequence is that they are speedily and, for the most part, completely burnt down and reduced to impalpable oxides. It is thus that, by the seemingly feeble resistance of the atmosphere, Providence secures us effectively from a bombardment which would destroy all animated nature exposed to its influence*.

The experiments to carry out and verify our previous results on the thermal effects which belong to friction on large surfaces at low velocities were made as follows:—A disk of zinc or cardboard was attached to the revolving axis; an ether-thermometer was attached to this disk, the bulb being near the circumference and describing a circle with a radius of about $1\frac{1}{2}$ foot. On rotating the disk at the velocity of $1\frac{3}{4}$ foot per second, so much as a rise of $\frac{1}{30}$ of a degree Cent. was observed.

On the Intensity of Light during the recent Solar Eclipse. By J. P. Joule, *LL.D., F.R.S., &c.* [In a letter to the Editors of the 'Philosophical Magazine.']

['Philosophical Magazine,' ser. 4. vol. xv., April 1858.]

Gentlemen,

Desiring to obtain an image of the annulus in the late solar eclipse, I took a camera to the Werrington Junction on the Great Northern Railway. A few minutes before the central eclipse, however, it became evident that the sky would remain obscured with cloud; I therefore employed the camera simply to obtain, if possible, a measure of the intensity of the light. The country about the Werrington Junction is an extensive plain. I placed the camera on the ground, directing it to the south-east horizon. The sensitive plate was exposed during the five minutes which preceded concentricity, and the image of the landscape was developed

* See p. 272.

immediately afterwards. The next day, the weather appearing very similar, possibly a little more cloudy, I exposed, two miles on the south of Manchester, plates prepared with the same collodion and nitrate of silver, and developed them with the same solution of sulphate of iron, the direction in which the camera was placed, the time of the day, and every other circumstance being as nearly as possible the same as before. In this latter case, a picture, judged by Mr. Dancer (a gentleman of great experience in photography) and also by myself to be of considerably greater intensity than that procured during the eclipse was obtained in two seconds.

The ratio of luminous influence, as measured by the camera, was therefore at the greatest only 1 to 150; but the average exposed area during the five minutes preceding concentricity, compared with the entire solar disk, was about 1 to 24. I therefore infer that the circumference of the sun's disk gives out a very weak luminous radiation (at least as measured by a sensitive plate) in comparison with the central part. This observation is quite in accordance with the experience of Mr. Dancer, who finds that in the photography of the sun the central part is always much more rapidly depicted than the circumference—so much so, that he finds it impossible to obtain in the same image a satisfactory delineation of both parts.

On the eclipse becoming central, the darkness suddenly increased, remained for a few seconds apparently constant, then as suddenly cleared up. The light at the darkest was evidently enormously greater than that of the full moon; but, from my experience in photography, I have no doubt that, with it, two hours at least would have been necessary to produce an effect equal to that attainable by one second's exposure had the sun been uneclipsed.

Yours respectfully,

James P. Joule.

On an Improved Galvanometer.
By J. P. Joule, *LL.D., F.R.S., &c.*

['Philosophical Magazine,' ser. 4. vol. xv. p. 432.]

The important experiments required in carrying out the gigantic projects of electric-telegraph engineers having rendered a delicate and portable galvanometer an essential piece of apparatus, I am induced to hope that the following description of one recently made from my design will interest at the present time. In figure 85, A A represents the frame

Fig. 85. Scale $\frac{1}{3}$.

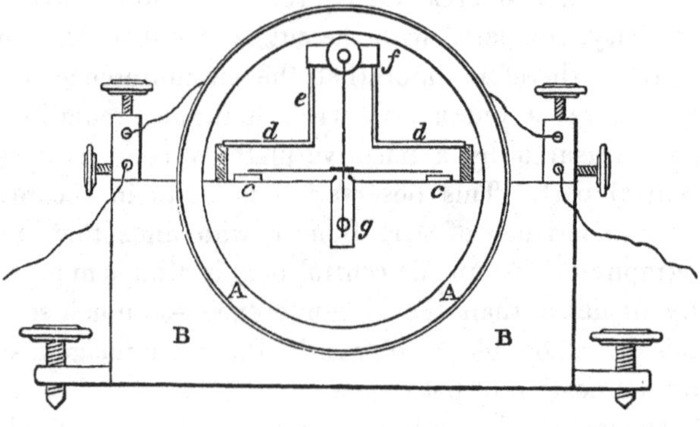

on which the wire forming the coil is wound; it is inserted in a groove cut into the block of wood B B. Another similar block, not shown in the figure, is fastened to the first by clasps, so as to hold the coil firmly in its place. *c c* shows the section of the graduated circle, enclosed in a box whose glass lid *d* is fitted with a glass chimney *e*, surmounted by a cap and roller *f*. Over this roller is thrown the filament of silk which supports the needle (a piece of magnetized sewing-needle a quarter of an inch long) and the glass index to which the needle is attached. A small piece of fine copper wire attached to the needle hangs within a groove cut into the wooden block. By means of a hole pierced horizontally from the back of the instrument, entering about midway down the groove, a stud can be made to press the fine copper wire against the anterior wooden block (that not seen in the figure). By doing this the needle and pointer (previously let down by turning the roller so as to slacken the silk filament) are held securely. The instrument can then travel safely, and the experimenter is saved the otherwise inevitable trouble of suspending the needle afresh at the journey's end.

In the instrument I have got fitted for Mr. Gordon, 2798 yards of no. 40 silked copper wire are wound on a reel 4 inches in diameter, the object being to obtain an effect accurately measurable with a very small quantity of current. But coils of different lengths and sizes of wire can be readily attached to the instrument. Although the silk filament is only $1\frac{1}{2}$ inch long, and the needle a quarter of an inch only, the torsion is so trifling that an entire twist of the filament deflects the needle only one degree from the magnetic meridian*. The resistance presented by the air to the motion of the glass pointer stops the oscillations of the needle in about half a dozen seconds.

On the Thermo-electricity of Ferruginous Metals; and on the Thermal Effects of stretching Solid Bodies. By J. P. Joule, *F.R.S.*†

['Proceedings of the Royal Society,' January 29, 1857.]

The experiments on the above subjects were made with a thermo-multiplier placed in the vacuum of an air-pump. Its sensibility was such, that with the junction antimony and bismuth a thermometric effect of $\frac{1}{8000}$ of a degree Centigrade could be estimated. In determining the thermo-electric position of the metals, it was necessary to increase

* By employing the thread of a diadema spider the torsion would be very much further reduced.—*Note*, 1882.

† The experiments were made at Oak Field, Whalley Range, near Manchester.

the resistance of the instrument a hundredfold, by placing in the circuit a coil of fine wire. In thermo-electric arrangement *steel* was found to be nearer copper than iron was. By hardening, steel was raised to the place of copper. *Cast iron* was found to surpass copper; so that the junction cast iron and copper is reverse to that of wrought iron and copper, and the arrangement cast iron and wrought iron is much more powerful than copper and wrought iron. A new test of the quality and purity of ferruginous metals is thus indicated, which will probably be found of value to the arts.

The experiments on the stretching of solids, showed in the case of the metals a decrease of temperature when the stretching weight was applied, and a heating effect when the weight was removed. An iron wire $\frac{1}{4}$ inch in diameter was cooled $\frac{1}{8}$ degree Centigrade when stretched by a weight of 775 lb. Similar results were obtained with cast iron, hard steel, copper, and lead. The thermal effects were in all these cases found to be almost identical with those deduced from Professor Thomson's theoretical investigation, the particular formula applicable to the case in question being $H = \frac{t}{J} \times Pe$, where H is the heat absorbed in a wire one foot long, t the absolute temperature, J the mechanical equivalent of the thermal unit, P the weight applied, and e the coefficient of expansion per 1° With gutta-percha also a cooling effect on extension was observed; but a reverse action was discovered in the case of vulcanized india-rubber, which became *heated* when the weight was laid on, and *cooled* when the weight was removed. On learning this curious result, Professor Thomson, who had already intimated the probability of a reverse action being observed under certain circumstances with india-rubber, suggested to me whether vulcanized india-rubber loaded with a weight would not be shortened by increasing its temperature. Accordingly, on trial, I found that this material, when stretched by a weight capable of doubling its length, had that length diminished by one tenth when its temperature was raised 50° Centigrade. This shortening effect was found to increase rapidly with the stretching weight employed, and, being exactly conformable with the heating effects of stretching, entirely confirmed the theory of Professor Thomson.

On the Thermal Effects of Longitudinal Compression of Solids. By J. P. JOULE, *Esq., F.R.S. With an Investigation on the Alterations of Temperature accompanying Changes of Pressure in Fluids. By Professor* W. THOMSON, *F.R.S.*

['Proceedings of the Royal Society,' June 15, 1857.]

IN the further prosecution of the experiments of which an outline was given in the 'Proceedings' for January 29, 1857, the author has verified the theory of Professor Thomson, as applied to the thermal effects of laying weights on and taking them off metallic pillars and cylinders of vulcanized india-rubber. Heat is evolved by compression, and absorbed on removing the compressing force in every substance yet experimented on. In the case of metals, the results agree very closely with the formula in which e, the longitudinal expansion by heat under pressure, is considered the same as the expansion without pressure. It was observed, however, that all the experimental results were a little in excess of the theoretical; and it became therefore important to inquire whether the force of elasticity in metals is impaired by heat. In the first arrangement for this purpose, the actual expansion of the bars employed in the experiments was ascertained by micrometric apparatus—1st, when there was no tensile force, and 2nd, when a weight of 700 lb. was hung to the extremity of the quarter-inch rods. The results, reliable to less than one hundredth of their whole value, did not exhibit any notable effect of tensile force on the coefficient of expansion by heat. An experiment susceptible of greater delicacy was now tried. Steel wire $\frac{1}{90}$ of an inch in diameter was wound upon a rod of iron $\frac{1}{4}$ of an inch in diameter. This was heated to redness. Then, after plunging in cold water, the spiral was slipped off. The number of convolutions of the spiral was 420, and its weight 58 grains. Its length, when suspended from one end, was 6·35 inches; but on adding to the extremity a weight of 129 grains, it stretched without sensible set to 14·55 inches. The temperature of the spiral thus stretched was raised or lowered at pleasure by putting it in or removing it out of an oven. After several experiments it was found that, between the limits of temperature 84° and 280° Fahr., each degree Centigrade of rising temperature caused the spiral to lengthen as much as 00337 of an inch, and that a contraction of equal amount took place with each degree Centigrade of descending temperature. Hence, as Mr. James Thomson has shown that the pulling out of a spiral is equivalent to twisting a wire, it follows that the force of torsion in steel wire is decreased 00041 by each degree of temperature.

An equally decisive result was obtained with copper wire, of which an elastic spiral was formed by stretching out a piece of soft wire, and then rolling it on a rod $\frac{1}{4}$ of an inch in diameter. The spiral thus formed consisted of 235 turns of wire $\frac{1}{40}$ of an inch in diameter, weighing altogether 230 grains. Unstretched, it measured 6·7 inches, but with a weight of 1251 grains attached it stretched, without set, to 10·05 inches. Experiments made with it showed an elongation of ·00157 of an inch for each degree Centigrade of elevation of temperature, and an equal shortening on lowering the temperature. The diminution of the force of torsion was in this case 00047 per degree Centigrade*.

---

* Since writing the above, I have become acquainted with M. Kupffer's researches on the influence of temperature on the elasticity of metals ('Compte-Rendu Annuel,' Petersburg, 1856). He finds, by his method of twisting and transverse oscillations, that the decrease of elasticity for steel and copper is ·000471 and ·000478. Very careful experiments recently made by Prof. Thomson indicate a slight increase of expansibility by heat in wires placed under tension.—J. P. J., Aug. 1, 1857.

Professor Thomson has obligingly furnished me with the following investigation :—

On the Alterations of Temperature accompanying Changes of Pressure in Fluids.

Let a mass of fluid, given at a temperature t and under a pressure p, be subjected to the following cycle of four operations in order :—

(1) The fluid being protected against gain or loss of heat, let the pressure on it be increased from p to $p+\varpi$.

(2) Let heat be added, and the pressure of the fluid maintained constant at $p+\varpi$, till its temperature rises by dt.

(3) The fluid being again protected against gain or loss of heat, let its pressure be reduced from $p+\varpi$ to p.

(4) Let heat be abstracted, and the pressure maintained at p, till the temperature sinks to t again.

At the end of this cycle of operations, the fluid is again in the same physical condition as it was at the beginning, but, as is shown by the following considerations, a certain transformation of heat into work or the reverse has been effected by means of it.

In two of these four operations the fluid increases in bulk, and in the other two it contracts to an equal extent. If the pressure were uniform during them all, there would be neither gain nor loss of work; but inasmuch as the pressure is greater by ϖ during operation (2) than during operation (4), and rises during (1) by the same amount as it falls during (3), there will, on the whole, be an amount of work equal to $\varpi\, dv$, done by the fluid in expanding, over and above that which is spent on it by pressure from without while it is contracting, if dv denote a certain augmentation of volume which, when ϖ and dt are infinitely small, is infinitely nearly equal to the expansion of the fluid during operation (2), or its contraction during operation (4). Hence, considering the bulk of the fluid primitively operated on as unity, if we take

$$\frac{dv}{dt}=e$$

to denote an average coefficient of expansion of the fluid under constant pressure of from p to $p+\varpi$, or simply its coefficient of expansion at temperature t and pressure p, when we regard ϖ as infinitely small, we have an amount of work equal to

$$\varpi\, e\, dt$$

gained from the cycle. The case of a fluid such as water below 39°·1 Fahr., which contracts under constant pressure, with an elevation of temperature, is of course included by admitting negative values for e, and making the corresponding changes in statement.

Since the fluid is restored to its primitive physical condition at the end of the cycle, the source from which the work thus gained is drawn must be heat; and since the operations are each perfectly reversible, Carnot's principle must hold; that is to say, if θ denote the excess of temperature of the body while taking in heat above its temperature while giving out heat, and if μ denote "Carnot's function," the work gained, per unit of heat taken in at the higher temperature, must be equal to

$$\mu\, \theta.$$

But while the fluid is giving out heat, that is to say during operation (4), its temperature is sinking from $t+dt$ to t, and may be regarded as being on the average $t+\frac{1}{2}dt$; and while it is taking in heat, that is during operation (2), its temperature is rising from what it was at the end of operation (1) to a temperature higher by dt, or on the average exceeds by $\frac{1}{2}dt$ the temperature at the end of operation (1). The average temperature while heat is taken in consequently exceeds the average temperature while heat is given out by just as much as the body rises in temperature during operation (1). If therefore this be denoted by θ, and if $\mathrm{K}dt$ denote the quantity of heat taken in during operation (2), the gain of work from heat in the whole cycle of operations must be equal to $\mu\,\theta\,\mathrm{K}dt$; and hence we have

$$\mu\,\theta\,.\,\mathrm{K}dt=\varpi\, e\, dt.$$

From this we find

$$\theta=\frac{e}{\mu \mathrm{K}}\varpi,$$

where, according to the notation that has been introduced,

θ is the elevation of temperature consequent on a sudden augmentation of pressure from p to $p+\varpi$; e is the coefficient of expansion of the fluid, and K its capacity for heat, under constant pressure; and μ is Carnot's function, being, according to the absolute thermodynamic scale of temperature, simply the reciprocal of the temperature multiplied by the mechanical equivalent of the thermal unit. If, then, t denote the absolute temperature, which we have shown by experiment* agrees sensibly with temperature by the air-thermometer Cent. with 274° added, and if J denote the mechanical equivalent of the thermal unit Centigrade, we have

$$\theta = \frac{t\,e}{\mathrm{JK}}\varpi.$$

This expression agrees in reality, but is somewhat more convenient in form than that first given, Dynamical Theory of Heat, § 49, Trans. R.S.E. 1851.

Thus for water, the value K, the thermal capacity of a cubic foot and under constant pressure, is 63·447, and e varies from 0 to about $\frac{1}{2200}$ for temperatures rising from that of maximum density to 50° Cent.; and the elevation of temperature produced by an augmentation of pressure amounting to n times 2117 lbs. per square foot (that is to say, to n atmospheres) is

$$\frac{t\,e \times 2117}{1390 \times 63{\cdot}447}n.$$

For mercury, we have

$$\frac{t\,e \times 2117}{1390 \times 28{\cdot}68}n.$$

If, as a rough estimate, we take

$$e = \frac{t-278}{46} \times \frac{1}{2200},$$

this becomes

$$\frac{t(t-278)}{420000}n.$$

* See Part II. of our Paper "On the Thermal Effects of Fluids in Motion," Philosophical Transactions, 1854.

If, for instance, the temperature be 300° on the absolute scale (that is, 26° of the Centig thermometer), we have

$$\frac{n}{636}$$

as the heating effect produced by the sudden compression of water at that temperature; so that ten atmospheres of pressure would give $\frac{1}{64}$ of a degree Cent., or about five divisions on the scale of the most sensitive of the ether-thermometers I have as yet had constructed.

Thus if we take $\frac{1}{5500}$ as the value of e, this becomes

$$\frac{t}{103600}n;$$

and at temperature 26° Cent. the heating effect of ten atmospheres is found to be $\frac{1}{34}$ of a degree Cent.

TABLE giving the thermal effects of a pressure of ten atmospheres on water and mercury*.

Temperature.	Increase or decrease of temperature in water.	Increase of temperature in mercury.
0°	·005 decrease	·026
3·95	·0	·0264
10	·006 increase	·027
20	·015 do.	·028
30	·022 do.	·029
40	·029 do.	·030
50	·035 do.	·031
60	·041 do.	·032
70	·047 do.	·033
80	·055 do.	·034
90	·065 do.	·035
100	·078 do.	·036

* Added August 1, 1857.

On some Thermo-dynamic Properties of Solids. By J. P. Joule, *LL.D., F.R.S., F.C.S., Hon. Mem. Phil. Soc. Cambridge, Vice-President of the Lit. and Phil. Soc. Manchester, Corresp. Mem. R.A. Sc. Turin, &c.**

['Philosophical Transactions,' 1859, vol. cxlix. p. 91.]

1. After finding the numerical relation between heat and work in 1843, it immediately occurred to me to investigate various phenomena in which heat is evolved by mechanical means, and of these one of the most interesting and important appeared to be the evolution of heat by the compression of elastic fluids. If the heat given out in this case proved to be the equivalent of the work spent, then the natural inference was that the elastic force of a gas and its temperature are owing to the motion of its constituent particles, both being proportional to the square of the velocity of the particles, and that the force of a falling body employed in compressing a gas is exhibited anew in the form of temperature. On the other hand it was possible, secondly, to conceive of an elastic fluid which would not give out any heat by compression. This would be the case if it were made up of mutually repelling particles, the temperature of which was that of the mass. The work required to compress the fluid would be the same on either hypothesis; but in one supposition the effect would be developed in actual energy, in the other in the potential form. Thirdly, we may suppose a fluid exhibiting as heat a portion of the force employed in its compression, and retaining the rest in the potential form. Or we may have a fluid giving out more heat than the equivalent of the work spent upon it when it is compressed and maintained at a constant temperature. Experiment proved that the heat actually evolved was very approximately that due to the compressing force. Nevertheless, it seemed

* The experiments were made at Oak Field, Whalley Range, near Manchester.

desirable to demonstrate the possibility of a gas so constituted that heat shall not be evolved by its compression. It occurred to me that a bag full of elastic metallic springs would illustrate such a gas. If the springs were properly formed, the elasticity of the bag would follow the laws of gaseous pressure. To the question, Would such a bag evolve heat on compression? I could readily answer, no; for in the bending of a spring one part is extended while the other is compressed, and thus it might be expected that the thermal effect on the whole would be neutral. Still it seemed desirable to decide the point by experiment.

Fig. 86.

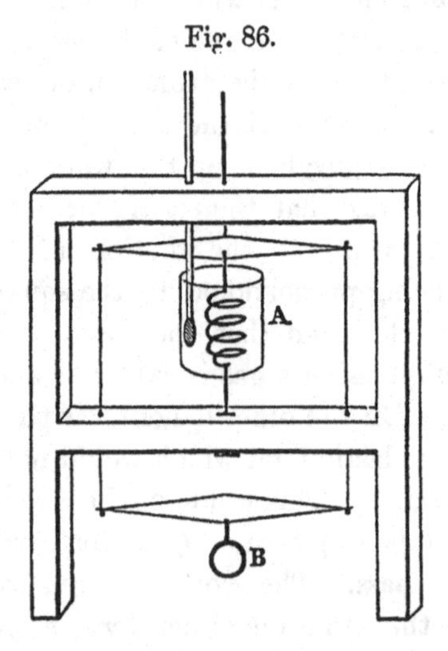

2. The apparatus I employed is represented by the adjoining sketch, where a spiral spring of tempered steel is seen immersed in the can A. By applying weights to a lever connected with the link B the spring could be compressed, and the heat evolved, if any, measured by the increase of temperature observed to take place in the water or mercury filling the can. The plan I pursued was to note the temperature successively—first, two minutes after the weight had been laid on; second, two minutes after the weight had been removed; and third, after two minutes more had elapsed. The mean of a number of these observations taken in succes-

sion gave me,—first, the thermal effect resulting from the laying on of the weight and the atmospheric influence; second, that of the removal of the weight and the atmospheric influence; and third, the atmospheric influence alone.

3. The pressure applied was 318 lb., which pushed down the spiral spring 1·136 inch. In the following summary of results, each number is the mean of eight or ten observations, given in terms of the graduation of a thermometer, of which each degree was equal to 0558 of a degree Centigrade.

First Series.—Spring immersed in 8 oz. of water.

	Weight laid on.	Weight taken off.	Atmospheric influence.
First experiment . .	−·548	−·568	−·560
Second experiment .	−·009	−·023	−·012
Third experiment . .	+·282	+·271	+·252
Mean	−·092	−·107	−·107

Second Series.—Spring immersed in 7 lb. of mercury.

	Weight laid on.	Weight taken off.	Atmospheric influence.
First experiment . .	+.180	+·175	+·169
Second experiment .	+·066	+·092	+·059
Third experiment . .	+·041	+·039	+·020
Fourth experiment .	−·060	−·059	−·053
Mean	+·057	+·062	+·049
Mean of both series	−·032	−·038	−·041

4. The capacity for heat of half a pound of water being about twice as great as that of 7 lb. of mercury, I have divided the mean result of the second series by two, before combining it with the first series in a general mean. The mean of both series therefore represents the thermal effect in the capacity of half a pound of water. It indicates, on subtracting the effects of the atmosphere, a heating effect of 009 after laying on the weight, and a heating effect of 003 on taking off the weight. The highest of these numbers

represents a temperature less than one thousandth of a degree Fahr.

5. Now the actual force expended in the compression of the spring was 14·268 foot-pounds, which is equivalent to 0°·037 Fahr. in half a pound of water, a thermal effect which would have been made manifest with the greatest facility. Hence it was obvious that a gas might be conceived as so constituted that the heat evolved by its compression would be in no respect the equivalent of the mechanical force employed, and that therefore we had no right to assume such equivalency except as a hypothesis to be tested by experiment. Accordingly, after this hypothesis had been proved by me approximately*, Professor Thomson devised the experiments† by which we have succeeded in defining the limits of its accuracy. The same philosopher has also applied his powerful analysis to the investigation of the thermo-electric properties of matter ‡. The results of the experiments I have just given an account of can only be considered as negative; but it has been decided by Professor Thomson that the compression of a spring gives a certain, though excessively small thermal effect, owing to the almost exact counterpoise of heating and cooling effects on the compressed and extended sides. The method of obtaining appreciable results was obviously to examine these opposite effects separately. I have therefore, on the suggestion of Professor Thomson, undertaken some experiments with a view to ascertain the heat developed by longitudinal compression, and that absorbed on the application of tensile force.

6. At the outset it was obvious that a very delicate test of temperature would be required, and no means appeared to offer so many advantages as that of thermo-electricity. Professor J. D. Forbes had constructed a thermo-multiplier capable of detecting temperatures not exceeding one thousandth of a degree Fahrenheit. Adopting some of the re-

* Proceedings of the Royal Society, June 20, 1844; and Philosophical Magazine, May 1845.

† Philosophical Magazine, 1852, Supplement; Philosophical Transactions, 1853, Part III. p. 357, and 1854, Part II. p. 321.

‡ Quarterly Mathematical Journal, April 1855.

finements introduced by Melloni and Forbes, I have simplified the instrument so as to render its construction and management very easy, and also increased its reliability by immersing it into the vacuum of an air-pump. My thermo-multiplier is represented by the adjoining sketch (fig. 87), where *a* is an air-pump firmly clamped to a strong stool, the legs of

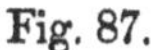

Fig. 87.

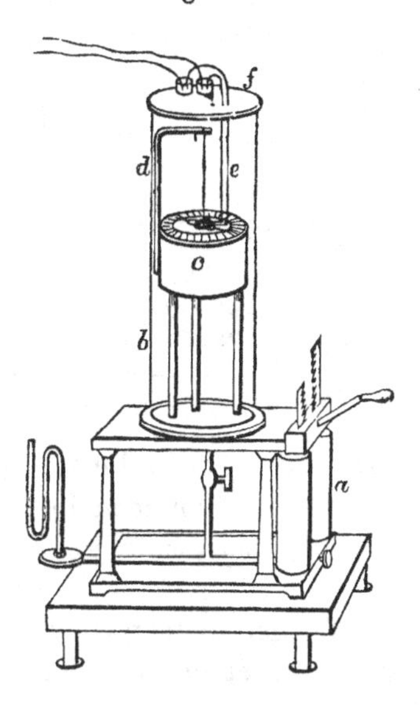

which pass through holes in the laboratory-floor and are driven into the ground beneath; *b* is a glass chimney-receiver; *c* a block of wood supported on feet which rest on the pump-plate; *d* a piece of glass rod fixed to the block, over which is thrown the filament which supports the astatic needles. Two thick copper wires (*e*) dip into mercury-cups formed in the block, and, being carried out of the receiver through holes drilled in the ground-glass plate *f*, are bent into two mercury-cups placed on the top of the instrument. Pitch was employed to close all orifices air-tight.

7. The details of the astatic needles, which are poised according to the plan first suggested by Professor Thomson,

will be better understood by inspecting fig. 88. The needles are parts of one sewing-needle magnetized to saturation, one part being a little longer than the other, so as to exceed it in magnetic moment and give direction to the system. A piece of glass tube drawn very fine, and bent as represented in the sketch, is attached at right angles to the upper magnetic needle and serves as the pointer. The lower needle is hooked to the pointer by means of the fine glass tube to

Fig. 88.

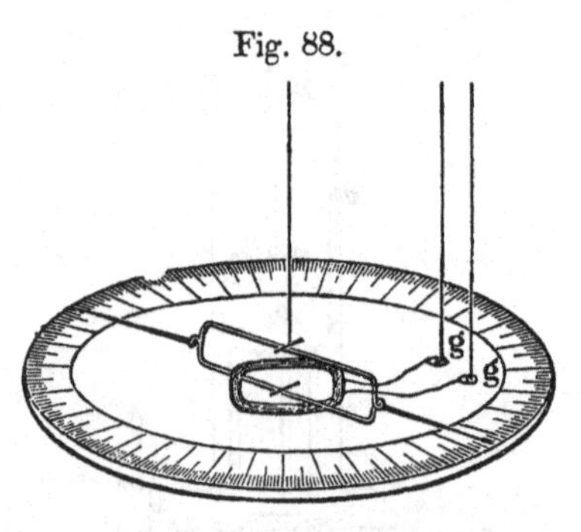

which it also is attached. The coil consists of twenty turns of silked copper wire, $\frac{1}{40}$ of an inch in diameter, the ends of which dip into the mercury-cups *g g* formed in the block of wood.

8. In order still further to increase the sensibility of the instrument, a steel magnet one yard long, the permanency of which had been tested, was placed so as to counteract and almost entirely overcome the action of the earth's magnetism in the locality of the needle. A small telescope placed at the distance of a few yards, and looking obliquely downwards through the chimney-glass at the graduated circle, completed the apparatus.

9. With air in the receiver at the atmospheric pressure, the mere standing at the distance of two yards on one side of the instrument would in a short space of time cause the needle to travel through 10°, in consequence of the currents of air produced by the unequal heating of the walls of the glass receiver. But when the air was reduced to a pressure of only half an inch in the mercury-gauge, this did not take

place, though still, when the hand was put in contact with the receiver, a very considerable deflection of the needle was speedily produced.

10. On working with my instrument, I was agreeably surprised to find that when the bar-magnet was placed so as to make the needle take up one minute in being deflected to a new position, no perceptible return swing of the needle took place, even when the rarefaction of the air was carried to half-an-inch pressure. If a small magnet was suddenly placed where it could deflect the needle 30°, the pointer would steadily travel towards that degree of deflection, and on arriving there would remain settled without any previous oscillation that could be discerned. When the time of a swing was reduced to 30 sec., a return swing was observed amounting to $\frac{1}{150}$, $\frac{1}{25}$, and $\frac{1}{18}$ of the first swing, according as the gauge was reduced to 1, $\frac{6}{10}$, and $\frac{1}{2}$ inch respectively.

11. As a test of the delicacy of the instrument arranged so as to give the swing in 45 sec., I may mention that, after increasing the resistance of the coil and its appendages a hundredfold by the addition of 100 yards of fine copper wire $\frac{1}{40}$ of an inch diameter, a single degree Centigrade communicated to the junction bismuth and antimony produced a deflection of 2° 57′. Therefore, as it was quite possible to estimate a deflection amounting to 2′, it followed that a change of temperature in the junction bismuth and antimony directly connected with the multiplier would be estimable, if it were only $\frac{1}{8800}$ of a degree Centigrade.

12. An objection which might be raised on account of an air-pump being permanently occupied, might be got rid of by the following arrangement:—A rim *a*, fig. 89, contains pitch or other cement to secure the joint between the chimney and top plate glass; a rim at *b* contains pitch to secure the orifices where the wires pass through the top plate. The bottom of the chimney rests upon a round metallic plate, into which a metal pipe *c* is screwed, which is attached by india-rubber tube to the glass tube *d*. This tube is hermetically sealed after the air of the receiver has been exhausted through it. All the lower joints are then rendered permanently tight by

putting the instrument in the shallow dish *e* filled with melted pitch.

Fig. 89.

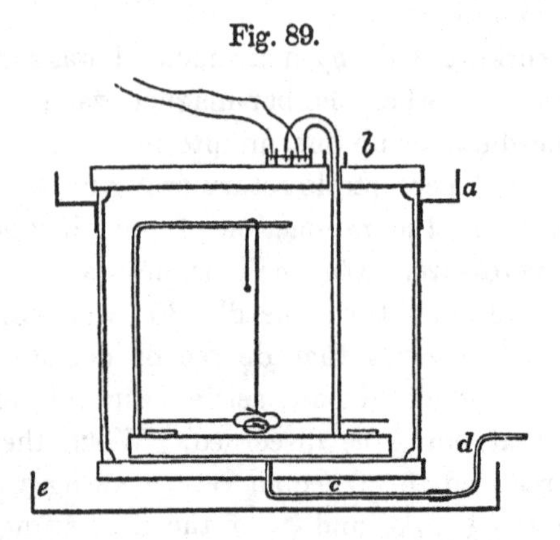

Thermo-electricity of Iron in different states.

13. In my earliest experiments on the thermal effect of stretching a steel bar, I placed a copper wire in contact with the steel, and completed the circuit by an iron wire in contact with another part of the steel bar not under tension. I found anomalous results, which I was ultimately able to refer to the strong thermo-electric relation between hard steel and iron. It was at once obvious that the existence of such marked differences between ferruginous metals in various states of aggregation or purity might render the thermo-multiplier a valuable test in the hands of practical men. I therefore allowed myself to be diverted awhile from the main object of inquiry in the endeavour to throw some light on so interesting a question.

14. Professor Thomson has described * the changes in thermo-electric position which are produced by the various conditions under which metals are placed. He has shown that the effects of lateral hammering are in most metals the reverse of those found for iron. I find that by hardening steel the change is in the same direction as with iron, but of enormously greater magnitude. In a softened state, the

* Philosophical Transactions, 1856, part iii. p. 722.

position of steel is about midway between copper and iron; but after hardening it by plunging it at a bright-red heat into water, I have in some instances found it to be on the bismuth side of copper, the alteration of the thermo-electric position of the same specimen amounting to as much as $\frac{1}{15}$ of the entire range between bismuth and antimony. In all the specimens of wrought and cast iron I have examined, there is a notable effect in the same direction produced by plunging at bright-red heat into water; but although a great change in the molecular state was thus occasioned, evinced by the iron bending with twice the difficulty it did when annealed, and the cast iron resisting the action of the file, either metal was only brought nearer bismuth by $\frac{1}{200}$ of the interval between bismuth and antimony. A fresh illustration of the extraordinary physical change produced in iron by its conversion into steel is thus afforded; and I believe that the excellence of the latter metal might be tested by ascertaining the amount of change in thermo-electric condition which can be produced by the process of hardening.

15. The different varieties of cast iron I have tried present a surprising range of thermo-electric intensities, extending almost from that of wrought iron on the one hand, to that of German silver on the other. By the kindness of Professor F. C. Calvert*, I have been enabled to examine several interesting specimens, of which he has furnished me with the analysis. The general conclusion arrived at is, that the metal is brought nearer bismuth as the quantity of carbon in combination is increased, but much more so than would be the case if cast iron exhibited merely the combination of thermo-electric intensities which carbon and iron separately possess, at least if the intensity generally assigned to carbon is correct. The intensity of the junction wrought iron and highly carbonized cast iron is as much as one fifth of the intensity of the junction antimony and bismuth.

* The late eminent Professor of Chemistry in the Manchester Royal Institution. I take pleasure in recording the assistance he was ever ready to supply from his laboratory.—*Note*, 1882. J. P. J.

Thermo-electric Intensities of Metals, Alloys, &c.

16. In constructing the following Table, I employed the thermo-multiplier already described, furnished with a variable extra resistance. I examined first those metals whose thermo-electric qualities were most widely separated, and then those which lay intermediate and differed less from each other. Small arcs of deflection were observed, so that the deviation of the needle was a sufficiently correct measure of the intensity of the current; and constantly repeated comparisons were made with a standard thermo-electric junction of copper and iron.

17. *Scale of Thermo-electric Intensities at* 12° *Centigrade.*

Antimony, specimen of commercial	100
Antimony, pure, prepared by Professor Calvert	98·06
Antimony, pure, prepared by Professor Calvert, not well annealed	95·95
Alloy consisting of 5 equivalents of antimony +1 equivalent of bismuth	83·75
Alloy consisting of 4 equivalents of antimony +1 equivalent of bismuth	80·50
Alloy consisting of 3 equivalents of antimony +1 equivalent of bismuth	78·94
Alloy consisting of 2 equivalents of antimony +1 equivalent of bismuth	68·79
Alloy consisting of 1 equivalent of antimony +1 equivalent of bismuth	45·51
Iron, thick wire, Professor Calvert's	80·0
Iron, thin wire	79·24
Iron, thick wire, very well annealed	78·24
Iron, thick wire, hardened by plunging at bright red into water	77·62
*Iron (Professor Calvert's No. 1) drawn into wire	77·23
†Iron (Professor Calvert's No. 1) after puddling, but previous to being drawn into wire	76·00
‡Iron (Professor Calvert's No. 1), cast, previous to puddling	65·21
Iron (Professor Calvert's No. 1), cast, previous to puddling, hardened by plunging it at bright red into water	63·89
Iron (Professor Calvert's No. 1), cast, white fracture, very hard	70·92
Iron (Professor Calvert's No. 1), cast, black fracture	57·77
Steel, small file, annealed	73·15
Steel, small file, hardened by plunging at bright red into water	65·8
Steel, large file, annealed	70·28

Steel, large file, hardened by plunging at bright red into water.	64·06
Iron, cast, annealed	63·25
Iron, cast, hardened by plunging at bright red into water......	62·7
Iron, cast, hardened by plunging at bright red into water, another specimen	60·83
Iron and copper wires, faggot of, in proportion 1 iron to 5 copper	68·29
Iron and copper wires, faggot of, in proportion 2 iron to 5 copper	69·83
Iron and copper wires, faggot of, in proportion 4 iron to 5 copper	71·17
Iron and copper wires, faggot of, in proportion 8 iron to 5 copper	72·71
Iron and copper wires, faggot of, in proportion 16 iron to 5 copper	75·53
Zinc	67·94
Zinc, amalgamated	67·77
Copper deposited by electricity (brittle)	67·9
Copper, thin wire........	67·14
Copper, another specimen	66·2
Copper, another specimen	65·92
Gold, pure........	67·71
Silver deposited by electricity (brittle)........	67·08
Silver, sheet	66·81
Tin, pure	65·46
Lead, commercial........	65·29
Lead, pure........	64·42
Platina, annealed	65·02
Platina, unannealed........	64·87
Platina, fine wire	64·15
Aluminium	64·68
Mercury........	61·74
German silver	51·88
German silver, hardened by hammering	49·51
Bismuth, pure, prepared by Professor Calvert........	4·7
Bismuth, specimen of commercial........	0

Professor Calvert's analysis of the specimens marked in the above table *, †, and ‡, gave for the

	Carbon.	Silicium.	Sulphur.	Phosphorus.	Iron.	Total.
Cast iron	2·275	2·720	0·301	0·645	94·059	100
Puddled bar ..	0·296	0·120	0·134	0·139	99·311	100
Iron wire	0·111	0·088	0·094	0·117	99·590	100

From the above digression I now return to the main subject of the present paper, by describing my

Experiments on the Thermal Effects of Tension on Solids.

18. All the metals employed, with the exception of lead,

were in the form of cylindrical bars about a foot long and a quarter of an inch in diameter. The upper end of the bar was screwed into a piece of metal supported by a wooden framework; the lower end was attached to a lever, to the extremity of which weights could be hung without approaching the apparatus. In the first method the thermo-electric junction was made, as represented in the adjoining sketch,

Fig. 90.

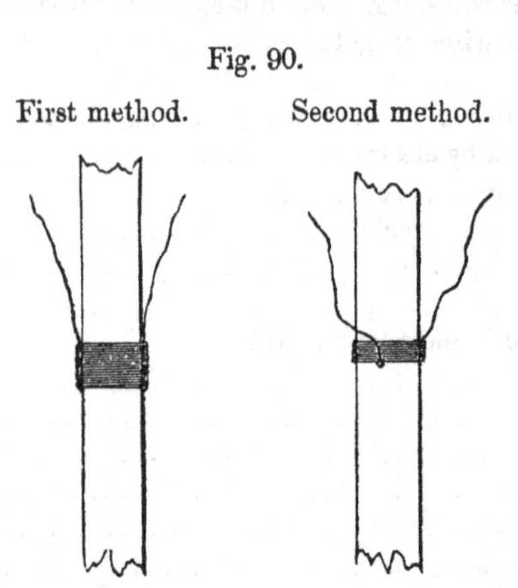

by binding to opposite sides of the bar copper and iron wires $\frac{1}{40}$ of an inch in diameter, hammered flat at the ends. In the second method, the junction of fine wires was inserted in a hole $\frac{1}{40}$ of an inch in diameter, bored through the centre of the bar, a small portion of wire being bound to the bar by means of cotton thread, but metallic contact prevented by an intervening slip of paper. The terminals of these wires were immersed in deep mercury-cups formed in a solid block of wood, whence thick copper wires proceeded to the thermo-multiplier.

19. Immediately after each experiment on the effect of tension, the thermometric value of the deflections was ascertained by immersing the bar, to within one third of an inch of the junction, in water of different temperatures. The deflections thus produced were about two thirds of those occasioned by the same changes of temperature when the junction was completely immersed. The diminished effect

in the former case is owing for the most part to the conduction of heat from the air by the thermo-electric wires. The experiments on tension were liable to be affected in the same way; but they were not subject to the loss arising from the conduction of heat from the surface of the bath to the junction. The error intervening from this latter circumstance could not be great, and was moreover, in all probability, almost exactly neutralized by a small error in the tension experiments, arising from the escape of $\frac{1}{14}$ of the thermal effect from the quarter-inch bars during the 40 seconds occupied by the swing of the needle.

20. *Iron.*—The weight of the bar per foot was ·1568 of a lb. The lever alone gave it a constant tension of 70 lb. The additional tensions successively given were 194 lb., 388 lb., 583 lb., and 1166 lb., or as 1, 2, 3, and 6. The mean of five trials gave 20′·6 as the deflection indicating cold on applying the tension of 194 lb., and 20′·2 as the deflection in the opposite direction, indicating heat, when the weight was removed. With 388 lb., I had 35′·8 and 42′·2; with 583 lb., 56′·4 and 57′·8; and with 1166 lb., 1° 55′·8 and 1° 58 ·6. Hence it appeared that the quantity of cold produced by the application of tension was sensibly equal to the heat evolved by its removal, and, further, that the thermal effects were proportional to the weights employed.

21. The above trials having been made without an accurate determination of the thermometric value of the deflections immediately afterwards (a practice I found it important to follow in order to obviate the effects of any alteration in the magnetic intensities of the earth, astatic needles, or controlling magnet, which might be taking place), I repeated the experiment, using the first method above described, and found a mean deflection of 1° 22′·8 with a tension of 775 lb. The thermometric value of this deflection, determined in the manner already explained, was −0°·115 Centigrade.

22. Professor Thomson, in his researches on the thermoelastic properties of metals, has demonstrated the following formula as applicable to the phenomenon in question :—

$$H=\frac{t}{J}\times\frac{p}{1}\times\frac{e}{1}\times\frac{1}{s}\times\frac{1}{w},$$

where H is the thermal increase in degree Centigrade, t the temperature Centigrade from absolute zero, J the mechanical equivalent of the thermal unit, p the pressure applied in pounds, which in the case of tension is of course a negative quantity, e the longitudinal expansion per degree Centigrade, s the specific heat, and w the weight in pounds of a foot length of the bar *. Applied to the above experiment the formula gives

$$H=\frac{276{\cdot}7}{1390}\times\frac{-775}{1}\times\frac{1\,\dagger}{81200}\times\frac{1\,\ddagger}{\cdot 11}\times\frac{1}{\cdot 1568}=-0^{\circ}{\cdot}11017.$$

23. In another experiment by the first method, but with another thermo-electric junction, I obtained a thermal effect of 0°·124. In this case the formula gave

$$H=\frac{276}{1390}\times\frac{-775}{1}\times\frac{1}{81200}\times\frac{1}{\cdot 11}\times\frac{1}{\cdot 1568}=-0^{\circ}{\cdot}1099.$$

24. A third experiment, in which the second method was used, and in which I had the advantage of Professor Thomson's assistance, gave a thermal effect of −0°·1007. In this case the formula gave

$$H=\frac{287}{1390}\times\frac{-725}{1}\times\frac{1}{81200}\times\frac{1}{\cdot 11}\times\frac{1}{\cdot 1568}=-0^{\circ}{\cdot}1069.$$

25. *Hard Steel.*—With stretching-weights of 194, 388, and 775 lb., I observed deflections of the needle amounting to 27′, 48′, and 1° 37′. Another experiment by the first method to obtain the absolute thermal effect gave 53, which was found to indicate a temperature of −0°·162. The theoretical result in this case was

* The formula signifies in fact that the heat evolved by compressing a solid is equivalent to the work required to compress the volume of it due to temperature, just as in the case of a perfect elastic fluid, applied to which the expression becomes simplified to $\frac{p}{Jsw}$. M. Clapeyron has given a theoretical estimate of the heat disengaged by the cubical compression of iron (Scientific Memoirs, vol iii. p. 73).

† Lavoisier and Laplace. ‡ Dulong and Petit.

$$H=\frac{275{\cdot}4}{1390}\times\frac{-775}{1}\times\frac{1\,*}{81633}\times\frac{1\,\dagger}{{\cdot}1024}\times\frac{1}{{\cdot}1499}=-0^{\circ}{\cdot}125.$$

26. *Cast Iron.*—The deflections produced by tensile forces of 194, 388, and 775 lb. were 17′, 31′, and 59′·9 respectively. The thermal effect indicated by the last was $-0^{\circ}{\cdot}1605$. The formula gives

$$H=\frac{277{\cdot}4}{1390}\times\frac{-775}{1}\times\frac{1\,\ddagger}{90139}\times\frac{1\,\S}{{\cdot}1198}\times\frac{1}{{\cdot}1281}=-0^{\circ}{\cdot}112.$$

27. In another experiment I obtained a thermal effect of −·1481, theory giving

$$H=\frac{283}{1390}\times\frac{-784}{1}\times\frac{1}{90139}\times\frac{1}{{\cdot}1198}\times\frac{1}{{\cdot}1281}=-0^{\circ}{\cdot}115.$$

28. *Copper.*—With tensile forces of 191·6, 383·3, and 766·6 lb., I obtained by the first method deflections of 21′, 41′, and 1° 12′·6 respectively, the absolute thermal effect indicated by the last being $-0^{\circ}{\cdot}174$. The formula gives

$$H=\frac{274{\cdot}9}{1390}\times\frac{-766{\cdot}6}{1}\times\frac{1\,\|}{58200}\times\frac{1\,\P}{{\cdot}095}\times\frac{1}{{\cdot}1781}=-0^{\circ}{\cdot}154.$$

29. With copper, and also with iron wires, I have observed the cooling effect of tension close up to the breaking-point. But whenever the pull occasioned permanent change of figure, the heat frictionally evolved overpowered the cooling effect.

30. *Lead.*—The bar employed was half an inch in diameter, and weighed 6308 grains to the foot. The stretching-weights were, first, 193 lb. with 70 lb., the weight of the lever, constant; and second, 263 lb., the weight of the lever itself being used along with the 193 lb. laid on it. Using the first method, I obtained with the above weights deflections of 21′·5 and 30′·7, indicating thermal effects of $-0^{\circ}{\cdot}0531$ and $-0^{\circ}{\cdot}0758$. The formula gives in the two cases,

$$H=\frac{278{\cdot}5}{1390}\times\frac{-193}{1}\times\frac{1\,**}{35100}\times\frac{1}{{\cdot}0303}\times\frac{1}{{\cdot}901}=-0^{\circ}{\cdot}0403,$$

* Smeaton. † Potter. ‡ Roy.
§ Myself. ‖ Lavoisier and Laplace. ¶ Dulong and Petit.
** Lavoisier and Laplace.

and

$$H=\frac{278\cdot5}{1390}\times\frac{-263}{1}\times\frac{1}{35100}\times\frac{1}{\cdot0303}\times\frac{1}{\cdot901}=-0^\circ\cdot0550.$$

31. *Gutta Percha.*—The piece used was half an inch in diameter, and weighed 1246 grains per foot. The thermo-electric junction, formed of very thin copper and iron wires, was inserted in a slit made two thirds through the substance of the gutta percha. The weights employed were, first, that of the lever alone, amounting to 70 lb.; and second, the lever with the further addition of 80 lb., making in all 150 lb. As with the metals, cold was produced when the gutta percha was stretched, and the heat restored when the stretching-weight was removed. With the first-named tensile force a mean deflection of 16′·4 was observed, and with the second a mean deflection of 30′·2. In order to turn the above into thermal measure, the gutta percha was plunged into water of different temperatures, deep enough to immerse the part in which the thermo-electric junction was imbedded, the former plan of operation being inapplicable in this instance, on account of the low conducting-power of the substance used. It was found in this way that the quantities of thermal effect due to the above deflections were $-0^\circ\cdot0284$ and $-0^\circ\cdot0524$—quantities, however, which are likely to err in defect by a very small quantity, owing to the nature of the method of finding the value of the deflections. The theoretical results, assuming, as I did for the metals, that the expansion of gutta percha by heat is the same whether suffering tension or not, are

$$H=\frac{276\cdot4}{1390}\times\frac{-70}{1}\times\frac{1}{6354}\times\frac{1}{\cdot402}\times\frac{1}{\cdot178}=-0^\circ\cdot0306,$$

and

$$H=\frac{276\cdot4}{1390}\times\frac{-150}{1}\times\frac{1}{6354}\times\frac{1}{\cdot402}\times\frac{1}{\cdot178}=-0^\circ\cdot0656.$$

32. The values just given for the expansion and specific heat of gutta percha are those which I arrived at by experiments on a part of the same sheet out of which the cylinder above used was made. The method pursued in obtaining the expansion was one I have frequently found very convenient.

It consists in weighing the body in water at two temperatures, one as much below as the other is above the point of maximum density. In this way I found the linear expansion between 2°·4 and 5°·8 to be $\frac{1}{6354}$ per 1°, and the specific gravity to be 1·00462 at 4°. The specific heat at 4°·5 was found by the method of mixtures to be ·402.

33. *India-rubber.*—The extraordinary physical properties of this substance appear to have been first remarked by Mr. Gough *. By placing a slip of it in slight contact with the edges of the lips, and then suddenly extending it, he experienced a sensation of warmth arising from an augmentation of the temperature of the rubber; and then by allowing the slip to contract again, he found that this increase of temperature could be destroyed in an instant. In his next experiment, he found that "if one end of a slip of caoutchouc be fastened to a rod of metal or wood, and a weight be fixed to the other extremity, in order to keep it in a vertical position, the thong will be found to become shorter with heat and longer with cold"†. The third experiment described by Mr. Gough has indicated the means of using this substance in the production of textile fabrics; he says:—"If a thong of caoutchouc be stretched, in water warmer than itself, it retains its elasticity unimpaired; on the contrary, if the experiment be made in water colder than itself, it loses part of its retractile power, being unable to recover its former figure; but let the thong be placed in hot water, while it remains extended for want of spring, and the heat will immediately make it contract briskly."

34. In addition to the above, my own experience has led me to the curious fact that a piece of india-rubber, softened by warmth, may be exposed to the zero of Fahrenheit for an hour or more without losing its pliability, but that a few days rest at a temperature considerably above the freezing-point will cause it to become rigid.

* Memoirs of the Literary and Philosophical Society of Manchester, 2nd series, vol. i. p. 288.

† Ibid. p. 292. From the context it appears that the weight was used to give tension, as well as to keep the slip vertical.

35. The ends of a piece of elastic india-rubber, about $\frac{4}{10}$ of an inch square, were attached to an apparatus by means of which it could be stretched by various weights. A thermo-electric junction of thin copper and iron wires was placed in an orifice pierced through the centre of the slip; and at either side, at $\frac{6}{10}$ of an inch distance from one another, pins were stuck which afforded the means of knowing the exact length to which the rubber was stretched. With small tensile forces no thermal effects were observed; but when the rubber was stretched by a weight of 6 lb., a sensible deflection of the needle took place, indicating heat; the needle returning to zero, and thus indicating absorption of heat, when the weight was removed. Experiments with different weights gave the following results:—

Stretching-weight in pounds.	Length of rubber in inches, when stretched.	Length of rubber in inches, after weight was removed.	Deflection of thermo-multiplier.	Thermal effect indicated by deflection.
			° ′	°
0	0·59	0·59	0	0
2	0·67	Not observed.	0	0
4	0·77	Not observed.	− 1	−0·002
6	0·94	Not observed.	11	0·020
8	1·13	Not observed.	32	0·059
10	1·45	0·71	1 15	0·139
12	1·75	0·9	2 47	0·370
14	Not observed.	Not observed.	4 50	0·537
16	2·14	Not observed.	5 55	0·657
18	2·36	1·02	6 45	0·750

36. The low temperature (about 6° Centigrade) at which the above experiments were tried considerably interfered with the perfection of the elasticity of the india-rubber, which in consequence became gradually elongated on the removal of successively increasing weights. This circumstance no doubt interferes slightly with the absolute quantity of thermal effect recorded in the last column of the Table. Still the essential characteristics of the phenomena are plainly indicated, viz. that little heat, or even a reverse effect, is produced by moderate stretching-weights, but that after a certain weight is reached, a rapid increase of thermal effect takes place.

37. The same piece of rubber was afterwards immersed in water in the manner represented by the adjoining sketch. A wire attached to the rubber at the upper extremity was made fast, and another wire affixed to the lower end passed through a cork in the glass tube, and, being bent into a hook, supported a scale for weights. The temperature of the water in the tube was kept uniform by constant additions and removals of water, which was kept well mixed by a current of bubbles of air blown from time to time through a narrow tube into the lower end. The positions of the pins, showing the length of the rubber, were ascertained by means of a graduated scale placed behind the tube, viewed through a telescope which was raised or depressed according to the level of the pins.

Fig. 91.

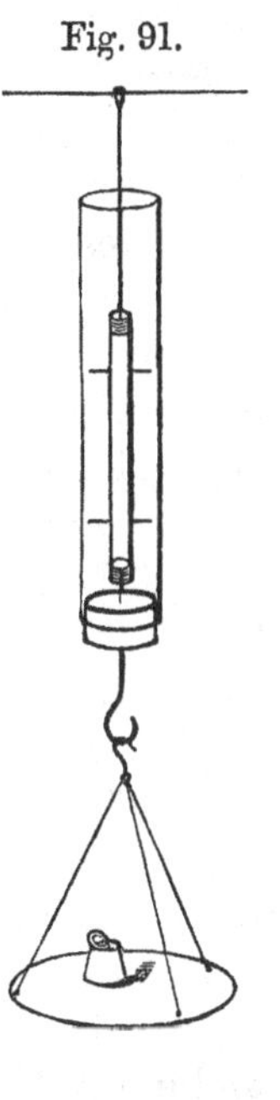

38. In my first experiment the stretching-weight, inclusive of the weight of the tube, was 3 lb., and the water in the tube being adjusted to 7° Cent., the length of the thong between the pins was 1·026 inch. The temperature of the water was then raised to 17°, and kept there for ten minutes, when the length was found to be reduced to 0·964 inch. A third observation at 28° gave the length 0·9. It was evident that this contraction was partly owing to the recovery by heat of the rubber from the constrained state in which it was left by the experiments of the preceding table; for the lengths observed on lowering the temperature successively to 16°·6 and 7°·4, namely 0·91 and 0·922, indicated a lengthening effect from the application of cold only one quarter of the amount of the previous shortening by heat. I now used a stretching-weight of 6 lb., which immediately increased the length to 1·22, and in the course of thirty-six hours further increased it to 2·548 inches at the temperature 5°·4. Raising the temperature successively to 12° and 17°, the length was diminished to 2·527 and 2·459. Then, on depressing it to

9°·5 and 4°·3, I observed 2·476 and 2·490. After leaving the same weight on for five hours longer, I observed lengths of 2·559, 2·514, and 2·47 with the temperatures 6°·2, 12°·4, and 20°·2, and 2·483 and 2·498 at temperatures successively depressed to 12° and 5°. After the lapse of twenty-two hours more, I found the lengths 2·624, 2·568, and 2·51 at the temperatures 6°·2, 13°·6, and 22°; and 2·523 and 2·548 on depressing the temperature to 13° and 6°·4. All these observations show the gradual elongation effected by tension, the tendency of heat to restore the original length, and the increase or decrease of elastic force with the elevation or depression of the temperature of the rubber.

39. I now reduced the stretching-weight to 1 lb., being that of the tube and the water contained by it. The observed length was 1·63, and in forty-five minutes further decreased to 1·621. Afterwards it began to increase gradually, in five hours attaining to 1·635, in seventeen more to 1·724, and after a further lapse of forty hours to 1·751. These observations were at 7°·1. Then raising the temperature to 21°·6, the length was reduced to 1·191; and on depressing it again to 7°·4, the length was further reduced to 1·181. Keeping the rubber at the same temperature it still continued to contract, and in two hours its length was only 1·17. Then, on raising the temperature to 23°, and afterwards depressing it to 7°·2, the length was successively reduced to 1·151 and 1·141. These and similar observations were continued for several days with like results, confirming my previous remarks, and also developing the curious fact, that when a piece of india-rubber, which has been previously gradually lengthened by stretching, is shortened by the temporary application of heat, the shortening effect continues to go on for some time, even if the rubber has been reduced to its previous temperature.

40. Although my experiments, already detailed, on the thermal effects of stretching india-rubber were not, on account of the imperfect elasticity of that substance at low temperatures, well calculated to give exact numerical results, the general conclusion derived from them is perfectly in accordance with Professor Thomson's formula, where, in this

instance *e*, the expansion, being negative, the result indicated is a rise of temperature on the application of the tensile force.

41. When, by keeping india-rubber at rest at a low temperature for some time, it has become rigid, it ceases to be heated when stretched by a weight, and, on the contrary, a cooling effect takes place as in the metals and gutta percha, The experiment by which I ascertained this fact was made with a piece of india-rubber $\frac{1}{4}$ of an inch square, which had been exposed for several days to a temperature not lower than 5° Cent. In this state it was found that the application of a tensile force of 14 lb. produced a deflection of 20′, indicating cold, and that a contrary deflection, indicating heat, was produced by removing the weight. After raising the temperature of the rubber to 15°, its elasticity was restored, and the reverse phenomenon of heat on stretching produced, indicated by a deflection of 15′.

42. *" Vulcanized" India-rubber.*—I observed the principal physical characters of this substance before I was acquainted with Mr. Gough's discovery of the properties of simple india-rubber, above noticed. The superior permanency of the elasticity of rubber in combination with sulphur, and its unimpaired existence at low temperatures, rendered it better qualified for experiments in which accurate numerical results were desired.

43. I determined the *specific heat* of vulcanized india-rubber from portions of the specimens used in the experiments about to be detailed. The method of mixtures was employed, a quantity in small bits being raised to a given temperature in a bath of air, and then plunged into a thin copper can partly filled with cold water. The result, reliable to at least one tenth of its amount, was 0·415.

44. The *expansion by heat* was found by weighing in water 2°·25 above and 2°·25 below the maximum density. The result showed a cubical dilatation of 0·000526 per degree Centigrade—an expansion greater than that of any other solid hitherto examined.

45. Mr. Gough has stated that the specific gravity of india-rubber is increased by stretching it; and to this he attributes the heat thereby evolved. The experiment, how-

ever, is a very delicate one, and it does not appear that this philosopher possessed the means of arriving at a reliable result. My experience with vulcanized india-rubber leads me to doubt the accuracy of Mr. Gough's conclusion. I weighed elastic bands alternately stretched and unstretched in water. The result of the first series of experiments was, that the bands unstretched had a specific gravity ·996057; but when pulled twice their natural length, ·994641. The second series gave me for the specific gravity of the unstretched bands ·990918; but when pulled to two and a half times the natural length, ·988483. The same vulcanized india-rubber bands were used in both series; and the diminution of the specific gravity in the second series is owing to the constant loss of sulphur taking place. On account, however, of the alternation of the experiments in each series, this circumstance does not affect the conclusion that the action of tension on a band of vulcanized india-rubber sensibly diminishes its specific gravity.

46. In my experiments on the thermal effects of stretching vulcanized rubber, a piece about $\frac{3}{8}$ of an inch square, and weighing 1452 grs. to the foot, was employed. The upper end was made fast, and to the lower weights could be applied and removed at pleasure. A thermo-electric junction of thin copper and iron wires was inserted in a slit made in the direction of the length of the thong and near its middle. Also pins were stuck into the rubber at 4 inches distance from each other, in order to measure its length under various tensile forces. The following is a Table of results :—

Stretching weight, in pounds.	Length of measured part, under tensile force, in inches.	Deflection of thermo-multiplier.	Thermal effect indicated by the deflection.
0	4	$\overset{\circ}{0}$	$\overset{\circ}{0}$
2	4·06	−1·4	−0·003
4	4·12	−2	−0·004
7	4·3	−2·4	−0·004
14	4·8	−0·5	−0·001
21	5·21	7·9	0·014
28	5·87	29·1	0·053
35	6·6	51·9	0·095
42	7·25	1 15·7	0·137
49	7·75	1 39·3	0·180

47. The data for turning the deflections in the third column into the thermal effects in the fourth were obtained by immersing the rubber, to a point above the place where the thermo-electric junction was inserted, in water of different temperatures.

48. These experiments developed the following facts:— 1st. That the effect of laying on the weights was not sensibly different in quantity from the reverse effect of removing them. 2nd That with light weights and a low temperature there was a slight cooling effect on the application of tensile force, which, first increasing with the weights laid on, ultimately changed into a heating effect, increasing much more rapidly than the stretching-weight.

49. The temperature of the thong during the experiments was $7^{\circ}{\cdot}8$. I found that at temperatures a few degrees higher, the reverse action with weak tensile forces did not take place, but that there was, on the contrary, a very slight heating effect.

50. Professor Thomson suggesting to me that in those circumstances in which the rubber was heated by being stretched, it would contract its length under tension when its temperature was raised, I arranged the same piece with which the foregoing experiments were made in the manner already described in the case of common india-rubber. The method pursued was to observe the lengths at several temperatures successively raised, and then at similar temperatures descending. The necessity for this arose from the circumstance that a gradual elongation took place, particularly at the higher temperatures; and by taking the mean of the ascending and descending series, I hoped to eliminate the error thus arising. I give the following experiments in the order in which they were made:—

Tensile force, in pounds.	Temperature successively raised.	Length, in inches.	Temperature successively lowered.	Length.	Mean temperature of the two series.	Mean length.	Shortening per degree Cent. from 0°.
14	13°·4	4·899	9°·2	5·008	11·3	4·953	$\frac{1}{2854}$
	22·5	4·894	21·5	4·979	22	4·936	
	32	4·888	30·8	4·952	31·4	4·920	
	48·6	4·888	....	....	48·6	4·888	
7	7·95	4·569	8	4·476	7·97	4·522	$\frac{1}{3536}$
	22	4·526	19·1	4·480	20·55	4·503	
	33·2	4·481	31·5	4·481	32·35	4·481	
	50·1	4·468	....	....	50·1	4·468	
21	2·8	5·497	6	5·661	4·4	5·579	$\frac{1}{2124}$
	18	5·468	17·4	5·612	17·7	5·540	
	33	5·458	32·4	5·540	32·7	5·499	
	50	5·459	....	....	50	5·459	
28	1·4	6·398	4·4	6·513	2·9	6·455	$\frac{1}{1001}$
	18	6·298	18	6·435	18	6·366	
	33·6	6·224	34	6·302	33·8	6·263	
	49·6	6·153	....	....	49·6	6·153	
35	1·8	7·467	3·4	7·589	2·6	7·528	$\frac{1}{687}$
	19	7·264	18	7·432	18·5	7·348	
	34	7·121	33	7·250	33·5	7·185	
	49·8	7·009	....	....	49·8	7·009	
42	– 0·4	8·604	2·6	8·792	1·1	8·698	$\frac{1}{628}$
	17	8·415	16·6	8·575	16·8	8·497	
	34	8·195	32	8·327	33	8·261	
	50	8·020	....	....	50	8·020	

51. The above results completely justify the anticipation of Professor Thomson, and, as we shall presently see, afford a remarkable confirmation of his theory. Before, however, instituting a comparison between theory and experiment, it will be proper to observe that the length of the rubber at the low temperatures beginning and ending each experiment shows a gradual increase. To this circumstance chiefly is it owing that so considerable a shortening effect took place with 7 lb. tension; for a tensile force of 14 lb. having been just previously tried, a certain amount of increased length of the nature of *set* was produced, which, being taken out by heating the rubber under less tension, caused a greater apparent contraction by heat. On lowering the temperature again, a small contraction took place, which, being corrected for the gradual elongation of the rubber, gave me $\frac{1}{9009}$ as the expansion by heat under the tension of 7 lb.

52. The Table on p. 438 gives the theoretical estimate, compared with the experimental result, for each additional tensile force of 7 lb.

53. In the last two columns of this Table are recorded the actual quantities of heat observed in the experiments given in (46), and the same corrected on account of the thong not having then been so much stretched by use. I am of opinion, however, that this circumstance would not cause any material difference in the thermal effect of a given strain, and that therefore the last column but one is the proper one to compare with theory, which it will be found to do quite as well as could have been expected.

54. After the experiment on the contraction by heat of the thong when it was stretched by 42 lb., I removed all the weights except 2 lb., and then a day or two afterwards observed as follows:—

Temperature successively raised.	Length.	Temperature successively lowered.	Length.	Contraction per degree of temperature raised.	Contraction per degree of temperature lowered.
$-0^{\circ}\!\cdot\!4$	4·445	$3^{\circ}\!\cdot\!2$	4·236		
25	4·339	26	4·251	$\frac{1}{1212}$	$\frac{1}{8260}$
50	4·26	50	4·26		

Weight laid on, in pounds.		$\frac{t}{J}$	p.	e, or mean expansion by heat between permanent and temporary tension.	$\frac{1}{s}$.	$\frac{1}{w}$, or reciprocal of mean weight per foot.	Theoretical result, or product of the preceding five columns.	Experimental result taken from Table, p. 434.	Experimental result corrected for elongation of rubber by use.
Permanent.	Temporary.								
0	7	$\frac{281}{1390}$	−7	$+\frac{1}{9009}$	$\frac{1}{\cdot 415}$	$\frac{1}{\cdot 1917}$	−0·002	−0·004	−0·004
7	14	$\frac{281}{1390}$	−7	$-\frac{1}{5003}$	$\frac{1}{\cdot 415}$	$\frac{1}{\cdot 1759}$	+0·004	+0·003	+0·003
14	21	$\frac{281}{1390}$	−7	$-\frac{1}{2435}$	$\frac{1}{\cdot 415}$	$\frac{1}{\cdot 1576}$	+0·009	+0·015	+0·016
21	28	$\frac{281}{1390}$	−7	$-\frac{1}{1360}$	$\frac{1}{\cdot 415}$	$\frac{1}{\cdot 1383}$	+0·018	+0·039	+0·043
28	35	$\frac{281}{1390}$	−7	$-\frac{1}{814}$	$\frac{1}{\cdot 415}$	$\frac{1}{\cdot 1194}$	+0·035	+0·042	+0·047
35	42	$\frac{281}{1390}$	−7	$-\frac{1}{656}$	$\frac{1}{\cdot 415}$	$\frac{1}{\cdot 1031}$	+0·050	+0·042	+0·050
0	42						+0·114	+0·137	+0·155

55. It will be observed that the effect of a rise of temperature in the above experiment * was to take out the *set* which had been given to the thong by previous experiments at higher tension. Afterwards, the effect of cooling, as shown in column 4, is a contraction so great as to prove that the *set* continued to be taken out after the heat which commenced its removal was withdrawn. That this was the case was proved by repeating the experiment several times, using the same weak tension of 2 lb.; when the lengths at 8°·8 and 52°·8 were found to be 4·241 and 4·254 respectively, indicating an expansion of only $\frac{1}{14376}$ per degree.

56. On removing all tensile force and applying a moderate warmth, the thong returned to the same length it had at the commencement of my experiments with it, viz. exactly 4 inches.

57. In order to try the thermal effects of stretching vulcanized india-rubber for a greater range of tensions, I inserted a very fine thermo-electric junction into the breadth of an elastic ring 2 inches in diameter and weighing 30 grains. The ring, when pulled sufficiently to make the two sides parallel, became a double band 3·3 inches long, which elongated further to 6·8 inches with a tension of 2½ lb. The results I arrived at are as follow:—

Permanent weight, in pounds.	Length with permanent weight, in inches.	Weight laid on in addition to the permanent weight.	Length with the additional weight.	Thermal effect in degree Centigrade.
0	3·3	2·5	6·8	0·110
2·5	6·8	2	10·9	0·242
4·5	11	2	14·6	0·330
6·5	14·8	2	17·2	0·132
8·5	17·8	2	18·5	0·088
10·5	18·9	2	19·4	0·068
12·5	19·7	2	20·1	0·004
14·5	20·5	4	20·9	0·001
18·5	21·1	2	21·3	0·009
20·5	21·7	4	22	0·044
24·5	22	4 band broke.		
			Sum......	1·028

58. The Table shows that the rise of temperature occasioned

* Bottom of p. 437.

by stretching vulcanized india-rubber bands up to their breaking-point is about 1° Centigrade—a quantity which I regard as erring somewhat in deficiency, owing to the attenuation of the band under high tension preventing the full communication of heat to the junction. The thermal effect is evidently greatest when the rubber receives the greatest elongation by an additional weight. When the band was stretched to six times its original length, little further elongation took place by increasing the tensile force, very little thermal effect being at the same time produced. I thought it possible that in this case a small cooling effect might exist; but after many trials did not succeed in finding it, owing probably to frictional generation of heat near the breaking-point.

59. *Wood.*—The physical properties of this substance have great interest, from its various applications to the wants and luxuries of mankind. On making experiments on the thermal effects of its extension and compression, I soon found anomalies which induced me to investigate at some length its expansion by heat, its elasticity, and the variations of these arising from such modifying influences as temperature, moisture, and tension.

Fig. 92.

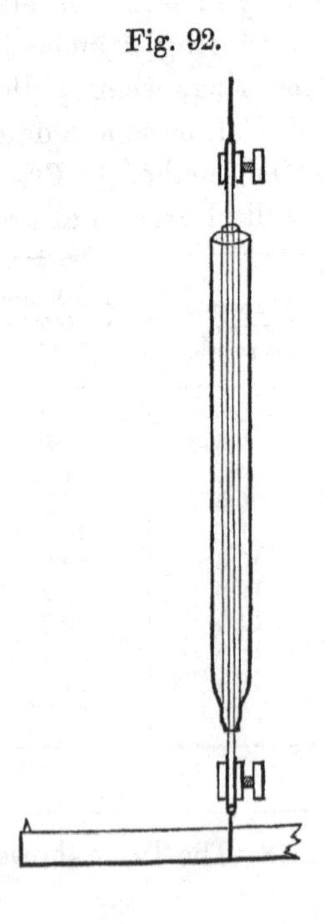

60. The apparatus I employed will be understood from the adjoining sketch, in which a wooden rod is represented as passing through a glass tube, which again passes through a wide tube of gutta percha. This latter fits tightly at its lower end on the glass tube, which again is made tight to the wood by means of a piece of india-rubber tube. Mercury or other fluid may be poured into the interstice between the glass tube and the wooden rod, and so serve as the means of convection of heat from the water with which the gutta-percha

tube is filled. A stopcock enables one to change this water when desired. The ends of the rod are held by suitable clamps, which are attached, one to a firm upper support, the other to a lever of which one extremity has a knife-edge resting on a steel plate, and the other is furnished with a micrometer consisting of a glass plate divided into $\frac{1}{200}$ of an inch, examined by a microscope. The weight of the lever alone produced a tension of 35 lb., which could be further increased by the addition of weights. When a very weak tension was desired, a lighter lever was employed.

61. Preliminary trials informed me of the time (never exceeding five minutes) required to bring the rod sensibly to the temperature of the water in the gutta-percha tube; I also found that the most ready means of obtaining uniform temperature in this water was to fill the tube twice, allowing a little intervening time. The observations of the micrometer were generally made at intervals of five minutes—the first observation being at a low temperature, the second at a high temperature, and the third at a low temperature again. The mean of the first and third observations gave me the length at the low temperature, the difference between which and the length at the high temperature gave me the expansion of the rod. When a high tension was employed, gradual elongation frequently took place for so long a time that I did not wait till it had entirely ceased. The system of observation employed was, however, such as prevented error arising from this circumstance; and since the gradual elongation was more rapid as the temperature was raised, the precaution was taken to expose the rod to the high temperature for the same space of time immediately before and after each micrometer-observation at the high temperature.

62. *Bay Wood.*—The piece selected was well seasoned and very straight in the grain. Its length was 46·63 inches, the part passing through the glass tube and exposed to change of temperature measuring 33 inches. The diameter of the rod was $\frac{3}{8}$ of an inch, and its weight, in the first instance, about 196 grains to the foot. Mercury poured into the inner glass tube formed the means of conducting heat

from the surrounding water to the rod. The following Table contains the observed expansions after various intervals of time:—

Interval of time between the successive determinations.	Tension on the rod, in pounds.	Expansion per degree Centigrade.
	35	·000004612
A few hours	435	·000005665
Two or three days	35	·000004372
A few hours	435	·000005685
Two or three days	35	·000003952
A few hours	435	·000004619
Two or three days	35	·000004003
A few hours	435	·000004850
Mean for	35	·000004235
Mean for	435	·000005225

63. The above experiments manifest most strikingly the effect of tension in increasing the expansibility of wood, that with 435 lb. being nearly one quarter greater than that with only 35 lb. The gradual diminution of the coefficient perplexed me a good deal at first, particularly as the mercury in contact with the wood prevented the access of air to its sides; but I was ultimately able to refer it to the gradual absorption of moisture through the pores in the direction of the length of the rod.

64. The rod was now exposed to a tension of 435 lb. for twenty-two days, at the end of which time its expansion by heat was found to be ·000004784. The weights were then removed so as to reduce the tension to 35 lb. After remaining two days in this state the first effect was found not to be expansion, but, on the contrary, a contraction amounting to ·000002408 per degree. Then the subsequent application of cold produced a contraction of ·000005705. On raising the temperature again after an interval of five minutes I found an expansion of ·000002170, and then on cooling a contraction of ·000005017. After twenty minutes I found that the expansibility by heat, indicated by the mean effect of raising and depressing the temperature alternately, was ·000003575.

65. From the above it appears that wood has to a certain extent the property possessed by india-rubber, of returning from a state of strain on the application of heat. Mr. Hodgkinson has observed this property in iron; I have noticed it in whalebone, hair, leather, and, to a slight extent, in copper.

66. The increase of expansibility with tension suggested that the force of elasticity of wood decreases with a rise of temperature. To try this I affixed a graduated glass plate to the top of the glass tube (see fig. 92), while parallel and close to it another divided glass plate was affixed to the wooden rod. The extension of the rod under various tensile forces was thus read off by means of a microscope in $\frac{1}{200}$ of an inch and fractions of the same appreciable to $\frac{1}{20}$ of the actual divisions. The mean of several series of trials gave me, at a low and high temperature, the following indications of the micrometer, with stretching-weights increased by 100 lb. at a time until 435 lb., and then decreased by the successive removal of the weights again.

Tension.	Micrometer-observations, the temperature of the wood being 15°·99.	Micrometer-observations, the temperature of the wood being 41°·78.
35	0	0
135	4·199	4·381
235	8·539	8·914
335	12·890	13·415
435	17·135	17·743
335	12·830	13·392
235	8·579	8·968
135	4·373	4·602
35	0·231	0·268

67. $17{\cdot}135 - \frac{{\cdot}231}{2} = 17{\cdot}02$, and $17{\cdot}743 - \frac{{\cdot}268}{2} = 17{\cdot}609$,

which are therefore the elastic strains in $\frac{1}{200}$ of an inch, produced by a tension of 400 lb. at the respective temperatures 15°·99 and 41°·78, the length of the rod being 33 inches. These results, reduced to unity of length and temperature, give the difference between ·000099953 and ·000103410, or 000003459, as the quantity by which the expansion by heat of the wood ought to be increased by adding a tension of

400 lb. Referring to p. 442, it will be seen that this increment is actually only about ·000001. I believe that two causes are concerned in producing this discrepancy. In the first place, I have reason to think that when tension is applied to wood and various other substances, a strain takes place of the nature of *set* besides the true elastic strain, which set is almost instantaneously taken out again on the removal of the tension; also that this temporary set increases with the temperature. Secondly, I think that when the expansion by heat of a body under tension is observed, the tendency of heat to take out *set* operating in a contrary direction to the expansive action, causes a slight apparent diminution of the latter. It will thus happen that the apparent decrease of elasticity by heat is greater than the real, while the apparent increase of expansion by heat, in consequence of the application of tension, is less than the real. These observations will be found to receive further confirmation from the phenomena of the elasticity of moistened wood, and also in the case of whalebone, a substance in which the characteristics of wood as regards *set* are very strongly developed.

68. Young's modulus of elasticity, deduced from the above data in terms of the length of a rod, weighing one lb. to the foot, which would be extended by unity by the tensile force of one lb., is 5539710 and 5354349 for the temperatures 15°·99 and 41°·78 respectively.

69. The same rod of bay wood was now employed to determine the effects of hygrometric condition on its expansion and elasticity. Mercury or water was employed as the medium for conducting heat between the glass tube and the wood. When mercury was employed, some inconvenience resulted from its entry into the pores of the wood. Ultimately I found that the heat was conveyed with sufficient rapidity when air only occupied the narrow space intervening between the glass and the wood.

Remarks.	Weight of rod per foot in grains.	Medium of communication of heat between the glass tube and the rod.	Tension in pounds.	Limits of temperature between which the experiments were made.	Linear expansion or contraction per degree Centigrade.	Young's modulus of elasticity, or length of rod, 1 lb. per foot, which would be extended unity by one lb. tension.
Moistened by immersion in water a short time. Experiment made half an hour afterwards	201	Mercury	35 435	42°·27 and 15°·2 42·3 and 16·01	·000002255 E. ·000004109 E.	
Moistened further by immersion under water forty hours. Experiment made shortly afterwards..	227	Mercury	35 35 35 435	21·6 and 10·8 33 and 13·3 49·6 and 11·7 37·33 and 13·4	0 ·000000176 E. ·000000436 E. ·000001140 E.	5721154 at 38°·2 6086956 at 12·4
Dried before a fire for twenty hours..	196	Mercury	35 435	40·6 and 12·3 40·6 and 13·6	·000004555 E. ·000005481 E.	7472530 at 13
Immersed in alcohol for two days....	215	Alcohol	35 435	31·5 and 12·8 37·05 and 14·72	·000002692 E. ·000003000 E.	
After immersion for one week in water, sp. gr. being then 0·82....	291	Water......	35 435	39·3 and 12·7 36·5 and 13·5	·000000388 C. ·000000327 E.	5948141 at 10·4 5242724 at 37
After it was kept warm before a fire for five days in frosty weather it became electrified on rubbing it with the hand, and retained its charge a long time	176	Mercury	35	39·3 and 11·7	·000004516 E.	7309582 at 13

70. *Deal.*—My next experiments were on a rod of St. John's Pine, perfectly free from knots and straight in the grain. Its diameter was $\frac{3}{8}$ of an inch, and the length subjected to thermal influence 33 inches, as before.

Remarks as to the condition of the wood.	Weight of rod per foot in grains.	Proximate specific gravity.	Medium of communication of heat between the glass tube and the rod.	Tension in pounds.	Limits of temperature between which the experiments were made.	Linear contraction or expansion per degree Centigrade.
Dried by remaining before a fire a few hours	132·4	....	Mercury	35 235	41°·7 and 13°·43 41·5 and 15·41	·000004277 E. ·000004379 E.
Moistened by immersion for a few hours in water	172	·557	Water	35	48·7 and 15·36	·000000768 E.
Boiled in water, and then left immersed in cold water some hours.	241·2	·761	Water	35 35	41 and 12·78 24·5 and 12·98	·000000377 C. ·000000360 C.
Left immersed in water three days longer	272·4	·854	Water	35	48·7 and 14·62	·000000636 C.
Left several days longer immersed in water	303·6	....	Water	35	12·7 and 2·7 36·4 and 7·5 40·8 and 28·8 63·6 and 38·6	·000000235 C. ·000000260 C. ·000000088 E. ·000000446 E.
Dried before a fire for several days, until it could be electrified by rubbing with the hand	128·8	....	Air	35	37·16 and 11·57	·000004342 E.
Covered with pitch	141·8	....	Air	35	32·2 and 10·83	·000004030 E.

71. *Deal cut across the grain.*—The specimens were cut at right angles to the pores of the wood, but obliquely in respect to the concentric rings which indicate the growth of the tree. The length subjected to thermal influence was 12 inches.

Remarks on the condition of the wood.	Weight of rod per foot in grains.	Proximate specific gravity.	Medium of communication of heat between the glass tube and the rod.	Tension in pounds.	Limits of temperature between which the experiments were made.	Linear contraction or expansion per degree Centigrade.	Young's modulus of elasticity, or length of rod, 1 lb. per foot, which would be extended unity by 1 lb. tension.
Dried before the fire for several days until it could be electrified by rubbing with the hand, and retain the charge a long time	204·2	428	Mercury	lb. oz. 4 5 16 5	36°·9 and 7°·35 38·5 and 8·9	·000041666 E. ·000045928 E.	203732
Boiled in water a few minutes, and then left immersed in water for one day................	593·2	1·122	Water.......	4 5 4 5	33·1 and 4·38 52 and 6·1	·000004497 C. ·000006455 C.	

72. The length of the wood saturated with water was 1·062 of its length in the dry state. In the saturated condition it broke after a tension of $17\frac{1}{3}$ lb. had been applied for half a minute. The specific gravity of the wood, considered apart from the air or water absorbed, was found to be 1·506.

73. The following Table contains the results of experiments with a rod of deal, similar to the last, also cut across the grain.

Remarks on the condition of the wood.	Weight of rod per foot in grains.	Extreme length in inches.	Medium of communication of heat between the glass tube and the rod.	Tension in pounds.	Limits of temperature between which the experiments were made.	Linear contraction or expansion per degree Centigrade.	Young's modulus of elasticity, or length of rod, weighing 1 lb. per foot, which would be extended unity by 1 lb. tension.
Dried before a fire until it could be electrified by friction	193·5	15·625	Mercury	4·3	37°·5 and 10°·9	·000042470 E.	188120
On immersion in water the wood absorbed in a few minutes sufficient moisture to make it weigh 243 grs. per foot. Its length was then 15·8; but after allowing it to rest two hours its length became 16·24, without receiving any additional water in the interim. Twelve hours afterwards its length was 16·3, and its weight per foot 228·2 grs.	228·2	16·3	Mercury	4·3	35·1 and 10·7	·000065153 E.	104600
Further immersion in water .	270·5	16·594	Mercury	4·3	41·3 and 10·91	·000000380 E.	60033

Immersed several hours in water; specific gravity then was found to be ·891	446	16·625	Mercury	4·3	43·2 and 10·95	·000005404 C.	38257
Boiled in the exhausted receiver of an air-pump, and afterwards left immersed in water several hours; sp. gr. 1·143	573·4	16·7	Water	4·3 7·3	34·5 and 13·25 27·2 and 12·12	·000002201 C. ·000001105 C.	29618
Dried a little by placing it a short time before a fire..	434·4 444·0	16·63 16·63	Air .... Water	4·3 4·3	39·03 and 12·8 34·9 and 12·7	·000003159 E. ·000001561 C.	
Dried still further........	358 370	16·62 16·62	Air .... Water	4·3 4·3	37·38 and 13·3 43·8 and 15	·000010390 E. ·000001356 C.	
Dried still further..........	231·5	16·48	Air	4·3	37·7 and 10·24	·000084212 E.	96170
Water poured between the glass tube and the rod. Then after a short time,	....	....	Water	4·3	37·2 and 10·3	·000058836 E.	82430
After ten minutes more had elapsed	....	....	Water	4·3	40·9 and 12·15	·000019457 E.	
After another ten minutes...	....	....	Water	4·3	37·6 and 12·3	·000004815 E.	
After another ten minutes...	....	....	Water	4·3	42 and 12·3	·000001017 C.	
After another ten minutes; sp. gr. ·5646	272·3	16·656	Water ...... Air	4·3 4·3	39·34 and 12·53 37·5 and 9·7	·000002951 C. ·000000817 C.	
Dried some hours in a moderately heated oven	206	15·55	Air	4·3	41·5 and 9·77	·000038620 E.	
Immersed in oil under the exhausted receiver of an air-pump	329	15·54	Air	4·3	38·8 and 10·6	·000043670 E.	

74. The specific gravity of the wood, considered apart from the air or water in its pores, was in the above specimen found to be 1·595.

75. The following are deductions from the above experiments: — 1st. That tension increases the expansibility of wood by heat. 2nd. The expansibility of dry wood cut across the grain is about ten times as great as its expansibility when cut in the direction of the grain. 3rd. The length of wood cut longitudinally is a little increased by the absorption of water; but when cut crossways a very great augmentation, as is well known, takes place. 4th. When water is absorbed into the pores of wood cut longitudinally, its expansibility decreases until ultimately it passes into contraction by heat. 5th. When water is absorbed into the pores of wood cut across the grain, its expansibility appears to increase in the first instance, then decreases until it changes into contraction, which, after increasing with the absorption of water to a certain point, seems to diminish as the wood becomes completely saturated. 6th. Cut lengthways, the elasticity of wood appears to be lessened by the presence of water; when, however, it is cut across the grain, the effect of moisture is very greatly to impair the elastic force. 7th. A short period of immersion enables wood cut crossways to take up sufficient water to make it contract by heat; but, on the contrary, wet wood partially dried on the surface is expanded by heat. So that we may have contraction or expansion of wood charged to the same extent with water, according probably to peculiarities in the distribution of the water among the pores.

76. When wet wood, cut across the grain, is dried, it gets shorter very gradually, in consequence of a tendency to retain its former dimensions. The strain thus arising may be removed by raising the temperature, which speedily reduces the wood to the length due to its state of dryness. The same kind of effect takes place, but in the reverse order, in wood which is taking up water. It does not immediately assume the length due to the state of hygrometry; but if

heated, the set is taken out, and the wood expands to its proper length.

77. Inasmuch as at a certain degree of humidity wood contracts in every direction with a rise of temperature, it might be inferred that on weighing it in water its specific gravity would be found to increase on the elevation of temperature. Such, however, is not the fact. A piece of saturated wood, being part of the specimen employed in the last series of experiments, was found to weigh 392·6 grs. in air, 42·26 grs. in distilled water at 0°, and 42·01 grs. in distilled water at 8°. Hence the cubical expansion was ·0000892 per degree; but calculated on the wood considered apart from the water in its pores, ·000396. Therefore on raising the temperature a decrease of specific gravity occurs simultaneously with a diminution in the external dimensions of the wood. This can only take place in consequence of a contraction of the surfaces of the walls of the cellular structure, while the actual bulk of the material of which they are composed is increased, a certain minute quantity of water exuding at the same time.

78. The phenomenon in question is therefore, I believe, owing to capillary attraction, the diminution of which with elevation of temperature has the effect of removing a part of the swelled condition due to that action. Dr. Young has given $\frac{1}{555}$ as the descent of water in a narrow tube due to a rise of temperature of 1° Cent. The fraction arrived at by M. C. Wolf is $\frac{1}{525\cdot67}$. To apply the latter estimate to the foregoing results, we find from the last Table that between extreme dryness and perfect humidity the wood increased in length from unity to 1·068. Hence, on the hypothesis that the contraction by heat is owing to diminution of capillary attraction, it ought to be equal to ·0001293 minus the expansion in the dry condition. An examination of the foregoing Tables will be found to afford ample support to this view.

79. *Ratan Cane.*—My observations of the expansion and elasticity of this substance are comprised in the following Table. The length exposed to thermal influence was 33 inches.

Remarks on the condition of the wood.	Weight of rod per foot in grains.	Medium of communication of heat between the glass tube and the rod.	Tension in pounds.	Limits of temperature between which the experiments were made.	Linear contraction or expansion per degree Centigrade.	Young's modulus of elasticity, or length of rod, weighing 1 lb. per foot, which would be extended unity by 1 lb. tension.
Dried by placing it before a fire for a short time . .	150·2	Mercury	35 91	25°·3 and 11°·67 28·05 and 12·09	·000024750 E. ·000027000 E.	1094242 at 10°·6 918179 at 34 956720 at 29 1048546 at 12
Wetted by immersion under water for a few hours	208	Water	35	26·6 and 11·04	·000001436 C.	
Dried by exposure before a fire for several hours .	145	Mercury	35	27·8 and 11·1	·000016740 E.	1214179 at 12

80. The following are the results of my experiments on rose, vine, and poplar. These were sprouts of recent growth, selected for their straightness. They were cut from the tree in November, directly before the experiments. I have also inserted in the Table the expansion and elasticity of wheat-straw.

Kind of wood, &c.	Weight of rod per foot in grains.	Medium of communication of heat between the glass tube and the rod.	Tension in pounds.	Limits of temperature between which the experiments were made.	Linear expansion or contraction per degree Centigrade.	Young's modulus of elasticity, or length of rod, weighing 1 lb. per foot, which would be extended unity by 1 lb. tension.
Rose-wood in the green state, the bark not removed; sp. gr. ·638	171·6	Water	35 91	46° and 14°·53 42 and 16	·000000240 E. ·000001101 E.	
Vine in the green state, bark not removed	116	Water	35	28·6 and 11·3	·000002121 E.	
Poplar in the green state, bark not removed	158·6	Water	35 40	23·2 and 10 39 and 9	·000000773 C. ·000000037 E.	
Wheat-straw dried before a fire	14·8	Air	5·2	33·4 and 11	·000004865 E.	3520625
Wheat-straw moistened by immersion in water	38·1	Air	5·2	16·8 and 12·13	·000000329 C.	982978

81. Cane, like wood cut in the direction of the grain, increases slightly in length in consequence of being moistened. The expansion by heat being diminished at the same time, it follows that the increment due to moisture is less when taken at a high than at a low temperature.

82. The Table on p. 455 records my observations on the expansion and elasticity of paper, leather, and whalebone. The length exposed to thermal influence was 33 inches.

83. It will be observed that an increase in the expansibility by heat of paper and leather was produced by moisture. It is probable, however, that had they been more thoroughly saturated the expansion would again diminish, as in the case of wood cut across the grain, in which, although the expansibility in the first instance increased with the application of moisture, it ultimately changed into contraction when the wood became more completely saturated. I think that a cause tending to increase the expansibility in consequence of the application of moisture exists in every case, but that it is modified, and in wood sometimes completely overcome, by the effect of decreased capillary attraction with rise of temperature, to which I have already adverted.

84. Upon the facility with which hot moistened whalebone can be moulded into form, and the permanency with which it keeps the shape in which it is cooled, much of its use in the arts depends. When heat is applied to whalebone thus cooled in a constrained state, it at once returns to its original shape. Even when cold moistened whalebone is strained, it returns almost immediately to its original dimensions on the application of heat. The slip of moistened whalebone employed in the experiments recorded in the Table, 46 inches long and weighing 71·8 grains to the foot, became rapidly stretched 7 inches longer by a tension of 64 lb. Left to itself after the tensile force was removed, it began to return gradually towards its original length, which, however, it at once assumed on immersion in hot water.

85. The apparent elasticity of whalebone is always somewhat less than the real, owing to the observed elongation, when tension is applied, consisting of a considerable amount

Remarks.	Weight per foot in grains.	Medium of communication of heat between the glass tube and the substance.	Tension in pounds.	Limits of temperature between which the experiments were made.	Linear expansion per degree Centigrade.	Young's modulus of elasticity, or length of strip, weighing 1 lb. per foot, which would be extended unity by 1 lb. tension.
A strip of cardboard one third of an inch broad, dried before a fire	35	Air..........	5·2	39°·4 and 9°·4	·000015780	
A strip of cardboard one third of an inch broad, moistened with water, which increased its length $\frac{1}{48}$	39·8	Air..........	5·2	40·3 and 11·5	·000026520	
A strip of cardboard one third of an inch broad, moistened still further, which increased its length still further, by $\frac{1}{120}$	47	Air..........	5·2	39·45 and 12·2	·000028950	
Cowskin leather dried before a fire. A narrow strip	17·55	Air..........	5·2	27·77 and 11·5	·000038720	227586
Cowskin leather moistened with water	19·95	Air..........	5·2	29 and 10·47	·000047116	167813
Cowskin leather moistened still further	25·8	Air..........	5·2	24·25 and 9·75	·000054401	118642
Whalebone, a strip one third of an inch broad, dry	60	Air..........	35 70	24·4 and 9·27 24·35 and 9·9	·000052200 ·000055020	1176554 at 10°
Whalebone, a strip one third of an inch broad, moistened with water	71·8	Air..........	5·2	23·7 and 11	·000024270	

of *set* as well as of elastic strain. Also when the tensile force is removed, the contraction which takes place consists of the return from *set* as well as of the elastic recoil. The interesting phenomena connected with imperfect elasticity might, I think, be very advantageously studied in this substance. The limits I had set myself did not, however, allow me to do more than obtain the following determinations of its elasticity in the dry state at two different temperatures, and with various intervals between the application or removal of tension and the corresponding observations.

Time allowed to elapse between laying on or taking off a tensile force of 35 lb., and making the observation of length.	Young's modulus of apparent elasticity, or length of rod, weighing 1 lb. per foot, which would be extended unity by 1 lb. tension.	
	At 10°·1.	At 30°·1.
15 sec.	1176554	962450
30 sec.	1147383	927100
1 min.	1119624	894545
2 min.	1080415	
4 min.	1043860	
15 min.	1028395	
1 hour.	922481	

86. *Thermal effect of Tension on Wood.*—My first experiment was tried on a square rod of straight-grained pine, dried, and weighing 132·4 grains to the foot. A thermo-electric junction of thin copper and iron wires having been inserted into it, its upper extremity was clamped to a firm support, and the lower attached to a lever giving a tension of 35 lb. On applying an additional tension of 200 lb., the motion of the needle of the thermo-multiplier indicated cold; on removing the tension, a reverse motion of equal extent indicated the evolution of heat. The mean of several trials gave me 9′·87, the thermal value of which, determined by immersing the junction in water of various temperatures, was 0°·01364. Owing to the large surface of the wood as compared with its capacity for heat, it was affected by the temperature of the surrounding atmosphere nearly three times as much as the quarter-inch metal bars. I found, in fact, that during the time occupied by a swing of the needle, viz. 45 sec., one fifth of the thermal

effect was lost. Adding, therefore, one quarter to $0^{\circ}{\cdot}01364$, we find $0^{\circ}{\cdot}01705$ for the cooling effect, which, it may be observed, errs a trifle in defect, owing to the conduction of a small portion of heat along the wires of the thermo-electric junction.

87. In order to arrive at the theoretical result, it was necessary to determine the specific heat of the wood in the condition it was in when used. This was done by raising the temperature of a faggot of it in a bath of mercury, and then plunging it into a thin copper can filled with mercury at the atmospheric temperature. The fall of temperature of the wood and the rise of temperature of the can and mercury, corrected for the atmospheric influence, indicated the specific heat—which, in the case of dried St.-John's pine, was found to be ·3962, and in that of dried bay wood ·3582.

88. The theoretical thermal effect of tension in the foregoing experiment, derived from the above data and the expansion by heat at the mean tension, is

$$H=\frac{279{\cdot}4}{1390}\times\frac{-200}{1}\times\frac{1}{231053}\times\frac{1}{{\cdot}3962}\times\frac{1}{{\cdot}0189}=-0^{\circ}{\cdot}02322.$$

89. *Bay Wood.*—The rod I employed was dried until it could be electrified by rubbing. A tensile force of 400 lb. produced a deflection of $25'{\cdot}8$, which the test experiment on the junction proved to indicate $0^{\circ}{\cdot}0475$. This, increased one quarter, gives $0^{\circ}{\cdot}0594$ as the actual cooling effect at the moment of the application of the tension.

90. The theoretical result, using my own determination of the specific heat of the wood, and its expansion by heat at the mean tension 235 lb., is

$$H=\frac{278}{1390}\times\frac{-400}{1}\times\frac{1}{196858}\times\frac{1}{{\cdot}3582}\times\frac{1}{{\cdot}01886}=-0^{\circ}{\cdot}06016.$$

91. *St. John's Pine, cut across the grain.*—This specimen was dried slowly until it became electrical by rubbing. A tensile force of 14 lb. produced a deflection of $3'{\cdot}4$, indicating a cooling effect of $0^{\circ}{\cdot}00494$. This, increased by one quarter, gives $0^{\circ}{\cdot}00617$ as the cold produced by the tension. The theoretical result is

$$H=\frac{276}{1390}\times\frac{-14}{1}\times\frac{1}{23256}\times\frac{1}{{\cdot}3962}\times\frac{1}{{\cdot}03214}=-0^{\circ}{\cdot}0093.$$

92. *Bay Wood saturated with water.*—The specific gravity of this specimen was ·933; it weighed 333 grains per foot, 157 grains being imbibed water. The mean of twenty experiments gave me 0°·0073 of *heat* on applying a tension of 200 lb., and 0°·0013, likewise of heat, on removing the tensile force. There could be no doubt that in this instance the imperfect elasticity of the moist wood caused a considerable quantity of heat to be generated frictionally. It may therefore be safely concluded that the thermal effect, considered apart from the result of friction, was, as in the case of india-rubber, *one of heat on the application of tension, and cold on its removal.* Its actual value would be overestimated by $\frac{\cdot 0073 - \cdot 0013}{2} = 0^\circ{\cdot}003$, on account of the set communicated by tension being always greater than that taken out when the tension is removed. The approximate theoretical result is

$$H = \frac{277}{1390} \times \frac{-200}{1} \times -\frac{1}{2000000} \times \frac{1}{\cdot 68} \times \frac{1}{\cdot 0476} = 0^\circ{\cdot}0006.$$

93. For the sake of ready comparison, I have collected in the following Table the results of the foregoing experiments on the thermal effects of tension, placing by their side the results of Professor Thomson's theory.

Material.	Experiment.	Theory.	Theoretical thermal effect of 1 lb. tension on a prism weighing 1 lb. to the foot, at the temperature of 0° Cent.
Iron	−·115	−·110	−0°·0000220
Iron	−·124	−·110	
Iron	−·101	−·107	
Hard steel	−·162	−·125	− ·0000235
Cast iron	−·160	−·112	− ·0000168
Cast iron	−·148	−·115	
Copper	−·174	−·154	− ·0000355
Lead	−·053	−·040	− ·0001847
Lead	−·076	−·055	
Gutta percha	−·028	−·031	− ·0000769
Gutta percha	−·052	−·066	
Vulcanized india-rubber	+·114	+·137	
Pine wood	−·017	−·023	− ·0000021
Bay wood	−·059	−·060	− ·0000028
Pine, cross-grained	−·006	−·009	− ·0000213
Wet bay wood	+·003	+·001	+ ·00000015

On the Thermal Effects of Longitudinal Compression on Solids.

94. *Wrought Iron.*—A pillar 2 inches long and one quarter of an inch in diameter had a fine hole bored through it, in which a thermo-electric junction of fine copper and iron wires was inserted, as described in § 18. It was placed under the lever which had served for the tension-experiments, the weight of which alone gave a pressure of 50 lb. When, in addition to this, a pressure of 1060 lb. was applied, a deflection of 21′·3, indicating heat, was produced; and a like deflection in the opposite direction, indicating cold, was produced by removing the pressing weight. The value of the deflection, ascertained by immersing the pillar in water of various temperatures, to within one eighth of an inch from the thermo-electric junction, was 0°·1517. The theoretical result is

$$H=\frac{287{\cdot}8}{1390}\times\frac{1060}{1}\times\frac{1}{81200}\times\frac{1}{{\cdot}11}\times\frac{1}{{\cdot}1491}=0^\circ{\cdot}1645.$$

95. My next experiment was with a pillar of the same length, but half an inch in diameter, using a pressure of 1060 lb. In pillars of this and greater diameters the experiment to test the value of the deflections was made by plunging the pillar into water, which rose a little above the orifice in which the thermo-electric junction was inserted. Thus in the present instance I found the deflection of 8′·94 to indicate a rise of temperature equal to 0°·0319. The theoretical result is

$$H=\frac{289}{1390}\times\frac{1060}{1}\times\frac{1}{81200}\times\frac{1}{{\cdot}11}\times\frac{1}{{\cdot}632}=0^\circ{\cdot}039.$$

96. Wishing to try higher pressures, I employed a very excellent hydraulic press, the pressure exerted by which I had ascertained by experiment as far as 2000 lb. The friction amounted to only one seventh of the exerted force. By supplying the correction thus indicated, I had a very convenient and tolerably accurate mode of applying pressure by simply placing weights on the handle of the pump. A pressure of 6458 lb., applied in this way, gave a deflection of 42′ at the

thermo-multiplier, placed at a distance of 40 yards, which deflection was found to indicate a thermal effect of 0°·2344. The theoretical result is

$$H=\frac{285\cdot5}{1390}\times\frac{6458}{1}\times\frac{1}{81200}\times\frac{1}{\cdot11}\times\frac{1}{\cdot632}=0^\circ\cdot235.$$

97. Next I tried a pillar 2 inches long and 1 inch in diameter. When a pressure of 1780 lb. was applied to this by the lever, the needle was deflected 10′·6, indicating a rise of temperature of 0°·01803. The theoretical result is

$$H=\frac{290\cdot6}{1390}\times\frac{1780}{1}\times\frac{1}{81200}\times\frac{1}{\cdot11}\times\frac{1}{2\cdot533}=0^\circ\cdot01642.$$

98. With the hydraulic press, pressures of 4154, 8762, and 20282 lb. produced deflections of 9′, 21′·6, and 1°15′·3 respectively, indicating temperatures of 0°·03157, 0°·07578, and 0°·2642. The theoretical results are

$$H=\frac{289\cdot8}{1390}\times\frac{4154}{1}\times\frac{1}{81200}\times\frac{1}{\cdot11}\times\frac{1}{2\cdot533}=0^\circ\cdot03822,$$

$$H=\frac{289\cdot8}{1390}\times\frac{8762}{1}\times\frac{1}{81200}\times\frac{1}{\cdot11}\times\frac{1}{2\cdot533}=0^\circ\cdot08063,$$

and

$$H=\frac{289\cdot8}{1390}\times\frac{20282}{1}\times\frac{1}{81200}\times\frac{1}{\cdot11}\times\frac{1}{2\cdot533}=0^\circ\cdot1866.$$

99. On applying a pressure of 47930 lb. the pillar was squeezed to the length of 1·88 inch, giving out about 6°, or nearly the thermal equivalent of the work thus done. Then using the lighter pressure of 37744 lb. I obtained a deflection of 2° 6′, indicating a thermal effect of 0°·442. The theoretical result is

$$H=\frac{290}{1390}\times\frac{37744}{1}\times\frac{1}{81200}\times\frac{1}{\cdot11}\times\frac{1}{2\cdot694}=0^\circ\cdot3266.$$

100. *Cast Iron.*—A pillar 2 inches long and one quarter of an inch in diameter being pressed by the lever to the extent of 730 lb. produced a deflection of 41 , which was proved to

indicate a rise of temperature equal to 0°·1123. The theoretical result is

$$H=\frac{287{\cdot}6}{1390}\times\frac{730}{1}\times\frac{1}{90139}\times\frac{1}{{\cdot}1198}\times\frac{1}{{\cdot}1281}=0^{\circ}{\cdot}1091.$$

Another experiment, with 1010 lb. pressure on another bar, gave a thermal effect of 0°·1667, theory giving

$$H=\frac{288{\cdot}6}{1390}\times\frac{1010}{1}\times\frac{1}{90139}\times\frac{1}{{\cdot}1198}\times\frac{1}{{\cdot}1346}=0^{\circ}{\cdot}1443.$$

101. Pillar, 2 inches long and half inch in diameter. A pressure of 1060 lb. applied by the lever produced a deflection of 6′·5, indicating a thermal effect of 0°·0454. The theoretical result is

$$H=\frac{285{\cdot}3}{1390}\times\frac{1060}{1}\times\frac{1}{90139}\times\frac{1}{{\cdot}1198}\times\frac{1}{{\cdot}57}=0^{\circ}{\cdot}0353.$$

102. Using the hydraulic press, pressures of 1850, 4154, and 8762 lb. produced thermal effects of 0°·0822, 0°·1883, and 0°·3423 respectively. The theoretical results are

$$H=\frac{285{\cdot}3}{1390}\times\frac{1850}{1}\times\frac{1}{90139}\times\frac{1}{{\cdot}1198}\times\frac{1}{{\cdot}57}=0^{\circ}{\cdot}0616,$$

$$H=\frac{285{\cdot}3}{1390}\times\frac{4154}{1}\times\frac{1}{90139}\times\frac{1}{{\cdot}1198}\times\frac{1}{{\cdot}57}=0^{\circ}{\cdot}1385,$$

and

$$H=\frac{285{\cdot}3}{1390}\times\frac{8762}{1}\times\frac{1}{90139}\times\frac{1}{{\cdot}1198}\times\frac{1}{{\cdot}57}=0^{\circ}{\cdot}2923.$$

Under a pressure of 20282 lb. the pillar broke diagonally, the deflection of the needle indicating an evolution of 4°·16.

103. Pillar, 2 inches long, 1 inch in diameter. A pressure of 1781 lb. applied by the lever gave a thermal effect of 0°·01365. The theory gives

$$H=\frac{290{\cdot}8}{1390}\times\frac{1781}{1}\times\frac{1}{90139}\times\frac{1}{{\cdot}1198}\times\frac{1}{2{\cdot}352}=0^{\circ}{\cdot}01466.$$

Using the hydraulic press, pressures of 4154, 8762, and

20282 lb. gave thermal effects of 0°·0364, 0°·0511, and 0°·1463 respectively, the theoretical results being

$$H=\frac{290}{1390}\times\frac{4154}{1}\times\frac{1}{90139}\times\frac{1}{\cdot1198}\times\frac{1}{2\cdot352}=0^\circ\cdot0341,$$

$$H=\frac{290}{1390}\times\frac{8762}{1}\times\frac{1}{90139}\times\frac{1}{\cdot1198}\times\frac{1}{2\cdot352}=0^\circ\cdot0719,$$

$$H=\frac{290}{1390}\times\frac{20282}{1}\times\frac{1}{90139}\times\frac{1}{\cdot1198}\times\frac{1}{2\cdot352}=0^\circ\cdot1666.$$

Under 47930 lb. the pillar was squeezed to the length of 1·78 inch. Then with the same pressure I obtained a thermal effect of 0°·4708. The theoretical result is

$$H=\frac{290}{1390}\times\frac{47930}{1}\times\frac{1}{90139}\times\frac{1}{\cdot1198}\times\frac{1}{2\cdot643}=0^\circ\cdot3505.$$

104. *Copper*.—A pillar 2 inches long and one quarter of an inch in diameter, being pressed by 717 lb., gave a deflection of 10′·1, indicating a thermal effect of 0°·1359. The theoretical result is

$$H=\frac{292}{1390}\times\frac{717}{1}\times\frac{1}{58200}\times\frac{1}{\cdot095}\times\frac{1}{\cdot1744}=0^\circ\cdot156.$$

105. Pillar, 2 inches long and half an inch in diameter. A pressure of 1325 lb. applied by the lever gave a deflection of 30′, indicating a rise in temperature of 0°·08316. The theoretical result is

$$H=\frac{291\cdot7}{1390}\times\frac{1325}{1}\times\frac{1}{58200}\times\frac{1}{\cdot095}\times\frac{1}{\cdot7207}=0^\circ\cdot06972.$$

Pressures of 1850 and 4154 lb., communicated by the hydraulic press, produced deflections of 24′·1 and 56′, indicating temperatures of 0°·1182 and 0°·2747 respectively. The theoretical results are

$$H=\frac{292}{1390}\times\frac{1850}{1}\times\frac{1}{58200}\times\frac{1}{\cdot095}\times\frac{1}{\cdot7207}=0^\circ\cdot0974,$$

and

$$H=\frac{292}{1390}\times\frac{4154}{1}\times\frac{1}{58200}\times\frac{1}{\cdot095}\times\frac{1}{\cdot7207}=0^\circ\cdot2188.$$

106. Pillar, 2 inches long and 1 inch in diameter. A pressure of 1792 lb. communicated by the lever gave a deflection of 9′·7, indicating a rise equal to 0°·0278. The theoretical result is

$$H=\frac{291\cdot2}{1390}\times\frac{1792}{1}\times\frac{1}{58200}\times\frac{1}{\cdot095}\times\frac{1}{2\cdot953}=0^{\circ}\cdot023.$$

Pressures of 4154, 8762, and 20282 lb., communicated by the hydraulic press, produced deflections of 10′·6, 33′, and 1° 9′, which were found to indicate elevations of temperature of 0°·0494, 0°·1538, and 0°·3216 respectively. The theory gives

$$H=\frac{291}{1390}\times\frac{4154}{1}\times\frac{1}{58200}\times\frac{1}{\cdot095}\times\frac{1}{2\cdot953}=0^{\circ}\cdot05322,$$

$$H=\frac{291}{1390}\times\frac{8762}{1}\times\frac{1}{58200}\times\frac{1}{\cdot095}\times\frac{1}{2\cdot953}=0^{\circ}\cdot1122,$$

and

$$H=\frac{291}{1390}\times\frac{20282}{1}\times\frac{1}{58200}\times\frac{1}{\cdot095}\times\frac{1}{2\cdot953}=0^{\circ}\cdot26.$$

A pressure of 57146 lb. squeezed the pillar to the length of 1·32 inch. Then, when the full amount of set had taken place, the application of the same weight gave 2° 18′ of deflection, indicating a rise of temperature equal to 0°·643. The theoretical result is

$$H=\frac{291}{1390}\times\frac{57146}{1}\times\frac{1}{58200}\times\frac{1}{\cdot095}\times\frac{1}{4\cdot474}=0^{\circ}\cdot4832.$$

107. *Lead.*—A pillar, 2 inches long and half an inch in diameter, being pressed by the lever with a force of 350 pounds, produced a deflection of 12′·4, indicating a rise of temperature of 0°·0805. The theory gives

$$H=\frac{292}{1390}\times\frac{350}{1}\times\frac{1}{35100}\times\frac{1}{\cdot0303}\times\frac{1}{\cdot8796}=0^{\circ}\cdot0786.$$

Pillar, 2 inches long and 1 inch in diameter. A weight of 1500 lb. applied by the lever, produced a deflection of 33′, indicating a thermometric rise of 0°·0944. Theory gives in this case,

$$H=\frac{290}{1390}\times\frac{1500}{1}\times\frac{1}{35100}\times\frac{1}{\cdot0303}\times\frac{1}{3\cdot717}=0^{\circ}\cdot0792.$$

4154 lb. laid on by the hydraulic press, squeezed the pillar to the length of 1·8 inch. Then, when this set was fully established, the same pressure gave a deflection of 41′·6, indicating a temperature equal to 0°·1835. The theoretical result is

$$H=\frac{288\cdot6}{1390}\times\frac{4154}{1}\times\frac{1}{35100}\times\frac{1}{\cdot0303}\times\frac{1}{4\cdot13}=0^{\circ}\cdot1964.$$

108. *Glass.*—A cylindrical pillar of flint-glass, 10 inches long and $\frac{9}{10}$ of an inch in diameter, had a thermo-electric junction of thin wires tightly bound to its side by cotton thread. A pressure of 900 lb. applied by the lever gave me a deflection of 10′·6, the thermometric value of which, estimated by immersing the pillar above the junction in water of various temperatures, proved to be 0°·01684. The theory gives

$$H=\frac{290\cdot8}{1390}\times\frac{900}{1}\times\frac{1}{124800}\times\frac{1}{\cdot19}\times\frac{1}{\cdot9576}=0^{\circ}\cdot0083.$$

In another experiment I obtained a deflection of 11′·3, indicating a rise of 0°·01774 when a weight of 1622 lb. was laid on by means of the lever. In this case the theoretical result is

$$H=\frac{292}{1390}\times\frac{1622}{1}\times\frac{1}{124800}\times\frac{1}{\cdot19}\times\frac{1}{\cdot9576}=0^{\circ}\cdot015.$$

109. *Wood.*—A pillar of seasoned pine, 13 inches long and 1·4 inch in diameter, had a junction of fine copper and iron wires inserted into its centre. When 869 lb. were laid on this pillar, a deflection of 6′·05 occurred, which was found to indicate a rise of temperature equal to 0°·0068. The theoretical result, taking my own results for expansion and specific heat, is

$$H=\frac{290\cdot4}{1390}\times\frac{869}{1}\times\frac{1}{238000}\times\frac{1}{\cdot4}\times\frac{1}{\cdot311}=0^{\circ}\cdot0061.$$

110. My next experiments were with a 3-inch cube of pine, which, being furnished with a junction of fine wires in its centre, had pressures of 4154 and 8762 lb. applied by the hydraulic apparatus. The deflections obtained were 3′·8 and 5′·9, indicating thermal effects of 0°·0093 and 0°·0145. Theory gives in these cases

$$H=\frac{290}{1390}\times\frac{4154}{1}\times\frac{1}{238000}\times\frac{1}{\cdot 4}\times\frac{1}{1\cdot 766}=0^{\circ}\cdot 00516$$

and

$$H=\frac{290}{1390}\times\frac{8762}{1}\times\frac{1}{238000}\times\frac{1}{\cdot 4}\times\frac{1}{1\cdot 766}=0^{\circ}\cdot 01088.$$

111. When the same block was pressed, in a direction perpendicular to the grain, with a weight of 1792 lb. communicated by the lever, a deflection of 37′·3 was produced; but when the pressure was removed, the deflection in the reverse direction amounted only to 31′·2. I believe that the excess in the result of pressure is owing to frictional evolution of heat through the imperfect elasticity of wood cut crossways to the grain. The mean, 34′·25, represents a thermal effect of 0°·0461. Theory gives

$$H=\frac{288\cdot 6}{1390}\times\frac{1792}{1}\times\frac{1}{20160}\times\frac{1}{\cdot 4}\times\frac{1}{1\cdot 766}=0^{\circ}\cdot 0261.$$

112. A second experiment gave a thermal effect of 0°·016, the theoretical result being 0°·0264. In this latter experiment the position of the block was reversed. I attribute the discordance between the two results to the difficulty I experienced in obtaining with my lever-apparatus a perfectly even distribution of pressure over so large a surface as 3 inches square.

113. On exposing the block, still crossways to the grain, to a greater pressure by means of the hydraulic press, I obtained a deflection of 40′·5, indicating a rise of 0°·1, whenever a weight of 4154 lb. was laid on, but observed no perceptible effect whatever when the pressure was removed. In this case the bulging of the wood under pressure extended to about half an inch. It was evident that frictional heat

increased to a great extent the thermal effect of pressure, and that a similar though smaller frictional effect diminished the cooling effect on removal of pressure. The theoretical result, if bulging had not taken place, and supposing perfect elasticity, is

$$H=\frac{290}{1390}\times\frac{4154}{1}\times\frac{1}{20160}\times\frac{1}{\cdot 4}\times\frac{1}{1\cdot 766}=0^{\circ}\cdot 0609.$$

114. *Vulcanized India-rubber.*—A pillar, 1·92 inch long, 1·22 inch in diameter, and weighing 692 grains, had a thermo-electric junction of thin copper and iron wires inserted into its centre. On applying pressure, the multiplier showed a rise of temperature; and when the pressure was removed, a depression took place sensibly equal to the previous rise. The following is a Table of the results:—

Weight laid on the pillar, in pounds.	Deflection.	Heat by laying on the weight, or cold by removing it.
28	° ′ 3	° 0·0058
33	3·7	0·0072
47	6·7	0·0131
62	16	0·0312
93	32	0·0625
124	1 4·3	0·1254

With the pressure of 124 lb. the pillar was compressed to the length of 1·05 inch, and returned to 1·78 when the weight was removed.

115. The expansion of the pillar by heat, in the direction in which the pressure was applied, was found to be about $\frac{1}{1660}$ per degree Cent. Hence the theoretical result for a pressure of 124 lb. would appear to be

$$H=\frac{287}{1390}\times\frac{124}{1}\times\frac{1}{1660}\times\frac{1}{\cdot 415}\times\frac{1}{\cdot 77}=0^{\circ}\cdot 05.$$

The excess of the actual result may be attributed to the central part, in which the junction was placed, bearing an undue share of pressure in consequence of the bulging of the waist of the pillar.

116. On examining the above Table, it will be remarked that the thermal effect increases more rapidly than the pressure. This is owing partly to the influence of a rise of temperature in increasing the elasticity of rubber under strain, to which we have already alluded, and partly to the unequal distribution of pressure to which I have just adverted. The longitudinal expansion of a pillar of vulcanized india-rubber, 2·21 inches long and 1·44 inch in diameter, under various pressures, was found to be

Pressure in pounds.	Expansion per degree Cent. on the length under pressure.
0	$\frac{1}{5700}$
55	$\frac{1}{3730}$
135	$\frac{1}{2228}$

117. Being desirous of ascertaining the thermal effects of higher pressures, I made the experiments tabulated below, on a pillar 1·7 inch long and 2·5 inches in diameter.

Pressure applied or removed, in pounds.	Heat on applying pressure, or cold on removing it, in degrees Centigrade.	Length of the pillar when under pressure.	Length of the pillar after the pressure was removed.
70	0°·011	....	1·7
113	0·015		
134	0·029		
162	0·036		
242	0·052		
275	0·078		
386	0·114		
547	0·130	1·43	
1426	0·384	1·00	
2742	0·750	·72	
4058	0·885	·64	
6692	1·192	·51	1·66
11958	1·463		
22400	1·426	·36	1·68

118. After a few experiments with the last pressure the rubber burst, being ruptured in a singularly symmetrical manner at the four quadrants of its circumference. Placed in hot water, it almost immediately regained its original shape.

119. I have collected in the following Table the results of the foregoing experiments on the thermal effects of pressing pillars. In reckoning the mean, the last result in each of the series for wrought iron, cast iron, copper, and lead is rejected on account of having been obtained with a pressure nearly up to the limit of strength, and after the form and structure of the pillar had been changed by the force to which it had been subjected.

Material.	Diameter of pillar.	Experimental thermal effect.	Theoretical thermal effect.
	inch.	°	°
Wrought iron	$\frac{1}{4}$	·152	·164
	$\frac{1}{2}$	·032	·039
	$\frac{1}{2}$	·234	·235
	1	·018	·016
	1	·032	·038
	1	·076	·081
	1	·264	·187
	1	·442	·327
	Mean	·115	·108
Cast iron	$\frac{1}{4}$	·112	·109
	$\frac{1}{4}$	·167	·144
	$\frac{1}{2}$	·045	·035
	$\frac{1}{2}$	·082	·062
	$\frac{1}{2}$	·188	·139
	$\frac{1}{2}$	·342	·292
	1	·014	·015
	1	·036	·034
	1	·051	·072
	1	·146	·167
	1	·471	·350
	Mean	·118	·107
Copper	$\frac{1}{4}$	·136	·156
	$\frac{1}{2}$	·083	·070
	$\frac{1}{2}$	·118	·097
	$\frac{1}{2}$	·275	·219
	1	·028	·023
	1	·049	·053
	1	·154	·112
	1	·322	·260
	1	·643	·483
	Mean	·146	·124

TABLE (*continued*).

Material.	Diameter of pillar.	Experimental thermal effect.	Theoretical thermal effect.
	inch.	°	°
Lead	$\frac{1}{2}$	·080	·079
	1	·094	·079
	1	·183	·196
	Mean	·087	·079
Glass	$\frac{9}{10}$	·017	·008
	$\frac{9}{10}$	·018	·015
	Mean	·017	·011
Wood	1·4	·007	·006
	3	·009	·005
	3	·014	·011
	Mean	·010	·007
Wood cut across the grain	3	·046	·026
	3	·016	·026
	3	·050	·061
	Mean	·037	·038
Vulcanized india-rubber	1·22	·125	·050

120. The above results, as in the case of the tension-experiments, indicate a slight excess of experiment over theory. I at first thought that this might be owing to the diminution of elastic force by heat in metals, of which I had not taken account as in wood; for in applying or removing tension the thermal effect would be increased in consequence of the increased expansion by heat in such a case. This cause, however, would diminish the thermal effect of the application or removal of pressure. The discrepancy must therefore be referred to experimental error, or to the incorrectness of the various coefficients which make up the theoretical results. Having, however, been led to believe that with a a rise of temperature a certain change of elasticity takes place in metals, although too minute to be appreciated in the

foregoing results, I made some experiments in which spirals weighted at one end were measured when exposed to cold air, and then again after they had been heated in the atmosphere of an oven. In the latter case there was considerable elongation, indicating in the case of steel a diminution of elasticity amounting to ·00041 per degree Centigrade, and in the case of copper to ·00047. After I had made these experiments, I became acquainted with M. Kupffer's valuable researches on this subject, by which, using the method of vibrations, he finds the decrement of elasticity per degree Centigrade in steel and copper to be ·00047 and ·00048 respectively.

121. Another source of error exists, which, although not of sufficient amount to make a sensible alteration in the thermal effects of tension and pressure on metals, yet ought not to be neglected in a complete view of the subject. Mr. Hodgkinson has long ago shown that any force, however small, is able to produce a certain permanent deflection in a bar, and that this deflection increases rapidly with the force which has produced it. Professor Thomson has added the observation, that even after a metal has been exposed to great tensile force, its elasticity is not thereby rendered perfect for smaller degrees of stress. Thus he finds that when weights are successively hung to a wire so as gradually to increase its tension, and then successively removed, the wire never assumes immediately its just length, but is always shorter during the putting on of the weights than during their removal. Hence work is done on the wire which must necessarily evolve a certain quantity of heat; and if, as is probable, a greater quantity of work is thus done whilst the tensile force is being removed than whilst it is being applied, the result will be that the cold of tension will not be diminished to the same extent as the heat consequent upon the removal of tension will be increased, and so the mean thermal effect will be increased. On the other hand, it is probable that in the act of compression less work is done on the wire than during the removal of the compressing force, the result being that the mean thermal effect of applying and removing

the pressure is lessened. The foregoing experiments do not afford sufficiently delicate tests to detect the excessively minute quantities of heat developed frictionally in the above manner.

---

122. Professor Thomson has pointed out that the dynamical theory of heat, with the modification of Carnot's principle introduced by himself and Clausius, shows that "if a spring be such that a slight elevation of temperature weakens it, and the full strength is recovered again with the primitive temperature, work done against that spring by bending or working in whatever way must cause a cooling effect." The quantity of cold expected was excessively small; yet I hoped to measure it by taking the mean of a large number of observations with the thermo-multiplier. I took a spiral of tempered steel wire $\frac{1}{9}$ of an inch thick, of which each convolution was $1\frac{1}{2}$ inch in diameter and one quarter of an inch distant from its neighbour. A thermo-electric junction was attached to one of the convolutions; and means were provided to compress or extend the spiral at pleasure without approaching it. In the case of a spiral stretched by a weight hung to it, the application of heat causes, as we have seen, a considerable elongation in consequence of diminution of elasticity; but in the case of a spiral compressed by a weight laid on the top of it, the effect of the same cause is to diminish its length. Hence either the pulling out or the compression of the spiral must cause the absorption of heat, and the return of the spiral to its normal state must be accompanied by the evolution of heat.

123. The above thermal effects of bending a spring are evidently proportional to the square of the pressing or tensile force; for if these be increased, the elastic spring and the alteration of length by rise of temperature will be also proportionally increased.

124. Having arranged the thermo-multiplier so as to give one swing in 30 sec., I pursued the experiments as follows:— A weight of 7 lb. was laid on the top of the spiral to com-

press it; and after 30 sec. had elapsed, the change in the position of the needle was noted. Then the weight was removed, and the needle observed after 30 sec. as before. One hundred such experiments were tried alternately. Afterwards I made another series of one hundred experiments, on the effect of stretching and removing the stretching force. The results are placed in the following Table, in which the signs + and – distinguish deflections indicating evolution and absorption of heat.

125. The thermal value of 11° 6′ deflection being found to be 1°·63 Cent., the deflections will indicate respectively ·00343 cold, ·00338 heat, ·00338 cold, and ·00215 heat, the average showing a quantitative thermal effect of 0°·00306. Using my own coefficient for the diminution of the elastic force of steel by rise of temperature, I find for the theoretical result, in the case of compression,

$$H=\frac{283}{1390}\times\frac{7}{1}\times\frac{-1}{2379}\times\frac{1}{\cdot 11}\times\frac{1}{1\cdot 35}\quad -0^{\circ}\cdot 00403,$$

and in the case of extension,

$$H=\frac{283}{1390}\times\frac{-7}{1}\times\frac{1}{7500}\times\frac{1}{\cdot 11}\times\frac{1}{\cdot 428}=-0^{\circ}\cdot 00403,$$

the results being necessarily the same in both cases. The deficiency of the actual result is not great, and is on the side of the probable error in consequence of the unavoidable loss of a portion of the thermal effect by conduction from the junction.

126. Thus even in the above delicate case is the formula of Professor Thomson completely verified. The mathematical investigation of the thermo-elastic qualities of metals has enabled my illustrious friend to predict with certainty a whole class of highly interesting phenomena. To him especially do we owe the important advance which has been recently made to a new era in the history of science, when the famous philosophical system of Bacon will be to a great extent superseded, and when, instead of arriving at discovery by induction from experiment, we shall obtain our largest accessions of new facts by reasoning deductively from fundamental principles.

Deflection in minutes by compressing with 7 lb.	Deflection on removal of compressing force.	Deflection in minutes by stretching with a force of 7 lb.	Deflection on removing the tensile force.
−6	−2	0	−4
+1	+2	−4	−2
−1	−2	−8	−1
−1	0	−2	+1
0	+1	−3	+1
−4	+3	−3	−4
−6	−3	−4	0
+6	0	−2	+2
+2	+1	−6	−2
−1	0	+1	+2
0	+2	+1	−2
−3	−1	0	−3
−1	0	+1	−4
−4	0	0	−6
0	+6	−5	−1
−3	−1	−1	+4
0	0	−2	−1
−6	0	0	+3
−1	0	0	+5
−3	−1	+4	+3
−3	+5	−10	−1
−4	−2	−4	+7
+3	+5	−5	−3
+3	+3	−5	+4
−2	0	+4	+1
−1	+4	−5	−1
0	+5	+1	−1
0	+8	−6	+3
+4	0	−2	+1
−2	−1	0	+4
+2	+4	−3	+1
−1	+4	−3	+3
+2	+3	−2	+1
−1	+6	+4	+4
+2	+6	+2	+4
−2	+1	+6	+6
−5	−2	+3	+7
−2	+1	−6	−1
−1	+1	+2	−1
−3	−2	+2	+4
−4	−3	0	+1
−4	+2	−2	+1
−4	0	+2	+2
+1	+4	−1	0
0	+2	−2	+2
0	+6	−2	−2
−4	0	−1	0
0	+4	+1	+3
−7	−1	−1	+2
−6	+1	−1	+2
Mean.... −1·4	+1·38	−1·34	+0·88

On the Thermal Effects of Compressing Fluids. By J. P. JOULE, *LL.D., F.R.S. F.C.S., Hon. Mem. Phil. Soc. Cambridge, Vice-President of the Lit. and Phil. Soc. Manchester, Corresp. Mem. R.A. Sc. Turin, &c.**

['Philosophical Transactions,' 1859, vol. cxlix. p. 133. Read November 25, 1858.]

PROFESSOR WILLIAM THOMSON has published† a theoretical investigation of the subject of the present paper, in which he arrives at the formula $\theta=\frac{\mathrm{T}ep}{\mathrm{JK}}$, where θ is the increase of temperature, T the temperature from absolute zero, e the expansibility by heat, p the pressure in pounds on the square foot, J the mechanical equivalent of the thermal unit in foot-pounds, and K the capacity for heat in pounds of water, of a cubic foot of the fluid employed. He has also given a Table of theoretical results for the compression of water and mercury. The investigation being established on the basis of well-ascertained principles and facts, the correctness of the Table could not be reasonably doubted. Nevertheless, believing that an experimental inquiry would be interesting (if not important), I have ventured to offer the following to the notice of the Royal Society.

The only previous experiments on the subject of which I am aware are those of M. Regnault. To his memoir on the Compressibility of Liquids, he appends a note ‡ on the heat disengaged by the compression of water. The method employed by this celebrated physicist, though less delicate, is similar to that which I have adopted. One set of the junctions of a thermo-electric pile was placed in a copper vessel

* The experiments were made at Oak Field, Whalley Range, near Manchester.

† Proceedings of the Royal Society, June 18, 1857, vol. viii. No. 27, p. 566.

‡ Mémoires de l'Académie Royale des Sciences, xxi. p. 462.

filled with water, to which a pressure of ten atmospheres could be instantaneously communicated by means of a reservoir of compressed air. The $\frac{1}{64}$ of a degree Centigrade could be detected by his thermo-multiplier. Nevertheless the conclusion arrived at was the negative one, that "the heat disengaged by a sudden pressure of ten atmospheres on water is unable to raise its temperature $\frac{1}{50}$ of a degree Centigrade."

In the absence of any statement to the contrary, we may consider that the temperature of the water compressed by Regnault was not above an ordinary one of the atmosphere, say 18° Centigrade. Thomson's formula gives a thermal effect of 0°·013 for a pressure of ten atmospheres at this temperature; and therefore the conclusion of Regnault above cited is strictly correct: indeed it is so as regards any temperature below 30°. It is to be regretted that he did not pursue his experiments a little further; for had he done so he would without fail have solved this, as he has done so many other more difficult problems, by showing the minute but nevertheless appreciable thermal effect which actually takes place at all temperatures but that of maximum or minimum density.

The apparatus I employed consisted of a strong vessel of copper, 12 inches long and 4 inches in diameter. This vessel was connected at the upper part with a cylinder of $1\frac{3}{8}$ inch internal diameter, furnished with a piston, by means of which the requisite pressure could be laid on or taken off at pleasure. A thermo-electric junction of iron and copper wires was placed in the centre of the copper vessel, the orifice through which it was passed being made tight by means of a plug of gutta percha. The outer junction was immersed in a bath of water; and the induced currents were measured by the thermo-multiplier *in vacuo*, described in my former paper.

The method employed was alternately to lay on and remove weights from the piston and to examine the consequent deflection of the needle of the multiplier, from which, by means of data derived from experiments made from time to time to determine the amount of deflection arising from a change of temperature of the outer junction, the thermal effect sought could be readily deduced.

It was found that the needle took rather more than half a minute to assume a new deflection. I therefore fixed upon 40 sec. as the time allowed to elapse between the application or removal of pressure and the thermo-electric observation. It was at first suspected that the small cooling effect in consequence of the dilatation of the copper vessel by internal pressure might interfere with the effect sought for; but it was found on trial that a sudden application of heat to the outside of the copper vessel did not sensibly affect the temperature of the central part of the liquid in which the junction was plunged, until an interval of time had elapsed equal to twice that occupied by a swing of the needle. This source of error was therefore disregarded.

Another possible source of error occurred to me. Was the thermo-electric relation of the metals employed to form the junction sensibly altered by the influence of pressure? Thomson has shown that such an alteration in the metals copper and iron accompanies the temporary strain produced by longitudinal tension *. This effect of temporary strain, however, is very minute, and in the case of pressure uniformly applied in every direction, which we are now considering, is probably far too small to be appreciated by the most delicate tests. However, I made an attempt to ascertain whether it existed, by applying pressure when the temperature of the outer junction was widely different from that of the inner, the needle of the multiplier being kept in range by means of a controlling magnet. It was then found that the effect of pressure remained the same as before; and therefore the conclusion was necessarily arrived at, that the effect of pressure, uniformly applied, in altering the thermo-electric relation of the metals, was, if it existed at all, too small to produce any sensible error in the present experiments.

In the case of oil, it occurred to me to inquire whether, in consequence of the imperfect fluidity of that liquid, the full

* Philosophical Transactions, 1856, p. 711.

thermal effect was communicated to the junction (composed of wires $\frac{1}{20}$ inch in diameter) in the 40 sec. allowed for the swing of the needle. To settle this point a long series of experiments was made, in which the deflections after 40 sec. were alternately observed with deflections after three minutes. It was found that in the latter case the amount of deflection was one tenth more than in the former; and therefore, in my experiments with this fluid, the deflections observed at intervals of 40 sec. were increased one tenth, as will be observed in the sequel.

As it was not convenient to apply a manometer to the apparatus during the experiments, I afterwards ascertained the pressure I had employed, by means of the indications of a carefully graduated air-gauge, when the piston was pressed down with the same weights I had employed in the experiments. The pressures thus obtained were nearly the same as those estimated from the weights laid on the piston, a small allowance being made for friction. The pressure given in the Table is that of the fluid after the weight had been laid on half a minute, minus the residuary pressure arising from friction of the piston after the weight was removed.

The specific heat of the oil (in Table II.) was found by the method of mixtures to be 0·5223 at 16°·5. Its expansion, determined by the weight of a volume meter filled with it at various temperatures, proved to be ·0007582 at 21°·3. Its specific gravity at 0° was 0·915. It was important to ascertain its specific heat and expansion at a temperature near that which it had in the experiments, because, though quite transparent, it became considerably more fluid when the temperature was much raised. Such a gradual change of state in any viscous substance is accompanied by the absorption of "latent heat" and an increase in the rate of expansion, which are greatest at the temperature at which the change of state is most rapid.

TABLE I.—Experiments on the Heat evolved by the Compression of Water.

Temperature in degrees Centigrade. t.	Expansibility by heat per degree Centigrade. e.	Pressure in pounds on the square foot. p.	Capacity for heat of a cubic foot of liquid. K.	Deflections of the needle, each the mean of ten observations.	Value of deflections.	Experimental thermal effect.	Theoretical effect, or $\frac{(273+t)ep}{\text{JK}}$.
1°·2	− ·000042	53634	62·43	− 5′·2, − 4·5; Mean − 4′·85	596′ = 1°·03	− 0°·0083	0°·0071
5	·000016	53634	62·45	2·6	596 = 1·03	0·0044	0·0027
11·69	·000112	53634	62·45	12·6, 15·8; Mean 14·2	596 = 1·03	0·0244	0·0197
11·69	·000112	53634	62·45	9·5, 9·8; Mean 9·65	596 = 1·03	0·0166	0·0197
18	·000185	53634	62·41	17·2, 20·9; Mean 19·05	890 = 1·46	0·0312	0·0333
18·76	·000193	53634	62·41	14·3, 17·1, 20·6, 19·0; Mean 17·75	1454 = 2·58	0·0315	0·0347
30	·000300	53634	62·29	3·6, 5·7, 3·2; Mean 4·17	204 = 2·66	0·0544	0·0563
31·37	·000303	33117	62·29	20·8, 14·4; Mean 17·6	554 = 1·24	0·0394	0·0353
40·4	·000396	33117	62·14	22·0, 22·9, 23·8; Mean 22·90	694 = 1·36	0·0450	0·0476

TABLE II.

Experiments on the Heat evolved by the Compression of Sperm Oil.

Temperature in degrees Centigrade. t.	Expansibility by heat per degree Centigrade. e.	Pressure in pounds on the square foot. p.	Capacity for heat of a cubic foot of the oil, in pounds of water. K.	Deflections of the needle, each the mean of ten observations.	Value of the deflections.	Experimental thermal effect increased by one tenth.	Theoretical effect, or $\frac{(273+t)ep}{JK}$.
16°	·0007582	16777	29·83	32′·3, 32·5 } Mean 32′·4	950′ = 2°·11	0°·0792	0°·0886
17·29	·0007582	33117	29·83	64·3, 66·0 } Mean 65·15	867 = 2·04	0·1686	0·1758
16·27	·0007582	53634	29·83	108·7, 109·3 } Mean 109	950 = 2·11	0·2663	0·2837

On a Method of Testing the Strength of Steam-Boilers. By J. P. Joule, *LL.D., F.R.S., &c.*

['Memoirs of the Literary and Philosophical Society of Manchester,' 3rd ser. vol. i. p. 97.]

In the course of my experiments on steam, I had to employ pressures which I did not consider absolutely safe unless the boiler was previously tested. The means I adopted being simple, inexpensive, and efficacious, may, I think, be recommended for general adoption. My plan is as follows:—The boiler is to be first entirely filled with water, care being taken to close all passages leading therefrom. A brisk fire must then be made under it, and, after the water has become moderately heated, say to 90° Fahr., the safety-valve must be loaded to the pressure up to which the boiler is intended to be tried. Bourdon's circular gauge, or other pressure-indicator, is then to be constantly observed; and if the pressure arising from the expansion of the water goes on increasing continuously, without sudden decrease or stoppage, until the testing pressure is attained, it may be inferred that the boiler has sustained it without having suffered strain.

In testing my own boiler, the pressure ran up from zero to sixty-two pounds on the inch in five minutes. It rose more rapidly at the commencement than towards the termination of the trial, owing to leakage, which was considerable and of course increased with the pressure. But as there was no sudden alteration or discontinuity in the rise of pressure, it was evident that no permanent alteration of figure or incipient rupture had taken place.

In the so-called testing by steam pressure it is impossible to be sure that a boiler has not thereby suffered strain, and there is therefore no guarantee that it will not burst if subsequently worked at the same or even a somewhat lower pressure. It is to be hoped that this practice, objectionable on account of its uselessness as well as its danger, will be immediately abandoned.

In the ordinary hydraulic test the water is introduced discontinuously, and therefore the pressure increases by successive additions, rendering it difficult to be sure that strain is not taking place. This system also requires the use of a special apparatus.

The plan I recommend is free from the objections which belong to the others; and the facility with which it may be employed will probably induce owners to subject their boilers to those periodical tests the necessity for which fatal experience has so abundantly testified.

Observations of Pressure every minute.

Experiment I.	Experiment II.
Temp. at commencement, 97° F.	Temp. at commencement, 126° F.
Pressure in pounds.	Pressure in pounds.
1·0	0
2·9	2·8
4·4	5·9
6·0	8·8
7·7	12·6
9·05	16·1
11	20·8
12·35	26·1
13·85	31·8
15·1	38
16·8	44
20·1	49·9
24·8	54·8
31	59·4
37·2	63·8
44·2	Temperature at conclusion, 139° F.
51·4	
58·2	
63	
Temperature at conclusion, 126° F.	

Experiments on the Total Heat of Steam. *By* J. P. Joule, *LL.D., F.R.S., &c.**

['Manchester Memoirs,' ser. 3. vol. i. p. 99.]

The total heat of steam is understood to mean that which is evolved when the steam is condensed into water of the freezing temperature. It is a mixed quantity, and consists of, 1st, the heat due to the change of state from vapour to water, or the true latent heat; 2nd (as I showed long ago †), the heat arising from the work done on the vapour in the act of condensation; and 3rd, the heat evolved by the water during its descent from the temperature of condensation to the fixed temperature chosen, viz. 0° Centigrade.

The importance of a correct determination of the total heat can hardly be overestimated; and it is fortunate that one of the most eminent physicists of modern times has made it the object of long and elaborate research. It is not my design to attempt to improve upon the experiments of M. Regnault; but having had the opportunity of making some determinations in a different manner from that he employed, I think my results may not be thought without interest.

In Regnault's experiments the steam was passed into globes and a worm immersed in the water of a calorimeter. By the use of an artificial atmosphere connected with the worm, the operator was enabled in all cases to operate under similar circumstances as to the relative pressures of the steam and atmosphere.

In my own experiments, a vulcanized india-rubber tube, eight inches long, was attached to the nozzle of a short pipe (furnished with a stopcock), connected with the top of an upright boiler. To the end of the india-rubber tube a brass

* The experiments were made at Oak Field, Whalley Range, near Manchester.

† 'Transactions of British Association,' Birmingham, 1849.

Time during which the steam was introduced, in minutes.	Weight of water in can, including can reduced to sp. heat of water. W	Temperature of water. Before experiment. t.	Temperature of water. After experiment. t'.	Weight of steam condensed, in grains.	Total pressure of steam in inches of mercury.		Total heat of steam. $\frac{W(t'-t)}{w}+t'$		Regnault's result.
2	140351	6°·362	43°·443	8700	40·0		641·64		
$2\frac{1}{2}$	140351	6·587	56·936	12089	36·4		641·48		
$2\frac{1}{2}$	140351	6·457	58·785	12577	36·95	37·25	642·73	638·43	638·77
2	140351	4·893	43·592	9205	38·25		633·61		
$2\frac{1}{2}$	140351	5·096	53·909	11686	36·1		640·16		
2	140351	5·139	52·814	11574	35·8		630·94		
2	140351	6·384	48·228	9835	57·3		645·37		
2	140351	6·54	58·325	12418	57·6		643·61		
3	140351	6·399	55·083	11609	52·2	57·52	643·66	644·77	642·87
2	140351	6·54	48·048	9775	54·6		644·03		
2	140351	5·096	58·347	12800	60·3		642·24		
2	140351	5·371	56·523	12103	63·1		649·70		
2	140351	6·529	63·043	13490	105·1		651·02		
2	140351	6·684	58·835	12272	117·37		655·27		
2	140351	6·684	55·285	11276	115·3	111·58	660·21	655·45	649·6
2	140351	5·342	53·257	11191	109·2		654·18		
2	140351	5·574	54·033	11266	112·6		657·83		
2	140351	5·516	64·891	14141	109·9		654·2		

nozzle was attached. The stopcock was left constantly open. In making an experiment the brass nozzle through which the steam was blowing was suddenly plunged into a can of water, and then, after two or three minutes, suddenly removed again. The weight gained by the can indicated the quantity of water condensed, which, with the observations of temperature before and after the experiment, afforded the means of computing the total heat of the steam.

The requisite corrections were readily made, and not of large amount. They arose from the heat lost by the steam by conduction in passing from the boiler to the can, the thermal effects of the atmosphere on the can itself, and the evaporation of the water from the can which took place before the weighing was accomplished. The data for these were derived from observations made after each experiment. The table on p. 483 comprises the results I obtained.

In the above experiments the steam was condensed at twice the rapidity it was in those of Regnault. I had also an advantage in the size of my boiler, which was eight feet high by two feet ten inches in diameter, whereas his was only two feet seven inches by two feet one inch in diameter. Owing, however, to the small number of my experiments, the results at which I have arrived can only be regarded as confirmatory of those of the French physicist. I believe, nevertheless, that the simple method I have adopted may be resorted to with advantage whenever it shall be required to obtain a further increase of accuracy.

Experiments on the Passage of Air through Pipes and Apertures in thin Plates. By J. P. Joule, *LL.D., F.R.S., &c.**

['Manchester Memoirs,' ser. 3. vol. i. p. 102.]

Sir Isaac Newton, Polenus, Daniel Bernoulli, and others have observed that water, when it is made to flow out of a vessel through a hole cut in a thin plate, becomes contracted in diameter and increased in velocity at a short distance from the hole, the ratio of the diameter of the stream at its narrowest part to the diameter of the hole being, according to Newton's experiments, as twenty-one to twenty-five. The phenomenon is occasioned by a concourse of the particles of water as they enter the orifice, and may, as Venturi has shown, be obviated by employing a short pipe instead of a hole in a thin plate.

Air and other fluids are known to comport themselves in the same manner as water. The subject is one of considerable importance; and as I have had an opportunity of trying some experiments on it, I trust they will be found of sufficient interest to warrant my offering them to the notice of the Society.

The principal part of my apparatus was a large organ-bellows, which, by means of weights laid on the top, could be worked at pressures varying from 1·44 to 5·65 inches of water, as indicated by the difference of level of water in a bent glass tube. A circular hole, two and a half inches in diameter, was cut in the chest; and in this could be placed, or to it affixed, thin plates with holes, or other means for the egress of air.

The method of experimenting was to note the time in seconds and tenths occupied by the running down of the

* The experiments were made at Oak Field, Whalley Range, near Manchester.

bellows. The capacity of the bellows being known to be 29,660 cubical inches, this observation gave the quantity of air issuing per second, plus the unavoidable leakage of the bellows. The amount of the latter was ascertained by observing the time in which the bellows ran down when the hole was made tight, and, being subtracted from the gross effect, gave the quantity which actually passed through the orifice.

Theoretically, the quantity of air emitted in a given time ought to be proportional to the size of the aperture in the thin plate multiplied by the square root of the pressure; or, in other words, the quantity emitted per square inch of aperture divided by the square root of the pressure ought to be a constant quantity. My observations to confirm this law were made with circular holes in thin tinned iron, measuring 0·535, 1·045, and 1·61 inch in diameter respectively. The pressure and temperature of the air in all the experiments were about 29·8 inches and 4° Cent.

Diameter of aperture.	Pressure, in inches of water.	Cubic inches of air discharged per second, reduced to one square inch aperture.	Cubic inches of air discharged per second, divided by square root of pressure.
0·535	1·44	496	413·3
	5·6	1033·3	436·8
1·045	1·44	541·4	451·2
	5·6	1058·4	447·5
1·61	1·44	589·5	491·2
	5·6	1132·7	478·7

The last column of the above Table shows the accuracy of the law so far as pressure is concerned, but seems to indicate a slight increase of the quantity issuing per square inch as the aperture becomes larger.

The law is also verified so far as pressure is concerned by the following tabulated experiments with tubes of various lengths and diameters, but all terminated by a short piece of wide pipe, three inches long by two and a half inches in diameter, which was inserted into the bellows:—

Length and diameter of tubes.	Pressure.	Cubic inches of air discharged per second per square inch aperture.	Cubic inches of air discharged per second per square inch aperture divided by square root of pressure.
44 and 0·875	1·44	562·9	469·1
	3·52	909·7	484·7
20 and 0·98	1·44	671·5	559·6
	3·52	1049·4	559·1
20 and 1·594	1·44	710·6	592·2
	3·52	1117·1	595·1

At an early period of the research it was found that a very slight bur or projection on the edge of the hole in a thin plate produced a remarkable change in the quantity of effluent air. I had holes of the respective least diameters 0·535, 0·75, and 1·61 inch cut out of a thin plate of tinned iron by a brace-bit. A slight bur projected to one fortieth of an inch beyond the plain surface. The following experiments were then made, using a pressure of air equal to 1·44 inch of water:—

Position of the bur with respect to the bellows.	Cubic inches of air per second, reduced to one square inch of aperture.		
	Hole of diameter 0·535 inch.	Hole of diameter 0·75 inch.	Hole of diameter 1·61 inch.
Outwards	597·8	579·2	647·4
Inwards	529·3	524·4	584·7

A hole one inch square, without any bur, gave 567 cubic inches per second.

The influence of a tube in increasing the quantity of effluent air has been already adverted to. It was a matter of considerable interest to determine the length of tube which would produce the maximum effect. In my first experiments to determine this point, I employed a tube 0·98 of an inch in diameter, and terminated at one end by a piece of wider tube three inches long and two and a half inches in diameter:—

Length of tube of 0·98 inch diameter.	Cubic inches of air per second, reduced to square inch of aperture of narrow tube.	
	Air entering by the short length of wide tube.	Air entering by the narrow tube.
40 inches.	642·7	
20 "	666·7	
10 "	714·2	
4 "	759·4	728
2 "	787·7	723·3
1 inch.	806·5	730·4
$\frac{1}{2}$ "	810·9	646·5
$\frac{1}{4}$ "	803·7	578 {flapping sound.
$\frac{3}{16}$ "	749·5 {flapping sound.	546
$\frac{1}{8}$ "	685·5	547·5
$\frac{1}{16}$ "	666·6	541·4

In an experiment in which a flange with a hole of one and a quarter inch diameter in its centre was placed on the wide end, the quantity of air entering by the narrow tube reduced to the length of three sixteenths of an inch was increased from 546 to 600.

The next experiments were made with a tube, 0·92 of an inch outside and 0·8 inch inside diameter, successively reduced in length. The inner sharp edge was removed at one end, and the outer edge at the other end of the tube. It will be seen that the greatest quantity of air flowed when it entered at the end from which the inner edge had been removed.

Length of tube of 0·8 inch diameter.	Cubic inches per second, per square inch of aperture.	
	Air entering the end from which the inner sharp edge was removed.	Air entering the end from which the outer sharp edge was removed.
44 inches.	513·5	473·2
24 "	564	538
12 "	589·6	573·4
4 "	660	637·4
2 "	685·2	668
1 inch.	726·2	663·2
$\frac{1}{2}$ "	699·6	594
$\frac{1}{4}$ "	594·8	526

My last experiment was with a hollow cone, the sides of

which formed an angle of 60°, and the opening at one end was three inches, and at the other 0·625 inch in diameter. Using a pressure of 1·44, the quantities of effluent air per second per square inch of narrowest aperture were, accordingly as the air entered at the broad or narrow apertures, 666·1 and 510·7 respectively.

The height of a column of air of the density and temperature of that used in the experiments, which would give a pressure of 1·44 inch of water, is 88·93 feet. The formula for very small pressures is $v = \sqrt{2gh}$. Thus the theoretical velocity in the absence of disturbing causes would be 75·64 feet per second, which gives 907·7 cubic inches issuing per second through an orifice one inch square. Calling this theoretical efflux unity, the above experiments give

For apertures in thin plates	0·6074
For a tube of the same diameter as length . .	0·7676
For a similar tube with a wide entrance-tube .	0·8933

I have not been able to detect any effect due to vibration of the issuing stream. By placing the end of a tube composed of thin metal, four feet long and one inch in diameter, at about half an inch distance from an aperture in a thin plate of one inch diameter, musical tones were produced, which by increasing the pressure gradually ascended in harmonics through a scale of many octaves. The same musical effects could be produced, using a constant pressure of air, by moving the tube nearer the aperture through the space of a tenth of an inch. Savart and Masson have adduced facts of this kind to prove that air rushing out of an aperture has a vibratory motion. Although I do not admit this conclusion, there can be no doubt that the vibration constituting sound, produced by such methods as above indicated, will be able to travel back to the air rushing through the orifice if its velocity be not greater than 1090 feet per second. I have failed, however, to discover any sensible influence from this cause on the velocity of efflux. It occurred to me also to try whether the air issued with a rotary motion; but such experiments as I have been able to make with vanes have led me to no decisive conclusion, although there can be no doubt that many circumstances might cause such vortices, the operation of which would be to diminish the velocity of efflux.

On some Amalgams.
By J. P. Joule, *LL.D., F.R.S., &c.**

['Manchester Memoirs,' ser. 3. vol. ii. p. 115.]

The experiments I am about to describe were made twelve years ago; but their publication was delayed to the present time in the hope of being able to extend them. Although I have not found an opportunity of doing so, I trust that these comparatively old observations will be deemed of sufficient interest to justify me in having submitted them to the Society.

My attention was first directed to the subject through my wish to discover a ready means of procuring a perfectly true and polished metallic surface. Since it was believed that mercury refused to enter into combination with iron, I thought that, by depositing the latter on mercury, a plate of it would be formed possessing a smoothness equal to that of the fluid metal. However, on making the experiment, I found that the iron entered into combination with the mercury, forming an amalgam †.

* The experiments were made at Acton Square, Salford.

† In consequence of iron possessing nearly the same affinity as hydrogen for oxygen, there is considerable difficulty in depositing it electro-chemically on a metallic plate. I have only once or twice obtained a good electrotype deposit on a polished surface, to which the iron adhered so firmly that it could only be removed by abrasion. Even in the process of amalgamating iron, the constant evolution of hydrogen from the mercury shows that decomposition of water takes place simultaneously with that of the salt of iron.

I find that Sir John Herschel has anticipated me in the production of the amalgam of iron.—*Note*, 1882.

One element of a Daniell's battery was amply sufficient for the purpose. Its zinc plate was connected by a wire with a globule of mercury covered by a solution of sulphate of iron, whilst an iron wire attached to its copper plate, and dipping into the solution, completed the circuit. The iron wire gradually dissolved, whilst an equal portion was taken up by the mercury, which, in doing so, by degrees lost its fluidity, until at length a mass of crystals of amalgam was formed having a greyish-white colour of metallic brilliancy. The time required to complete the operation was generally about one day; but a longer or shorter period was occupied in some instances, in consequence of variations in the quantity of mercury employed and in the efficiency of the voltaic arrangement. The following Table contains the results of most of the experiments made on the amalgam of iron. The analysis of this and other amalgams was made by heating them in a glass tube through which a current of hydrogen was passed.

No.	Composition.		Sp. grav.	Remarks.
	Mercury.	Iron.		
1.	100	0·143	....	Perfectly fluid.
2.	"	1·39	....	Fluid.
3.	"	2·97	....	Semifluid.
4.	"	11·8	12·19	Soft.
5.	"	18·3	....	Solid; colour greyish white.
6.	"	47·5	....	Solid; good metallic lustre.
7.	"	127·6	10·11	Solid; friable.
8.	"	14·74	....	The superfluous mercury pressed out from the semifluid amalgam by hand.
9.	"	79	....	Compressed rapidly, and with a force of fifty tons on the square inch.
10.	"	103·2	....	Ditto.

No. 5 of the above Table was a solid amalgam of a greyish white, approaching the colour of iron. It could be easily broken into powder. When dried and left undisturbed, it soon became covered with small globules of mercury, until ultimately it was entirely decomposed.

To obtain No. 6, I used a solution of chloride of iron instead of the sulphate which was used in all the other experiments.

No. 7 could be easily reduced to powder. It had a bluish colour, and was destitute of metallic lustre until it was rubbed. It remained some time under water without change, but when dried became speedily decomposed, whether it was exposed to the action of air, or was placed under the exhausted receiver of an air-pump.

The amalgam of iron, whether solid or fluid, is attracted by the magnet, and in the solid condition is capable of receiving a slight dose of permanent magnetization.

In No. 1, the iron, though apparently completely dissolved by the mercury, remained in the full possession of its magnetic virtue.

A portion of No. 2, weighing 87·69 grains, placed in a piece of quill, was attracted by a magnet with a force equal to 0·36 gr. 3·058 grains of iron wire, cut into small pieces and placed in the same quill, were attracted by a force of 0·94 gr. The quantity of iron contained by the amalgam was 1·2 grain. Hence it appeared that the iron had lost very little of its magnetic virtue by combination with the mercury.

The following observations were made to discover the position of the amalgam of iron in the electro-chemical series. The galvanometer which was employed had a coil 1 foot in diameter, composed of 400 convolutions of wire $\frac{1}{40}$ of an inch in diameter.

Positive Metal.	Negative Metal.	Deflection.
Amalgamated zinc.	Zinc.	10°
Zinc.	Iron.	42°
Zinc.	Copper.	65°
Amalgamated iron.	Copper.	15°
Iron.	Amalgamated iron.	5°

It appears therefore that the amalgamation of iron produces a contrary effect to the amalgamation of zinc. This is especially remarkable, as the amalgamated iron contained no carbon, which must have existed to a certain extent in the plate of iron with which it was associated.

When amalgam of iron is left under water for a few days, it becomes coated with rust. If shaken violently, it becomes almost immediately decomposed, the iron as a black powder floating on the surface of the liberated mercury.

When the amalgam is heated to the boiling-point of mercury, the liberated iron unites with the oxygen of the air, throwing off bright red sparks, and leaving a hard lump of oxide.

The experiments seem to indicate that the solid amalgam of iron which contains the largest quantity of mercury is a binary combination of the two metals.

The specimen marked No. 8 in the Table was procured by compressing by hand between folds of linen a quantity of amalgam in a soft state. There resulted a mass of white crystals of perfect metallic lustre. The mercury left was about two equivalents. It seemed probable that one of these was left uncombined among the pores of the amalgam.

The specimens Nos. 9 and 10 were obtained by hydraulic pressure acting on a piston of steel $\frac{3}{8}$ of an inch in diameter, working in a cylinder in which a silken bag filled with amalgam was placed. The resulting amalgam was so hard that it could only be broken by the smart blow of a hammer. Its black colour seemed to indicate nearly total decomposition.

Amalgam of Copper.—To form this amalgam, a small quantity of mercury was poured into a dish containing solution of sulphate of copper. A copper wire connected the mercury with the zinc of a Daniell's cell, whilst a coil of copper wire immersed in the cupreous solution completed the circuit. A mass of crystals was gradually formed, branching out to the distance of half an inch or more. Ultimately pure copper was deposited on the extremities of the crystals in a fringe of light red, the whole presenting the appearance of a beautiful flower. In the following Table I have collected the results of several such experiments :—

No.	Mercury.	Copper.	Sp. grav.	Remars.k
1.	100	22·5	13·32	Arborescent crystals; no pink deposit.
2.	„	24·73	13·260	Ditto ditto.
3.	„	25	13·185	Ditto ditto.
4.	„	27·76	13·17	Ditto ditto.
5.	„	29·92	....	Pink deposit on the extremities of the crystals.
6.	„	37·7	....	Pink deposit over the greatest part.
7.	„	31·35	13·51	The copper which was deposited on the outside of the crystals was constantly removed. The experiment was stopped when the central button of amalgam became pink in one or two places.
8.	„	29·08	....	Ditto ditto.
9.	„	29·0	13·76	Ditto ditto.
10.	„	34·19	13·01	From a hot solution of sulphate of copper. Hard and crystalline mass.
11.	„	39·64	12·99	Formed slowly in eight days. Pink in several places.
12.	„	41·5	....	This amalgam was continually pounded whilst it was being produced. Pink in several places.
13.	„	38·12	12·65	Sulphate of copper, kept at 100° Fahr. In two days the amalgam was covered with arborescent crystals tipped with pink.

On inspecting the above Table, it will appear that whenever the quantity of copper approaches nearly to an equivalent, a deposit of unamalgamated copper begins to take place. This seems to demonstrate that the solid amalgam containing the least quantity of mercury is a binary compound.

The mean of the specific gravities of the specimens possessing an equivalent (or a little less than an equivalent) of copper is 13·31.

The specific gravity of the other amalgams, containing excess of copper, is 12·82. It follows that, if we admit that the specific gravity of copper is not altered when it enters into combination with mercury, the specific gravity of the latter, in the amalgam, is 15·415.

In the following Table I give the analysis of amalgams

after pressure of various degrees of force had been applied during various lengths of time:—

No.	Pressure per square inch in tons.	Time.	Mercury.	Copper.	Sp. grav.
1.	$\frac{3}{4}$	12 hours.	100	20·3	
2.	$\frac{3}{4}$	12 hours.	"	17·28	
3.	1	36 hours.	"	20·5	
4.	$1\frac{1}{4}$	17 hours.	"	18·95	
5.	2	12 hours.	"	18·4	
6.	Gradually increased up to 1 ton.	$3\frac{1}{2}$ months.	"	39·0	12·76
7.	Ditto.	13 days, with intervals amounting to 54 days.		38·43	12·56
8.	9	A few minutes.	"	25·84	12
9.	15	"	"	28·57	
10.	18	"	"	28·4	13·01
11.	20	"	"	29·46	
12.	72	"	"	30·95	13·06
13.	72	"	"	32·82	12·93
14.	144	"	"	35·13	12·96
15.	144	"	"	34·87	12·57
16.	144	"	"	35·63	12·62
17.	20	30 minutes.	"	33·04	
18.	36	30 minutes.	"	30·25	12·88
19.	72	1 hour.	"	32·34	
20.	30	2 hours.	"	40·18	
21.	30	7 hours.	"	44·34	12·38

On inspecting the above Table, it will appear that a moderate pressure continued for a short time leaves a binary compound of the metals along with the quantity of mercury which may be supposed to be entangled among the crystals. When the pressure was very great, or was continued for a long time, the resulting amalgam invariably contained more than one equivalent of copper. I believe that this arises from a decomposition of the binary amalgam by the violent mechanical means adopted.

On the supposition that the copper retains its own specific gravity, the density of the above amalgams gives for the mercury a specific gravity of 14·985.

The *Amalgam of Silver* was generally produced by treating mercury with nitrate of silver. The action goes on until a hard mass of shining crystals is formed, consisting of about an equivalent of silver to one of mercury.

No.	Mercury.	Silver.	Sp. grav.	Remarks.
1.	100	52·6	14·68	From cold solution of nitrate of silver.
2.	"	100·3	....	Ditto ditto.
3.	"	115·4	....	Ditto ditto.
4.	"	115·2	13·25	Ditto ditto.
5.	"	155·8	12·34	Boiled in solution of nitrate of silver.
6.	"	106·4	12·49	From a hot concentrated solution of nitrate of silver.
7.	"	293·3	12·54	Button of amalgam formed by the electrolytic action of one cell of Daniell's battery.
8.	"	2614·0	11·42	Crystals formed on the edges of the above button of amalgam.

From the above Table it appears that the amalgam most readily formed by the action of nitrate of silver on mercury is a binary compound; for the average result gives the proportion of 107·6 silver to 100 mercury. It will be noticed that the specific gravity of the specimens indicates, as in the case of the amalgam of copper, a very considerable contraction of volume, principally referable no doubt to the assumption of the solid state by the mercury, the specific gravity of which comes out 16·5 from the above and succeeding experiments on the amalgam of silver.

In the next Table I give the composition of amalgams of silver after compression. Before placing it in the press, each specimen was mixed up with excess of mercury so as to form a thick paste. I should mention here, that, on making the analysis, it was found necessary to employ a temperature nearly sufficient to fuse the silver in order to drive from it the last portions of mercury.

No.	Pressure.	Mercury.	Silver.	Sp. grav.
1.	2½ tons for 1 day.	100	33·78	
2.	3 tons for 3 days.	”	37·76	
3.	72 tons for 1 hour.	”	40·13	13·61
4.	72 tons for 1½ hour.	”	40	13·78
5.	72 tons for 1½ hour.	”	51·55	13·44
6.	72 tons for 20 min.	”	43·15	

The mean composition of the amalgam, after being pressed with 72 tons on the inch, was therefore 43·71 silver to 100 mercury. Allowing for mercury remaining among the crystals in an uncombined state, we may conclude that the solid amalgam containing the largest quantity of mercury is composed of one equivalent of silver to two of mercury.

Amalgam of Platinum.—To obtain this amalgam, platinum was deposied on mercury by the electrolytic action of two or three voltaic cells on the bichloride.

No.	Mercury.	Platinum.	Sp. grav.	Remarks.
1.	100	15·48	14·29	Metallic lustre when rubbed.
2.	”	21·6	....	Solid. Dark grey colour.
3.	”	34·76	14·60	Dark grey; no metallic lustre.

An amalgam of 12 platinum to 100 mercury possesses a bright metallic lustre, and is soft and greasy to the touch. Pressed with a force of 72 tons to the square inch, a hard button of dark-grey amalgam is left, consisting of 43·2 parts of platinum to 100 of mercury. I infer therefore that the solid amalgam of platinum which contains the largest quantity of mercury is composed of two equivalents of mercury to one of platinum *.

The specific gravity of this amalgam appears to be nearly that which it would be on the supposition that no condensation of volume takes place on the union of the metals; but

* Amalgam of platinum in the form of a thick paste may be obtained by exposing mercury to the action of bichloride of platinum for a sufficient length of time.

the specimens were too small to make very accurate determinations of specific gravity.

Amalgam of Zinc was obtained electrolytically from sulphate of zinc; after some time the mercury lost its fluidity, and branching crystals began to be formed.

No.	Mercury.	Zinc.	Sp. grav.	Remarks.
1.	100	39·4	11·34	White and crystalline.
2.	"	122·8	8·935	Ditto ditto.
3.	"	184·9	8·349	Prepared from hot sulphate of zinc.

The first of the above three specimens, consisting of an equivalent of each metal, appears to be the amalgam which, containing the largest quantity of mercury, is yet solid. The specific gravity indicates a certain contraction of volume, though not nearly as much as that in the amalgams of silver and copper, but such as would place the specific gravity of the mercury at 14·1. Pressure seemed to have the effect of decomposing this amalgam, or at least of expelling mercury until the amalgam consisted of about one equivalent of mercury to three of zinc.

No.	Pressure.	Mercury.	Zinc.
1.	¾ ton for 1 day.	100	59·25
2.	1¼ ton for 1 day.	"	69
3.	50 tons for 1 hour.	"	76·7
4.	Ditto ditto.	"	79·6
5.	Ditto ditto.	"	75·9

Amalgam of Lead.—On making mercury negative in acetate of lead, a crystalline amalgam was gradually formed. The operation was stopped when the characteristic flat blue crystals of lead began to make their appearance. The amalgam was found to have a specific gravity of 12·64 (indicating 13·85 for its mercury), and to consist of 100 mercury to 69·83 lead, and, allowing for unavoidable excess of mercury, may be considered a binary compound.

To ascertain the effect of pressure, a liquid amalgam was

formed by heating the two metals together. It was then compressed with a force of three tons to the square inch for a day. A greater pressure than this would have caused the amalgam as well as the mercury to escape from the press. The result was a mass of bright crystals, easily fractured, which had a specific gravity of 12·11, and was composed of 100 mercury to 194 lead. I think there can be no doubt that the pressure had partly decomposed the binary compound. It appears that little or no contraction of volume is occasioned by the combination of the metals.

Amalgam of Tin was obtained by making mercury negative in a solution of chloride of tin.

No.	Mercury.	Tin.	Sp. grav.	Remarks.
1.	100	51·01	10·518	Beautiful crystalline amalgam.
2.	"	44·12	10·94	Ditto ditto.
3.	"	70·7	....	Some unamalgamated tin crystals at the extremities of the amalgam.

The amalgam formed by the electrolytic process appears therefore to be a binary compound. Its specific gravity, along with that given in the next Table, shows a specific gravity of 14·1 for the mercury in combination. Pressure of the amalgam gave the following results :—

No.	Pressure.	Mercury.	Tin.	Sp. grav.	Remarks.
1.	1440 lb. for 10 min.	100	75·9		
2.	1440 lb. for 2 days.	"	255·5		
3.	2724 lb. for 2 days.	"	392·4		
4.	5400 lb. during 30 days.	"	384·1	8·154	Pressure gradually increased.
5.	50 tons for 15 min.	"	402·3		
6.	2700 lb. during 30 days.	"	408·9	....	Pressure of 50 tons during 1 day did not afterwards drive out more mercury.

The above results show most decisively that pressure is able to decompose the amalgam of tin, the mercury left after long continued high pressure having a volume little more than one eighth of the entire mass.

I made an unsuccessful attempt to amalgamate hydrogen by developing it at a low temperature (4° Fahr.) on mercury. It did not appear that the smallest quantity of hydrogen was taken up. This appears, however, to be an experiment worth repeating. I think it highly probable that, by using intense cold and very great pressure, an amalgam of hydrogen might be formed.

As metals generally retain their specific gravities when they meet to form alloys, it may be inferred that the foregoing experiments indicate the specific gravity of mercury in the solid state. This value, from the average of the thirty six determinations of specific gravity above given, is 15·19.

On the Probable Cause of Electrical Storms. *By* J. P. Joule, *LL.D., F.R.S.*

[Proceedings of the Literary and Philosophical Society of Manchester, vol. ii. p. 218, March 18, 1862.]

The very close correspondence between the theoretical rate of cooling in ascending and the actual indicates a rapid transmission of the atmosphere from above to below, and *vice versâ*, continually going on. We may believe that during thunderstorms this interchange goes on with much greater than ordinary rapidity. At a considerable distance from the thundercloud, where the amosphere is free from cloud, the air descends, acquiring temperature according to the law of convective equilibrium in dry air. The air then traverses the ground towards the region where the storm is raging, acquiring moisture as it proceeds, but probably without much diminution of temperature, on account of the heated ground making up for the cold of evaporation.

Arrived under the thundercloud, the air rises, losing temperature, but at a diminished rate, owing to the condensation of its vapour to form part of the immense cumulus cloud which overcasts the sky on these occasions. The upward current of air carries the cloud and incipient rain-drops upwards, but presently, in consequence of the increased capacity of the mass from the presence of a large quantity of water, the refrigeration of the air, in consequence of its dilatation, will be so far diminished as to prevent the condensation of fresh vapour, and ultimately to redissolve the upper portion of the cloud. This phenomenon, which has been noticed by Rankine in the cylinder of the steam-engine, will account for the defined outline of the upper edges of cumulus clouds. The upward current no doubt extends occasionally to regions below the freezing temperature. If cloud be carried with it, snow or hail will be formed, which, if sufficiently abundant, will pass through the cloud and fall to the ground before it is melted. Now, the dry cold air in which the snow and hail are formed is a perfect insulator. Ice has also been proved, by Achard of Berlin, to be a non-conductor and an electric. Even water, in friction against an insulator, is known from the experiments of Armstrong, explained by himself and Faraday, to be able to produce powerful electric effects; and this fact has been suggested by Faraday to explain powerful electric effects in the atmosphere. Sturgeon has noted the remarkable development of electricity by hail-showers. Few heavy thunderstorms occur without the fall of hail. Hail, whether in summer or winter, is almost, if not invariably, accompanied with lightning. In the presence of these facts, it seems not unreasonable to consider the formation of hail essential to great electrical storms; although, as has been pointed out by Professor Thomson, very considerable electrical effects might be expected from the negatively charged air on the surface of the earth being drawn up into columns, and although, as the same philosopher has observed, every shower of rain gives the phenomena of a thunderstorm in miniature. The physical action of insulators and electrics in mutual friction must certainly produce very marked effects on the grand scale of nature. If we suppose that the falling hail is electrified by the air it meets, the electrification of the cloud into which the hail falls might thus be constantly increased until the balance between it and the inductively electrified earth is restored by a flash of lightning. If the hail is negatively electrified by the dry air with which it comes into contact, the latter will float off charged with positive electricity, which may account for the normal positive condition of the atmosphere in serene weather, as well as the electrification of the upper strata evidenced by the aurora borealis. The friction of wind has been supposed by Herschel to contribute to the intense electrification of the cloud which overhangs volcanoes during eruption.

On the Surface-condensation of Steam. By J. P. Joule, *LL.D., F.R.S., President of the Literary and Philosophical Society of Manchester, &c.**

[' Philosophical Transactions' 1861, vol. cli. p. 133. Read December 13, 1860.]

The laws which regulate the transmission of heat through thin plates of metal under various circumstances, although of extensive practical application, and although their elucidation would necessarily involve scientific conclusions of great interest, have hitherto received little of the attention of natural philosophers. Two great divisions of the inquiry are, first, the communication of heat from the products of combustion to a boiler, and, second, the application of cold to a vessel employed for the condensation of steam. With a view to supply some information on the latter subject I have, with the assistance of a grant from the Royal Society, undertaken the present research.

* The experiments were made at Oak Field, Whalley Range, near Manchester.

The adjoining sketch will explain my apparatus. B is a steam-boiler, into the side of which a pipe P furnished with a stopcock T is screwed. Jointed to this by a caoutchouc tubulure *t* is the condensing pipe *s*, connected at the lower end to a short pipe *q*, which in turn is connected with the copper receiver R, closed at the bottom by a screw-nut *n* furnished with a washer of india-rubber. The refrigerating water is transmitted through the channel E D C, consisting of a pipe 1¼ inch in diameter, and the concentric space between the steam-condensing pipe and an exterior pipe of larger diameter. The refrigerating water on flowing away is collected in V, the vessel in which it is afterwards weighed. In order to avoid the necessity of applying a large correction to the temperature of this water, it is, when its quantity is not very great, received in the first instance by the small can U, into which a thermometer is plunged. A branch pipe *p*, screwed into the main pipe, is connected to the barometer-tube *b* in order to measure the degree of vacuum.

The pipe P enters the boiler at 8 inches above the surface of the water. Separate experiments showed that no water came up to this height by "priming." On the other hand, the arrangement of the boiler, the flue of which is entirely below the level of the water, prevented the steam being surcharged with heat to any notable extent.

By careful experiments I found that a thermometer, of which the bulb was held six inches above the water of the boiler, indicated exactly the same temperature, whether the boiling was carried on very slowly or very rapidly. But when the bulb was immersed 3 inches below the surface, the temperature with slow boiling was $0^\circ \cdot 532$ higher than that of the steam, which difference was further increased to $0^\circ \cdot 538$ by rapid boiling. This would lead to the belief that the steam must have been a little overcharged with heat by passing through superheated water; but as there was a trifling cooling effect by the influence of the atmosphere on the pipe P, the steam passing through the stopcock might be safely considered neither superheated nor mixed with water.

Fig. 93. Scale $\frac{1}{8}$.

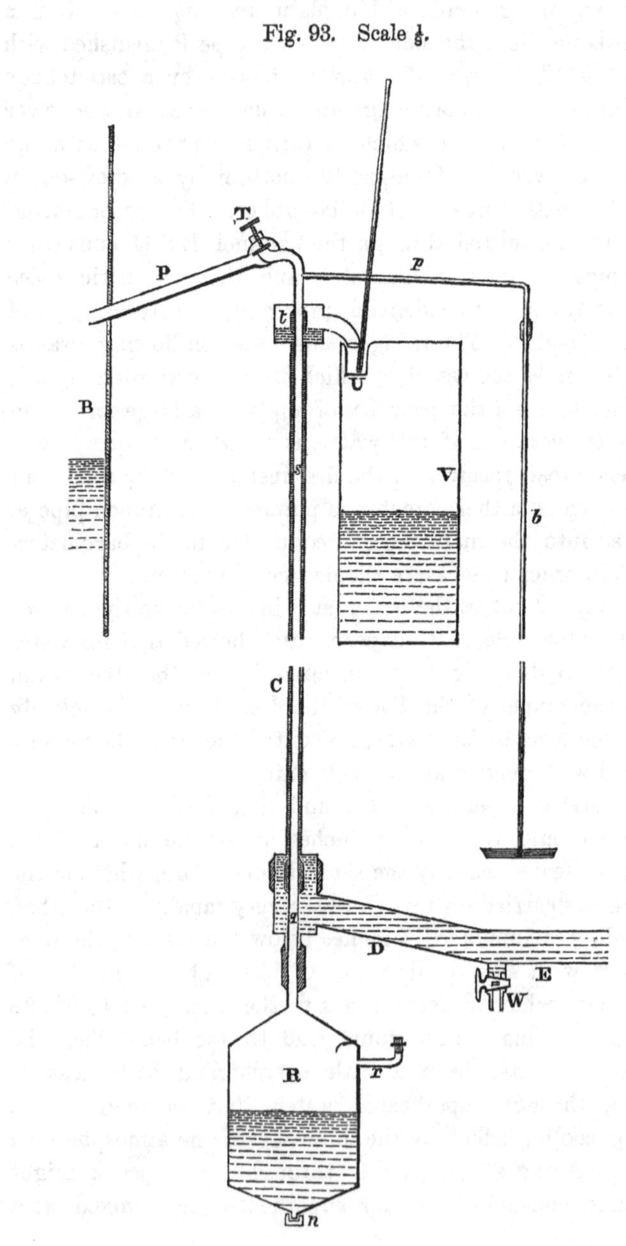

Up to the stopcock T the temperature of the pipe may be considered that of the boiler; beyond it the temperature becomes gradually that of the condenser. A certain, though very small, quantity of heat is thus conducted along the tube from the stopcock T as far as the india-rubber junction *t*. Any water condensed in P falls back again into the boiler; that between the stopcock and *t* falls into the receiver; so that the small quantity of conducted heat just mentioned is probably compensated by the trifling cooling effect of the atmosphere between T and the refrigerating water.

The short continuation pipe *q* exposes to the water an effective length of 3 inches, which, on account of the wideness of the channel there, cannot generally have had an effect greater than that due to 2 inches in the narrower part. As, however, a length amounting to an inch and a half of the ends of the condensing tube is overlapped by the vulcanized tubing, the entire amount of condensation may, without appreciable error, be laid to the account of the condensing pipe.

The receiver R and the pipes C, P, and *p* are enveloped by a thick coating of cotton-wool and flannel, so as to prevent, as far as possible, the refrigerating effect of the atmosphere.

Great pains were taken to make every part of the apparatus in which the pressure is below that of the atmosphere perfectly air-tight. It will be seen that the form of the stopcock T effectually prevents any leakage except from the high-pressure side into the atmosphere, which is of no consequence. The india-rubber junctions were at first made by simply binding on the ends of the tubes short lengths of vulcanized caoutchouc; but it was soon found that enough air passed to vitiate the experiments, which were consequently rejected. The method afterwards adopted was to smear the ends of the tubes with melted vulcanized caoutchouc before the short india-rubber tubes were bound on. This plan was found to be so efficacious that air appeared to be perfectly excluded, and the vacuum wholly unimpaired, however long an experiment was carried on.

The vacuum-gauge glass tube is 0·45 of an inch in internal

diameter. It is plunged into a wide dish of mercury, from the surface of which the height of the column is measured. The temperature of the mercury in the gauge was always nearly that of the barometer which registered the atmospheric pressure. During each experiment a small quantity of condensed water settled by degrees on the top of the mercury, the length of which, divided by 13·56, gave the correction to be applied to the height of the column.

It will be observed that the pipe leading to the vacuum-gauge is inserted near the stopcock which admits the steam. It was important to ascertain whether the gauge would stand at the same level if it were connected with other parts of the vacuous space. To determine this, a pipe was attached to the receiver at *r*, and connected with a gauge placed side by side with the first gauge, and dipping into the same dish of mercury. The gauges were observed during rapid and slow condensation, at different and at varying pressures; but the height of the columns appeared to be in general exactly the same: if any difference could be observed at any time, I should say that the receiver-gauge indicated the less perfect vacuum of the two; the difference, however, amounted in no case to more than $\frac{1}{30}$ inch.

The following is my method of experimenting:—The nut *n* being unscrewed, the dish of mercury removed from under the gauge-tube, and the water being completely discharged from the tap W, the cock T is partly opened, and the steam is blown through the steam-pipe *s*, the gauge *b*, and the receiver R until they are completely freed from air. The nut *n* is then screwed on, W closed, and the water let on, the three operations being performed as simultaneously as possible. At the moment when the steam is about to cease issuing from the gauge-pipe, its end is introduced into the dish of mercury. After an interval of time, varying from half a minute to three minutes, the condensation goes on with perfect regularity, and the mercury in the vacuum-gauge remains steady. The temperature of the water flowing away and the gauge are observed every two or three minutes. The experiment is terminated by simultaneously shutting off the steam and the water, and opening the tap W to let off the water remaining in the pipe. The nut n is then removed; and a quantity of air having entered the receiver, the condensed water is caught by a small can (held close and containing a thermometer), which overflows into a larger vessel, in which the water is immediately afterwards weighed.

The values of several small corrections which had to be applied to the observations were obtained from data derived from separate experiments. Of the thermometers employed, one was made by Fastré, in which each division is equal to $0^{\circ}{\cdot}225$ Cent.; the two others were from Kew Observatory, and have for each division the values $0^{\circ}{\cdot}1$ and $0^{\circ}{\cdot}0994$ respectively. A correction had generally to be applied in consequence of the non-immersion of the stems.

The cooling effect of the atmosphere on the receiver R operates partly to condense steam and partly to cool condensed water. The correction on the former account was found to be equal to the product of the time in minutes, the proportion of acting surface, and the difference between the temperatures of the receiver and atmosphere, divided by 77 times the difference between 640 and the temperature of the condensed water: the result had to be subtracted from the weight of condensed water. The correction on the latter account is equal to the product of the time, acting surface, and difference of temperature, divided by 77 times the weight of condensed water: it had to be added to the observed temperature of the condensed water.

The correction on account of the cooling of the refrigerating water on flowing through C into the vessel U, was found to be equal to the difference of temperature between the water and the atmosphere, multiplied by 0·51, and divided by the quantity of water flowing per hour. This rule applies to the case in which the external pipe C was 4 feet long and 1 inch in diameter. Corrections in the instances in which other tubes were used were made by calculation without express experiments, inasmuch as they were of very trifling amount.

The slight loss of water by evaporation before and during

the process of weighing, was allowed for in the weighing both of the refrigerating and condensed water.

The metal of the steam-pipe and receiver is necessarily at 100° at the commencement of an experiment, and therefore communicates some heat during the first few moments. On the other hand, the small quantity of water drawn off at W at the termination of an experiment is always more or less heated. Corrections on both these accounts were easily applied.

I had at first some doubts whether the vacuum would not become gradually impaired by air coming over from the boiler; for it has been frequently asserted that water becomes perfectly free from air only after long-continued boiling. I found, however, that after boiling had taken place for only two or three minutes, the air was entirely expelled, and that, even if condensation were afterwards carried on until the receiver was entirely filled with water, no change took place in the height of the gauge. Hence, by blowing off steam for ten minutes at the commencement of a day's experimenting, I effectually secured myself against any risk of the interference of air*.

The Table of experiments (facing p. 512) requires little explanation. It will be seen that column 5 contains some numbers with the negative sign. This might be expected where a small quantity of water was used, on account of its being raised in temperature during its ascent. When the water was intended to go in the same direction as the steam,

* I could not discover any alteration in the composition of the air after it had remained in the boiler some days. There appears to be no truth in the hypothesis which ascribes boiler-explosions to the formation of hydrogen. The obvious cause is over-pressure; and it is not wonderful that, when multitudes of boilers are worked at a very considerable proportion of the pressure calculated to burst them when new, accidents occasionally occur. I have repeatedly insisted upon the absolute necessity of periodical testing, and have proposed a method requiring no extra apparatus or expense, which consists simply in lighting a fire under the boiler when completely filled, and so producing the proof-pressure by the expansion of water by heat. I try my boiler every six weeks by this process, which appears to answer the end in view in every respect.

it was poured in at the upper end of the outer tube, and flowed away at the lower end, the pipe E being removed. Each number in the 14th column is the average of all the observations of the pressure in the condenser after it became constant; and column 17 contains the averages of all the observations of the temperature of the refrigerating water at its overflow made at the moments of gauge-observation. Hence this column contains numbers generally a little different from those of column 7, which, being taken for the purpose of deducing the total heat of steam, are the averages of all the temperature-observations of the overflow water in the several experiments.

In order to explain the principle on which the 18th column is based, I cannot do better than give textually the extract of a letter I received from Professor Thomson, to whom at the outset I communicated my design, and who, with his usual zeal and kindness, immediately offered me very valuable suggestions.

"Steamer Venus, August 10th, 1859.

"If the resistance to equalization of temperature between the steam and water depended on *conduction* through the separating metal alone, the heating effect would take place according to the law you name. The formula would be thus found,

$$wdv = -k\frac{\mathrm{A}dx}{a}v,$$

where w is the mass of water passing per unit of time, dv the augmentation of the difference of temperatures inside and outside in a length from x to $x+dx$, v the difference itself at any point P, k the conducting-power of the metal, A the area of the tube per unit length, a its thickness. By integrating, we find

$$\log\frac{\mathrm{V}}{v} = \frac{k\mathrm{A}x}{aw},$$

where V denotes the difference of temperatures at the entrance end; A will be the area corresponding to a mean diameter calculated by the formula $\dfrac{2a}{\log \dfrac{\mathrm{D}}{\mathrm{D}-2a}}$ when the outer diameter D and the inner $\mathrm{D}-2a$ differ so much that it will not do to use one or the other indifferently. For all practical purposes, with such tubes as are actually used, it will do to take as the mean diameter the arithmetic mean $\mathrm{D}-a$.

"The truth, however, is that, except with a very great velocity of the water, there will be a heated film close to the metal much higher in temperature than the average temperature of the water in the same section, and the abstraction of heat will be much slower than according to the preceding formula. It is not improbable, however, that some law of variation will still hold from point to point in the direction of flow; and if so, the same formula would apply, only that for k something much smaller than the true conductivity of the metal must be substituted. Thus, supposing k to be a function of w, smaller the smaller is w and increasing to a limit (the true conductivity of the metal), your experiments might give values of k for different rates of the flow of the water by the expression

$$k=\frac{aw}{\mathrm{A}x}\log\frac{\mathrm{V}}{v}.$$

It would be necessary to ascertain by experiment how nearly the geometrical law of decrease of the difference of temperatures along the tube holds, as there is no sufficient theory for convection to give any decided indication.

"As the results would probably depend but little on the thickness and quality of the metal, it would be better perhaps to take $\frac{k}{a}$ as the thing to be determined: calling it C, we have

$$\mathrm{C}=\frac{w}{\mathrm{A}x}\log\frac{\mathrm{V}}{v}, \text{ or } v=\mathrm{V}\epsilon^{\frac{-\mathrm{CA}x}{w}}.$$

ϵ being the base of the nap. log, $\epsilon^{\frac{-\mathrm{CA}}{w}}$ is the fraction expressing the reduction of the difference per unit of length, and therefore $\left(1-\frac{\text{CA}}{w}\right)$ 100 is the percentage of difference lost per unit of length. If this be called θ, we have

$$v=\text{V}(1-\theta)^x, \text{ or } \log\frac{1}{1-\theta}=\frac{1}{x}\log\frac{\text{V}}{v},$$

where log denotes any kind of log. These are, in fact, the compound-interest formulæ, and are perhaps the most convenient for numerical reductions."

The results of my experiments were quite in conformity with Professor Thomson's view as to the smallness of the resistance to conduction through the thickness of the metal compared with the resistance at the surfaces of the tubes through the closely adhering film of liquid. I therefore sought to discover in each instance the entire conductivity by the formula

$$\text{C}=\frac{w}{\text{a}}\log\frac{\text{V}}{v},$$

where, a being the area of the tube in square feet, and w the quantity of refrigerating water transmitted per hour, C represents the number of units of heat, in pounds of water raised 1°, which would be conducted through a surface of 1 foot area, the opposite sides of which differ from one another by 1°. The determination of C in each instance will be found in column 18.

I generally obtained observations of the vacuum-gauge directly after the stoppage of the condensation. The results of these, reduced to the value they would have had at the precise time of the closing of the stopcock, are given in column 15 of the Table. The effect of stopping the condensation was generally a diminution of pressure, which took place rapidly at first, and afterwards slowly and with great regularity. I believe that this diminution of pressure is owing to the water collected in the receiver, which, having fallen somewhat in temperature during an experiment, governs the vacuum as soon as the fresh hot condensed water ceases to be supplied to its surface. In some few instances the mercury in the gauge was observed to *fall* immediately on the stoppage of the condensation. In these the vacuum appeared to be more perfect while the condensation was being carried on than was due to the temperature of the condensed water. It was long before I was able to form any conjecture as to the cause of this anomalous circumstance. I now think that it might have been occasioned by a stricture in the india-rubber junction which connected the gauge with the steam-tube *p*. It is not, however, easy to see how this can account for the sudden fall of the gauge at the moment of the stoppage of the condensation. In the Table, I have marked those results which I suspect to have been influenced by a contraction at the junction by a note of interrogation. I may observe that the india-rubber tubulures were frequently renewed, in order to prevent the chance of a stricture, which, moreover, I always endeavoured to detect at its first approach, by observing whether the mercury descended instantaneously on the admission of the first bubble of air into the receiver when the nut was unscrewed.

Great care was always taken to keep the flow of steam and refrigerating water as constant as possible during each experiment. If this had not been done, the temperature of the water collected in the receiver during the former part of an experiment would have influenced to a certain extent the vacuum observed at the latter part. It was easy, by first condensing rapidly, and afterwards slowly by partially closing the steam-cock, to maintain for some time a vacuum much more perfect than that due to the temperature of the water in the receiver. In this case "bumping boiling" took place in the receiver, whilst the pressure gradually decreased to the value due to the new conditions.

On a cursory examination of the Table, it will become evident that the numbers in column 18, representing the

TABLE OF EXPERIMENTS ON THE

1. Description.	2. No.	3. Duration of experiment, in minutes.	4. Total pressure of steam in the boiler, in inches of mercury.	5. Head of refrigerating water above its overflow, in inches.	6. Mean temperature of refrigerating water. At its entrance (t).	7. Mean temperature of refrigerating water. At its exit.
Copper steam-tube, *s*, 4 feet long, exterior diameter ·75 inch, interior ·63 inch, mean area a=·7225 sq. ft. Outer tube C 1·4 inch in diameter. Refrigerating water moving in a direction contrary to that of the steam. In the experiments 10–16 the receiver was in communication with the atmosphere.	1.	60	48·2	0·2	5·18°	20·21°
	2.	60	41·88	−0·1	5·18	40·38
	3.	30	46·23	1·13	5·15	19·21
	4.	30	48 36	1·2	4·96	17·63
	5.	45	91·47	0·47	4·78	16·13
	6.	37	120 14	0·66	4·81	19·19
	7.	50	114·27	0·51	4·93	15 23
	8.	60	39·29	0·47	4·67	13·62
	9.	52	35 68	0 54	4·7	14·17
	10.	60	51·98	0 12	4·94	11·58
	11.	30	45·22	0·12	5·17	31·6
	12.	30	48·02	0·48	5·12	22·02
	13.	20	50·31	0·97	5·39	22·21
	14.	30	54·4	−0·1	5·37	48·35
	15.	20	45·6	1·35	5·12	26·1
	16.	18½	50·9	1·04	5·37	29·62
The same copper steam-tube. The outer tube ·87 inch in interior diameter. Experiments No. 30, 31, 32, and 33 were made when the steam-tube had been recently cleaned by dilute sulphuric acid. Refrigerating water moving in a direction contrary to that of the steam.	17.	60	44·91	4·37	6·57	16·5
	18.	60	48·8	−1·4	6·22	81·08
	19.	120	47·77	−0·13	6·62	50·51
	20.	120	45·27	0·6	6 86	25·61
	21.	50	48 71	−0·49	6·36	88·08
	22.	60	48 61	1·19	6·42	52·07
	23.	44	45·25	6·45	6·24	27·32
	24.	19	47·1	14·07	6·04	34·48
	25.	555	40·78	0·08	6·2	16·7
	26.	26½	49·86	14·73	5·36	26·94
	27.	27	46·68	21·45	5·22	22·67
	28.	20	51·5	23·29	5·22	26·75
	29.	435	46·27	− 0·63	5·22	53·73
	30.	24	53·13	12·9	8·5	32·135
	31.	18½	52·09	20·12	8·4	30·014
	32.	30	52·09	14·15	8·46	13·518
	33.	30	53·81	11·48	8·44	20·66
Refrigerating water moving in the same direction as the steam.	34.	30	48·99	48	8·655	22·55
	35.	15	46·51	48	8·62	22·955
	36.	12	46·51	48	8·62	27·78
	37.	11½	49·07	48	8·63	29·739
	38.	10	48·09	48	8 64	29·817
The same copper steam-tube. The outer tube 0·8 inch in internal diameter. Experiments 53–61 inclusive, were made when the steam-tube had been recently cleaned with dilute sulphuric acid. Experiments 62, 63, and 64 were made with the steam-tube greasy by rubbing it with oil. Refrigerating water moving in a direction contrary to that of the steam.	39.	60	43·81	12·8	6·73	38·69
	40.	60	41·32	− 0·1	7·27	89·146
	41.	60	42·77	37·16	7·13	22·852
	42.	57	43·05	35·61	6·87	47·62
	43.	60	45·78	18·33	6·95	56·784
	44.	120	44·75	0·5	6·95	54·01
	45.	60	45·33	9·92	6·61	35·29
	46.	60	45·06	28·58	6·67	18·353
	47.	60	47·16	210·2	6·51	14·317
	48.	23	50·72	232	6·47	36·732
	49.	17	48·52	206·3	6·84	45·282
	50.	15	51·97	211·7	6·82	47·496
	51.	14	51·67	237·2	6·82	47·371
	52.	30	53·76	292·06	6·82	24·312
	53.	60	62·4	28·74	7·09	25·343
	54.	30	55·06	28·6	6·7	41·49
	55.	30	53·47	26·66	6·88	63·893
	56.	15	49·16	34·66	6·88	84·222
	57.	30	51·97	231·4	6·82	13·54
	58.	30	57·88	211·3	6·82	20·574
	59.	20	58 82	233	6 82	31·987
	60.	13½	60·36	235·9	6·82	48 266
	61.	10	58·41	223·5	6·82	51·442
	62.	20	50·03	211	6·075	16 563
	63.	15	49·62	193·5	6·075	34·808
	64.	10	47·65	216	6·075	51·417
The same tubes. The steam-tube freshly cleaned. Refrigerating water going in the same direction as the steam.	65.	60	50·89	48	6·97	26·808
	66.	30	51·84	48	6·97	47·73
	67.	15	49·49	48	6·97	71·317
Lead steam-tube 4 feet long, exterior diameter 0·77 inch, interior 0·52 inch, mean area a=0·6503 sq. ft. The outer tube 0·87 inch internal diameter. Refrigerating water moving in a direction contrary to that of the steam. In experiments 81, 82, and 83 the receiver was in communication with the atmosphere.	68.	30	43·24	1·57	9·4	67·44
	69.	30	41·73	4·96	9·14	60·235
	70.	20	42·18	13·35	9·14	41·71
	71.	30	36 5	1·47	7·34	70·65
	72.	30	37 08	13 3	7·3	28·73
	73.	20	36·7	25·4	7·3	30·75
	74.	30	35·17	1·24	7·28	79·66
	75.	30	38 08	26·5	7·24	10·305
	76.	30	44·12	29	6·44	11·1

SURFACE-CONDENSATION OF STEAM.

8.	9.	10.	11.	12.	13.	14.	15.	16.	17.	18.	19.
eight of refrigerating water, in pounds.		Weight of condensed water, in pounds.		Temperature of the condensed water.	Total heat of steam.	Barometer minus vacuum-gauge, or pressure in the condenser, in inches of mercury.	Pressure in the condenser immediately after the conclusion of the experiment.	Temperature due to the pressure in the condenser (column 14) per Regnault (t_2).	Temperature of the refrigerating water at its exit, at the times the vacuum was observed (t_1).	Conduction of heat, per square foot of the surface of the steam-pipe, or $\frac{w}{a}\log\left(\frac{t_2-t}{t_2-t_1}\right)$.	No.
In the xperiment.	Per hour (w).	In the experiment.	Per hour.								
97·64	97·64	2·374	2·374	33·08°	651·41	1·56	1·49	34·03°	20·074°	98·13	1.
98·73	98·73	5·818	5·818	55·27	652·6	5·0	4·66	56·63	40·493	158·46	2.
379·45	758·9	9·467	18·934	78·87	640·83	17·992	12·53	86·37	18·956	195·68	3.
399·08	798·16	8·914	17·828	75·24	640·48	15·32	11·65	82·3	17·29	191·86	4.
443·26	591·01	8·410	11·213	61·58	656·49	9·094	6·074	69·79	15·82	152·24	5.
389·64	631·85	9·556	15·496	69·95	654·04	13·332	......	78·87	18·882	184·3	6.
495·39	594·47	8·611	10·332	58·94	644·58	7·865	6·57	66·48	15·152	149·44	7.
597·01	597·01	9·128	9·128	56·65	638·63	6·346	4·373	61·73	13·492	138·79	8.
570·26	657·99	9·284	10·712	59·12	637·72	7·342	5·4	64·95	13·94	151·6	9.
430·64	430·64	4·418	4·418	12·68	645·15	29·38	......	99·49	11·42	......	10.
218·34	436·68	9·321	18·642	25·62	638·76	29·618	......	99·72	31·165	......	11.
309·42	618·84	8·200	16·399	20·78	651·19	29·618	......	99·72	21·72	......	12.
261·26	783·78	7·027	21·081	29·08	647·86	29·915	......	100	21·924	......	13.
33·95	67·9	2·18	4·36	23·9	693·27	29·9	......	99·98	47·8	......	14.
283·3	849·9	10·6	31·8	97·0	657·72	29·68	......	99·77	26·1	......	15.
227·4	737·5	9·8	31·78	79·0	641·7	29·92	......	100	30·66	......	16.
189·47	189·47	2·985	2·985	22·75	652·85	0·96	......	25·58	16·5	193·77	17.
20·41	20·41	2·983	2·983	84·31	596·4	18·326	......	86·85	83·47	89·62	18.
87·14	43·57	6·54	3·27	54·36	639·16	4·965	4·965	56·48	51·9	143·92	19.
122·32	61·16	3·66	1·83	25·71	652·48	1·185	1·185	29·18	25·64	155·91	20.
63·72	76·5	9·55	11·46	87·25	632·77	25·126	18·806	95·19	88·87	279·85	21.
124·5	124·5	9·838	9·838	67·39	645·09	10·835	10·505	73·88	52·56	198·48	22.
251·6	343·1	9·938	13·552	60·6	594·28	8·2	7·95	67·42	27·46	202·28	23.
194·06	612·82	9·671	30·54	84·43	654·33	23·22	15·0	93·06	34·48	335·65	24.
167·16	18·07	2·652	0·287	9·18	671·01	0·77	......	21·91	17·03	29·26	25.
261·6	592·3	9·771	22·123	68·77	644·99	11·76	7·83	75·83	26·8	297·38	26.
332·91	739·8	9·77	21·711	59·24	651·8	7·46	5·414	65·3	22·3	342·5	27.
255·72	767·16	9·679	29·037	73·61	641·1	13·45	9·4	79·08	26·14	353·6	28.
118·66	16·37	9·966	1·375	38·92	616·85	5·03	......	56·75	55·3	80·9	29.
242·53	606·33	9·89	24·725	71·663	649·82	13·784	8·124	79·08	31·805	332·85	30.
245·85	797·34	9·145	29·66	69·825	649·23	11·92	9·59	76·15	29·376	408·89	31.
345·85	691·69	2·665	5·33	27·08	669·8	1·145	1·145	28·59	13·188	256·3	32.
305·85	611·69	6·059	12·119	46·782	659·2	3·271	3·26	47·97	20·248	300·41	33.
328·62	657·25	7·403	14·806	39·636	652·34	2·161	......	40	22·343	522·1	34.
311·0	1244	7·710	30·84	60·411	636·08	7·876	6·67	66·51	22·371	466·96	35.
258·44	1292·18	8·853	44·264	79·88	638·08	19·012	15·81	87·8	27·065	474·32	36.
249·56	1302·06	9·576	49·962	87·677	637·15	25·624	......	95·72	29·42	491·56	37.
216·0	1296	8·475	50·85	89·732	628·85	27·074	21·67	97·23	29·43	479·77	38.
57·53	57·53	3·023	3·023	39·54	647·76	2·186	1·986	40·21	38·858	255·6	39.
21·4	21·4	3·629	3·629	89·45	572·29	24·49	22·49	94·5	89·476	84·55	40
133·45	133·45	3·399	3·399	25·07	642·26	1·037	0·976	26·89	23·068	303·43	41.
147·29	155·04	9·992	10·518	49·51	650·19	6·064	......	60·74	47·62	303·08	42.
99·45	99·45	8·737	8·737	61·96	629·15	6·934	5·33	63·67	56·744	289·44	43.
22·432	11·216	1·957	0·978	55·92	595·34	4·762	4·21	55·6	54·01	53·1	44.
45·276	45·276	2·151	2·151	33·97	637·64	1·662	1·55	35·16	34·688	257·1	45.
91·197	91·197	1·742	1·742	17·58	629·21	0·698	0·7	20·31	18·258	239·09	46.
455·01	455·01	5·559	5·559	15·625	645·23	1·589	0·964	34·36	14·041	198·52	47.
200·531	523·12	10·191	26·585	61·80	654·95	8·97	6·12	69·47	36·703	473·3	48.
144·807	511·08	9·66	34·094	74·563	649·19	15·47	10·32	82·55	44·731	490·99	49.
133·31	533·23	9·594	38·376	80·073	643·97	20·54	13·82	89·81	47·563	498·64	50.
133·588	572·52	9·551	40·933	82·86	648·92	21·315	......	90·79	46·82	512·66	51.
310·978	621·95	8·811	17·622	41·145	654·37	4·864	2·51	56·05	24·038	370·49	52.
103·32	103·32	3·073	3·073	20·507	634·23	1·092	1·09	27·78	25·343	305·89	53.
58·712	117·42	3·4	6·8	43·703	644·5	2·946	2·95	45·91	41·49	354·74	54.
64·603	129·21	6·598	13·195	67·608	625·86	9·108	9·11	69·83	63·53	411·66	55.
44·447	177·79	6·164	24·656	89·197	646·9	24·947	21·95	95	84·51	523·85	56.
252·385	504·77	2·732	5·464	17·437	619·46	0·831	0·831	23·17	13·035	334·13	57.
248·947	497·9	5·567	11·135	31·841	639·27	1·622	1·592	34·73	20·136	446·82	58.
181·822	545·47	7·623	22·87	50·362	646·56	3·975	3·57	51·88	21·233	589·21	59.
132·56	589·14	9·687	43·055	81·526	647·47	19·83	14·83	88·89	47·807	564·26	60.
97·79	586·75	7·828	46·969	86·612	642·98	25·01	......	95·06	50·857	561·39	61.
158·2	474·59	2·466	7·398	25·952	680·13	1·115	1·065	28·13	15·8	381·96	62.
123·29	493·16	5·948	23·793	58·839	650·11	7·288	5·09	64·78	34·16	444·27	63.
90·42	542·5	7·458	44·747	87·877	636·57	27·27	19·97	97·42	50·112	494·05	64.
150·75	150·75	5·014	5·014	28·287	624·68	1·817	1·81	36·78	26·808	228·48	65.
92·44	184·88	6·524	13·047	56·1	633·66	5·448	4·848	58·44	48·162	412·24	66.
53·187	212·75	6·451	25·805	85·906	616·42	22·77	19·37	92·54	71·317	410·55	67.
14·785	29·57	1·608	3·216	65·113	598·77	10·06	......	72·13	67·185	115·51	68.
77·582	155·16	6·691	13·382	78·951	671·41	18·22	......	86·7	60·235	256·55	69.
139·27	417·82	7·394	22·183	81·184	694·63	21·53	......	91·05	42·08	330·53	70.
16·883	33·77	1·943	3·886	74·253	624·33	11·45	9·45	75·19	71·69	145·75	71.
197·03	394·07	7·271	14·542	66·22	644·6	9·28	9·28	70·25	28·73	252·19	72.
218·47	655·42	9·249	27·748	81·92	634·83	21·07	17·1	90·48	30·89	336·18	73.
16·109	32·218	2·268	4·535	84·59	598·78	16·56	14·26	84·26	79·53	138·2	74.
312·03	624·07	1·323	2·647	18·526	710·4	1·18	......	29·11	9·955	127·19	75.
328·03	656·07	2·133	4·266	15·613	712·49	1·145	1·425	28·59	10·65	212·67	76.

Description	Conditions	No.					
		77.	20	42·34	30	6·26	26·1
		78.	27½	38·29	7·4	6·2	57·67
		79.	30	37·1	0·63	6·22	89·19
		80.	60	36·7	1·0	5·0	60·7
		81.	30	35·77	13·4	5·8	15·14
		82.	30	34·8	18·7	5·32	13·46
		83.	30	35·04	1·3	5·0	90·68
Iron steam-tube 4 feet long, exterior diameter 0·74 inch, interior diameter 0·602 inch, mean area a=0·7026 sq. ft. Outer tube 0·87 inch internal diameter. Refrigerating water moving contrary to the direction of the steam.		84.	15	43·5	14·8	13·54	38·85
		85.	20	45·5	11	13·54	42·3
Copper steam-tube 4 feet long, area ·7225 sq. ft. Outer tube 0·87 inch interior diameter. A taper glass rod was placed in the axis of the steam-tube; its length was 40 inches, diameter at thick end ·55 inch, at thin end ·3 inch.	The taper glass rod with its thin end uppermost. Refrigerating water moving in the same direction as the steam.	86.	15	52·71	48	8·52	28·313
		87.	15	57·39	48	8·52	22·157
		88.	9½	58·91	48	8·47	34·233
		89.	10	56·43	48	8·42	28·909
	The taper glass rod with its thin end uppermost. Refrigerating water moving in a direction contrary to that of the steam.	90.	20	48·6	10·8	8·08	21·82
		91.	15	52·85	30·25	8·02	17·96
		92.	15	52·15	27·66	8·02	27·832
		93.	10	52·61	32·5	7·92	32·999
	The taper glass rod with its thick end uppermost. Refrigerating water moving in a direction contrary to that of the steam.	94.	15	48·25	32	7·67	19·308
		95.	15	49·12	25·66	7·67	13·325
		96.	11	48·6	25	7·54	37·22
Copper steam-tube 4 ft. long, area ·7225 sq. ft. A spiral consisting of 30 turns of copper wire $\frac{1}{17}$ of an inch diameter was wound round it. Half of the spiral was right-handed, the other half left-handed. Outer tube 1·4 inch diameter. Refrigerating water moving in a direction contrary to that of the steam.		97.	30	53·86	1·5	16·875	29·501
		98.	20	60·27	1	16·65	40·625
		99.	15	58·64	3	16·425	43·99
Copper steam-tube 4 feet long, area ·7225 sq. ft. Outer tube 1·4 inch diameter. Spiral of 45 turns of copper wire ·21 inch thick between the tubes. Refrigerating water moving contrary to the direction of the steam.		100.	30	47·42	1·95	15·547	25
		101.	30	51·8	2·06	15·66	32·456
		102.	20	59·45	1·43	15·547	44·634
		103.	14	55·96	4·7	15·48	45·034
Copper steam-tube 4 feet long, area ·7225 sq. ft. Outer tube 1 inch interior diameter. Between the tubes there was a spiral of 103 convolutions, composed of copper wire ·105 inch thick.	Refrigerating water moving in a direction contrary to that of the steam.	104.	30	39·99	21·94	14·085	36·025
		105.	30	42·01	24·36	14·04	61·013
		106.	20	45·15	245	13·95	37·583
		107.	15	42·1	270	13·59	48·768
		108.	12	45·63	257	13·59	51·97
		109.	50	50·35	66·9	13·567	33·782
		110.	80	40·6	58	13·545	46·27
	Refrigerating water moving in the same direction with the steam.	111.	30	50·14	48	12·44	23·135
		112.	30	56·3	48	12·44	40·29
		113.	30	57·66	48	12·6	54·32
Copper steam-tube 2 feet long. Interior diameter ·63 inch, exterior ·75 inch, mean area a=·3612 sq. ft. Outer tube, interior diameter 1 inch. Between the tubes there was a spiral consisting of 50 convolutions of copper wire ·105 inch thick.	Refrigerating water moving in the same direction as the steam.	114.	30	52·48	24	12·29	26·8
		115.	30	45·42	24	10·935	27·46
		116.	30	48·34	24	11·07	23·17
		117.	30	50·9	24	11·07	28·08
		118.	30	44·02	24	9·88	41·1
		119.	30	44·125	24	10·17	45·73
		120.	30	42·61	24	10·17	59·27
	Refrigerating water moving in a direction contrary to that of the steam.	121.	15	44·32	176	8·01	12·93
		122.	14	46·15	207	8·01	22·856
		123.	15	37·96	193	8·01	29·825
		124.	30	44·41	8·4	7·2	23·07
		125.	30	43·84	4·47	7·2	48·74
		126.	30	45·22	5·8	7·72	66·31
Copper steam-tube 4 feet long. Interior diameter ·63 inch, exterior diameter ·75 inch, mean area a=·7225 sq. ft. Outer tube, interior diameter 1 inch. Between the tubes there was a spiral consisting of 96 convolutions of copper wire ·105 inch thick.	Refrigerating water moving in a contrary direction to the steam.	127.	30	42·14	274	3·64	16·48
		128.	30	45·37	248	3·64	29·93
		129.	20	46·37	218	3·65	51·19
	Refrigerating water moving in the same direction as the steam.	130.	30	46·38	48	3·51	32·02
		131.	30	52·24	48	3·51	42·02
		132.	20	49·07	48	3·51	71·81
Copper steam-tube 6 feet long. Interior diameter ·63 inch, exterior diameter ·75 inch, mean area a=1·0837 sq. ft. Interior diameter of the outer tube 1 inch. Between the tubes there was a spiral consisting of 143 convolutions of copper wire ·105 inch thick.	Refrigerating water moving in a direction contrary to that of the steam.	133.	30	39·2	22·1	5·53	82·25
		134.	30	38·43	24	5·65	57·77
		135.	30	41·06	22·6	5·76	38·22
		136.	30	47·75	341	4·95	29·72
		137.	30	45·43	315	4·905	32·84
		138.	30	51·01	279	3·87	42·52
		139.	20	44·8	261	3·87	67·27
		140.	15	45·6	250	3·85	77·45
		141.	30	42·67	301	2·65	32·07
		142.	30	42·02	301	2·65	32·96
		143.	30	39·34	327	3·28	37·69
		144.	30	41·62	337	3·28	28·22
	Refrigerating water moving in the same direction as the steam.	145.	30	43·88	72	4·3	41·8
		146.	30	43·28	72	4·3	50·32
		147.	30	41·11	72	4·3	76·47
Iron steam-tube 4 feet long. Exterior diameter ·74 inch, interior diameter ·602 inch. Interior diameter of the outer tube ·87 inch. A spiral consisting of 55 convolutions of copper wire ·055 inch thick was placed between the tubes.	Refrigerating water moving in the same direction as the steam.	148.	30	37·65	48	4·72	31·11
		149.	30	37·84	48	4·72	61·27
	Refrigerating water moving in a direction opposite to that of the steam.	150.	30	40·33	282	4·2	21·32
		151.	20	40·57	265	4·2	42·7

264·66	793·98	9·063	27·19	78·337	656·49	15·73	......	82 96	26·1	365·43	77.
101·35	221·12	9·608	20·96	88·926	631 81	27·73	18·7	97 89	57·67	280·2	78.
15·472	30·94	2·635	5·27	88·67	575·78	25·27	18·07	95·34	89·19	127·19	79.
18·16	18·16	1·743	1·743	60·687	640·94	6·87	......	63·46	60·7	85·27	80.
162·72	325·44	2·366	4·73	20·088	662·31	29·79	......	99·88	15·14	......	81.
237·66	475·32	2·991	5·982	18·907	652·18	29·82	......	99·9	13·46	......	82.
11·472	22·944	1·642	3·284	35·552	634·12	30·02	......	100·1	90·68	......	83.
139	556	6·608	26·432	94·07	626·47	28·52	23·02	98·66	38·85	279·24	84.
152·2	456·6	7·917	23·751	89·09	641·98	29·54	22·64	99·64	42·3	264·16	85.
137·812	551·25	4·628	18·512	50·701	634·76	4·329	3·93	53·63	28·131	435·25	86.
323·68	1294·72	7·804	31·218	54·454	617·12	5·979	5·31	60·44	22·037	540·42	87.
204·25	1290	9·718	61·379	88·654	629·53	26·244	......	96·37	33·955	611·33	88.
222·94	1337·64	8·023	48·136	75·579	643·42	15·601	13·1	82·75	28·441	581·02	89.
195·1	585·3	4·408	13·224	30·844	631·12	4·354	3·4	53·74	21·346	278·07	90.
246·41	985·64	3·997	15·987	39·706	644·99	2·543	1·84	43·07	17·564	433·62	91.
240·97	963·88	8·176	32·706	67·187	649·06	11·122	9·39	74·49	27·434	460·81	92.
205·04	1230·24	9·166	54·993	86·363	646·65	23·188	20·19	93·03	32·607	583·32	93.
251·97	1007·88	4·682	18·726	38·804	658·63	6·201	2·15	61·23	18·828	325·87	94.
216·1	864·4	1·825	7·301	23·365	671·96	2·375	0·775	41·77	12·834	196·48	95.
172·67	941·84	9·021	49·208	81·147	651·34	25·779	19·28	95·88	36·556	519·07	96.
281·89	563·78	5·905	11·81	49·357	645·62	4·041	......	52·216	29·273	337·12	97.
220·546	661·64	9·103	27·31	75·963	654·81	14·45	......	80·84	40·6	427·66	98.
195·827	783·31	9·503	38·01	88·09	655·18	24·49	......	94·5	44·09	474·29	99.
301·116	602·232	4·377	8·754	33·382	670·11	1·545	......	33·86	24·792	586·25	100.
288·678	577·356	7·891	15·783	48·75	657·31	4·222	......	53·11	32·523	478·15	101.
182·162	546·486	9·125	27·375	74·834	653·01	13·65	......	79·44	45·14	470·53	102.
186·022	797·24	9·692	41·537	86·059	652	23·214	......	93·06	45·16	532·07	103.
86·996	173·99	3·15	6·3	36·208	642·14	1·343 ?	1·74	31·37 ?	36·441	......	104.
98·037	196·07	7·628	15·256	60·985	664·7	6·566	6·53	62·47	60·931	935·8	105.
208·775	626·32	7·965	23·894	43·349	657·79	2·287 ?	3·18	41·06 ?	37·136	......	106.
161·15	644·6	9·617	38·47	66·068	653·06	9·715	7·14	71·31	48·176	815·87	107.
132·525	662·62	8·68	43·4	73·931	657·8	13·518	11·42	79·2	51·715	798·15	108.
147 9	177·48	4·736	5·683	34·968	666·26	1·084 ?	1·784	27·65 ?	36·113	......	109.
145·15	108·86	7·836	5·877	41·66	647·84	2·391 ?	3·69	41·9 ?	48·23	......	110.
87·375	174·75	1·648	3·296	29·64	596·68	0·798 ?	0 898	22·5 ?	23·755	......	111.
73·343	146·68	3·698	7·396	45·52	597·88	1·57 ?	3·01	34·14 ?	41·13	......	112.
98·25	196 5	7·456	14·912	59·77	609·53	4·85 ?	6·95	56 ?	55·824	......	113.
84·44	168·88	2·108	4·216	31·9	613·1	0·92 ?	......	24·86 ?	26·856	......	114.
100·94	201·88	2·612	5·224	32·71	671·51	1·496	1 496	33·28	27·2	727·52	115.
100·87	201·74	1·902	3·804	29·99	671·7	1·116	1·686	28·14	23·27	700·44	116.
55·19	110·38	1·456	2·912	31·92	676·7	1·286	1·34	30·6	27·87	601·2	117.
86·82	173·64	4·413	8·826	48·46	662·67	3·44	4·34	48·97	41·07	768·58	118.
109·63	219·26	6·368	12·736	58·018	670·22	5·741	6·84	59·56	43·57	684·5	119.
102 74	205·48	8·708	17·416	79·96	659·25	15 924	17·124	83·27	59·39	636·38	120.
172·9	691·6	1·105	4·42	22·21	756·76	0·67	0·9	19·65	12·22	859·4	121.
199·65	855·6	4·845	20·704	54 14	661·32	4·57	......	54·75	22·406	872·28	122.
194·28	777·1	7·399	29·596	78·14	649·52	15·61	......	82·77	29·553	731·1	123.
67·9	135·8	1·732	3·464	27·87	650·02	1·126	0·946	28·3	23·17	531·59	124.
58·53	117·06	4·155	8·31	60·86	646·05	5·837	5·83	59·92	49·41	522·6	125.
67·03	134·06	7·295	14·59	80·14	618·59	17·033	17·03	84·97	66·56	532·54	126.
166·59	333·18	3·454	6 908	14·03	633·32	0·682	0·682	19·95	16·48	713·7	127.
173·59	347·18	7·395	14·79	28·68	645·81	1·699	1·94	35·56	29·93	833·78	128.
108·9	326·7	8·721	26·163	56·12	649·76	5·555	5·205	58·86	51·19	892·52	129.
65·44	130·88	3·141	6·282	33·32	627·3	1·722	1·722	35·8	32·02	388·56	130.
68	136	4·433	8·866	42·27	632·99	2·717	2·717	44·34	42·38	571·56	131.
47·87	143·61	5·906	17·718	74·47	628·06	12·173	11·173	76·65	71·81	539·75	132.
37·26	74·52	5·26	10·52	79·24	622·7	16·377	15·852	83·98	82·25	262·27	133.
35·14	70·28	3·099	6·198	56·4	647·72	5·159	......	57·29	57·77	......	134.
31·57	63·14	1·593	3·186	35·8	678·93	1·855	1·875	37·16	38·22	......	135.
133·01	266·02	5·266	10·532	23·82	649·47	1·345	1·245	21·4	29·72	676·6	136.
123·45	246·9	5·54	11·08	22·09	644·58	1·594	1·694	34·41	32·8	662·57	137.
124·64	249·28	7·894	15·788	33·17	643·42	2·634	3·034	43·74	42·52	802·02	138.
76·88	230·64	8·266	24·798	64·68	654·34	8·458	8·258	68·13	67·27	918·05	139.
60·01	240·04	7·627	30·508	72·49	651·58	13·293	12·49	78·79	77·45	891·3	140.
129·2	258·4	6·149	12·298	23·58	641·74	1·468	1·48	32·94	32·07	846·46	141.
125·57	251·14	6·116	12·232	28·55	650·86	1·596	1·55	34·44	32·96	710·76	142.
128·76	257·52	7·21	14·42	28·743	643·25	2·106	2·206	39·51	37·69	710·74	143.
125·82	251·64	5·018	10·036	19·973	645·31	1·2	1·2	29·4	28·22	719·16	144.
65·25	130·5	4·157	8·314	40·9	629·53	2·649	2·75	43·86	41·8	355·84	145.
63·18	126·36	5·043	10·086	49·6	626·15	3·922	3·72	51·6	50·32	420·87	146.
68·87	137·74	9·133	18·266	75·17	619·37	13·11	11·11	78·45	76·47	460·48	147.
59·5	119	2·571	5·142	32·82	643·56	1·601	1·6	34·49	31·11	368·43	148.
68·92	137·84	6·817	13·634	65·61	637·33	8·392	8	67·95	61·27	440·92	149.
169·64	339·28	4·857	9·714	28·05	626	1·412	1·41	32·25	21·32	455·08	150.
116·39	349·17	7·786	23·358	62·45	637·97	7·192	6·69	64·48	42·7	505·87	151.

conducting-power, increase as the space between the tubes, which serves to convey the refrigerating water, is contracted. It will also be noted that an increase of conduction likewise takes place when the quantity of water transmitted between the same tubes is augmented. I will begin by arranging the results so as to show the effect of altering the velocity of the refrigerating water.

Series 1.—Copper steam-tube. Water-space between tubes 0·325 inch.

No. of experiment.	Quantity of refrigerating water.		Conductivity.	
1.	97·6	345·5	98·1	139·6
2.	98·7		158·4	
5.	591		152·2	
7.	594·4		149·4	
8.	597	688·8	138·8	172·5
6.	631·8		184·3	
9.	658		151·6	
3.	758·9		195·7	
4.	798·2		191·9	

Series 2.—Copper steam-tube. Water-space between tubes 0·06 inch.

No. of experiment.	Quantity of refrigerating water.		Conductivity.	
29.	16·37	99·24	80·9	152·67
25.	18·07		29·26	
18.	20·41		89·62	
19.	43·57		143·9	
20.	61·16		155·9	
21.	76·5		279·8	
22.	124·5		198·5	
17.	189·47		193·8	
23.	343·1		202·3	
26.	592·3	689·04	297·4	334·31
30.	606·3		332·8	
33.	611·7		300·4	
24.	612·8		335·6	
32.	691·7		256·3	
27.	739·8		342·5	
28.	767·2		353·6	
31.	797·3		408·9	

Series 3.—Copper steam-tube. Water-space between tubes 0·025 inch.

No. of experiment.	Quantity of refrigerating water.		Conductivity	
44.	11·2	95·19	53·1	281·8
40.	21·4		84·5	
39.	57·5		255·6	
45.	45·3		257·1	
46.	91·2		239·1	
43.	99·4		289·4	
53.	103·3		305·9	
54.	117·4		354·7	
55.	129·2		411·7	
41.	133·4		303·4	
42.	155		303·1	
56.	177·8		523·8	
47.	455	536·17	198·5	457·06
63.	493·2		444·3	
58.	407·0		446·8	
57.	504·8		334·1	
49.	511·1		491	
48.	523·1		473·3	
50.	533·2		498·6	
59.	545·5		589·2	
51.	572·5		512·7	
61.	586·7		561·4	
60.	589·1		564·3	
52.	622		370·5	

Series 4.—Lead steam-tube. Water-space between tubes 0·05 inch.

No. of experiment.	Quantity of refrigerating water.		Conductivity.	
80.	18·2	74·42	85·27	164·1
68.	29·6		115·51	
79.	30·9		127·2	
74.	32·2		138·2	
71.	33·8		145·7	
69.	155·2		256·5	
78.	221·1		280·2	
72.	394·1	590·24	252·2	270·7
70.	417·8		330·5	
75.	624·1		127·2	
73.	655·4		336·2	
76.	656·1		212·7	
77.	794		365·4	

We deduce from the averages in Series 1......... $C \propto w^{\frac{1}{3\cdot26}}$.

" " " 2......... $C \propto w^{\frac{1}{2\cdot47}}$.

" " " 3......... $C \propto w^{\frac{1}{3\cdot57}}$.

" " " 4......... $C \propto w^{\frac{1}{4\cdot14}}$.

Suppose we take the average index, then $C \propto w^{\frac{1}{3\cdot25}}$ will express the influence of the quantity of refrigerating water on the conductivity with sufficient accuracy. But it is evident that this relation can only be relied on between certain limits, indicated pretty plainly by the experiments. The influence of a change in the quantity of refrigerating water is doubtless gradually lessened as the flow is increased; and ultimately, at a very high velocity, the conductivity must necessarily reach a constant value.

To find the influence of the extent of the water-space, successively narrowed by diminishing the diameter of the outside tube, we will select those experiments in which the flow of water was nearly the same in quantity.

Width of water-space between the tubes.	No.	Quantity of refrigerating water.		Conductivity.	
0·325 inch.	5.	591·01	614·46	152·24	155·27
	6.	631·85		184·3	
	7.	594·47		149·44	
	8.	597·01		138·79	
	9.	657·99		151·6	
0·060 inch.	24.	612·82	622·96	335·65	304·52
	26.	592·3		297·38	
	30.	606·33		332·85	
	32.	691·67		256·3	
	33.	611·69		300·41	
0·025 inch.	48.	523·12	554·22	473·3	488·34
	49.	511·08		490·99	
	50.	533·23		498·64	
	51.	572·52		512·66	
	52.	621·95		370·49	
	57.	504·77		334·13	
	59.	545·47		589·21	
	60.	589·14		564·26	
	61.	586·75		561·39	

Reducing the conductivity in each case to the flow of 618 lb. of water, by the rule just found, we deduce for the spaces ·325, ·06, and ·025, the conductivities 156, 303·7, and 504·4 respectively. Whence, for the circumstances of the experiments, it follows that

$$C \propto \frac{1}{S^{\frac{1}{2\cdot185}}}.$$

The above laws are neither exact, nor universal in their application, but they afford the means of estimating the probable amount of benefit to be anticipated from increasing the rapidity of the refrigerating stream in such tubes as I have employed, which are indeed of the dimensions most likely to be practically adopted.

I pass now to the consideration of the effect of cleanliness of surface. In the experiments 62, 63, and 64 the outside of the copper steam-tube was made greasy by rubbing it with oil. In the five immediately preceding these the tube was kept perfectly clean, so that the water readily adhered to it.

State of surface.	No.	Quantity of refrigerating water.		Conductivity.	
Clean.	57.	504·77	544·81	334·13	499·16
	58.	497·9		446·82	
	59.	545·47		589·21	
	60.	589·14		564·26	
	61.	586·75		561·39	
Greasy.	62.	474·59	503·42	381·96	440·09
	63.	493·16		444·27	
	64.	542·5		594·05	

The conductivity with the oiled tube, reduced to 544·8 lb. of refrigerating water by means of the relation we have deduced, will be 450·6. The closeness of this number to 499·16 shows that the influence of a greasy surface is inconsiderable.

The experiments 86 to 96 inclusive are proper to determine whether any effect can be produced by placing a solid in the axis of the steam-tube.

Description.	No.	Quantity of refrigerating water.		Conductivity.	
Thin end of the tapered rod uppermost.	90.	585·3	941·27	278·07	438·95
	91.	985·64		433·62	
	92.	963·88		460·81	
	93.	1230·24		583·32	
Thick end of the rod uppermost.	94.	1007·88	938·04	325·87	347·14
	95.	864·4		196·48	
	96.	941·84		519·07	

Selecting similar experiments, with the exception that the core was not present, we have:—

No.	Quantity of refrigerating water.		Conductivity.	
27.	739·8	721·76	342·5	339·39
28.	767·16		353·6	
31.	797·34		408·89	
32.	691·69		256·3	

The conductivity in the last instance, reduced to 940 lb. of refrigerating water, will be 367·1, a number which does not differ sufficiently from 439 and 347 to lead us to expect any practical advantage from narrowing the steam-space.

Let us now inquire into the effect of changing the direction in which the refrigerating water was transmitted. Its usual direction was contrary to that of the steam and condensed water; but by removing the pipe E (see figure) and pouring the water into the upper part of the outer tube C, it could be made to flow in the same direction. The experiments suitable for ascertaining the effect of changing the direction of flow are collected in the following Tables:—

Series 1.—Thickness of water-space 0·06 inch.

Direction of water.	No.	Quantity of refrigerating water.		Conductivity.	
Contrary to the steam.	24.	612·82	689·55	335·65	332·89
	27.	739·8		342·5	
	28.	767·16		353·6	
	30.	606·33		332·85	
	31.	797·34		408·89	
	32.	691·69		256·3	
	33.	611·69		300·41	
The same as that of the steam.	34.	657·25	1158·3	522·1	486·94
	35.	1244		466·96	
	36.	1292·2		474·32	
	37.	1302·1		491·56	
	38.	1296		479·77	

Series 2.—Thickness of water-space 0·025 inch.

Direction of water.	No.	Quantity of refrigerating water.		Conductivity.	
Contrary to the steam.	41.	133·45	136·04	303·45	367·11
	42.	155·04		303·08	
	53.	103·32		305·89	
	54.	117·42		354·74	
	55.	129·21		411·66	
	56.	177·79		523·85	
The same as that of the steam.	65.	150·75	182·79	228·48	350·42
	66.	184·88		412·24	
	67.	212·75		410·55	

Thus with the refrigerating water flowing in a direction opposite to that of the steam, we have the conductivities 332·89 and 367·11, whilst with the water flowing in the same direction as the steam we have the conductivities (referred to the same quantities of refrigerating water) 417·3 and 320·96. The means for the two directions are 350 and 369·13; whence we may conclude that the conductivity is little influenced by the direction in which the water flows.

We will now consider the influence of the kind of metal of which the steam-tubes were made. In the Table will be found results obtained with tubes of copper, iron, and lead:—

Metal.	No.	Refrigerating water.		Conductivity.	
Copper.	23.	343·1	640·25	202·28	314·43
	24.	612·82		335·65	
	26.	592·3		297·38	
	27.	739·8		342·5	
	28.	767·16		353·6	
	30.	606·33		332·85	
	31.	797·34		408·89	
	32.	691·69		256·3	
	33.	611·69		300·41	
Iron.	84.	556	506·3	279·24	271·7
	85.	456·6		264·16	
Lead.	70.	417·82	590·24	330·53	270·7
	72.	394·07		252·19	
	73.	655·42		336·18	
	75.	624·07		127·19	
	76.	656·07		212·67	
	77.	793·98		365·43	

The water-spaces around the copper, iron, and lead tubes were respectively ·06, ·065, and ·05 inch wide. By reducing all the mean results to the space ·06 and 640·25 lb. of water by means of the formulæ we have already deduced, we obtain for the conducting-power with the three tubes the numbers 314·4, 302·2, and 255·1 respectively. Taking into account the thickness of the metal, which was ·06 in the copper, ·069 in the iron, and ·125 in the lead tube, we arrive at the conclusion that the resistance to conduction through the metal itself is so small in comparison with the resistance at the bounding surface of the metal and through the adhering films of water (inside as well as outside of the steam-tube) as to be almost inappreciable.

We have seen that the tendency of the water flowing between the tubes is to adhere to their sides, and that a head of water of considerable height is required in order to give the water sufficient velocity to remove the adhering film rapidly. It seemed possible that part of the force due to the head might be employed for the purpose of agitating the water. I have not yet found an opportunity to construct an apparatus for this purpose; but it occurred to me that the same object might be attained by placing a wire bent into the form of a spiral between the tubes. By this means the water would be impelled in a spiral direction, which would contribute largely to the rapid intermixture of the partieles of water as they advanced. Accordingly, in experiments 97, 98, and 99 this arrangement was tried for the first time. The spiral (in these three experiments only) was half of it left-handed, and the other half right-handed, so that the rotatory motion produced by the first half was reversed in the second. Although the thickness of the wire which formed the spiral was only one third of the width of the water-space in which it was placed, the effect it produced was marked, as the following results testify :—

No.	Head of water.		Quantity of refrigerating water.		Conductivity.	
97.	1·5		563·78		337·12	
98.	1·0	1·83	661·64	669·58	427·66	413·02
99.	3·0		783·31		474·29	

If we contrast these results with those obtained with the same tubes unfurnished with spirals, we shall find :—

No.	Head of water.		Quantity of refrigerating water.		Conductivity.	
3.	2·27		758·9		195·68	
4.	2·4		798·16		191·86	
5.	0·94		591·01		152·24	
6.	1·33	1·43	631·85	661·34	184·3	166·27
7.	1·03		594·47		149·44	
8.	0·94		597·01		138·79	
9.	1·08		657·99		151·6	

proving that a great increase of conductivity was obtained by the use of the spiral, without entailing the necessity of a much higher head of water.

The effect of increasing the velocity of the spirally directed refrigerating water will appear from the following experiments :—

No.	Head of water.		Quantity of refrigerating water.		Conductivity.	
124.	8·4		135·8		531·59	
125.	4·47	6·22	117·06	128·97	522·6	528·91
126.	5·8		134·06		532·54	
121.	176		691·6		859·4	
122.	207	192	855·6	774·77	872·28	820·93
123.	193		777·1		731·1	

whence we find $C \propto (w)^{\frac{1}{4\cdot078}}$.

By classifying the experiments so as to show the comparative effect of transmitting the refrigerating stream in the same direction with, and opposite to, the steam and condensed water, we obtain the following Table :—

Description.	No.	Quantity of refrigerating water.		Conductivity.	
Copper steam-tube 2 feet long. Water in the same direction with the steam.	115.	201·88		727·52	
	116.	201·74		700·44	
	117.	110·38	185·4	601·2	686·44
	118.	173·64		768·58	
	119.	219·26		684·5	
	120.	205·48		636·38	
Copper steam-tube 2 feet long. Water moving in the opposite direction to the steam.	121.	691·6		859·4	
	122.	855·6		872·28	
	123.	777·1	451·87	731·1	674·92
	124.	135·8		531·59	
	125.	117·06		522·6	
	126.	134·06		532·54	
Copper steam-tube 4 feet long. Water in the same direction as the steam.	130.	130·88		388·56	
	131.	136	136·83	571·56	499·29
	132.	143·61		539·75	
Copper steam-tube 4 feet long. Water in the contrary direction.	127.	333·18		713·7	
	128.	347·18	335·69	833·78	813·33
	129.	326·7		892·52	
Copper steam-tube 6 feet long. Water in the same direction as the steam.	145.	130·5		355·84	
	146.	126·36	131·53	420·87	412·4
	147.	137·74		460·48	

TABLE (*continued*).

Description.	No.	Quantity of refrigerating water.		Conductivity.	
Copper steam-tube 6 feet long. Water in the contrary direction.	136.	266·02	250·17	676·6	770·85
	137.	246·9		662·57	
	138.	249·28		802·02	
	139.	230·64		918·05	
	140.	240·04		891·3	
	141.	258·4		846·46	
	142.	251·14		710·76	
	143.	257·52		710·74	
	144.	251·64		719·16	
Iron steam-tube 4 feet long. Water in the same direction as the steam.	148.	119	128·42	368·43	404·67
	149.	137·84		440·92	
Iron steam-tube 4 feet long. Water in the opposite direction.	150.	339·28	344·22	455·08	480·47
	151.	349·17		505·87	

The above mean results are collected and averaged as follows :—

Direction of stream.	Quantity of water.		Conductivity.	
With the steam.	185·4	145·54	686·44	500·7
	136·83		499·29	
	131·53		412·4	
	128·42		404·67	
Contrary to the steam.	451·87	345·49	674·92	684·89
	335·69		813·33	
	250·17		770·85	
	344·22		480·47	

The conductivities for the different directions of the flow of refrigerating water will therefore be

$$500{\cdot}7 \text{ and } \left(\frac{145{\cdot}54}{345{\cdot}49}\right)^{\frac{1}{4{\cdot}078}} \times 684{\cdot}89 = 554{\cdot}06.$$

The difference between the two values is not great. If we average them with the results obtained when the tubes were not furnished with spirals, we shall obtain the following result :—

Tubes employed.	Conductivity. Water going in the same direction.	Conductivity. Water going opposite to the steam.	Ratio of conductivities.
Plain	369·13	350	0·9482
Furnished with spirals..	500·7	554·06	1·1065
Mean	......	......	1·0273

showing a trifling advantage on the side of the arrangement in which the refrigerating water goes in a contrary direction to the steam and condensed water, which, however, is too small to be attributed to any thing beyond experimental errors.

The quantity of transmitted water being, *cæteris paribus*, nearly proportional to the square root of the height of the head, it is evident that the limit to the economical increase of the conductivity by diminishing the thickness of the water-space, or by increasing the velocity of the stream, is soon attained. Hence, as I have already observed, the importance of any method which promotes the rapid removal of the adhering film of water without necessitating a great initial pressure. I have arranged my results with reference to the head of water in the following Tables, so as to enable a comparison to be readily made in this respect between the plain tubes and those furnished with spirals.

Table I.—Plain Tubes.

Description.	No.	Head of water.		Conductivity.	
Copper steam-tube 4 feet long. Thickness of water-space 0·325 inch.	2.	− 0·15	0·56	158·46	157·83
	1.	0·2		98·13	
	5.	0·47		152·24	
	8.	0·47		138·79	
	7.	0·52		149·44	
	9.	0·54		151·6	
	6.	0·66		184·3	
	3.	1·13		195·68	
	4.	1·2		191·86	

TABLE I.—Plain Tubes (*continued*).

Description.	No.	Head of water.		Conductivity.	
Copper steam-tube 4 feet long. Thickness of water-space 0·06 inch.	18.	– 1·4		89·62	
	21.	– 0·49		279·85	
	19.	– 0·13	0·69	143·92	176·92
	20.	0·6		155·91	
	22.	1·19		198·48	
	17.	4·37		193·77	
	23.	6·45		202·28	
	90.	10·8		278·07	
	33.	11·48		300·41	
	30.	12·9	12·08	332·85	286·13
	24.	14·07		335·65	
	32.	14·15		256·3	
	26.	14·73		297·38	
Copper steam-tube 4 feet long. Thickness of water-space 0·06 inch.	31.	20·12		408·89	
	27.	21·45		342·5	
	28.	23·29		353·6	
	96.	25·0		519·07	
	95.	25·66	26·44	196·48	402·68
	92.	27·66		460·81	
	91.	30·25		433·62	
	94.	32·0		325·87	
	93.	32·5		583·32	
	86.	48		435·25	
	87.	48		540·42	
	88.	48		611·33	
	89.	48		581·02	
	34.	48	48	522·1	511·41
	35.	48		466·96	
	36.	48		474·32	
	37.	48		491·56	
	38.	48		479·77	
Copper steam-tube 4 feet long. Thickness of water-space 0·025 inch.	40.	– 0·1		84·55	
	44.	0·5	3·11	53·1	131·58
	45.	9·92		257·1	
	39.	12·8		255·6	
	43.	18·33	21·6	289·44	327·86
	55.	26·66		411·66	
	54.	28·6		354·74	
	46.	28·58		239·09	
	53.	28·74		305·89	
	56.	34·66	32·95	523·85	335·07
	42.	35·61		303·08	
	41.	37·16		303·43	
	65.	48		228·48	
	66.	48	48	412·24	350·42
	67.	48		410·55	

Table I.—Plain Tubes (*continued*).

Description.	No.	Head of water.		Conductivity.	
Copper steam-tube 4 feet	63.	193·5		444·27	
long.	49.	206·3		490·99	
Thickness of water-space	47.	210·2		198·52	
0·025 inch.	62.	211	208·6	381·96	422·18
	58.	211·3		446·82	
	50.	211·7		498·64	
	64.	216		494·05	
	61.	223·5		561·39	
	57.	231·4		334·13	
	48.	232		473·3	
	59.	233	240·7	589·21	486·49
	60.	235·9		564·26	
	51.	237·2		512·66	
	52.	292·06		370·49	

Table II.—Tubes furnished with Spirals.

Description.	No.	Head of water.		Conductivity.	
Copper steam-tube 4 feet long.	100.	1·95		586·25	
Water-space 0·325 inch.	101.	2·06	2·53	478·15	516·75
Spiral of 45 turns of wire	102.	1·43		470·53	
0·21 inch thick.	103.	4·7		532·07	
Copper steam-tube 2 feet	125.	4·47		522·6	
long.	126.	5·8	6·22	532·54	528·91
Water-space 0·125 inch.	124.	8·4		531·59	
Spiral of 50 turns of wire	115.	24		727·52	
0·105 inch thick.	116.	24		700·44	
	117.	24		601·2	
	118.	24	24	768·58	686·44
	119.	24		684·5	
	120.	24		636·38	
	121.	176		859·4	
	122.	207	192	872·28	820·93
	123.	193		731·1	
Copper steam-tube 4 feet	130.	48		388·56	
long.	131.	48	48	571·56	499·96
Water-space 0·125 inch.	132.	48		539·75	
Spiral of 96 turns of wire	129.	218		892·52	
0·105 inch thick.	128.	248	246·66	833·78	813·33
	127.	274		713·7	

TABLE II.—Tubes furnished with Spirals (*continued*).

Description.	No.	Head of water.		Conductivity.	
Copper steam-tube 6 feet long.	133.	22·1		262·27	
	145.	72		355·84	
Water-space 0·125 inch.	146.	72	72	420·87	412·4
Spiral of 143 turns of wire	147.	72		460·48	
0·105 inch thick.	140.	250		891·3	
	139.	261		918·05	
	138.	279		802·02	
	141.	301		846·46	
	142.	301	301·33	710·76	770·85
	137.	315		662·57	
	143.	327		710·74	
	144.	337		719·16	
	136.	341		676·6	

The averaged results of the preceding Tables are collected together as follows :—

Description.	Head of water.		Conductivity.	
	0·56		157·83	
Plain tube..............	0·69	1·44	176·92	155·44
	3·11		131·58	
Tube with spiral		2·53		516·75
Plain tube..............		12·08		286·13
Tube with spiral		12·44		528·91
Plain tube..............	48	48	511·4	430·91
	48		350·42	
	48		686·44	
Tube with spiral	48	48	499·96	532·93
	48		412·4	
Plain tube..............		240·7		486·49
	384		820·93	
Tube with spiral	246·66	277·18	813·33	801·7
	200·88		770·85	

The cause of the inferiority of the plain tubes may be attributed in some measure to a want of perfect concentricity and truth in the pipes, resulting in an irregular action of the refrigerating water, the greatest quantity of which would thus be transmitted through the widest part of the water-space. In the arrangement with spirals, the width of the water-space was too great for any such circumstance to have a sensible influence. I think, however, that the imperfections of the tubes and of their concentricity were not such as to account for the great advantage which appeared to be produced by the spirals in my experiments; and I therefore attribute it to the continuous intermixture of the particles of water favoured by that arrangement.

The following is a summary of the principal foregoing results:—

1st. The pressure in the vacuous space is sensibly equal in all parts.

2nd. In the arrangement in which the steam is introduced into a tube whilst the refrigerating water is transmitted along a concentric space between the steam-tube and a larger tube in which it is placed, it is a matter of indifference in which direction the water is transmitted. Hence,

3rd. The temperature of the vacuous space is sensibly equal in all parts.

4th. The resistance to conduction is to be attributed almost entirely to the film of water in immediate contact with the outside and inside surfaces of the tube, and is little influenced by the kind of metal of which the tube is composed, or by its thickness within the limits of ordinary tubes, or even by the state of its surface as to greasiness or oxidation.

5th. The narrowing of the steam-space by placing a rod in the axis of the steam-tube does not produce any sensible effect.

6th. The conductivity increases as the rapidity of the stream of water is augmented. In the circumstances of my experiments, the conduction was nearly proportional to the cube root of the velocity of the water; but at very low velocities it evidently increases more rapidly than according to this law, whilst at high velocities it increases less and less rapidly as it gradually approaches a limit determined by the resistance of the metal and of the film of water adhering to the inside surface of the tube.

7th. The conductivity increases so slowly in relation to

TABLE III.—Atmospheric Air, the refrigerating agent,

1.	2.	3.	4.	5.	6.	7.
Description.	No.	Duration of experiment, in minutes.	Total pressure of steam in the boiler, in inches of mercury.	Pressure required to propel the air, in inches of water.	Mean temperature of the refrigerating air. At its entrance (t).	At its exit (t_1).
Copper steam-tube 4 feet long. Exterior diameter 0·75 inch, interior 0·63 inch. Outer tube 0·8 inch interior diameter.	1.	60	73·3	231	13·83	94·12
	2.	60	72·16	201	13·83	90·49
	3.	60	82·1	228	19·03	99·4
The same copper steam-tube. Outer tube 0·87 inch interior diameter.	4.	48	62·74	31·8	13·18	81·64
	5.	60	73·16	31·48	14·4	86·3
The same copper steam-tube. Outer tube 1 inch interior diameter.	6.	60	51·23	1·36	10·94	80·64
	7.	60	43·58	3·5	12·53	76·83
	8.	60	41·64	3·5	13·86	73·7
	9.	60	41·51	5·52	11·74	72·84
	10.	60	46·53	5·52	11·38	72·86
The same copper steam-tube. Outer tube interior diameter 1·4 inch.	11.	48	43·67	5·52	10·26	42·07
	12.	60	53·16	5·52	8·8	43·44
	13.	60	48·9	5·52	9·47	44·2
	14.	60	42·33	1·28	10·48	48·47
	15.	60	49·05	1·3	10·93	49·88
The same tubes. A spiral of 30 turns of copper wire $\frac{1}{10}$ inch thick was wound round the steam-tube. Half of this spiral was right-handed, the other half left-handed.	16.	60	43·54	1·3	10·57	73·58
	17.	60	42·32	5·32	14·87	67·29
Copper steam-tube 2 feet long. Exterior diameter 0·75 inch. Outer tube 1·4 inch interior diameter. A spiral of 20 turns of copper wire 0·21 inch thick between the tubes.	18.	60	41·76	1·44	9·13	69·13
	19.	60	46·88	3·55	8·46	63·33
Copper steam-tube 1 foot long. Exterior diameter 0·75 inch. Outer tube 1·4 inch interior diameter. A spiral of 10 turns of copper wire 0·21 inch thick between the tubes.	20.	60	45·04	1·44	6·43	52·87
	21.	60	45·06	3·55	8·23	46·73

propelled in a direction contrary to that of the Steam.

8.	9.	10.	11.	12.	13.	14.	15.	16.	17.
Quantity of water, in pounds, equal in capacity for heat to the refrigerating air.		Weight of condensed water, in pounds.		Temperature of condensed water.	Total heat of steam.	Barometer, minus vacuum-gauge, or pressure in the condenser, in inches of mercury.	Temperature due to the pressure in the condenser, per Regnault's tables (t_2).	Conduction of heat per square foot of the surface of the steam-pipe, or $\frac{w}{a} \log \left(\frac{t_2-t}{t_2-t_1}\right)$	No.
In experiment.	Per hour (w).	In experiment.	Per hour.						
6·614	6·614	1·09	1·09	95·29	582·48	25·88	96	34·58	1.
5·622	5·622	0·754	0·754	93·55	665·15	21·19	90·63	49·08	2.
6·244	6·244	0·85	0·85	71	661·39	30·5	100·56	36·75	3.
5·28	6·6	0·69	0·86	71·68	595·55	22·744	92·5	18·16	4.
6·707	6·707	0·996	0·996	84·07	568·22	27·268	97·42	18·66	5.
4·562	4·562	0·661	0·661	90·31	571·35	27·26	97·41	10·36	6.
8·298	8·298	0·948	0·948	90·4	653·23	27·857	98·01	16·03	7.
8·3	8·3	0·865	0·865	83·97	658·16	25·446	95·52	15·17	8.
10·157	10·157	1·129	1·129	90·55	640·23	26·328	96·45	17·94	9.
10·157	10·157	1·133	1·133	92·22	643·37	26·25	96·38	18·06	10.
25·208	31·51	1·375	1·719	68·68	651·86	27·33	97·49	19·78	11.
32·085	32·085	2·076	2·076	98·43	633·8	30·01	100·08	21·19	12.
32·14	32·14	2·08	2·08	97·73	634·37	30·05	100·11	21·49	13.
14·97	14·97	1·156	1·156	98·25	590·21	30·03	100·1	11·43	14.
13	13	1·006	1·006	97·21	600·56	30·146	100·21	10·31	15.
8·4	8·4	1·109	1·109	97·37	574·63	30·09	100·16	14·13	16.
18·5	18·5	1·92	1·92	99·39	604·48	30·366	100·42	24·29	17.
4·64	4·64	0·548	0·548	102·61	610·64	30·06	100·13	13·83	18.
7·744	7·744	0·748	0·748	102·55	670·61	30·08	100·15	19·55	19.
5·578	5·578	0·506	0·506	96·86	608·81	30·04	100·11	21·14	20.
9·122	9·122	0·629	0·629	99·12	657·47	30·06	100·13	27·42	21.

the height of the head of water, that the limit to the economical increase of the latter is soon attained.

8th. By means of a contrivance for the automatical agitation of the particles of the refrigerating stream, such as the spirals I have employed, an improvement in the conductivity for a given head of water takes place.

9th. The total heat of steam above 0° Cent., determined by the average of the 151 experiments, is 644°·28 for a pressure of 47·042 inches.

The experiments above recorded in which air was employed as the refrigerating agent were made in a similar manner to those in which water was used. At high pressures the air was propelled by the condensing pump used by Professor Thomson and myself in our experiments; and at low pressures a large organ-bellows was employed. The temperature of the air at its exit was obtained by placing the thermometer immediately over the concentric space between the tubes, varying its position from time to time so as to obtain an average result for the entire section of the channel.

On examining the Table of results with air as the refrigerating agent, we may remark,—

1stly. That a film of air does not *adhere* to the surface of the tube so tenaciously as a film of water does. This is evident from a comparison of Nos. 1, 2, and 3 with 4 and 5; from which it appears that for the spaces ·025 and ·06 inch the pressures able to propel equal quantities of air were as 7·66 to 1, or nearly as the squares of the velocities. When water was employed in the same tubes, these pressures were as 18·8 to 1.

2ndly. That the velocity of the elastic fluid appears to exercise a much more considerable influence on the conductivity than it does in the case of water.

3rdly. That spirals exercise a beneficial influence. This will be noted on comparing Nos. 6 to 15 with Nos. 16 and 17.

The very small conductivity when air is the refrigerating agent will probably prevent its being employed for the condensation of steam, except in very peculiar cases.

I must remark, in conclusion, that the above research, however laborious, has left much to be accomplished. One of my chief objects was to obtain figures which might prove useful to practical men; and I have therefore confined myself to such tubes as were most likely to be generally used. In taking up the subject afresh, greater accuracy might be attained by the use of a sheaf of tubes, so as, by condensing a larger quantity of steam, to diminish the amount of temperature-corrections. It would also be desirable to employ tubes of great thickness, so as to obtain the conductivity of metals after eliminating the resistance of the fluid film. The effect of irregularities in the water-space might also be more exactly determined, and the action of arrangements for agitating the refrigerating water might be more exhaustively treated.

Notice of a Compressing Air-Pump.
By J. P. Joule, *LL.D. &c.*

[Proceedings of the Manchester Literary and Philosophical Society, vol. iii. p. 5. October 21, 1862.]

The author referred to the difficulties of realizing in practice the theoretical advantages of the air, or the superheated steam-engine. The abrasion which takes place when metal rubs against metal, without an intermediate lubricator, speedily destroys the cylinder. He believed that the necessity of using elastic packing would not exist if the length of the channel along which the elastic fluid must pass in order to arrive at the opposite side of the piston were sufficiently increased. This might be accomplished by increasing the depth of the piston, or by placing on the rim of the piston concentric rings to enter, at the beginning and ending of each stroke, corresponding concentric grooves in the cover of the cylinder.

The principle of great depth of piston, as a substitute for packing, has been successfully carried out in the pump which was the subject of this communication. The cylinders, two in number, are 20 inches long and 2 inches in diameter. The pistons are solid cylinders of iron, 10 inches long, fitting as accurately to the cylinders as is consistent with freedom of motion. The depth of each piston, as compared with its diameter, renders the usual guide or parallel motion unnecessary; so that the connecting rod is simply jointed to the top of the piston. Air is readily compressed by this pump to sixteen atmospheres, the quantity passing the sides of the cylinders being very trifling.

Note on a Mirage at Douglas. *By* J. P. Joule, *LL.D., F.R.S.*

[Read before the Manchester Literary and Philosophical Society, January 13, 1863. Proceedings of the Society, vol. iii. p. 39.]

Dr. Joule described a peculiar kind of mirage which he had witnessed from the northern extremity of Douglas Bay, Isle of Man, and at an elevation of 20 feet above the surface of the water. At about 9 a.m. a number of chimneys in the town were fired; and the products of combustion were driven out to sea by a very gentle breeze. Presently the wind changed and drove the smoke at right angles to its former tract. The steamer for Liverpool had in the meantime attained a distance of about three miles, and, although somewhat obscured, was distinctly visible through the smoke. Noticing something extraordinary in her appearance, he viewed her through a telescope, and then observed that nearly the whole of the hull was obscured by the horizon, the uppermost part of the paddle-boxes, the bowsprit, and tafferel being alone visible, whilst at the same time the masts and funnel appeared considerably elongated. A quarter of an hour afterwards, when the smoke had cleared away and the steamer was about seven miles distant, the hull was seen as usual, quite unobscured by the horizon. He attributed the phenomenon to the gradual increase of the density of the refracting gases with elevation from the sea-level to that of the smoke.

Dr. ANGUS SMITH said:—I saw once a very remarkable instance of the diversion of the rays of light from the straight line. Although fogs are common on our hills, I have met no instance of similar exaggeration of effect among us. I went with some friends up Skiddaw, and near the top entered a cloud which prevented us from seeing many yards before us, although it was not extremely dark. When moving over those loose stones which form a highway to the summit for a considerable distance, we observed a building which appeared to us about 14 feet high. The side towards us seemed to be a wall nearly square; and we took it for granted that the top of the hill was attained, and that the foremost of the party would be found there; but as we approached the building sank, and in the course of a few steps it went downwards, until we found that there were only three layers of stones not very thick, and the whole under 2 feet high. The disappearance must have occurred within the space of 20 feet; my present impression is that it was less. Further on we attained the man at the summit, and the cloud became denser; we could not see many feet, and I did not think it safe to move from the spot for some time, not knowing the locality. Sitting there, I saw four perpendicular lines on the cloud: I looked up, and found that they terminated in the body of an animal which was moving slowly towards us. The distance could not be measured, except by the number of steps taken by the animal; and from these I concluded that when first seen it was from 12 to 20 feet from us; I looked on 12 as being most probable. I do not exaggerate when I say that the height was in appearance 30 feet. Wondering what was to be the end of this strange vision, I called the attention of all; but the animal diminished in size so rapidly that only one or two, who instantly attended to my call, could perceive the monstrousness of the exaggeration of form now presented by a moderate-sized pointer. On coming up to the more advanced portion of our party, they were enlarged and distorted, as we often hear described; but such effects are comparatively common, and they are not to be compared with the two instances mentioned. The first was seen with my face towards the sun, nearly. It was about midday. The second phenomenon was seen twenty minutes later, when looking nearly north.

Although I have lived much in sight of Morven, where one might expect to see the "spirit of the mist on the hills," I had not seen any thing similar before, nor have I learned from the shepherds that such things are common. Perhaps, if they were common, Ossian's ghosts would be less imposing.

On a Sensitive Barometer.

By J. P. Joule, *LL.D., F.R.S.*

[Read before the Manchester Literary and Philosophical Society, February 10, 1863. Proceedings of the Society, vol. iii. p. 47.]

Dr. Joule described a barometer for measuring small atmospheric disturbances. It consists of a large carboy connected by a glass tube with a miniature gasometer formed by inverting a small platinum crucible over a glass of water. The crucible is attached to the short end of a finely suspended lever multiplying its motion six times. When the apparatus was raised 2 feet the index moved through 1 inch; hence, in serene weather, he could observe the effect corresponding to an elevation of less than an inch. The barometer is placed in a building, the unplastered slated roof of which affords, without perceptible draught, free communication with the external atmosphere. In this situation it was found that the slightest wind blowing outside caused the index to oscillate, a gale occasioning oscillations of 2 inches, an increase of pressure being generally observed when the gusts took place.

On a Sensitive Thermometer. *By* J. P. Joule.

[Read before the Manchester Literary and Philosophical Society.
Proceedings of the Society, vol. iii. p. 73.]

Dr. Joule made the following communication respecting a new and extremely sensitive thermometer:—"Some years ago I remarked the disturbing influence of currents of air on finely suspended magnetic needles, and suggested that it might be made use of as a delicate test of temperature. I have lately carried out the idea in practice, and have obtained results beyond my expectation. A glass vessel in the shape of a tube, 2 feet long and 4 inches in diameter, was divided longitudinally by a blackened pasteboard diaphragm, leaving spaces at the top and bottom, each a little over 1 inch. In the top space a bit of magnetized sewing-needle, furnished with a glass index, is suspended by a single filament of silk. It is evident that the arrangment is similar to that of a 'bratticed' coal-pit shaft, and that the slightest excess of temperature on one side over that on the other must occasion a circulation of air, which will ascend on the heated side, and, after passing across the fine glass index, descend on the other side. It is also evident that the sensibility of the instrument may be increased to any required extent, by diminishing the directive force of the magnetic needle. I purpose to make several improvements in my present instrument; but in its present condition the heat radiated by a small pan containing a pint of water heated 30° is quite perceptible at a distance of 3 yards. A further proof of the extreme sensibility of the instrument is obtained from the fact that it is able to detect the heat radiated by the moon. A beam of moonlight was admitted through a slit in the shutter. As the moon (nearly full) travelled from left to right the beam passed gradually across the instrument, causing the index to be deflected several degrees, first to the left and then to the right. The effect showed, according to a very rough estimate, that the air in the instrument must have been heated by the moon's rays a few ten-thousandths of a degree, or by a quantity no doubt the equivalent of the light absorbed by the blackened surface on which the rays fell."

Note on the Meteor of February 6, 1818. *By* J. P. Joule, *LL.D.*

[Read before the Manchester Literary and Philosophical Society, December 1, 1863. Proceedings of the Society, vol. iii. p. 203.]

This meteor is noticed in Mr. Greg's very complete list of these phenomena in the British-Association Report for 1860. The account of it is given at greater length, however, in the 'Gentleman's Magazine,' a periodical which I think is not included in the large number of works consulted by Mr. Greg. This meteor is one of the few which have been seen in the daytime, and is also interesting as having been one of the first whose observation afforded materials for the estimation of its altitude. In the work just referred to it is described as follows:—

"At 2 o'clock P.M. a large and luminous meteor was seen descending vertically from the zenith towards the horizon, in the northern part of the hemisphere, by persons in the neighbourhood of Cambridge. The most remarkable circumstance attending the phenomenon is, that it was thus visible in broad daylight, the sun shining at the same time in great splendour, in a cloudless sky. The same meteor was seen at Swaffham, in Norfolk, at the same hour. It was seen also at Middleton Cheney, near Banbury, in the county of Northampton, not in the zenith, but perhaps 45° from it, in the north-eastern quarter of the heavens, shooting along towards the north. It seemed to be divided into two before it became extinct."

The distance between Middleton Cheney and Cambridge in a straight line is 61 miles, which therefore will be the elevation of this meteor above the surface of the earth according to the above observation.

The phenomena presented by meteors have for a long time occupied a good deal of my attention. Many years ago I published my opinion that they arise from the resistance of the atmosphere to the motion of bodies which, wandering through space, become entangled by the earth's attraction. I endeavoured to show that at even a very considerable elevation there would be sufficient air to cause the liquefaction and even vaporization by heat of bodies moving at a velocity so enormous as must be assigned to them if arrested in their motion through space. Latterly new facts have been collected by the Committee of the British Association, and also new experiments in physics have been made, bearing on the subject. My object is now to investigate these to see how far they confirm, oppose, or modify my original opinion.

In the Report of the Association for 1862 we have the following results for five remarkable meteors :—

Date.	Height at Beginning.	Height at End	Velocity.
1861—July 16.	195	65	55
Aug. 6.	126	21	35
Nov. 12.	95	20	48
Nov. 19.	55	30	24
Dec. 8.	110	45	23
1862—Jan. 28.	44	47	46
Feb. 2.	190	15	39
Average	116	35	39

Simultaneous observations at Hawkhurst and Cambridge of the meteors of August 10th of the present year give for the average of 10 :— 81 ...... 66 34

The above results give us a very good idea of the height and velocity of meteors. The observations are attended by some difficulty, so that errors are liable to occur in individual instances ; but we may rely on the general statement that the larger meteors have their origin at a higher elevation than the smaller ones, and that they attain a nearer distance from the earth at the end of their course.

In addition to the most valuable facts collected by Mr.

Herschel, we have a telescopic observation by Schmidt, of the Observatory of Athens, of a splendid meteor on the 19th of October, at 2.55 A.M. Mr. Schmidt was occupied in observing shooting-stars, and saw the one in question at its commencement. At first it appeared like a star of the fourth magnitude; after two seconds it was of the second magnitude; at the third and fourth second it surpassed the splendour of Sirius. It slowly passed towards the west, appearing a dazzling meteor of 10′ to 15′ diameter. At this moment M. Schmidt followed the meteor with his telescope for fourteen seconds. Its appearance was most remarkable. There were two brilliant bodies of a yellowish green in the form of elongated drops, each followed by a well defined tail of reddish colour. These were followed by smaller luminous bodies of the same shape, each followed by its red trace. The meteor disappeared at an elevation of 1° above the horizon. Four minutes afterwards M. Schmidt still observed the remains of the meteoric train, of a yellowish white and covering an area of nearly 5°.

The above observation is extremely important. It shows, firstly, that meteors are probably for the most part of a compound nature, consisting of a greater or less number of bodies associated together by the force of gravitation—small systems in fact; and, secondly, that they present evident marks of fusion, either in the whole or in part, after passing some time through the rarer portions of our atmosphere.

The researches by Professor Thomson and myself enable us to arrive at reliable conclusions with respect to the temperature acquired by bodies moving rapidly through the air. The law is very simple, viz.:—The temperature ultimately acquired by the moving body is the equivalent of the force with which the particles of air come into contact with it. It is, in fact, the temperature acquired by each particle of air on being caught and suddenly dragged on. This temperature, cleared from the effects of radiation, is 1° C. for a velocity of 145 feet per second, and goes on increasing with the square of the velocity.

Hence we find that the ultimate temperature acquired by

a body moving through air of whatever density is for the velocity of 39 miles per second 2000000° C. The question to be solved is, whether at the known height of meteors, as above stated, the density of the air is sufficient to give rise to effects in quantity sufficient to account for the actual phenomena.

Now, if we reckon the density to decrease to one quarter per elevation of seven miles, we shall find the quantity of air in a column of, say, a mile long and one square foot section, to be about ·0003 of a grain at the height of 116 miles, the elevation at which meteors in general are first observed. The temperature acquired at the surface of a meteorite of a foot section and of the specific heat ·23, moving at the average velocity at the average highest elevation through one mile, will be ·0003 gr. raised 2000000° C., or otherwise ·2 gr. raised 3000° C., which would be doubtless able to fuse any known substance, and bring it to a condition of dazzling brilliancy. The meteorite of a foot section might, I believe, have $\frac{1}{5}$ of a grain of its surface brought to this condition in its passage of one mile in the $\frac{1}{39}$ of a second. I do not think that the luminous effect would be neutralized by conduction of heat to the interior of the meteoritc, as it is very likely that the spheroidal condition would be produced. I can therefore easily believe that the $\frac{1}{5}$ grain at 3000° C. would give sufficient light to attract an observer at the distance of 100 or 200 miles.

From data given in Herschel's 'Outlines of Astronomy' I find that, at the elevation and velocity above stated, viz. 116 and 39, and supposing as before the entire effect to be given out by radiation, a meteorite of 5 feet diameter would have the brilliancy of *α* Centauri.

As the meteorite descends towards the earth its brilliancy will increase to a certain point, when, from the quantity of fused matter, a longer tail will be left behind. This process will of course be sustained for a longer time by the larger meteorites, which are thus enabled to penetrate to a nearer distance from the earth's surface.

On a Method of Hardening Steel Wires for Magnetic Needles. *By* J. P. Joule, *LL.D.*

[Read before the Manchester Literary and Philosophical Society. Proceedings of the Society, vol. iv. p. 28.]

Dr. Joule described the process he employed to harden steel wires for magnetic needles. The wire is held stretched between the ends of two iron rods bent into a semicircular shape. The free ends of the iron rods can be placed in connexion with a voltaic battery by means of mercury-cups. Underneath the steel wire a trough of mercury is placed. When the ends of the iron rods dip into the cups the current passes through the wire, heating it to any required extent. Then when these ends are lifted the current is cut off, while at the same instant the heated wire becomes immersed in the trough of mercury.

On an Instrument for showing Rapid Changes in Magnetic Declination. *By* J. P. Joule, *LL.D., F.R.S.*

[Read before the Manchester Literary and Philosophical Society, March 21, 1865. Proceedings of the Society, vol. iv. p. 131.]

The adjoining sketch represents a section half the real size of this instrument*. *a* is a column of small magnetic needles suspended by a filament of silk. Attached to its lower end is a glass lever *b*, with a hook at its end. A fine glass lever *c* is suspended by a single filament of silk, its shorter arm being connected with the first lever by means of a small hook attached to the fibre *d*. The whole is enclosed in a stout copper box, into which light is admitted through a

* A piece of watch-spring magnetized in the direction of its breadth was first employed, see vol. iv. p. 37.

lens *e*, cemented into an orifice immediately under the object-glass of the microscope *f*.

The microscope magnifies about 300 linear, and has a micrometer in its eyepiece, with divisions corresponding to

Fig. 94.

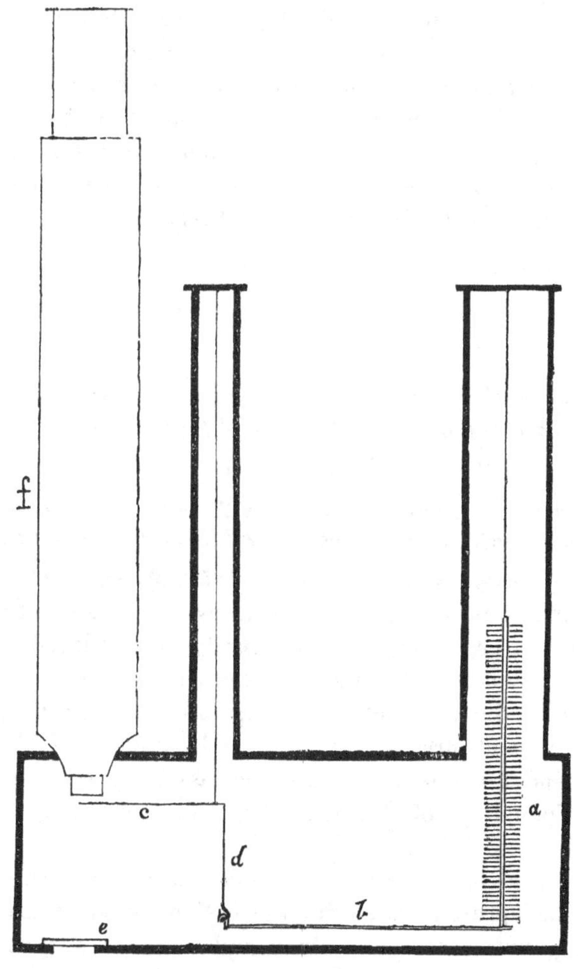

$\frac{1}{2000}$ of an inch. One division corresponds to a deflection of the needle of $4\frac{1}{2}''$; and as a tenth of a division can be very readily observed, the instrument measures deflections to within half a second. So rapid is the action, that on applying a small magnetic force the index takes up its new position steadily in two seconds of time. Besides being a damper to the motion of the needle, the copper box, by its conducting power, equalizes the temperature rapidly, so that the indications are not to any considerable extent disturbed by currents of air. The success of the present instrument encourages the hope that very much greater delicacy may be obtained by a further multiplication of the motion and the use of a more powerful microscope. Dr. Joule stated that he had observed an extensive magnetic disturbance the preceding evening, the index being driven entirely out of the field of view*

Determination of the Dynamical Equivalent of Heat from the Thermal Effects of Electric Currents. By J. P. Joule, *D.C.L., F.R.S., &c.*

[Report of the British Association for the Advancement of Science Dundee, 1867. Committee on Standards of Electrical Resistance.]

Sir W. Thomson, as long ago as 1851, showed that it was desirable to make experiments such as are the subject of the present paper. They have necessarily been delayed until a sufficiently accurate method of measuring resistance was discovered. Such a method having been described by Sir William, and carried out in practice by Professor J. C. Maxwell and his able coadjutors, the task assigned to me by the Committee of Electric Standards was comparatively simple.

My experiments were commenced nearly two years ago; and the apparent ease with which they could be executed gave promise of their early completion. It was, however

* Further instances of the use of the instrument are given in the Proceedings, vol. v. p. 15. On one occasion thirty-six changes of deflection, varying from 10″ to 1′ 40″ occurred per minute.

found essential that careful observations of the earth's hori zontal magnetic intensity should be frequently made; and these required the construction of apparatus whereby this element could be determined with accuracy and rapidity.

The apparatus finally adopted for this purpose consists of a suspended horizontal flat coil of wire between two fixed similar coils. A current of electricity can be made to traverse all three, communication with the suspended coil being made by the suspending wires themselves according to Sir W. Thomson's plan. The strength of a current is found by observing the sum of the forces of attraction and repulsion by which the suspended coil is urged. The strength of a current can in this manner be determined in absolute measure; for, the area of each of the three equal coils being called a, the weight required to counterpoise the force with which the suspended one is urged w, the force of gravity g, and the length of wire in each of the coils l, the current $c=\frac{1}{2l}\sqrt{\frac{a\,g\,w}{\pi}}(1+\text{correction})$, the correction being principally due to the distance between the fixed coils. In my instrument, in which this distance is 1 inch, the diameter of the coils being 12 inches, and their interior core 4 inches, this correction was proved by experiment to be ·1185.

There was, however, considerable difficulty in obtaining an exact measure of the distance between the fixed coils; and I therefore judged that the measure of the currents used in the experiments would be most accurately obtained by means of a tangent galvanometer, the above-described current-meter being employed to determine the horizontal intensity.

This determination was effected as follows:—Many careful observations of the horizontal intensity by an improved method on Gauss and Weber's system were made alternately with observations of the deflections of a tangent galvanometer and the weighings of the current-meter when the same currents traversed both instruments in succession. Then, calling the horizontal intensity H, the angle of deflection θ, and the weighing w, there was obtained a constant

$c = \frac{H \tan\theta}{\sqrt{w}} = \cdot 17676$. Hence with these instruments

$$H = \frac{\cdot 17676\sqrt{w}}{\tan\theta}.$$

The experiments for the determination of horizontal intensity by the use of this formula could be effected in a few minutes, and did not require an alteration in the disposition of any part of the apparatus. It was satisfactory to find that, although the presence of masses of iron at only a few yards distance made the field in which I worked considerably more intense than that due to the latitude, and although I worked at different times of the day, the highest intensity, out of upwards of seventy observations distributed over a year, was 3·6853, and the lowest 3·6607, indicating a much greater degree of constancy than might have been expected.

The galvanometer above mentioned was that employed in the thermal experiments. It had a single circle of $\frac{1}{10}$-inch copper wire, the diameter of which, being measured in many places by a standard rule, gave a radius of ·62723 of a foot. The needle was half an inch long, and furnished with a glass pointer traversing a divided circle of 6 inches diameter. In the experiments the deflections were not far from 26° 34′, the angle at which the influence of the length of the needle within certain limits is inappreciable. It was easy by a magnifier, arranged so as to avoid parallax, to read to one minute. The torsion of the fibre gave only 3′·5 for an entire twist. The trifling correction thus required is applied to the recorded observations of deflection.

The calorimeter first used was a copper vessel upwards of a gallon in capacity, filled with distilled water. It had a conical lid, attached by screws, in which were two tubulures, one for the introduction of a copper stirrer, the other for the thermometer, around the immersed stem of which a wire of platinum-silver, having a resistance nearly equal to that of the Association unit, was coiled.

The resistance of the wire was found by comparing it with the Association unit, sent me by the Committee, using Ohm's formula, $x=\frac{C_2}{C_1}\left(\frac{C_3-C_1}{C_3-C_2}\right)$, where C_3, C_2, and C_1 are the tangents of deflection with the battery and connexions only, with these and the unit, and with the coil respectively. This, though by no means so delicate a method as that of the Wheatstone balance improved by Thomson, was able to give a final result certainly accurate to the two-thousandth part. The results for the resistance of the coil in the first series of experiments are as follow. They were obtained before and after those experiments. A large galvanic cell, consisting of cast-iron and amalgamated zinc plunged in dilute sulphuric acid, was the source of electricity, which was measured by a galvanometer with a coil of nine turns, 17 inches in diameter.

C_3.	C_2.	C_1.	Temperature of unit.	Temperature of coil.	Resistance of coil in terms of unit.
tan 55° 6′·75	tan 28° 18′	tan 28° 1′·3	63°·7	62°·65	1·01901
tan 59 32·5	tan 32 39 6	tan 32 22	59·24	58·39	1·01825

The average resistance 1·01863 being reduced from the temperature 14°·5 Cent., at which the unit was adjusted, to 69°·9 Fahr., the average temperature of the calorimeter in the first series of experiments, becomes 1·0191, which, multiplied by 32808990, gives 33435640 as the resistance in British absolute measure.

A delicate thermometer was placed at a few inches distance from the calorimeter, for the purpose of registering the temperature of the air. In the Tables its indications are reduced to the scale of the instrument plunged in the calorimeter. A string attached the handle of the stirrer to a stick, so that the water could be effectually stirred without communicating the heat of the hand. A wooden screen separated the observer from the apparatus.

In the experiments of the first series a battery of five large Daniell's cells, arranged in series, transmitted the current through the coil for 40 minutes exactly, determined by chronometer. During this time twenty-eight observations of deflection were obtained, seven at each end of the pointer directed N.E. and S.W., and seven when it was directed N.W. and S.E., by reversing the current in the galvanometer for the latter half of the time. The water was stirred twenty-eight times. Its temperature was taken at the beginning, middle, and end of an experiment. There were also fourteen observations of the temperature of the air.

Immediately after each experiment the horizontal intensity of magnetic force was obtained by observing the deflection of the galvanometer and the weighing of the current-meter produced by the same current.

Before and after each experiment, two others were made in precisely the same manner, but excepting the current, in order to discover the influence of radiation and the conducting-power of the atmosphere.

First Series of Thermal Experiments.

Date.	Deflection.	$\tan^2$ Deflection.	Temperature of air.	Temperature of water.	Rise of temperature.	Horizontal intensity.
1866.						
Aug. 22 ...	32° 46·86	·414719	492·36	497·42	23·55	3·6763
„ 23 ...	34 0·29	·455133	494·77	493·27	25·65	3·6815
Sept. 8 ...	32 24·83	·403156	400·4	401·8	22·8	
„ 10 ...	31 50·22	·385542	441·11	433·85	22·21	3·6737
„ 11 ...	31 31·02	·376024	367·0	392·89	18·51	3·6758
„ 12 ...	31 14·42	·367944	344·33	344·45	21·9	3·6656
„ 13 ...	30 57·51	·359850	361·54	358·47	20·95	3·6671
„ 15 ...	30 24·86	·344607	346·7	330·01	21·98	3·6638
„ 15 ...	30 20·51	·342610	381·41	367·56	21·07	3·6711
„ 18 ...	30 34·34	·348982	342·64	324·32	22·29	3·6607
Average ...	....	·379857	397·226	394·406	22·0914	3·67073

First Series of Radiation Experiments.

Date.	Temperature of air.	Temperature of water.	Rise of temperature of water.
1866. Aug. 22	495·93	469·14	2·88
,,	502·22	477·83	3·15
Aug. 23	476·37	458·96	3·08
,,	490·81	499·22	−0·55
Sept. 8	393·5	382·75	2·0
,,	395 82	414·15	−1·7
Sept. 10	444·31	419·4	2·9
,,	437·15	396·96	4·83
Sept. 11	373·07	384·72	−0·63
,,	367·14	391·76	−1·75
Sept. 12	334·0	332·42	0·44
,,	365·34	360·2	1·6
Sept. 13	352·82	343·11	1·83
,,	366·65	369·16	−0·08
Sept. 15	330·78	315·41	2·78
,,	381·47	347·14	3·72
,,	378·93	350·67	3·34
,,	381·05	379·51	0·22
Sept. 18	326·99	309·28	2·55
,,	339·9	338·35	0·04
Average	373·058	364·686	1·3806

In applying the preceding Table for the purpose of correcting the results of the thermal experiments, it must be first observed that the external influences on the calorimeter do not amount to zero when the temperature of the air-thermometer coincides with the indication of that immersed in the calorimeter. This might arise partly from the locality of the two instruments not being the same, but was, I found, principally owing to the different radiating and absorbing powers of the air-thermometer bulb and of the surface of the calorimeter. Taking, then, the number of instances in which the temperature of the air appeared to exceed that of the water, there are fifteen, with a total excess of 259·63 and a resulting gain of temperature of 35·36. Also those in which the air appeared to be colder than the water were five, giving a total deficiency of 65·5, with a loss of temperature 4·71. Hence $\frac{65{\cdot}5-5x}{4{\cdot}71}=\frac{259{\cdot}63+15x}{35{\cdot}36}$, whence $x=4{\cdot}418$, which must be added to the indications of the thermometer registering the temperature of the air. After this correction has been made, it will be found that the effect of a difference of temperature, between the air and water, of 9·216 is unity.

4·418 added to 397·226 gives 401·644 for the corrected temperature of the air in the thermal experiments; and this being 7·238 in excess of the temperature of the calorimeter, the corrected thermal effect will be $22{\cdot}0914 - \frac{7{\cdot}238}{9{\cdot}216}$ $= 21{\cdot}306$, which, after applying the needful correction for the immersed portion of the thermometer-stem, becomes ultimately 21·326.

The thermal capacity of the calorimeter was made up of 95525 grains of distilled water, 26220 grains of copper, equivalent to 2501 grains of water, and the thermometer and coil equivalent to 80 grains, giving a total capacity equal to 98106 grains of water. 12·951 divisions of the thermometer are equivalent to one degree Fahr.

The dynamical equivalent is the quotient of the work done, by the thermal effect, or

$$\frac{\left\{\frac{k}{2\pi}\mathrm{H}\right\}^2 \tan^2\theta \mathrm{R}t}{\mathrm{T}} =$$

$$\frac{\left\{\frac{\cdot 62723}{6{\cdot}2832} \times 3{\cdot}67073\right\}^2 \times \cdot 379857 \times 33435640 \times 2400}{\frac{21{\cdot}326}{12{\cdot}951} \times 98106} = 25335.$$

It appeared to be desirable to diminish the atmospheric influence; I therefore commenced a second series, in which the calorimeter was covered with two folds of cotton wadding. The bulb of the air-registering thermometer was also placed in a small bag made of the same material. In this fresh series each experiment occupied one hour, as I had learned by experience that with my battery-arrangement the current would be sufficiently uniform. In fact the highest reading in an experiment was not more than $\frac{1}{30}$ higher than the lowest. There were, evenly distributed through the hour, forty observations of deflection, twenty of the air-, and three of the water-phenomenon; and the water was stirred forty times. Two minutes were allowed for the complete equalization of temperature previous to the final thermometer-reading. The experiments on radiation were also similarly extended.

The coil was the same as that used in the first series. It had a coat of shellac varnish. Five determinations of its resistance were made, using a single Daniell's cell with various resistances included in the circuit. The galvanometer had a coil 17 inches in diameter consisting of nine turns. The results are as follows :—

C_3.	C_2.	C_1.	Temperature of unit.	Temperature of coil.	Resistance of coil in terms of unit.
tan 79° 39′·5	tan 52° 33′·3	tan 52° 9′·3	59°·25	58°·6	1·0192
tan 71 39·5	tan 47 17·06	tan 46 55·6	48·6	48·5	1·0198
tan 70 16	tan 46 18·11	tan 45 57·4	54·68	57·4	1·0194
tan 71 54·33	tan 47 7·66	tan 46 45·93	......	......	1·0198
tan 62 6	tan 41 30·43	tan 41 13·46	......	......	1·0187
Average	.. ...	......	......	......	1·01938

The average temperature of the calorimeter in the experiments being 13°·55 Cent., and that at which the unit was adjusted 14°·5, the resistance during the experiments must have been 1·01906, which is equal to 33434330 in British measure.

Second Series of Thermal Experiments.

Date.	Deflection.	tan² Deflection.	Temperature of air.	Temperature of water.	Rise of temperature.	Horizontal intensity.
1866.						
Sept. 21 ...	29° 51′·68	·329623	397·4	363·42	30·38	3·6668
„ 22 ...	28 58·4	·306585	362·51	348·06	26·95	3·6707
„ 25 ...	29 14·63	·313472	345·19	306·94	29·75	3·6724
„ 26 ...	29 51·46	·329525	370·84	350·64	29·92	3·6644
„ 27 ...	28 54·78	·305064	365·91	361·71	25·88	3·6665
Oct. 5 ...	29 5·05	·309393	380·66	387·57	24·90	3·6612
„ 6 ...	28 22·54	·291761	426·55	392·77	27·40	3·6688
„ 8 ...	28 8·74	·286198	338·49	335·54	24·04	3·6595
„ 19 ...	28 42·81	·300074	398·56	332·35	31·08	3·6659
„ 20 ...	27 40·13	·274910	395·18	361·90	26·08	3·6654
„ 22 ...	26 40·5	·252409	371·72	388·63	19·12	3·6702
„ 23 ...	27 28·1	·270252	320·07	318·09	22·55	3·6638
„ 25 ...	27 9·63	·263230	275·65	286·25	20·98	3·6620
„ 26 ...	27 42·56	·275855	249·75	257·54	22·15	3·6623
„ 27 ...	28 7·84	·285838	245·96	247·27	23·57	3·6641
Average ...	....	·292946	349·63	335·912	25·65	3·6656

Second Series of Radiation Experiments.

Date.	Temperature of air.	Temperature of water.	Rise of temperature of water.
1866. Sept. 21	378·84	344·95	3·0
,, ...	390·13	381·34	0·32
Sept. 22	326·32	334·37	−0·43
,, ...	360·71	361·13	−0·41
Sept. 25	330·67	287·94	4·05
,, ...	347·56	326·13	1·59
Sept. 26	352·15	333·12	2·12
,, ...	377·56	368·12	0·70
Sept. 27	355·81	347·9	0·74
,, ...	388·0	375·69	1·31
Oct. 5	376·9	375·04	0·
,, ...	385·8	396·95	−1·15
Oct. 6	402·94	376·47	2·13
,, ...	433·28	411·33	1·52
Oct. 8	319·5	323·51	−0·29
,, ...	356·02	347·79	0·33
Oct. 19	365·08	303·94	5·95
,, ...	398·49	356·29	3·57
Oct. 20	357·9	344·01	1·61
,, ...	395·66	377·40	1·43
Oct. 22	371·24	380·45	−0·95
,, ...	362·7	392·44	−3·18
Oct. 23	297·96	305·0	−0·50
,, ...	334·07	329·05	0·5
Oct. 25	261·67	277·01	−1·26
,, ...	277·59	294·31	−1·86
Oct. 26	233·31	247·61	−1·40
,, ...	264·37	265·97	−0·66
Oct. 27	237·05	234·85	0·1
,, ...	251·15	257·24	−0·65
Average	343·011	335·245	0·6083

The correction to be applied to the thermometer immersed in air as deduced from the above Table is given by

$$\frac{123{\cdot}66-12x}{12{\cdot}74}=\frac{356{\cdot}65+18x}{30{\cdot}99},$$

whence $x=-1{\cdot}1835$. It appears also that a difference between the temperatures of the calorimeter and air-registering thermometer so corrected, equal to 10·822, gives the unit effect on the former.

Hence the corrected indication of the air-thermometer in the second series of thermal experiments will be

$$349{\cdot}63-1{\cdot}1835=348{\cdot}4465.$$

This being 12·5345 in excess of the temperature of the calorimeter, the corrected thermal effect will be

$$25{\cdot}65-\frac{12{\cdot}5345}{10{\cdot}822}=24{\cdot}4917,$$

which, after a small further correction for the immersed stem, becomes 24·512.

The thermal capacity in this second series was made up of 95561 grains distilled water, copper as water 2501, thermometer and coil as water 80, and cotton-wool as water 200 grs., giving a total of 98342 grains.

The equivalent, as deduced from the second series, is therefore

$$\frac{\left\{\frac{\cdot 62723}{6{\cdot}2832}\times 3{\cdot}6656\right\}^2\times \cdot 292946\times 33434330\times 3600}{\frac{24{\cdot}512}{12{\cdot}951}\times 98342}=25366.$$

The equivalents obtained in the two foregoing series of experiments are as much as one-fiftieth in excess of the equivalent I obtained in 1849 by agitating water. I therefore instituted a strict inquiry with a view to discover any causes of error, so that they might be avoided in a fresh series. The most probable source of error seemed to be insufficient stirring of the water of the calorimeter. Although agitated so frequently as forty times in the hour, there could be no doubt that, during any intervals of comparative rest, a current of heated water would ascend from the coil, and that, if a thin stratum of it remained any time at the top, some loss of heat would result. I resolved therefore to use a fresh calorimeter, and to introduce into it a stirrer which could be kept in constant motion by clockwork.

Another source of error (which, though it would be finally eliminated by frequent repetition of the experiments, it seemed to be desirable to avoid) was the hygrometric quality of the cotton-wool which enveloped the calorimeter in the second series of experiments. I therefore sought for a material which did not present that inconvenience. The plan finally adopted was to cover the calorimeter first with tinfoil, to place over that two layers of silk net (tulle), and to finish with a second envelope of tinfoil.

A third source of possible error was the circumstance that the silver-platinum alloy, when made positively electrical in distilled water, is slowly acted upon, an oxide of silver as a bluish-white cloud arising from the metal, while hydrogen escapes from the negative electrode. On this account the coil in the experiments of the last series, as well as the subsequent, was well varnished. But it was found at the conclusion of the experiments that the varnish had in a great measure lost its protecting power. This circumstance gave me considerable anxiety. I was, however, ultimately able, by the following facts arrived at after the thermal experiments were completed, to satisfy myself that no perceptible influence had been produced by it on the results:—

1st. The resistance of the coils, after long-continued use had deteriorated the varnish, was not sensibly less than it was after they had been freshly varnished.

2nd. The coil of the 3rd series was, in the unprotected state, immersed in distilled water, and compared with many hundred yards of thick copper wire, unimmersed, having nearly equal resistance. The result showed that the resistance to the current was sensibly the same whether a single cell or five cells of Daniell in a series were used. Now, had any considerable leakage by electrolytic action taken place, it would have been very much less in proportion in the former than in the latter instance.

3rd. When the coils of the second and third series, in the unprotected state, were placed in distilled water, and made the electrodes of a battery of five cells, the deflection was 40 of a degree on a galvanometer with a coil of 17 inches diameter composed of 18 turns of wire. This deflection indicates a current of about $\frac{1}{400}$ of the average current in the thermal experiments. In this case the chemical action was distinctly visible, but quite ceased to be so when the electrodes were connected by a wire of unit resistance, so as to reduce the potential to that in the thermal experiments.

4th. The coil of No. 2 series being used as a standard, that of No. 3 series, in the unprotected condition, was immersed, first in water, then in oil. The resistance to the current of five Daniell's cells was found to be sensibly equal in the two cases.

Hence there could be no doubt that the loss of heat during the experiments by electrolytic action could not possibly in any instance have been so great as one thousandth of the entire effect, and was probably not one quarter of that small quantity; whilst in the larger number of experiments, when the varnish was fresh, it must have been *nil*.

The coil used in the third series of experiments was made by bending four yards of platinum-silver wire double, and then coiling it into a spiral which was supported and kept in shape by being tied with silk thread to a thin glass tube. The terminals were thick copper wires; and the whole was coated with shellac and mastic varnish. The following results were obtained for its resistance. In the first three trials the current was measured by a galvanometer with a circle of nine turns 17 inches in diameter, and in the last six with an instrument with eighteen turns of wire. In the first six there was an extra unit of resistance included in the circuit:—

Battery.	Unit.	C_3.	C_2.	C_1.	Temp. of unit.	Temp. of coil.	Resistance in terms of my unit.
One cell, Daniell ...	Mine ...	tan 52° 53′	tan 37° 3′·15	tan 37° 10′·6	63°·27	62°·78	·98963
Ditto	" ...	tan 52 24·12	tan 36 29·02	tan 36 37·27	59·03	60·07	·98823
Ditto	Jenkin's	tan 52 3·62	tan 36 6·45	tan 36 14·79	60·88	60·57	·98752
Daniell's cell. Positive metal iron ...	"	tan 50 25·8	tan 35 21·88	tan 35 29·27	59·78	60·46	·98818
Ditto	Mine ...	tan 49 48·12	tan 34 57·36	tan 35 5·62	60·03	60·30	·98754
Ditto	" ...	tan 48 17·62	tan 34 5·48	tan 34 12·24	60·50	60·88	·98816
Ditto	" ...	tan 75 28	tan 49 58·6	tan 50 11·98	61·27	61·08	·98863
Ditto	Jenkin's	tan 75 17·25	tan 49 44·93	tan 49 57·51	61·96	61·27	·98871
Ditto	Mine ...	tan 75 59·6	tan 49 18·97	tan 49 33·08	69·35	70·28	·98820
Average	......	....	....	....	......	......	·98831

The above average resistance, reduced to 18°·63 C., the mean temperature in the third series, is ·98953 of the Association unit, or in British measure 32465480.

In the third series the experiments for the heat of the current, of radiation, and for horizontal magnetic intensity were alternated in such a manner that each class occupied the same portions of the day that the others did. I sought in this way to avoid the effects of any horary change in the humidity &c. of the atmosphere or in the magnetic force. Of the thirty experiments comprising each class, six were performed at about each of the several hours, 11 A.M., 12¼ P.M., 1½ P.M., 4 P.M., and 5½ P.M.

Third Series of Thermal Experiments.

Date.	Deflection.	$\tan^2$ Deflection.	Temperature of air.	Temperature of water.	Rise of temperature.	Fall of weight.
1867.						in.
June 28, 12.54 P.M.	28° 18′·25	·290024	488·660	494·17	25·1	30
„ 28, 5.36 „	30 56·37	·359310	534·155	524·214	32·08	26
„ 29, 1.30 „	28 55·45	·305345	509·172	490·13	27·82	27
July 1, 10.30 A.M.	29 41·1	·324949	428·81	425·67	28·52	27
„ 1, 4.24 P.M.	30 19·4	·342107	508·78	467·214	33·05	26
„ 2, 12.45 „	30 10·12	·337891	405·343	450·73	25·13	26
„ 2, 6. 0 „	30 30·98	·347424	401·822	458·104	24·99	28
„ 4, 1.20 „	31 23·4	·372299	516·992	452·97	57·98	27
„ 20, 11.11 A.M.	30 21·72	·343170	385·622	394·0	28·98	28
„ 20, 3.45 P.M.	31 37·55	·379241	454·19	430·97	34·92	28
„ 22, 12.36 „	32 0·6	·390765	482·44	460·621	35·48	30·5
„ 22, 5.21 „	32 23·47	·402470	493·087	498·573	34·47	28·4
„ 23, 1. 7 „	31 18·43	·369881	465·238	473·167	31·27	28·7
„ 24, 11. 0 A.M.	31 4·75	·363299	430·688	448·043	30·24	27·9
„ 24, 4. 5 P.M.	30 49·15	·355900	439·007	470·954	28·14	28·2
„ 25, 12.15 „	32 39·5	·410832	465·354	432·45	38·48	29·4
„ 25, 4.55 „	33 10	·427129	521·569	486·049	39·72	28·4
„ 26, 12.58 „	32 33·95	·407920	445·009	464·267	33·61	30
„ 27, 11.13 A.M.	33 1·6	·422590	391·0	419·21	34·46	30
„ 27, 4.14 P.M.	32 58·22	·420777	418·11	446·623	34·09	29·4
Aug. 2, 12.31 „	31 52·98	·386923	385·876	390·911	33·1	30
„ 2, 5.18 „	31 53·77	·387325	407·781	422·843	32·25	28
„ 3, 12.56 „	31 37·18	·379056	453·66	421·948	35·37	29·75
„ 6, 11.18 A.M.	26 34·35	·250162	439·906	435·699	22·32	29·7
„ 6, 3.55 P.M.	28 42·8	·300070	457·145	462·056	25·67	29·6
„ 8, 12.17 „	29 29·25	·319773	465·586	443·204	29·6	29·7
„ 8, 5.45 „	29 39·25	·324137	499·874	480·564	29·67	28
„ 9, 1.27 „	29 33·2	·321491	478·658	469·296	28·8	26·4
„ 10, 11. 9 A.M.	29 12·65	·312625	468·344	455·304	28·21	27·4
„ 10, 3.56 P.M.	28 14·47	·288500	519·082	493·136	27·28	28·4
Average	......	·3547795	458·699	455·436	31·02666	28·362

The calorimeter, in the foregoing experiments, was supported on the edges of a light wooden frame. It was carefully guarded against draughts by screens coated with tinfoil placed at a foot distance. The stirrer consisted of a vertical copper rod, to which vanes, on the plan of a screw propeller, were soldered at four equidistant places. The rod extended 2 inches above the calorimeter, and was there affixed to a light wooden shaft 2 feet long, attached at the upper end to the last spindle of a train of clock-wheels. The weight was 35 lb., which, falling about 2 feet per hour, produced a continuous revolution of the stirrer at a rate of about 200 in the minute. The action of the stirrer left nothing to be

Third Series of Radiation Experiments.

Date.	Temperature of air.	Temperature of water.	Rise of temperature.	Fall of weight.
1867.				in.
June 28, 10.38 A.M.	460·527	481·990	−1·48	31
„ 28, 3.53 P.M.	513·687	506·770	0·75	28·2
„ 29, 11.55 A.M.	493·088	473·930	1·82	28
„ 29, 4.40 P.M.	526·185	508·480	1·88	28·5
July 1, 1.23 „	469·368	442·114	2·46	27·5
„ 2, 10.58 A.M.	404·842	439·790	−2·82	27
„ 2, 4. 5 P.M.	401·779	450·930	−4·1	28·5
„ 4, 11.46 A.M.	492·210	427·517	5·97	28
„ 4, 4.42 P.M.	541 007	484·927	5·1	26·5
„ 20, 1. 0 „	416·237	409·044	1·03	28·75
„ 22, 1. 5 A.M.	474·393	439·140	3·32	30
„ 22, 13.50 P.M.	486·267	480·106	0·8	28·75
„ 23, 11.41 A.M.	451·029	456·947	−0·1	28·4
„ 23, 4.49 P.M.	475·319	486·113	−0·65	28·5
„ 24, 12.54 „	441·677	460·780	−1·48	26·5
„ 25, 10.40 A.M.	435·863	410·237	2·43	28
„ 25, 3.27 P.M.	515·653	460·939	5·03	28·8
„ 26, 11.29 A.M.	441·256	447·526	−0·2	28·5
„ 26, 4.49 P.M.	435·776	472·503	−3·0	29
„ 27, 1. 7 „	404·58	433·444	−2·28	29·8
Aug. 2, 10.55 A.M.	369·966	374·18	−0·15	29·75
„ 2, 3.50 P.M.	407·34	406·42	0·17	27·8
„ 3, 11.30 A.M.	435·813	401·187	3·24	28·6
„ 3, 4.33 P.M.	476·691	446·393	2·9	27
„ 6, 1.15 „	457·87	447·843	1·05	28·9
„ 8, 10.46 A.M.	442·403	426·304	1·68	29
„ 8, 4.17 P.M.	480·901	463·143	2·42	29·7
„ 9, 11.51 A.M.	466·428	453·149	1·27	26·5
„ 9, 5.37 P.M.	490·308	484·753	0·66	27·9
„ 10, 1.20 „	502·96	472·469	2·82	28·6
Average	460·6808	451·6356	1·018	28·498

desired. It was started five minutes before an experiment commenced, and kept going until the last observation of the thermometer had been made.

Each experiment, as in the second series, lasted one hour, during which were made eight observations of the thermometer immersed in the calorimeter, twenty of the temperature of the air, and forty of the deflection of the galvanometer.

The correction to be applied to the air-registering ther-

Determinations of Horizontal Magnetic Intensity.

Date.	Galvanometer-deflection, θ.	Weighing by current-meter, w.	$H=\frac{\cdot 17676\sqrt{w}}{\tan\theta}$.
1867.	° ′	grs.	
June 28, 1.30 P.M.	37 21·42	253·04	3·68334
„ 29, 10.50 A.M.	26 43·06	109·28	3·67114
„ 29, 3.50 P.M.	25 12·56	96·04	3·67964
July 1, 12.25 „	38 23·56	272·35	3·68144
„ 1, 5.20 „	38 59·25	284·95	3·68634
„ 2, 1.40 „	38 49·94	280·9	3·68034
„ 4, 10.45 A.M.	26 24·55	106·25	3·66894
„ 4, 3.45 P.M.	26 10·55	104·99	3·68474
„ 20, 12 Noon.	39 18·9	289·875	3·67484
„ 20, 4.40 P.M.	41 11·35	332·825	3·68504
„ 22, 1.30 „	41 21·4	335·13	3·67594
„ 23, 10.45 A.M.	32 5·1	169·616	3·67194
„ 23, 3.45 P.M.	31 56·15	168·608	3·68224
„ 24, 11.51 A.M.	39 52·95	301·591	3·67364
„ 24, 5 P.M.	40 24 9	315·092	3·68474
„ 25, 1.10 „	41 27·95	338·391	3·67964
„ 26, 10.30 A.M,	34 40·45	206·658	3·67324
„ 26, 3.33 P.M.	33 25·5	188·675	3·67864
„ 27, 12 Noon.	43 19·55	386·0	3·68194
„ 27, 5.12 P.M.	42 48·53	372·658	3·68414
Aug. 2, 1.30 „	41 15·35	332·733	3·67584
„ 3, 10.25 A.M.	34 13·9	198·99	3·66464
„ 3, 3.33 P.M.	33 40·3	191·983	3·67628
„ 6, 12.12 „	35 9·8	214·117	3·67156
„ 6, 4.50 „	37 8·1	248·258	3·67784
„ 8, 1.11 „	37 44·55	259·867	3·68110
„ 9, 10.53 A.M.	31 23·65	160·708	3·67186
„ 9, 4.42 P.M.	30 43·4	152·75	3·67590
„ 10, 12.12 „	36 25·4	235·433	3 67557
„ 10, 4.50 „	34 49·5	209·608	3·67864
Average	....	....	3·67771

mometer, as deduced from the radiation experiments of this third series, is found from

$$\frac{217{\cdot}452 - 10x}{16{\cdot}26} = \frac{488{\cdot}807 + 20x}{46{\cdot}8},$$

whence x, the quantity to be added to the observed temperature of the air in the thermal experiments, $= 2{\cdot}81$. The temperature of the air was therefore virtually 6·073 higher than that of the water. The results also show that the unit of effect on the calorimeter was produced by a difference of temperature of 11·645.

Hence $31{\cdot}0266 - \frac{6{\cdot}073}{11{\cdot}645} = 30{\cdot}5051$; and adding ·077 for the unimmersed part of the thermometer-stem, the corrected thermal effect in the third series is found to be 30·5821.

The average capacity of the calorimeter was equal to that of 93859 grs. of water, being made up of 91531 grs. distilled water, 22364 grs. of copper, 486 grs. of tin (the weight of the coating next the calorimeter), 52 grs. silk net (half that employed), the thermometer, coil, and corks.

The equivalent deduced from the third series is therefore

$$\frac{\left\{\frac{62723}{6{\cdot}2832} \times 3{\cdot}6777\right\}^2 \times {\cdot}35478 \times 32465480 \times 3600}{\frac{30{\cdot}5821}{12{\cdot}951} \times 93859} = 25217.$$

The equivalents above arrived at are :—

From Series 1. Average of 10, 25335.
From Series 2. Average of 15, 25366.
From Series 3. Average of 30, 25217.

The extra precautions taken in the last Series entitle the last figure to be taken as the result of the inquiry. Reduced to weighings *in vacuo* it becomes 25187.

Observations on the Alteration of the Freezing-point in Thermometers. By Dr. J. P. Joule, *F.R.S. &c.*

[Read before the Manchester Literary and Philosophical Society, April 16, 1867. Memoirs of the Society, 3rd series, vol. iii. p. 292.]

Having had in my possession, and in frequent use, for nearly a quarter of a century, two thermometers, of which I have from time to time taken the freezing-points, I think the results may offer some interest to the Society. Both thermometers are graduated on the stem, and are, I believe, the first in this country which were accurately calibrated. Thirteen divisions of one of them correspond to one degree Fahrenheit. It was made by Mr. Dancer, in the winter of 1843–44. My first observation of its freezing-point was made in April 1844. Calling this zero, my successive observations have given :—

0	April 1844,	8·8	February 1853.
5·5	February 1846.	9·5	April 1856.
6·6	January 1848.	11·1	December 1860.
6·9	April 1848.	11·8	March 1867*.

The total rise has been therefore ·91 of a degree Fahrenheit. The other thermometer is not so sensitive, having less than four divisions to the degree. The total rise of its freezing-point has been only ·6 of a degree; but this is probably owing to the time which elapsed between its construction and the first observation being rather greater than in the case of the other thermometer. The rise of the two thermometers has been almost identical during the last nineteen years.

A projection of the observations given above is shown in the following diagram :—

* To the above are now added 12·1, Feb. 1870; 12·5, Feb. 1873; 12·71, Jan. 1877; 12·92, Nov. 1879; and 13·26, Dec. 1882. See Proceedings Lit. & Phil. Soc. Feb. 22, 1870, and March 4, 1873.

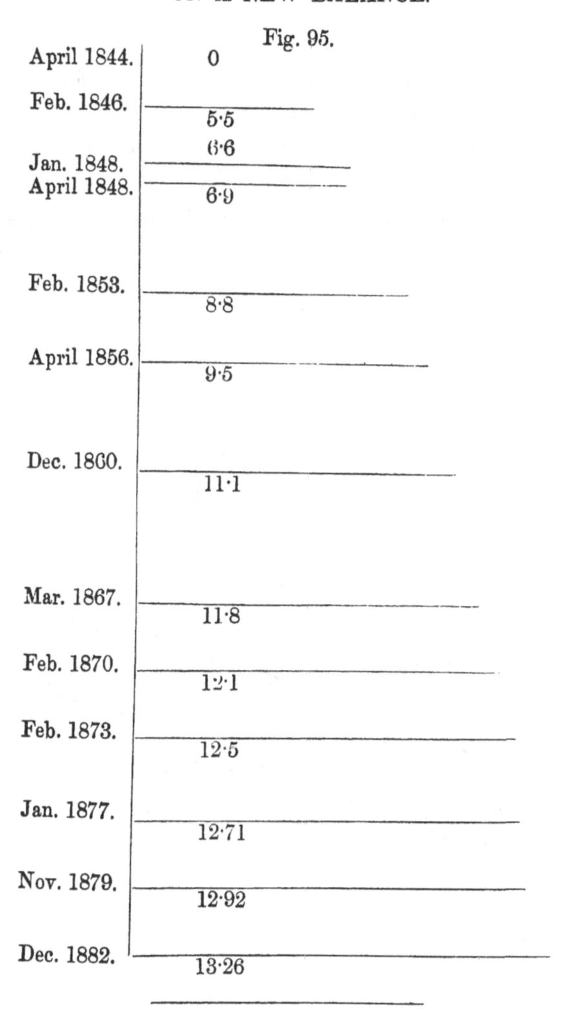

On a New Balance. *By* J. P. Joule, *D.C.L., F.R.S.*

[Read before the Manchester Literary and Philosophical Society, March 20, 1866. Proceedings of the Society, vol. v. pp. 145 & 165.]

Dr. J. P. Joule exhibited a balance which he had constructed on the principle which had been introduced by Professor Thomson, and employed by him in weighings for a long time. The following figure will fully explain the instrument. The beam has a leaden weight let into its extremity *b*. It is supported by a wire *a a* stretched between the sides of the box containing the balance. This wire is led round so as to form the suspender *a a'* of the scale. The beam is limited in its movements by fixed supports. Silk threads, *c c*, *c c*, hanging from the cross pieces, form a gimbal system by which the scale is supported in such a manner

Fig. 96.

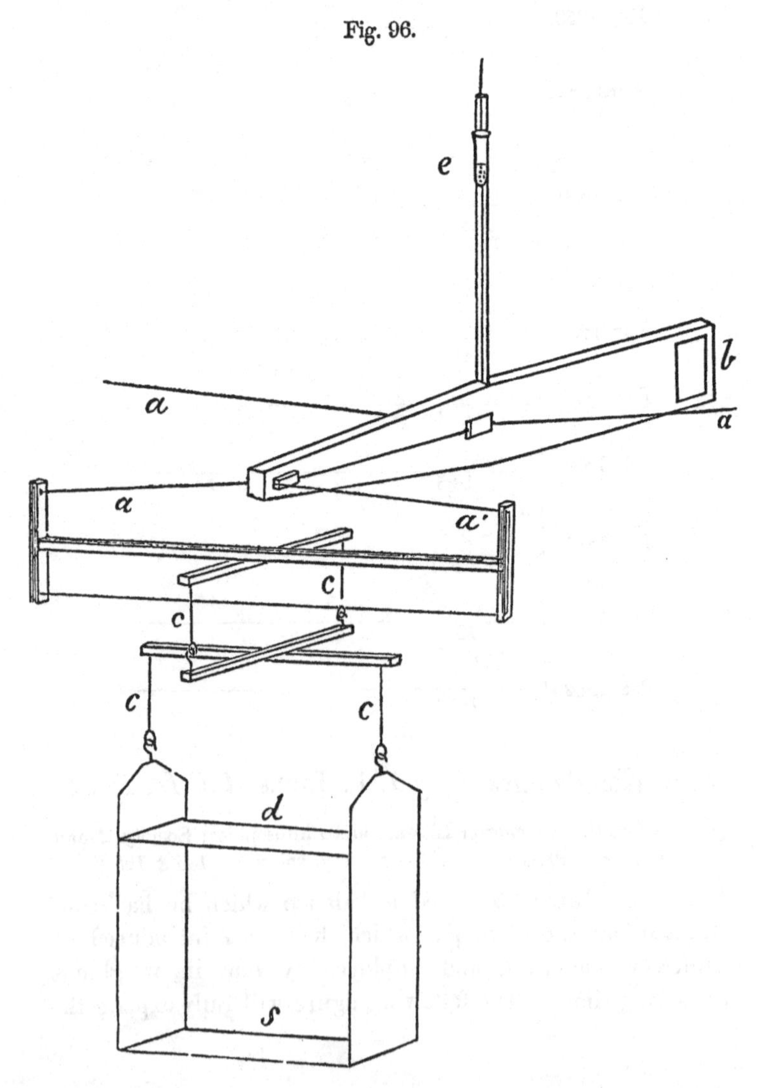

that any variation in the position of the weights does not alter the torsion of the suspender. A counterpoise is placed on the stage *d*. When an article is to be weighed it is placed in the lower part of the scale *s*, and then, the counterpoise being removed, weights are placed on the stage to effect the counterpoise in the new condition. The difference between the first and second counterpoises of course gives the weight required. The upper edge of the beam is furnished with an index for showing minute effects; and attached to this is a small bottle *e* for holding shot or sand, by the addition of which the stability of the beam may be decreased to any required extent. The instrument exhibited was able to weigh articles of upwards of 3000 grains to one hundredth of a grain. Dr. Joule stated that he had also employed Professor Thomson's principle of weighing in the construction of a galvanometer for the absolute measure of electrical currents. In this instrument a flat coil is suspended between two fixed flat coils, one of which attracts while the other repels the suspended coil, to which last the current is conducted by means of the suspending copper wires. This electrical balance is sensitive to one part in two millions.

On an Apparatus for determining the Horizontal Magnetic Intensity in Absolute Measure. By J. P. Joule, *LL.D., F.R.S., &c.**

[Read before the Manchester Literary and Philosophical Society, March 19, 1867. Proceedings of the Society, vol. vi. p. 129.]

Many years ago Weber described a small portable apparatus for ascertaining the horizontal component of the intensity of the earth's magnetism. In it a magnetic bar, about four inches long, was vibrated under the influence of the earth's magnetism, and afterwards employed to deflect a compass-needle. The moment of the deflecting bar was determined

* The experiments were made at Thorncliff, Old Trafford, near Manchester.

from its dimensions and weight; the compass-needle was supported on a point in the ordinary manner; and no allowance was made for the influence of the earth in increasing the strength of the magnet while vibrating in the magnetic meridian. Notwithstanding these defects, the results obtained were of considerable accuracy, and justified the employment of the portable apparatus by travellers. I have endeavoured to improve Weber's apparatus, so as to give it an accuracy comparable to that of the instruments employed in stationary observatories, without increasing its expense or diminishing its portability. I will, after these preliminary observations, at once commence by describing the various parts of the apparatus I have constructed for my own use.

I employ two deflecting-bars, for reasons which will afterwards appear. It is important that the length of the bars should be decreased as far as possible. By so doing, there is greater probability of obtaining perfect homogeneity and accuracy of dimension, whilst they can be more readily and thoroughly magnetized. I have, after much trial and consideration, fixed upon two inches as the most convenient length, and on a quarter of an inch as the section of the bar. The bars are first filed, then thoroughly hardened, and afterwards ground to an exact figure, the operations being finished by rubbing the bars together with a little cutting powder between them. The bars may easily be made so true that, when dry and perfectly polished, they will adhere strongly together. When finished, they must be measured by a standard rule under a microscope—the means of observations in many places being taken for the respective dimensions. Their weight must also be accurately determined.

The moment of a homogeneous parallelopiped in vibration is found by multiplying the square of the diagonal of the upper surface by the weight of the bar and dividing by 12. If it be thought desirable to test the accuracy of the moment thus deduced, the best plan is to fasten small cubes of lead to the ends of the bar by an elastic band. These cubes must be carefully wrought and measured; and their moment may be found by subtracting the moment of a bar of lead two inches long from that of a bar two inches long plus the length of the two pieces. When the moment of the pieces of lead is determined, that of the steel bar may be found by vibrating it first by itself, and then with the lead cubes attached. Another plan is to hang pieces of lead from the ends of a filament of silk thrown over the bar; but in this case great care is requisite that the filaments have very little force of torsion, and that the vibration is very slow. The moments of the four bars I have finished, as determined by the above means, are as follows :—

	By measurement.	By fixed lead weights.	By suspended lead weights.	Average.
No. 1	86·246	86·173	86·225	86·215
No. 2	86·424	86·330	86·435	86·396
No. 3	85·248	85·070	85·283	85·200
No. 4	84·843	84·773	84·906	84·841

The results show that the moments of steel bars, when carefully finished, may be very correctly deduced from their linear dimensions and weight.

It is important that the bars should be thoroughly and uniformly magnetized. This is best effected by bringing the ends of the bar into contact with the ends of two straight electromagnets of considerable power. In this way the confused polarity which more or less exists in all bars magnetized by stroking is avoided—the result being, that the magnetic virtue is more permanent, and at the same time much more effective.

The needle to be deflected should be of small dimensions, not exceeding half an inch in length. It may be conveniently made of several perfectly hard pieces of common sewing-needle mounted on perforated card. A glass index should be affixed to it, the ends being furnished with very fine bits of copper wire. These show well when traversing the graduated circle.

A circle of six inches diameter will be found a convenient size. It should be divided to thirds of a degree. The deflections may be read off with sufficient accuracy by the aid of an eye-glass, a piece of looking-glass being placed on the table to ensure the correct perpendicularity of the eye. It is easy after a little practice to read off the deflections to half a minute or less. The needle should be protected from currents of air by being placed, with its circle, in a shallow box, covered by a glass plate, in the centre of which there is a hole for the purpose of suspension.

A filament of silk of six inches length, and strong enough to carry the needle with safety, will be found to give a quantity of torsion so small that it may be neglected, not amounting to more than a small fraction of a minute in a deflection of 45°. For suspending the deflecting-bars a stronger thread is required; but by increasing its length to four or five feet, the effect of torsion in this latter case also becomes too small to require notice.

The remaining part of the apparatus gives the means of placing the deflecting-bars at accurately measured distances from the needle. This has generally been done by means of a divided metal rod; but I find that a plain board of well-seasoned deal, with a circular hollow excavation in the centre for the graduated circle, answers the purpose very well. It is needful to be able to place the bars rapidly to an exact position on the board. This is best done for each distance by driving pins into the board, and cutting them off pretty close to the wood, one pin regulating the distance and two others the direction. It is only necessary to push the bar up to its defined position. The distances of the pins on the east side of the suspended needle from the corresponding ones on the west side are in my own apparatus 15, 20, 25, and 30 inches, half those distances being the mean distances of the bar from the needle. The first two distances, however, are all that are actually required.

In making the observations of deflection two bars are used, one on the east side, the other on the west of the needle. They must be carried to their respective places by wooden pincers, and on no account touched by the fingers or subjected to any violence. The deflection marked by the pointer at both ends of the needle must be observed; and then the bars must be reversed and the deflection in the other direction also observed. The bars must then be removed to the next distances and the observations repeated. Finally, the bars must be made to change sides from east to west, and all the former observations repeated in the reverse order. Thus eight separate observations are obtained, of which the average gives the deflection due to any given distance.

The only further observation required for finding the earth's horizontal intensity is that of the number of vibrations of the bars under the influence of the earth's magnetism. It is clear that when a bar is vibrating about the magnetic meridian its magnetism is slightly increased by the inductive influence of the earth, and that thus the number of vibrations is increased. The use of two bars enables me to get rid of this difficulty in a very simple manner. I find the distance at which one bar must be held below the other so as to counteract the earth's magnetism; then I suspend the two bars, one beneath the other, at that interval, and vibrate both together. The result gives the number of vibrations due to the magnetism of the bars as uninfluenced by induction.

The method of suspension I have found most convenient is as follows:—A fine wire is twisted round the centre of one of the bars, forming a loop which is caught by a small hook fastened to the extremity of the suspending fibre. Then a filament of silk of the proper length, tied at the ends so as to form a long loop, is thrown along the length of the suspended bar, and the ends of the other bar are placed in the ends of the loop. There is some slight difficulty in making the bars hang quite horizontally; but this may be overcome by a little practice. The proof of accurate horizontal position is made by looking at the bar and a spirit level simultaneously from a distance with a telescope.

The bars being arranged as above in one system, and having nearly the same amount of magnetism in each, vibrate with perfect regularity. To ascertain the time occupied by the vibrations, I look at the system through a telescope placed at a distance of two or three yards, a clock beating seconds being within hearing. Then I have a slate, divided into ten columns; and each time that I note the end of a vibration I make a mark on the slate, counting at the same time the ticking of the clock. Every time that the end of a vibration appears to be sensibly coincident with the beat of the clock, I write the number of seconds on the slate alongside the usual mark. After observing the vibrations (which ought not to exceed 3° of amplitude) for about ten minutes, materials will have been collected for obtaining a sufficiently accurate determination of the time of vibration by dividing the times between the several coincidences by the number of intervening vibrations and taking the mean of the whole.

As has been shown by Gauss and Weber, if M be the force of the magnetic bars in absolute measure, θ the angle of deflection produced at the distance R, and T the absolute horizontal intensity of the earth's magnetism, $\frac{M}{T}=\frac{R^3 \tan\theta}{2}$, but approximately only, because the deflecting-power of a bar decreases in a more rapid ratio than the inverse cube of the distance, on account of its own magnetic length. Hence the necessity of a correction, which may be found by observing the deflections at two distances R and R_1, from which we obtain $L^2=\frac{\delta\, R^3 \tan\theta - R_1{}^3 \tan\theta_1}{2(R\tan\theta - R_1 \tan\theta_1)}$, where L signifies the virtual half length of the bar. Then we obtain the closely correct value $\frac{M}{T}=\frac{R^3 \tan\theta}{2} - L^2 R \tan\theta$. Calling this quantity r, as in Gauss and Weber's equation, we have finally $T=\frac{1}{t}\sqrt{\frac{12C}{r}}$, where t is the time of vibration in seconds, C is the moment of inertia of the bars multiplied by π^2, and 12 is the coefficient to reduce from inches to feet.

The following were the data for determining the horizontal intensity on four several occasions:—

	Jan. 29th, 1867.	Feb. 5th, 1867.	Feb. 11th, 1867.	March 16th, 1867.
C	1691·80	1690·04	1690·04	1690·04
R	7·5	7·5	7·5	10
R_1	10	10	10	12·5
tan θ	·913087	·97852	·97520	·40610
tan θ_1	·379294	·40607	·40480	·20656
t	2·8407	2·7445	2·7483	2·7423
L^2	·9679	1·028	1·0122	·9007
r	185·976	198·861	198·303	199·395
T	3·6779	3·6795	3·6796	3·6776

The above determinations were obtained in a room in which there are several pieces of iron in fixed positions, and in the neighbourhood of which are some considerable masses of that metal. I have ascertained that these had an influence in increasing the intensity by about a fortieth. My current-meter, by which the horizontal intensity is immediately deduced from a comparison of its weighings with the deflections of a tangent galvanometer included in the same circuit, shows a gradual increase of the intensity since last September, conformably with the results obtained in past years at Kew.

Note on the Tangent Galvanometer.
By J. P. Joule, *LL.D., F.R.S., &c.*

[Proceedings of the Manchester Literary and Philosophical Society, vol. vi. p. 135. See also a valuable paper by Professor W. Jack, M.A., ibid. p. 147.]

It is well known that a current circulating round a magnetic needle produces a deflection of the latter, the tangent of which is approximately proportional to the strength of the current. Pouillet made the approximation much more close by using a broad strip of copper bent into a circle, instead of the usual coil, and thus obtained the instrument commonly known as the tangent galvanometer. The arrangement of Pouillet ensures a close approximation, it is true, but it does not, after all, exactly fulfil the law of tangents; and in absolute measurements it has the disadvantage that the diameter of the circle of ribbon cannot be taken as the virtual diameter. It appeared to me that the best plan for a galvanometer would be to use a wire of about one tenth of an inch diameter surrounding a short needle, and to supply the small correction needed to the tangents of the deflections of the latter.

Calling, then, θ the angle of deflection, l the magnetical length of the needle (generally about four fifths of the actual length), and d the diameter of the coil, we have

$$\{(2\sin\theta)^2-(\cos\theta)^2\}\left(\frac{l}{d}\right)^2\tan\theta,$$

or, more elegantly and simply, as suggested to me by Professor Jack,

$$\frac{1}{2}(4\tan^2\theta-1)\frac{l^2}{d^2}\,.\sin 2\theta,$$

for the correction to be supplied to the tangent of the angle observed. This correction is additive at great deflections, and subtractive at small ones, whilst at 26° 34′ the correction vanishes.

It seems therefore desirable that in exact researches a needle of small length should be employed, and the apparatus so arranged as to give a deflection somewhere about the above-indicated angle, where the correction, if needed at all, would be of trifling amount.

Additional Note on the Galvanometer.

By Dr. J. P. Joule, *F.R.S.*

[Read before the Manchester Literary and Philosophical Society, April 2, 1867. Proceedings of the Society, vol. vi. p. 151.]

The current-meter has enabled me to find experimentally the diminution of deflection consequent on an increase of the length of the needle. This was done by comparing the square roots of the weighings of the meter with the deflections occasioned by a coil 15·2 inches diameter, in the centre of which needles of 1, 2, and 4 inches length were successively placed. I find the fractions representing the loss on the tangents of deflection as follow:

	From 1-in. to 2-in. Needle.	From 1-in. to 4-in. Needle.
68° 21′	·02239 loss	·11258 loss.
42°	·00993 loss	·04870 loss.
27° 2′	·00023 gain	·00254 loss.
19° 1′	·00287 gain	·01646 gain.

Hence the total fractions gained or lost by the two needles are:—

	For 2-in. Needle.	For 4-in. Needle.
68° 21′	·02985 loss	·12009 loss.
42°	·01324 loss	·05194 loss.
27° 2′	·00032 gain	·00230 loss.
19° 1′	·00383 gain	·01756 gain.

Since my last communication I have made trial of many bars and needles, and find that the magnetic length is about $\frac{7}{8}$ of the actual in most cases. If, therefore, instead of l we take the actual length of the needle n, the correction to be applied to the tangent of deflection will be nearly

$$\frac{1}{4}\left(4\tan^2\theta-1\right)\frac{n^2}{d^2}\sin 2\theta.$$

The fractions of loss or gain calculated from this formula for the deflections experimented with are as follow:—

	For 2-in. Needle.	For 4-in. Needle.
68° 21	·02873 loss	·11493 loss.
42°	·01072 loss	·04290 loss.
27° 2′	·00029 loss	·00116 loss.
19° 1′	·00406 gain	·01625 gain.

Very curious phenomena present themselves as the needle is prolonged. There ultimately becomes a point of stable equilibrium on the other side of the meridian. The following rough results obtained with a coil of $7\frac{1}{4}$ inches diameter will give an idea of these changes:—

Length of Needle. inch.	Deflection on the usual side.	Deflection on the opposite side.	Unstable Equilibrium on the opposite side.
1	74°	—	—
2	72°	—	—
3	68°	—	—
3½	65½°	—	—
4	62°	—	—
4½	56°	—	—
5	50°	—	—
5½	41½°	—	—
6	34°	—	—
6½	23°	35°	35°
6¾	21°	47°	26°
7	17°	57°	14°
7¼	13°	59°	13°
7½	8	59°	6
8	—	63°	
9	—	62°	
10	—	61°	
11	—	59°	
12	11°	58°	
13	17°	56°	
14	20°		
15	23°		

On a Self-Acting Apparatus for Steering Ships. *By* Dr. J. P. Joule, *F.R.S.*

[Read before the Institution of Engineers in Scotland, June 18, 1865. Transactions of the Institution, vol. viii. p. 56.]

Some investigations in which I have been recently engaged have led me to the construction of magnetic needles having considerably greater directive power than those in common use. It has occurred to me that it might be possible to apply the power, thus increased, to the purpose of the automatical steering of ships. My idea is, to suspend a large compound system of needles or magnetic bars in the way first described by Professor Thomson, viz. by threads or fine wires attached above and below the system. By means of an electromagnetic relay it would be possible to start a powerful machine in connexion with the tiller whenever the ship deviates from a prescribed course.

Suppose a system to be composed of a thousand 4-inch bar magnets, each $\frac{1}{4}$ of an inch in diameter, arranged in a vertical column, say 5 in breadth and 200 in height. According to a rough estimate I have made of the directive force of such a system, I find it to be equal, at one inch from the axis of revolution, to 300 grains when at right angles to the magnetic meridian. This corresponds to 31 grains at 6° deflection, and 5 grains at 1°. Five grains would be amply sufficient to overcome any resistance to motion offered by a mercury commutator; and 30 grains would be more than sufficient with a properly constructed solid metallic commutator.

I would have a bent wire affixed to the lower end of the system of magnetic bars, one extremity of which should be immersed in a central cup of mercury, and the other should dip in one or the other of two concentric semicircular troughs of mercury exterior to the central cup.

I would place the central cup in connexion with one of the poles of a voltaic battery. The other pole must be in connexion with a branched conductor leading to two electro-magnets. The free wires of these electromagnets should be put in connexion with the semicircular troughs.

By this arrangement it is obvious that, accordingly as the wire carried by the magnetic system is immersed in the one or the other of the semicircular troughs, one or the other of the electro-magnets will be excited.

An armature should be placed between the two electro-magnets, so as to reciprocate between them whenever the wire passes from one semicircular trough to the other.

In so doing I would have the armature, suitably connected by levers &c., to operate on easily-acting valves (throttle-valves for instance) placed in steam-pipes proceeding from a steam-boiler to opposite ends of a cylinder. A similar arrangement might be made for working the exit valves.

The piston of this cylinder should be connected with the tiller in such sort that whenever the ship turns to the right the helm will be put to port, so as to bring her back to her course.

It is obvious that, if the dipping bent wire is in the direction of the magnetic axis of the compound system of magnets, and the division between the semicircular troughs is in the direction of the ship's length, the ship will be kept directed to the magnetic north. By turning the commutator in the direction of the hands of a clock, the ship will at once change to a course the same number of degrees west of north.

The use of such an apparatus as I have described would of course be limited to very extraordinary circumstances. In general practice it will be impossible advantageously to displace the intellectually-guided hand of the steersman, whose art consists in a great part in anticipating the motion of the ship, and, in heavy seas, in directing his ship so as to encounter them with safety.

Note by the Author, received 25th Jan. 1865.—I find, on trial, that a much smaller number of magnets than that I have mentioned is able to work a mercurial commutator. Fifteen 4-inch bars would be amply sufficient to overcome the adhesiveness of the mercury to the wires dipping into it when the deflection is one degree. A similar observation applies to the metallic commutator. Professor Thomson, however, has shown me a far more delicate mode than either of the above. In his plan a single bar magnet is suspended by a fine platinum wire. To one arm of the magnet a platinum wire is attached vertically. Two horizontal parallel fixed wires are placed on either side of the suspended one. Whenever either of the fixed wires is, by the motion of the ship, brought into contact with the wire carried by the magnet, a current passes to it from the suspending wire. This current excites an electromagnetic relay, by which another current is thrown upon an electromagnet powerful enough to work the valves of the steam-cylinder. Experiments conducted in the Physical Laboratory of Glasgow College are quite conclusive as to the practicability of this plan, and demonstrate the possibility of directing a ship by the agency of a needle much less powerful than that of an ordinary compass.

On a Thermometer unaffected by Radiation. *By* Dr. J. P. Joule, *F.R.S. &c.*

[Read before the Manchester Literary and Philosophical Society, Nov. 26, 1867. Proceedings of the Society, vol. vii. p. 35.]

Fig. 97.

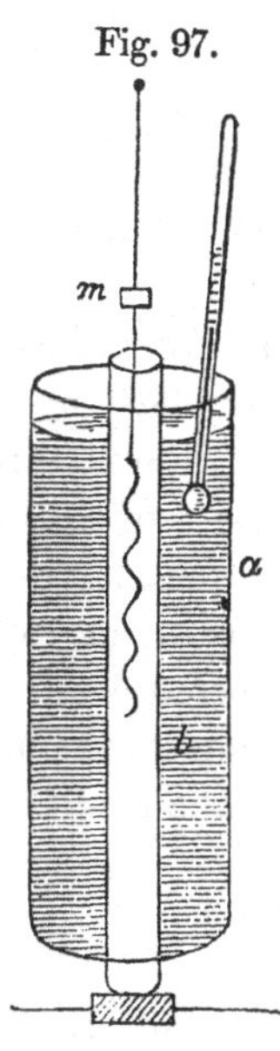

In the annexed figure *a* is a copper tube about one foot long, and has a tube open at both ends in the centre. Water is poured into the space between the two tubes. In the centre tube there is a spiral of fine wire suspended by a filament of silk, and having a mirror at *m*. There is a lid at *p*, which can be removed at pleasure from the lower end of the tube. When *p* is situated as in the figure, there can be no draught, and consequently the spiral with its mirror is at zero of the scale. But when *p* is removed, there is a current of air, which turns the spiral if the air in the tube has a different temperature from that of the outside atmosphere. In my apparatus, one degree Fahr. produces an entire twist of the filament. I find that the temperature of the water is generally warmer than that of the outside atmosphere, which must be owing to the conversion of light and other radiations into heat on coming into contact with the copper tube. I have tried the apparatus in the open air on a still day, with the same result. Of course when there is wind the effect is masked; but I feel confident that by increasing the length of the tube, making it 30 feet for instance, and using certain precautions, this difficulty may be overcome.

Note on the Resistance of Fluids. *By* Dr. J. P. Joule, *F.R.S. &c.**

[Proceedings of the Manchester Literary and Philosophical Society, vol. vii. p. 223.]

Coulomb appears to have been the first who pointed out clearly the two sources of resistance to bodies moving in fluids—the first varying in the simple ratio, the second in the duplicate ratio of the velocity. He found the simple ratio to prevail only at the very slow velocity of not more than half an inch per second. I have long had in view experiments for the purpose of further illustrating this subject, and possess an apparatus which, although constructed so long ago as 1849, I have not experimented with before the last few weeks. It consists of two vertical disks of turned and polished steel, each 18 inches in diameter, which revolve between corrugated iron plates at about one quarter of an inch distance. The axle passes through a leather collar; and the weight of the disks, which is considerable, is supported by friction-wheels. Revolution is effected by the agency of weights and pulleys. A train of wheels in connection with the axis, by which a papered drum is moved under a pen worked by a pendulum, furnishes the means of measuring the velocity. The following are the results at which I have already arrived:—

Fluid.	Velocities of rim, in feet per second, between which the experiments were made.	Index of the power of the velocity which is proportional to the resistance.
Air	0·7 and 5·36	0·64
	5·36 „ 13·2	0·98
	13·2 „ 23·03	1·7
Water	0·47 „ 5·2	1·084
	5·2 „ 13·44	1·78
Oil	0·19 „ 1·78	1·24

* The experiments were made at Thorncliff, Old Trafford, near Manchester.

I infer from the above that in the circumstances of these experiments the resistance of liquids up to five feet per second is nearly in the simple proportion of the velocity. The following are my results as to the resistance due to given velocities :—

Fluid.	Resistance referred to rim, in grains.	Velocity of rim, in feet per second.
Air	853	23
Water	6267	13·44
Oil	6715	1·79

When the liquids were exposed to three atmospheres' pressure, the resistance appeared to be slightly increased. I hope to clear up this and other points, and to obtain more exact results, when the apparatus shall have received the improvements which have suggested themselves in the course of this preliminary inquiry.

On a New Magnetic Dip-Circle.

[Read before the Manchester Literary and Philosophical Society, April 6, 1869. Proceedings of the Society, vol. viii. p. 171.]

Dr. Joule, F.R.S., gave an account of his endeavours to improve the instrument known as the *Dip-Circle*, which, notwithstanding what had been done by Lloyd and others, was not comparable in delicacy to the declinometer. The ordinary mode of causing the axis of the needle to roll on agate planes first claimed attention. He found it possible to obtain a steel cylinder of beautiful accuracy as follows :— A steel wire, stretched by a weight hanging from one extremity, being heated to redness, draws out a certain length, and in so doing becomes perfectly straight. The wire is then divided into pieces each about two inches long, which are ground true and then polished by rolling them against one another. If the operation has beencare fully conducted, one wire laid across two others will roll noiselesssly down an almost imperceptible gradient. The plan of suspension thus indicated appeared better than the use of agate planes, inasmuch as dust and moisture were less likely to interfere with the delicacy of the indications. Nevertheless it was impossible altogether to avoid the effects of these impediments to free rotation, and the more so as the obstruction to rotation is proportional to the square root of the height of any small particle between the rolling surfaces. In fact, on narrowly watching the periods of oscillation they were invariably found to become sensibly quicker when the arc was very small, showing that the needle was rocking on two points.

The suspension by inclined silk threads was then tried, but soon abandoned, as it was found that the violent torsion at the points of attachment could not be certainly allowed for, owing to the viscosity of the threads.

The system brought now before the Society was free from the above-named evils. In it each end of the axis of the needle is suspended by a fibre of silk, on which it rolls. Small washers on the axis serve to keep the fibres in a definite position. The ends of the fibres are supported by the extremities of a delicate balance-beam placed at the top of the instrument. Small pins &c. are used for adjusting the length of the fibres and to regulate the centre of gravity of the beam. The needle itself is a piece of softened watch mainspring, sufficiently long to extend completely across the graduated circle. It is seven inches long, and weighs 18 grains. A glass plate fastens before the instrument by a notch. By the reflection of the eye of the observer from this glass, parallax is avoided, while the position of each edge of the needle is read off by an eye-glass to a minute of arc.

Notes on the above, with Experiments.
By Dr. J. P. Joule, *F.R.S.*

[Added August 1881.]

The Dip-Circle described in the previous paper received successive improvements until it assumed the form shown by a drawing exhibited at South Kensington in the Loan Collection of Scientific Apparatus, 1876, and which is reproduced on a smaller scale in the adjoining figure. *a a a a* is a wooden frame into which a graduated circle is sunk so as to be flush with the surface. It is suspended by a brass wire to a clamp which can be rotated on the top of the instrument through an angle of 180° as determined by the fixed pins *c c*. *d d* is a light balance-beam suspended by the tension wires *e e*; from this, attached to four pins, proceed the two loops of split silk thread on which the gold axis of the needle *g* rolls. *h h* are holes for the insertion of pins to hold the needle on occasion. *i* is a clamp held down by a caoutchouc band *j*, by which the needle is held whenever its magnetism is to be reversed. *f f* is a piece supported at one end by a hinge, on the other by a screw, whereby the needle can be raised or depressed to a small extent, when, by their altered hygrometric condition or temperature, the silk or spider filaments become altered in length. *k* is a paddle plunged in a vessel full of castor oil to rapidly arrest the oscillations of the frame when the wire by which it is supported is turned through 180°. The outer frame *l l l l* is enclosed behind and in front by plate-glass doors not seen in the figure.

The positions of the pointed ends of the needle are observed with a telescope revolving on an axis in front of the circle. A peculiarity of this telescope is that its object-glass is narrowed to a slit parallel to the needle, by which clearness of definition of the graduation is secured in spite of the distance intervening between the needle and the circle.

The two component parts of a single thread of the silkworm are each able to sustain for some little time a weight of 80 grains; the thread of the diadema spider about 20 grains. A fifth of the breaking-strain being deemed suffi-

Fig. 98. Scale ¼.

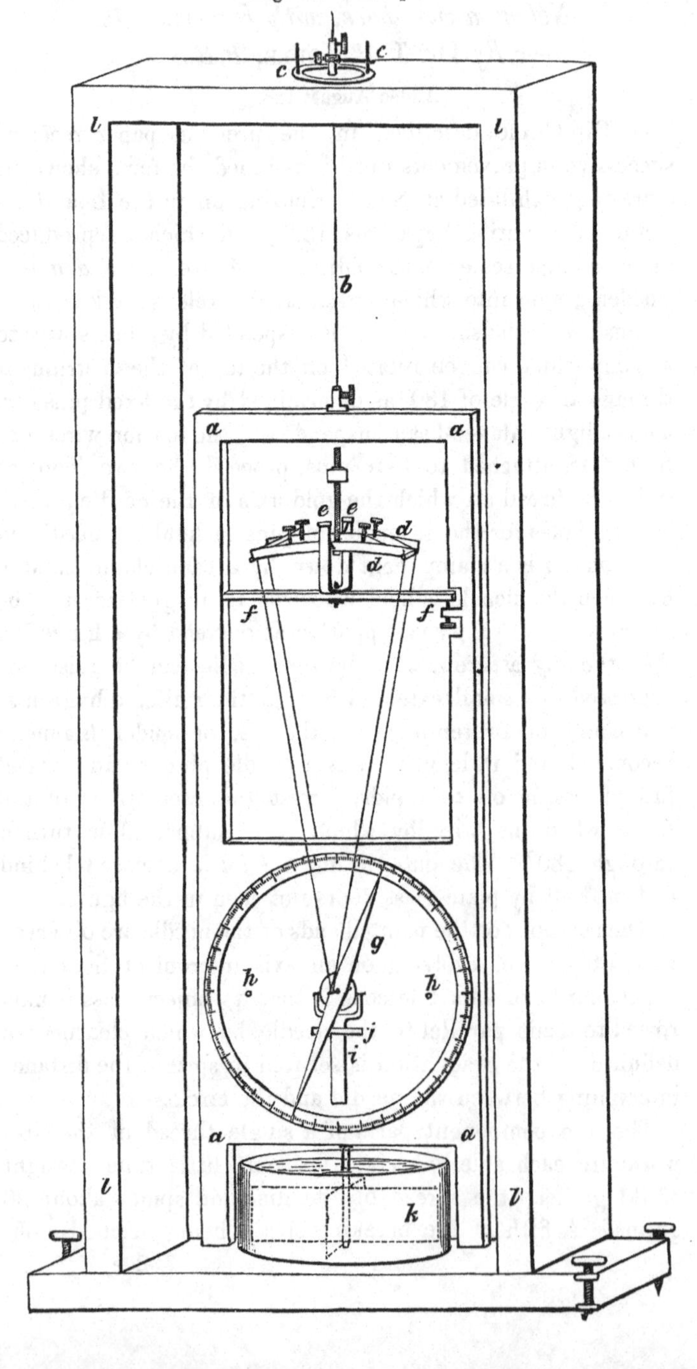

cient, a needle weighing 63 grains may be hung on silk, and a needle of 16 grains on spider threads. The wide margin of strength is needful on account of the great diminution of elasticity which occurs by increase of the hygrometric condition, to which I shall subsequently allude. The needle is of softened steel spring in order that it may be thoroughly and uniformly magnetized by two strokes, one on each end, by a magnet protected from actual contact by a covering of silk or leather.

In using the instrument both ends of the needle are in succession observed. Then the rotator is moved through 180°, and the observations are repeated on the other side. The magnetism of the needle is then reversed, and a third double observation made. Lastly the rotator is brought back to its first position, and the final readings taken. There are thus altogether eight readings for one complete observation of the dip, which when done leisurely occupies about ten minutes. The difference between any two consecutive determinations is hardly ever greater than the fraction of a minute, or such as may be fairly attributed to a change in the direction of the earth's magnetism which has occurred during the interval.

After the needful correction for any slight displacement of the centre of gravity of the needle is supplied by many observations, reversal of its magnetism is not absolutely necessary, and the dip may be obtained with sufficient accuracy in three or four minutes.

The viscosity of the suspending fibres has a slight tendency to increase the value of the dip, but to so trifling an extent that its influence may be neglected.

The following experiments on the strength of silk and spider filaments were made by me in 1870. The breaking-weights of the two threads were respectively 80 and 18 grs. Having suspended a weight of 20 grs. to the silk and 5 grs. to the spider line, I placed them along with a thermometer in a glass vessel, and then observed their lengths to experience the following changes :—

Date.	Tempe-rature.	Atmospheric condition.	Length of Silk. inches.	Length of Spider-line. inches.
May 23	62°	Moist.	12·0	7·0
„ 24	63·4	Getting gradually drier till the 30th.	12·03	7·015
„ 25	62·3		12·062	7·027
„ 26	63·7		12·07	7·031
„ 27	67·0		12·074	7·051
„ 28	65·1		12·077	7·05
„ 29	69·0		12·073	7·058
„ 30	66·5	Moist with rain.	12·081	7·07
„ 31	61·7		12·09	7·07
June 1	59·0		12·09	7·07
„ 2	60·7		12·092	7·071

On heating the filaments by holding a hot iron near them each became shortened by about $\frac{1}{35}$ of an inch. Then on moistening them with water they were elongated to 12·33 inches and 7·6 inches respectively. After drying, the observations were renewed as follow :—

Date.		Tempe-rature.	Length of Silk.	Length of Spider-line.
June 5	9.0 A.M.	66°	12·33	7·35
	4.30 P.M.	78	12·32	7·345
„ 6	6.45 A.M.	64	12·324	7·352
	4.30 P.M.	79	12·32	7·342
„ 7	8.20 A.M.	67	12·314	7·368
	5.0 P.M.	80	12·317	7·348
„ 8	9.0 A.M.	71·5	12·332	7·356
	3.2 P.M.	77·6	12·318	7·352
„ 9.	9.15 A.M.	64	12·327	7·356

On adding 20 grs. to the load on the silk filament, making on the whole 40 grs., its length was increased to 12·407, stretching out gradually till June 11, when it broke. The spider-thread had 3 grs. added, making 8 grs. altogether. Its length then became 7·386 on the 11th June, on which day I gradually increased the load to 15 grs., which broke it after stretching it a short time.

In December 1870 I resumed these experiments. I obtained a very strong diadema-spider thread whose breaking-weight was 27 grs. Having attached to it a weight of 10 grs., I suspended it in a glass tube having some chloride of calcium at the bottom in order to dry the air. This tube was placed in a larger beaker to act as a bath.

Date.				Temperature.	Length.	
				°	in.	
Dec. 9.	12.0	NOON	..	39·5	23·125	0·150 shortening for 83°·5 rise.
„	12.5	P.M.	..	71·8	23·032	
„	12.10	„	..	93·5	22·994	
„	12.15	„	..	123·0	22·975	0·172 lengthening for 85°·0 fall.
„	12.20	„	..	81·5	23·022	
„	12.25	„	..	44 0	23·113	
„	12.30	„	..	38·0	23·147	0·125 shortening for 87°·5 rise.
„	12.35	„	..	74·2	23·085	
„	12.40	„	..	125·5	23·022	0·178 lengthening for 87°·5 fall.
„	12.45	„	..	76·0	23·100	
„	12.50	„	..	38·0	23·200	
Dec. 10.	10.0	A.M.	..	40	23·185	0·105 shortening for 74° rise.
„	10.5	„	..	76·5	23·119	
„	10.10	„	..	114·0	23·080	0·140 lengthening for 73°·8 fall.
„	10.15	„	..	85·3	23·120	
„	10.20	„	..	40·2	23·220	
Dec. 11.	10.0	„	..	40	23·180	
Dec. 12.	10.0	„	..	38·8	23·172	0·092 shortening for 93°·2 rise.
„	10.5	„	..	79·5	23·123	
„	10.10	„	..	132·0	23·080	0·144 lengthening for 94°·5 fall.
„	10.15	„	..	83·7	23·130	
„	10.20	„	..	37·5	23·224	0·119 shortening for 88°·7 rise.
„	10.25	„	..	80·0	23·150	
„	10.30	„	..	126·2	23·105	0·144 lengthening for 91°·7 fall.
„	10.35	„	..	83·2	23·156	
„	10.40	„	..	42·0	23·231	
„	10.45	„	..	37·3	23·240	
„	10.50	„	..	34·5	23·249	0·111 shortening for 113°·5 rise.
„	10.55	„	..	55·3	23·215	
„	11.0	„	..	84·2	23·166	
„	11.5	„	..	102·0	23·147	
„	11.10	„	..	125·1	23·125	
„	11.15	„	..	148·0	23·138	0·262 lengthening for 112°·8 fall.
„	11.20	„	..	145·8	23·153	
„	11.25	„	..	116·0	23·188	
„	11.30	„	..	114·0	23·196	
„	11.35	„	..	87·5	23·231	
„	11.40	„	..	53·8	23·319	
„	11.45	„	..	35·2	23·400	0·115 shortening in 4 days.
„	12.15	P.M.	..	36·3	23·402	
„	12.45	„	..	37·5	23·400	
„	2.0	„	..	39·3	23·375	
Dec. 13.	11.30	A.M.	..	42·0	23·289	
Dec. 14.	10.0	„	..	46·8	23·285	
Dec. 16.	11.45	„	..	42·0	23·285	0·013 shortening for 9° rise.
Dec. 23.			..	33·5	23·280	
„			..	42·5	23·267	0·026 lengthening for 20°·8 fall.
Dec. 24.			..	21·7	23·293	0·021 shortening for 13°·3 rise.
„			..	35	23·272	
1871.						
April 4.			..	48	23·288	
June 14.			..	66	23·300	

On the last-named date I poured a little water into the tube, which had the immediate effect of lengthening the filament half an inch. Then, on reclosing the tube, I obtained the following measurements :—

Date.	Temperature.	Length.
June 15	68°	23·96
„ 16	68·5	24·04
„ 17	67·5	24·09
„ 18	67·0	24·14
„ 19	66·0	24·125
„ 20	64·5	24·105
„ 21	60·7	24·09
„ 22	69·0	24·092
„ 23	55·0	24·09

I now removed the calcium chloride, and left the tube with both ends open, so as to expose the filament to the full influence of the atmosphere.

Date.	Temperature.		Length.
	Dry-bulb Ther.	Wet-bulb Ther.	
June 23	56°·88	52°·9	23·979
„ 24	59·5	55·14	23·994
„ 25	58·82	55·12	23·994
„ 26	60·95	56·82	23·990
„ 27	62·0	57·62	23·992
„ 28	61·35	58·0	24·010
„ 29	61·5	58·1	24·016

On the 30th and four following nights the window of the laboratory was left open in damp weather, with the effect of lengthening the filament to 24·060. It was found broken on the morning of July 5.

I suspended a weight of 8 grs. from another diadema-spider thread on July 8. Its original breaking-weight was 24 grs. Its length was examined daily until August 8, when it broke. In this experiment the filament was exposed alternately to the atmosphere in its natural state and when dried by sulphuric acid. In the former case it was, *cæteris paribus*, about $\frac{1}{1000}$ longer than in the latter. The filament increased in length from 11·62 inches to 11·93 in the first half of the month, and then to 11·99 in the last half.

I took a remnant of the last broken thread and immersed

it alternately in air and water, each period of immersion being about three minutes. I added the stretching-weights grain by grain till it broke.

Stretching-weight.	Length in Air.	Length in Water.
1 gr.	3·5	3·63
2 „	3·75	
3 „	3·75	
4 „	3·87	3·94
5 „	3·94	4·00
6 „	4·00	4·00
7 „	4·00	4·04
8 „	4·05	4·125
9 „	4·11	
10 „	4·09	4·15
11 „	4·15	broke.

From the foregoing observations it will appear that both the silkworm- and the spider-filaments comport themselves in a similar manner to that I have long ago observed in caoutchouc. They assume a greater tensile force when they are heated, and when strained they have a tendency to return to their first condition. Used in an ordinary atmosphere, they are able to sustain for an indefinite time a load of one third of their immediate breaking-weight. Hence, although the large margin of strength already named may be advisable when the instrument is exposed in very damp weather, a needle of 27 grs. may be safely suspended on moderately strong diadema-loops in the ordinary atmospheric condition.

Observations with the new Dip-Circle.
By Dr. Joule, *F.R.S.*

[Read before the Manchester Literary and Philosophical Society, Nov. 1, 1870. Proceedings of the Society, vol. x. p. 15.]

The author exhibited a series of curves obtained by Dr. Stewart, F.R.S., from the self-recording instruments at the Kew Observatory, showing the large amount of disturbance of the magnetic declination and horizontal force during the progress of the Aurora of the 25th of October. He also showed a curve of the changes which took place in the magnetic dip as observed on the same day by himself at Broughton. The most remarkable variation occurred during the interval between 6^h 15^m and 6^h 23^m G. M. T., when the dip increased from 69° 8′ to 70° 30′.

Notice of, and Observations with, a new Dip-Circle. *By* J. P. JOULE, *LL.D., F.R.S.*

[Report of the British Association, Edinburgh, 1871, p. 48.]

THE method of suspension of the needle, which formed the principal feature of the new instrument, was explained. The increased facilities of observation had enabled the author to trace the diurnal variation of inclination with greater accuracy than he believed had hitherto been done. At Manchester, about the summer solstice, the greatest inclination was found to occur at 21^h 40^m mean local time, and the mean range extended 5 of a degree. The simultaneous variation of the horizontal force was such as to indicate that the total intensity was sensibly a constant quantity.

Description of an Electric-Current Meter.* *By* J. P. JOULE, *D.C.L., F.R.S.*

THIS instrument is similar in general form to that described in the Report on the heat developed by currents of electricity read in 1867 before the meeting of the British Association.

* Constructed at Thorncliff, Old Trafford, near Manchester. The term was adopted after some consideration. "*Meter* (from Mete), a measurer" (Johnson), has too long been in use as an English word to allow of its combination with Greek or Latin words. Such hybrid words as *barometer*, *hygrometer*, *chronometer*, &c., are unfortunate, and tend to destroy the elegance and purity of the English language.

The dimensions of the new meter which I am about to describe are, however, lessened by the use of coils composed of copper riband instead of round wire. The dimensions of the coils and their distances asunder are exactly three fourths of those of the original instrument. So that the two are strictly comparable.

In fig. 99, *a a* represents a brass tube or balance-beam, at the far end and at the top of which are screws with

Fig. 99. Scale $\frac{1}{8}$.

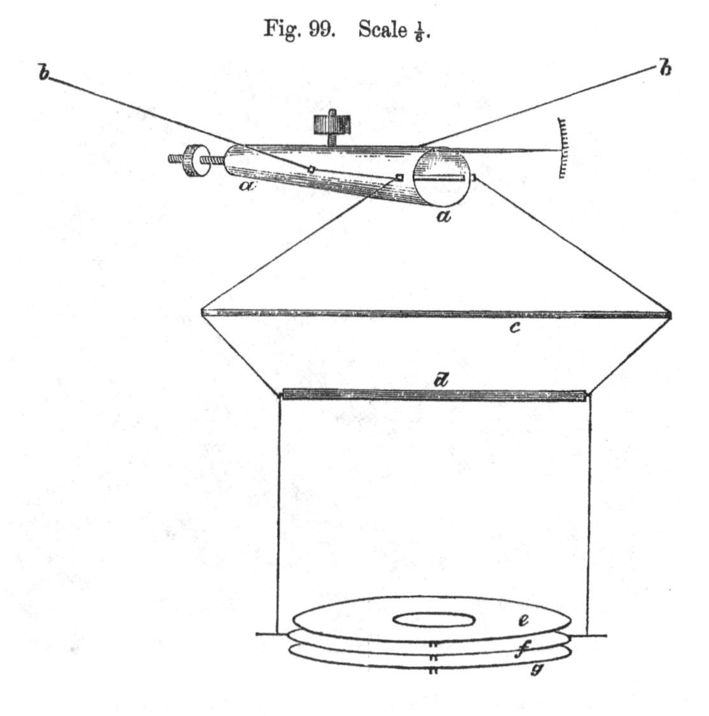

weights on them for the purpose of adjusting its equilibrium. On each side are copper wires *b b*, by which the balance-beam is sustained. These wires are insulated from the brass tube by ivory pieces. Passing, without touching the brass, to the front of the beam, they are stretched by a glass tube *c*, then support the glass shelf *d*, and finally descend perpendicularly to the movable coil *f* placed between the fixed coils *e* and *g*. These three coils are shown by the explanatory figures 100, 101, and 102. The actual number of convolutions in each coil is 66, though for the sake of clearness only 11 are represented in the figures. These coils or spirals are constructed of ribands of good conducting copper, $\frac{1}{6}$ of an inch broad and about $\frac{1}{30}$ of an inch thick, insulated by being bound with silk tape. Figs. 101 and 102 are the fixed coils attached to wooden boards, and kept at the distance of $\frac{3}{4}$ of an inch asunder, measured from the centre of the

Fig. 100. Scale $\frac{1}{3}$.

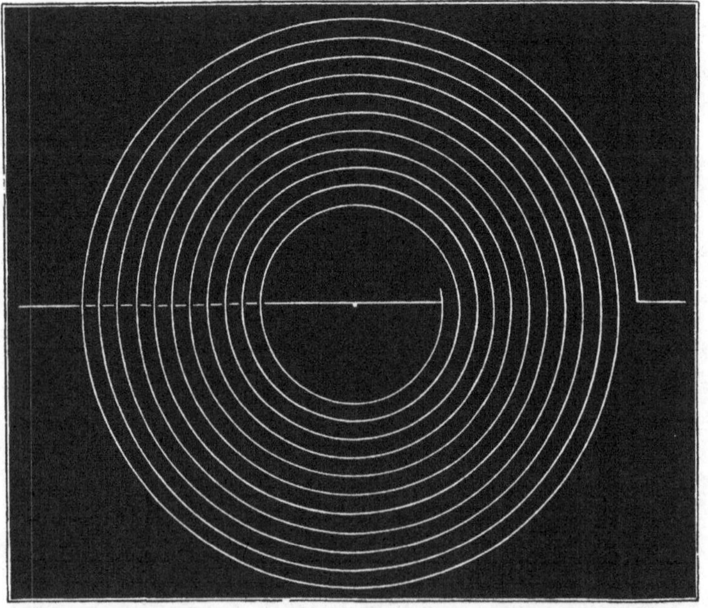

breadth of the wires. The movable coil, fig. 100, is kept from distortion by attachment to a thin sheet of glass. Its play between the two fixed coils is little under half an inch.

The current from the voltaic apparatus first traverses the fixed coils, then is led to one of the suspenders at *b*, whence it passes through the system of wires to the movable coil and thence to the other suspender, and so returns to the battery again. Inasmuch as the current in the movable coil goes in the same direction as in the upper fixed coil, and in a

Fig. 101. Scale $\frac{1}{3}$.

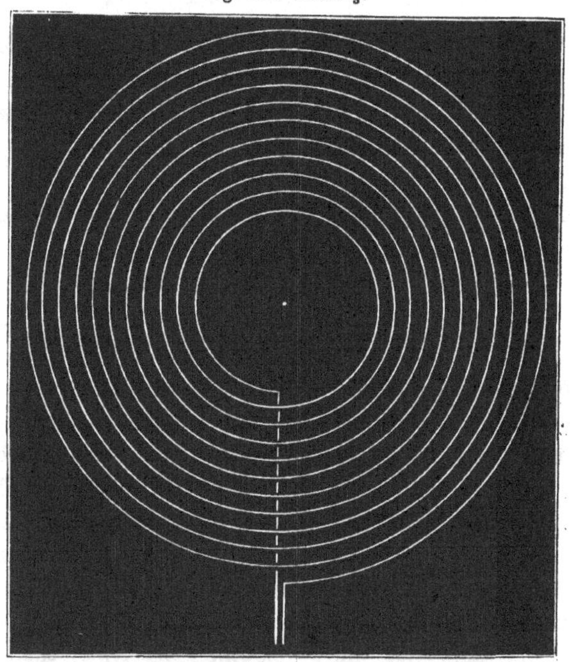

Fig. 102. Scale $\frac{1}{3}$.

contrary direction to it in the lower fixed coil, it follows that in whatsoever direction the current is passed through the instrument, its strength may be determined by the weight required to be laid on *d* to keep the suspended coil in its central position.

The strength of the current as determined by the meter is represented by the equation

$$C = \frac{1}{2l}\sqrt{\frac{agw}{\pi}}\left\{1 + \text{correction}\right\},$$

where l is the length of wire in any one of the three equal coils, a the area of a coil, and w the weight required to keep the movable coil in the central position. The correction depends upon the distance between the movable coil and the fixed ones; and being very difficult to estimate with the requisite exactness, I have determined it by passing a current through the meter and a tangent galvanometer having a single circle of known diameter, ascertaining at nearly the same time the horizontal intensity of the earth's magnetism at the place. In this way I have found the value of the correction to be ·1185 for both the old and the new instruments.

The current-meter supplies also a ready and accurate method of ascertaining the horizontal intensity of the earth's magnetism in absolute measure. For this purpose it is only necessary to pass a current in one stream through the meter and a tangent galvanometer. The formula by my old instrument along with a tangent galvanometer, furnished with a single circle 15·0535 inches in diameter, is $H = \frac{\cdot 17676\sqrt{w}}{\tan\theta}$.

In the new one $H = \frac{\cdot 11514\sqrt{w}}{\tan\theta}$. The first formula was employed in my paper on the heat evolved by the standard wire of the British Association; but the last is considerably more sensitive, accurate results being obtained using a single cell of Bunsen's battery. I exhibited the new instrument at the Manchester Literary and Philosophical Society on January 11, 1870; and the observations with it, which occupied only a few minutes, showed that the absolute horizontal intensity of the earth's magnetism in the hall of the Society was 3·83. The meter in connexion with a dip circle galvanometer may also be conveniently employed to ascertain the total terrestrial magnetic intensity.

Account of Experiments on Magnets begun in 1864 *at Old Trafford.* *By* J. P. Joule, *F.R.S.*

[Written in 1882–83.]

In the endeavour to improve some of the instruments used in magnetical observations, I have been led to seek the means of procuring magnets of high intensity and great permanency.

Scoresby * appears to have been the first who made a careful investigation of the magnetical properties of steel under the varied conditions of quality, mass, shape, and temper. In order to obtain the highest retentive power, Scoresby found that it was necessary to harden his bars to the greatest possible extent and along their entire length, and, in compound magnets, to separate the constituent plates by intervening spaces.

Having early recognized the importance of using small magnets in the galvanometer, I was guided by Scoresby's results to the following investigation, in which, on the small scale, I fully corroborated the principles he had enunciated.

Common sewing-needles, cut into uniform lengths and magnetized, were made up into systems by inserting them into the holes of perforated cardboard. But it was found that the action of the needles on each other very greatly reduced the intensity they possessed before combination, some of them even having their polarity reversed. I therefore procured a quantity of needles of various sizes in an early stage of their manufacture, immediately after the hardening

* Magnetical Observations, Part 1, 1839, and Part 2, 1843; Longman and Co.

process, in which condition they were found to be brittle, never bending permanently before breaking.

I constructed seven systems of one-inch needles, and six systems of half-inch needles. The one-inch systems were arranged either as shown in fig. 103 or as in fig. 104, the

Fig. 103. (Full size.) Fig. 104. (Full size.) Fig. 105. (Full size.) Fig. 106. (Full size.) Fig. 107. (Full size.)

former consisting of 71 needles, the latter of 34. The half-inch systems were arranged as shown in fig. 105 and fig. 106, in the former case containing 36 needles and in the latter 16.

Fig. 107, which represents a side view of one of the half-inch systems, shows the method of mounting. Two pieces of perforated cardboard, after receiving the needles, were bound to a piece of wood, and to avoid the chance of the needles being displaced, the whole was varnished by dipping it into a solution of shellac.

Table of Compound Wire-Magnets.

Distinguishing No.	Number of needles.	Their length.	Trade description of wire.	Weight of wire per foot.	Total weight of steel in the compound needle.
				grs.	grs.
1.	71	inch.	No. 6	13·3	78·67
2.	34	"	No. 6	13·3	37·67
3.	71	"	No. 7	10·8	63·9
4.	34	"	No. 7	10·8	30·6
5.	71	"	No. 8	8·36	49·48
6.	71	"	No. 9	6·79	40·18
7.	71	"	No. 10	4·9	29·0
8.	36	half inch.	No. 6	13·3	19·94
9.	16	"	No. 6	13·3	8·86
10.	36	"	No. 8	8·36	12·55
11.	16	"	No. 8	8·36	5·57
12.	36	"	No. 10	4·9	7·35
13.	16	"	No. 10	4·9	3·27

In addition to the above I procured a loadstone 0·6 inch long, weighing 53 grs., which I designated as No. 14.

The method of magnetizing, which I invariably used, was to apply the poles of two powerful electromagnets simultaneously, one at each end of the compound needle.

I have examined the deflecting powers of the above magnets during an interval of eighteen years. It would be unprofitable to publish the mass of figures obtained, as they present very uniform features which are easily summarized. In the first place I will give the average of the deflections of the inch-long magnets, and the average deflections of the half-inch long magnets at various times since 1864.

Date.	Inch-needles' deflection at 6 inches mean distance.		Half-inch needles' deflection at 3 inches mean distance.	
	°	′	°	′
1864. Aug. 21	13	5	13	52
1865. June 15	11	37	12	20
„ Nov. 13	11	20	11	58
1866. Aug. 11	11	16	12	1
1868. April 1	10	47	11	8
1874. Aug. 31	9	31	9	33
1882. Aug. 25	8	47	9	1

The horizontal intensity of the earth's magnetism must have increased by only $\frac{1}{90}$ during the 18 years, so that this element may be disregarded. Inasmuch as both the half-inch and the inch needles were magnetized to saturation in the commencement, the result that they lost the same proportion of intensity might have been anticipated. It will be seen that the rate of deterioration has by degrees become very considerably reduced.

The gradual deterioration of the magnets seemed to point to some active cause, and it occurred to me that it might be found in the changes of temperature to which the needles were incessantly exposed from day to day. To investigate this I plunged them alternately into baths of the respective temperatures 212° and 62°. After ten immersions at each temperature it was found that the average diminution of the deflecting power in the inch needles at 6 inches distance was from 12° 56′ to 12° 1′, whilst that in the half-inch needles at 3 inches was from 13° 51′ to 12° 57′.

I made similar experiments using baths of the respective temperatures 132° and 62°, the result being that the deflecting power of the inch compound needles was reduced from 13° 5′ to 12° 53′, and that of the half-inch needles from 13° 52′ to 13° 42′. Using smaller variations, viz. from 100° to 60°, the amount of deterioration was little more than 1′. It appears therefore to be nearly in proportion to the square of the amount of variation of the temperature.

On Nov. 13, 1865, having placed them in bottles, I buried two of the compound needle-magnets, viz. No. 1 and No. 8, under the surface of the ground. Their respective deflecting powers—No. 1 at six inches, and No. 8 at three inches—were 12° 11′ and 14° 48′. On the 1st of April 1868 these compound needles were again examined, and found to give deflections of 11° 49′ and 14° 31′. These on Sept. 12, 1868, were found further reduced to 10° 47′ and 12° 19′. Again, on the 21st July, 1877, the deflections observed were 10° 34′ and 11° 57′.

Comparing these results with those when the magnets were left exposed to the ordinary vicissitudes of daily temperature, it does not appear that the maintenance of a sensibly constant temperature secures the absolute constancy of the magnetic intensity; so that the cause of the gradual decline of power has yet to be discovered.

The chief violence to which the compound needles were subjected arose from the mutual induction of the individual needles. The extent of this action in the several magnets will appear by the following Table, which gives their powers after being charged to the greatest extent.

	No. of compound needle.	Thickly or thinly distributed.	Distance of deflected compass.	Total weight of steel in compound needle.	Deflection of compass-needle.	Tangent of deflection.	Weight of steel per unity of tangent.
				grs.	° ′		grs.
Inch needles.	1.	71 thickly.	6 inch.	78·67	14 27	0·257	306
	2.	34 thinly.	„	37·67	12 14	0·217	174
	3.	71 thickly.	„	63·9	14 43	0·262	244
	4.	34 thinly.	„	30·6	11 52	0·210	146
	5.	71 thickly.	„	49·48	13 18	0·236	210
	6.	71 thickly.	„	40·18	12 59	0·230	175
	7.	71 thickly.	„	29·0	11 59	0·212	137
Half-inch needles.	8.	36 thickly.	3 inch.	19·94	17 22	0·312	64
	9.	16 thinly.	„	8·86	13 38	0·242	37
	10.	36 thickly.	„	12·55	16 8	0·289	44
	11.	16 thinly.	„	5·57	12 29	0·221	25
	12.	36 thickly.	„	7·35	13 7	0·233	32
	13.	16 thinly.	„	3·27	10 27	0·184	18

We see from the above table that the inductive influence is such as to render much crowding of the needles in the formation of a compound system useless, and that even in the case of the hardest steel the limit to the increase of power with mass is soon attained.

To further illustrate the strain on the individual needles of a compound system, I took No. 1 when it had arrived at a magnetic age of 18 years. Its deflecting-power at six inches mean distance was then 10° 6′. I then pulled it to pieces, carefully examining the deflecting-power of each component needle. This was found to vary from about 16′* for the needles near the centre of the system, to 27′ for the needles near the outside, and averaged 22′ for the entire number of 71. The whole deflecting-power of the separated needles was therefore by separation increased more than 2½ times. The needles were then bound up into a bundle afresh, with the result that a deflecting-power of 9° 53′ was obtained, *i. e.* nearly equal to what it was at first.

It is evident that great permanency can be secured to slender magnetic needles by first magnetizing them to satu-

* One was found to be as low as 9′; but not one had its polarity reversed.

ration, and then reducing them by exposure to a uniform field of similar polarity. In large magnets this is difficult or impossible because different parts of the metal will be differently magnetized, and then it may very readily happen (as I have frequently observed) that violence of any kind will increase the apparent polarity of the mass by diminishing the intensity of portions of the bar magnetized contrary to the general direction.

In order further to study the effects of thermal alterations in reducing the intensity of magnets, I procured six bars of chilled cast iron, each about 3·81 inches long and weighing 353 grains. Having reduced the power of one of these from saturation to $\frac{4}{5}$, by mechanical violence, as will shortly be explained, I observed its deflecting-powers at 6 inches mean distance after the specified sudden variations of its temperature.

	°	′
Deflecting-power at commencement	35	30
„ after 10 alternations from 45° to 212° and back	29	22
„ „ 20 „ „ „	29	9
„ „ 20 „ „ „	29	0
„ „ 120 „ „ „	28	38
„ „ 160 „ „ „	28	27

Another of the chilled iron bars which had not been reduced gave me the following results:—

	°	
Deflecting-power at commencement	42	7½
„ after single alternation from 45° to 212° and back	37	34
„ after 2 alternations „ „	36	1
„ „ 4 „ „ „	35	36
„ „ 8 „ „ „	35	12
„ „ 16 „ „ „	34	55
„ „ 32 „ „ „	34	43
„ „ 64 „ „ „	34	30
„ „ 128 „ „ „	34	8

The average deflecting-power of the six chilled cast-iron magnets after the deteriorating treatment they had been subjected to was, in August 1866, 31° 14′. After sixteen years had elapsed, viz. on Sept. 1882, it was found to be 30° 53′, which declension is almost exactly accounted for by the increase of horizontal intensity during the time between the observations.

I also procured four hardened-steel bars, each of which was 4 inches long, 0·28 in diameter, and weighed 480 grains. One of these, which originally had a deflecting-power 66° 55′, was exposed to variations of temperature as follows:—

	°	′
Deflecting-power at commencement of the trials	61	26
„ after single alternation from 48° to 212° and back	61	25
„ after 2 alternations „ „	61	34
„ „ 4 „ „ „	61	27
„ „ 8 „ „ „	61	3
„ „ 16 „ „ „	60	50
„ „ 32 „ „ „	60	35
„ „ 64 „ „ „	60	23
„ „ 128 „ „ „	59	17
„ „ 256 „ „ „	57	50
„ „ 512 „ „ „	56	12
„ „ 768 „ „ „	54	31

In six years this bar, left to itself, was found to have further declined to 51° 54′.

Another of the steel bars had its deflecting-power at 6 inches mean distance examined from time to time as follows:—

	°	′
Soon after magnetization, in December 1864	70	30
August 1866	66	0
„ 1874	59	28
September 1882	58	11

Another, found to have the deflecting-power of 66° 20′ in Dec. 1864, had declined to 52° 45′ in July 1877.

The above-described facts do not encourage the hope that we may be able to procure magnets of absolute stability, although by subjecting them to alternations of temperature and other violence for considerable time we may put them into such a state that the gradual lowering of their intensity can be safely relied on in the reduction of magnetic observations.

I made many experiments with a view to ascertain the effect

of mechanical violence on the intensity of magnetic bars. If the bar is regularly magnetized, this is always a degradation. Otherwise it may well happen that, owing to parts of greater intensity overcoming parts of less intensity in the same bar, the total effect may be one of increase instead of decrease. In the following trials one of the 4-inch hardened-steel bars (p. 595) was let fall horizontally* on a sheet of plate-glass, with the results as below:—

	°	′
Intensity at beginning shown by deflection at 6 inches mean distance	66	55
Intensity after 1 fall from 6 inches high	65	0
Intensity after 36 falls ”	64	20

Experiments with the chilled iron bar used (p. 594):—

	°	
Intensity at commencement shown by deflection at 6 inches	41	42
Intensity after 1 fall from 6 inches	38	20
Intensity after 2 more falls	37	0
Intensity after 36 falls	35	26

Experiments with a glass-hard, a hard, and a soft steel bar. Attraction at 6 inches.

	Glass-hard bar.		Hard.		Soft bar.	
	°	′	°	′	°	′
At commencement	51	33	36	28	27	58
After one 6-inch fall	51	19	35	31	17	38
” another 6-inch fall	51	28	34	55	13	38
” ” ”	51	33	34	15	12	58
” ” ”	50	58	34	3	12	28
” ” ”	50	33	33	49	12	8
” ” ”	50	26	34	13	11	29
” 4 more falls	50	28	33	49	10	13
” 4 ”	50	15	33	55	9	48
” 8 ”	50	22	33	49	8	58
” 8 ”	50	22	33	43	8	23
” 22 ”	50	23	33	45		

In the above table we see, what might have been anticipated, how greatly the effects of percussion are increased by the softness of the metal. This is still further exemplified by the following results obtained with a 4-inch bar of

* The bars struck the glass nearly at right angles to the magnetic meridian, but the terrestrial magnetism was not sufficiently intense to have much influence; hence, in my experiments with perpendicular falls, it mattered little which end of a bar struck the ground.

specially softened iron, 4 inches long and $\frac{1}{4}$ by $\frac{1}{8}$ in. transverse dimensions. Notwithstanding its softness, this soft iron wire held a very good charge of magnetism.

Falls from one inch high :—

	°	′
At commencement	11	42
After one fall	4	48
„ another fall........	3	49
„ another fall........	3	34
„ three falls more....	3	14
„ six „	3	0
„ six „	2	47

After remagnetizing, the effects of falls of 6 inches were observed :—

	°	′
At commencement	9	47
After one fall	1	32
„ another fall	1	9
„ another fall	0	58

After heating the same bar to redness, and then cooling in air :—

	°	′
At commencement	6	20
After one fall of 6 inches ..	0	40

After heating again to redness, and plunging into cold water :—

	°	′
At commencement	6	40
After one fall	2	0

In passing from this part of my subject I would suggest the study of the effects of musical vibrations on the intensities of magnetic bars.

How far the intensity of a magnet is a function of its temperature has been the subject of several experiments. On raising the temperature of a magnet it loses intensity, as is well known, and then on cooling it it regains much of its power. The loss on heating is greater than the gain on cooling, because the mechanical violence occasioned by alteration of temperature produces a loss which increases the former, but decreases the latter effect.

In the following experiments the magnets were plunged into a bath of water, the temperature of which was made to alternate between about 55° and 200°, but all the results are reduced to 180°, the interval between the freezing- and boiling-points. The fractions represent the ratio of gain or loss to the mean intensity of the bar in the respective trials.

System of stout inch needles (No. 1). Deflecting-power at 6 inches, 13°.

Loss on raising the temperature 180° .... $\frac{1}{10}$.

Gain on lowering „ „ .... $\frac{1}{18}$.

System of less-crowded needles (No. 2). Deflecting-power at 6 inches, 10° 45′.

Loss on raising the temperature 180° .... $\frac{1}{21}$.

Gain on lowering „ „ .... $\frac{1}{36}$.

System of fine close-set needles (No. 7). Deflecting-power at 6 inches, 13° 8′.

Loss on raising the temperature 180° .... $\frac{1}{10 \cdot 7}$.

Gain on lowering „ „ .... $\frac{1}{22 \cdot 1}$.

Loss on raising „ „ .... $\frac{1}{18 \cdot 6}$.

Gain on lowering „ „ .... $\frac{1}{21 \cdot 1}$.

Four-inch steel bar (hard). Deflecting-power at 6 inches, 34° 17′.

1. Loss on raising the temperature 180° .... $\frac{1}{4}$.

1. Gain on lowering „ „ .... $\frac{1}{12}$.

2. Loss on raising „ „ .... $\frac{1}{8 \cdot 1}$.

2. Gain on lowering „ „ .... $\frac{1}{10 \cdot 7}$.

Remagnetized. Deflecting-power at 6 inches, 33° 44′.

1. Loss on raising the temperature 180° .... $\frac{1}{4 \cdot 6}$.

1. Gain on lowering „ „ .... $\frac{1}{11 \cdot 8}$.

2. Loss on raising „ „ .... $\frac{1}{9 \cdot 1}$.

2. Gain on lowering „ „ .... $\frac{1}{10 \cdot 6}$.

3. Loss on raising the temperature 180°			$\frac{1}{9\cdot1}$.
3. Gain on lowering	,,	,,	$\frac{1}{10\cdot6}$.
4. Loss on raising	,,	,,	$\frac{1}{10\cdot2}$.
4. Gain on lowering	,,	,,	$\frac{1}{10\cdot7}$.
5. Loss on raising	,,	,,	$\frac{1}{9\cdot2}$.
5. Gain on lowering	,,	,,	$\frac{1}{10\cdot4}$.
6. Loss on raising	,,	,,	$\frac{1}{9\cdot8}$.
6. Gain on lowering	,,	,,	$\frac{1}{10\cdot4}$.
7. Loss on raising	,,	,,	$\frac{1}{9\cdot8}$.
7. Gain on lowering	,,	,,	$\frac{1}{10\cdot1}$.
Average loss, excluding the first			$\frac{1}{9\cdot5}$.
Average gain	,,	,,	$\frac{1}{10\cdot3}$.

Same bar annealed and then remagnetized. Deflecting-power at 6 inches, 26° 51′.

1. Loss on raising the temperature 180°			$\frac{1}{4\cdot3}$.
1. Gain on lowering	,,	,,	$\frac{1}{7\cdot6}$.
2. Loss on raising	,,	,,	$\frac{1}{6}$.
2. Gain on lowering	,,	,,	$\frac{1}{6\cdot4}$.
3. Loss on raising	,,	,,	$\frac{1}{6\cdot6}$.
3. Gain on lowering	,,	,,	$\frac{1}{6\cdot8}$.
4. Loss on raising	,,	,,	$\frac{1}{6\cdot5}$.
4. Gain on lowering	,,	,,	$\frac{1}{7}$.
Average loss, excluding the first			$\frac{1}{6\cdot36}$.
Average gain	,,	,,	$\frac{1}{6\cdot74}$.

The same bar hardened and then remagnetized. Deflecting-power at 6 inches, 35° 47′.

1. Loss on raising the temperature 180°			$\frac{1}{3\cdot4}$.
1. Gain on lowering	,,	,,	$\frac{1}{13\cdot4}$.
2. Loss on raising	,,	,,	$\frac{1}{12\cdot7}$.
2. Gain on lowering	,,	,,	$\frac{1}{34\cdot2}$.

Magnetized afresh.

1. Loss on raising the temperature 180°	$\frac{1}{7\cdot8}$.
1. Gain on lowering „ „	$\frac{1}{19\cdot2}$.
2. Loss on raising „ „	$\frac{1}{16\cdot6}$.
2. Gain on lowering „ „	$\frac{1}{19}$.
3. Loss on raising „ „	$\frac{1}{18\cdot3}$.
3. Gain on lowering „ „	$\frac{1}{22\cdot4}$.
4. Loss on raising „ „	$\frac{1}{17\cdot3}$.
4. Gain on lowering „ „	$\frac{1}{19\cdot4}$.
Average loss, excluding the first in each set	$\frac{1}{17\cdot4}$.
Average gain „ „ „	$\frac{1}{20\cdot2}$.

The same bar glass-hardened and magnetized by an electric coil. Deflecting-power at 6 inches, 52°.

1. Loss on raising the temperature 180°	$\frac{1}{8\cdot2}$.
1. Further loss on lowering the temperature 180°	$\frac{1}{70\cdot4}$.
2. Loss on raising the temperature 180°	$\frac{1}{20\cdot1}$.
2. Gain on lowering „ „	$\frac{1}{65}$.
3. Loss on raising „ „	$\frac{1}{25\cdot2}$.
3. Gain on lowering „ „	$\frac{1}{51\cdot5}$.
4. Loss on raising „ „	$\frac{1}{23\cdot4}$.
4. Gain on lowering „ „	$\frac{1}{48\cdot7}$.
Average loss, excluding the first	$\frac{1}{22\cdot7}$.
Average gain „ „	$\frac{1}{50\cdot4}$.

The same bar hardened again, by being plunged at a high temperature into mercury. It was magnetized by an electric coil. Deflecting-power at 6 inches, 54° 42′.

1. Loss on raising the temperature 180°	$\frac{1}{4\cdot1}$.
1. Further loss on lowering the temperature 180°	$\frac{1}{74}$.
2. Loss on raising the temperature 180°	$\frac{1}{19\cdot1}$.
2. Gain on lowering „	$\frac{1}{67}$.

3. Loss on raising the temperature 180° .... $\frac{1}{20 \cdot 1}$.
3. Gain on lowering „ „ .... $\frac{1}{53}$.
4. Loss on raising „ „ .... $\frac{1}{21 \cdot 3}$.
4. Gain on lowering „ „ .... $\frac{1}{44}$.
5. Loss on raising „ „ .... $\frac{1}{23 \cdot 6}$.
5. Gain on lowering „ „ .... $\frac{1}{61 \cdot 7}$.
Average loss, excluding the first .... $\frac{1}{20 \cdot 9}$.
Average gain „ „ .... $\frac{1}{55}$.

Steel bar 2 inches long, hardened. Deflecting-power at 6 inches, 11° 12′.

1. Loss on raising the temperature 180° .... $\frac{1}{9 \cdot 9}$.
1. Gain on lowering „ „ .... $\frac{1}{37 \cdot 6}$.
2. Loss on raising „ „ .... $\frac{1}{6 \cdot 5}$.
2. Gain on lowering „ „ .... $\frac{1}{93}$.
3. Loss on raising „ „ .... $\frac{1}{10 \cdot 5}$.
3. Gain on lowering „ „ .... $\frac{1}{300}$.
4. Loss on raising „ „ .... $\frac{1}{14 \cdot 6}$.
4. Gain on lowering „ „ .... $\frac{1}{73 \cdot 5}$.
5. Loss on raising „ „ .... $\frac{1}{7 \cdot 8}$.
5. Gain on lowering „ „ .... $\frac{1}{73 \cdot 4}$.
6. Loss on raising „ „ .... $\frac{1}{29}$.
6. Gain on lowering „ „ .... $\frac{1}{133}$.
Average loss, excluding the first .... $\frac{1}{9}$.
Average gain „ „ .... $\frac{1}{97}$.

Same bar tempered.

1. Loss on raising the temperature 180° .... $\frac{1}{2 \cdot 9}$.
1. Gain on lowering „ „ .... $\frac{1}{5 \cdot 9}$.
2. Loss on raising „ „ .... $\frac{1}{3 \cdot 7}$.
2. Gain on lowering „ „ .... $\frac{1}{5 \cdot 5}$.

Same bar remagnetized. Deflection at 6 inches, 6° 0′.

1. Loss on raising the temperature 180° .... $\frac{1}{3\cdot2}$.

1. Gain on lowering ,, ,, ... $\frac{1}{4\cdot9}$.

2. Loss on raising ,, ,, ... $\frac{1}{5\cdot1}$.

2. Gain on lowering ,, ,, ... $\frac{1}{5\cdot3}$.

3. Loss on raising ,, ,, ... $\frac{1}{3\cdot8}$.

3. Gain on lowering ,, ,, ... $\frac{1}{4}$.

4. Loss on raising ,, ,, ... $\frac{1}{4\cdot4}$.

4. Gain on lowering ,, ,, ... $\frac{1}{4\cdot6}$.

Average loss, excluding the first .... $\frac{1}{4\cdot37}$.

Average gain ,, ,, ... $\frac{1}{4\cdot57}$.

Same bar annealed and remagnetized. Deflecting-power at 6 inches, 5° 41′.

1. Loss on raising the temperature 180° .... $\frac{1}{1\cdot9}$.

1. Gain on lowering ,, ,, ... $\frac{1}{6\cdot1}$.

2. Loss on raising ,, ,, ... $\frac{1}{4}$.

2. Gain on lowering ,, ,, ... $\frac{1}{5}$.

3. Loss on raising ,, ,, ... $\frac{1}{5}$.

3. Gain on lowering ,, ,, ... $\frac{1}{5\cdot6}$.

4. Loss on raising ,, ,, ... $\frac{1}{5\cdot2}$.

4. Gain on lowering ,, ,, ... $\frac{1}{51}$.

Average loss, excluding the first .... $\frac{1}{4\cdot67}$.

Average gain ,, ,, ... $\frac{1}{5\cdot23}$.

Very well annealed iron bar 4 inches long, $\frac{3}{16}$ broad, $\frac{1}{8}$ thick. Deflecting-power at 6 inches, 11° 40′.

1. Loss on raising the temperature 180° .... $\frac{1}{9\cdot5}$.

1. Gain on lowering ,, ,, ... $\frac{1}{61}$.

2. Loss on raising ,, ,, ... $\frac{1}{19}$.

2. Gain on lowering ,, ,, ... $\frac{1}{29\cdot1}$.

3. Loss on raising ,, ,, ... $\frac{1}{311}$.

3. Gain on lowering ,, ,, ... $\frac{1}{60}$.

4. Loss on raising the temperature 180° .... $\frac{1}{157}$.

4. Gain on lowering " " .... $\frac{1}{22}$.

5. Loss on raising " " .... $\frac{1}{16}$.

5. Gain on lowering " " .... $\frac{1}{18 \cdot 5}$.

Average loss, excluding the first .... $\frac{1}{32 \cdot 1}$.

Average gain " " .... $\frac{1}{26 \cdot 6}$.

The same iron bar, after being plunged red-hot into cold water, and afterwards remagnetized. Deflecting-power at 6 inches, 5° 4′.

1. Loss on raising the temperature 180° .... $\frac{1}{7 \cdot 3}$.

1. Gain on lowering " " .... $\frac{1}{74}$.

2. Loss on raising " " .... $\frac{1}{13 \cdot 4}$.

2. Gain on lowering " " .... $\frac{1}{13 \cdot 3}$.

3. Loss on raising " " .... $\frac{1}{19 \cdot 2}$.

3. Gain on lowering " " .... $\frac{1}{19 \cdot 3}$.

4. Loss on raising " " .... $\frac{1}{11 \cdot 1}$.

4. Gain on lowering " " .... $\frac{1}{15 \cdot 4}$.

Average loss, excluding the first .... $\frac{1}{13 \cdot 8}$.

Average gain " " .... $\frac{1}{15 \cdot 6}$.

The same iron bar, plunged at a white heat into mercury, then magnetized afresh. Deflecting-power at 6 inches, 6° 23′.

1. Loss on raising the temperature 180° .... $\frac{1}{8 \cdot 9}$.

1. Gain on lowering " " .... $\frac{1}{23 \cdot 4}$.

2. Loss on raising " " .... $\frac{1}{13 \cdot 7}$.

2. Gain on lowering " " .... $\frac{1}{17 \cdot 1}$.

3. Loss on raising " " .... $\frac{1}{17 \cdot 1}$.

3. Gain on lowering " " .... $\frac{1}{15 \cdot 6}$.

Average loss, excluding the first .... $\frac{1}{15 \cdot 2}$.

Average gain " " .... $\frac{1}{16 \cdot 3}$.

I have examined two loadstones, and they appear to have lost no appreciable amount of magnetic virtue during 18 years. However, they had previously been subjected to the effects of ten alternate plunges into cold and hot water, which in one of them produced a diminution of intensity amounting to $\frac{1}{7}$. The changes of intensity by raising the temperature 180° and lowering the same amount were as follows:—

Loadstone $\frac{5}{8}$ inch long.

1. Loss on raising the temperature 180° .... $\overline{10\cdot5}$.
1. Gain on lowering " " .... $\frac{1}{54\cdot2}$.
2. Loss on raising " " .... $\frac{1}{15\cdot2}$.
2. Gain on lowering " " .... $\frac{1}{18\cdot9}$.

Loadstone 1 inch square.

1. Loss on raising the temperature 180° .... $\frac{1}{9\cdot6}$.
1. Gain on lowering " " .... $\frac{1}{26\cdot8}$.
2. Loss on raising " " .... $\frac{1}{20\cdot4}$.
2. Gain on lowering " " .... $\frac{1}{26\cdot8}$.

The temperature of the bath, by promoting currents of air, might have exercised some influence on the needle, but as this was well screened off, the irregularities in the foregoing results cannot be accounted for in this way. We may perhaps draw the general conclusion that in steel bars thoroughly hardened the temperature-coefficient is smaller than in those which are annealed. But this seems not to be the case with wrought-iron magnets. Mr. Whipple's observations (Proceedings R. S. vol. xxvi.) were all made with steel of the best quality and perfectly hardened, and yet they, like my own, present great irregularities. It is probable that, whatever process is employed for hardening steel, inequalities in its physical condition will manifest themselves in altering the temperature-coefficient for the magnetic moment; and these effects will doubtless be studied with much advantage by future enquirers.

On some Physical Properties of Bees'-wax. By J. P. Joule, *F.R.S., in a Letter to Wm. Fairbairn, F.R.S., dated Salford, August* 24, 1853.*

[Life of Sir W. Fairbairn, Bart., by W. Pole, F.R.S., p. 296.]

I transmitted to Mr. Hopkins yesterday an account of some experiments on the physical properties of bees'-wax†, which may perhaps serve to throw some light on the experiments on the alteration of the point of liquefaction by pressure. The results I arrived at are as follows:—

Specific heat	between	48°·88	Cent.	and	18°·74	Cent.	=	0·991
"	"	39·8	"	"	19·1		=	0·923
"	"	32	"	"	18·34		=	0·647

The wax softened gradually until the point of absolute fluidity, viz. 54° Cent., was reached. The increase of the specific heat at high temperatures was evidently owing to the heat due to the change of state being mixed therewith.

I found in one experiment that the specific heat of the wax in a perfect fluid condition was 0·506. Another experiment gave me 0·509.

Taking the specific heat to be 0·5 both in the states of perfect solidity and perfect fluidity, I find the heat absorbed in changing the state of one grain of wax from perfect solidity to perfect fluidity to be 34°·2 Cent. per one grain of water.

The expansion of 61·828 grains of bees'-wax I found to be

From	16°·6	Cent. to	26°·8	Cent.	=	Volume of	0·793	grain	of water.
"	26·8	"	37·4	"	=	"	2·739		"
"	37·4	"	49·0	"	=	"	4·255		"
"	49·0	"	53·4	"	=	"	1·609		"
"	54·3	"	67·4	"	=	"	0·578		"

The total expansion of 61·828 grs. of wax between 26°·8 and 53°·4 is therefore equal to the volume of 8·603 grs. of water. The volume of the wax at the former temperature is

* The experiments were made at Acton Square, Salford.

† The specimen was bleached.

64·795 and at the latter 73·398. Professor Thomson's formula gives, with these data, 24° Cent. as the theoretical elevation of the temperature of change of state for the greatest pressure used in our experiments, the actual result which we obtained being 17° Cent.

Yours very truly,
J. P. JOULE.

Wm. Fairbairn, Esq.

On some Photographs of the Sun. *By* J. P. JOULE.

[Proceedings of the Manchester Literary and Philosophical Society, vol. x. p. 132. Read before the Society, March 7, 1871.]

THE author exhibited three photographs of the sun which he had taken on the 1st December, 1858. The images, 0·43 inch in diameter, were produced by the achromatic object-glass of a telescope with half-inch stop. The exposure, effected by an apparatus completely detached from the camera, occupied about $\frac{1}{200}$ of a second. He had been induced to examine them afresh after seeing the beautiful photograph of the late eclipse taken by Mr. Brothers. In all three a nebulous appearance appears on three quarters of the circumference, the remainder being quite free. There are also indications of a radial structure; so that he thinks it highly probable that the representations are actually those of the corona.

Since communicating the above he has examined two other photographs of the sun, which he took early in November 1858. These, one of which must have been exposed at about 2 hours 20 minutes after the other, present nothing remarkable to the naked eye; but when viewed through a glass of moderate power, a thin crescent-shaped envelope is observed on each, with this remarkable circumstance, viz. that on the two it appears on opposite limbs, suggesting the idea of a semi-revolution in the above interval of time at a velocity not much different from that due to Kepler's law of planetary motion. In one of the photographs there is, under the crescent and apparently on the rim of the sun itself, a narrow band, in breadth about $\frac{1}{300}$ of the diameter of the disk, and of at least double the intensity of the neighbouring surface of the sun.

On Sunset seen at Southport. *By* Dr. J. P. Joule.

[Proceedings of the Manchester Literary and Philosophical Society, vol. ix. p. 1.]

I enclose a rough drawing of the appearance of the setting sun. Mr. Baxendell noticed the fact that at the moment of the departure of the sun below the horizon, the last glimpse

Fig. 108.

is coloured bluish green. On two or three occasions I have noticed this, and also near sunset an appearance like what I have rudely depicted. Just at the upper edge, where bands of the sun's disk are separated one after the other by refraction, each band becomes coloured blue just before it vanishes.

On the Alleged Action of Cold in rendering Iron and Steel brittle. *By* J. P. Joule, *D.C.L.*, *F.R.S.*, *&c.*

[Read before the Manchester Literary and Philosophical Society, Jan. 10, 1871. Proceedings of the Society, vol. x. p. 91*.]

As is usual in a severe frost, we have recently heard of many severe accidents consequent upon the fracture of the tires of

* The experiments were made at Cliff Point, Higher Broughton.

the wheels of railway-carriages. The common-sense explanation of these accidents is, that the ground being harder than usual, the metal with which it is brought into contact is more severely tried than in ordinary circumstances. In order apparently to excuse certain Railway Companies, a pretence has been set up that iron and steel become brittle at a low temperature. This pretence, although put forth in defiance, not only of all we know of the properties of materials, but also of the experience of every-day life, has yet obtained the credence of so many people that I thought it would be useful to make the following simple experiments:—

1st. A freezing-mixture of salt and snow was placed on a table. Wires of steel and of iron were stretched so that a part of them was in contact with the freezing-mixture and another part out of it. In every case I tried the wire broke outside of the mixture, showing that it was weaker at 50° F. than at about 12° F.

2nd. I took twelve darning-needles of good quality, 3 in. long, $\frac{1}{24}$ in. thick. The ends of these were placed against steel props, $2\frac{1}{8}$ in. asunder. In making an experiment, a wire was fastened to the middle of a needle, the other end being attached to a spring weighing-machine. This was then pulled until the needle gave way. Six of the needles, taken at random, were tried at a temperature of 55° F., and the remaining six in a freezing-mixture which brought down their temperature to 12° F. The results were as follows:—

Warm Needles.	Cold Needles.
64 oz. broke.	55 oz. broke.
65 „ „	64 „ „
55 „ „	72 „ „
62 „ „	60 „ bent.
44 „ „	68 „ broke.
60 „ bent.	40 „ „
Average $58\frac{1}{3}$	Average $59\frac{5}{6}$

I did not notice any perceptible difference in the perfection of elasticity in the two sets of needles. The result, as far as it goes, is in favour of the cold metal.

3rd. The above are doubtless decisive of the question at

issue. But as it might be alleged that the violence to which a railway wheel is subjected is more akin to a blow than a steady pull, and as, moreover, the pretended brittleness is attributed more to cast iron than any other description of metal, I have made yet another kind of experiment. I got a quantity of cast-iron garden nails, inch and a quarter long, and $\frac{1}{8}$ in. thick in the middle. These I weighed, and selected such as were nearly of the same weight. I then arranged matters so that by removing a prop I could cause the blunt edge of a steel chisel, weighted to 4 lb. 2 oz., to fall from a given height upon the middle of the nail as it was supported from each end, $1\frac{1}{16}$ in. asunder. In order to secure the absolute fairness of the trials, the nails were taken at random, and an experiment with a cold nail was always alternated with one at the ordinary temperature. The nails to be cooled were placed in a mixture of salt and snow, from which they were removed and struck with the hammer in less than 5″.

Up to Series 10, each set of sixteen nails was made up of those of the previous set which were left unbroken, added to fresh ones to make up the number.

Series 1. Temperature of eight cold nails 10°. Of eight warm 36°. Height of fall of hammer 2 inches.

Result. No nails broke.

Series 2. Temperature of eight cold nails 14°. Of eight warm ones 36°. Fall of hammer $2\frac{1}{2}$ inches.

Result. No nails broke.

Series 3. Temperature of eight cold nails 2°. Of eight others 36°. Fall of hammer 3 inches.

Result. One cold nail broke. No warm one broke.

Series 4. Temperature of eight cold nails 2°. Of eight others 36°. Fall of hammer $3\frac{1}{2}$ inches.

Result. Two cold nails broke. One warm one broke.

Series 5. Temperature of eight cold nails 2°. Of eight others 36°. Fall of hammer 4 inches.

Result. One broke of each sort.

Series 6. Temperature of eight cold nails 0°. Of eight others 38°. Fall of hammer $4\frac{1}{2}$ inches.

Result. One broke of each sort.

Series 7. Temperature of eight cold nails 2°. Of eight others 36°. Fall of hammer 5½ inches.

Result. No cold nail broke. One warm nail broke.

Series 8. Temperature of eight cold nails 2°. Of eight others 40°. Fall of hammer 6½ inches.

Result. Two cold nails broke. One warm nail broke.

Series 9. Temperature of eight cold nails 2°. Of eight others 40°. Fall of hammer 7½ inches.

Result. Three cold nails broke. Three warm nails broke.

Series 10. Experiment with the ten left in the last. Temperature of five cold nails 2° Of the five others 40°. Fall of hammer 8½ inches.

Result. Two cold nails broke. One warm nail broke.

Series 11. Experiment with the six left from the last. Temperature of three cold nails 3°. Of the other three 40°. Fall of hammer 10 inches.

Result. Two cold nails broke. Three warm nails broke.

Series 12. Experiment with fresh nails. Twelve cooled for four hours to 3°. Twelve others 41°. Fall 7 inches.

Result. Seven cold nails broke. Eight warm nails broke.

The collective result is that 21 cold nails broke and 20 warm ones.

The experiments of Lavoisier and Laplace, of Smeaton, of Dulong and Petit, and of Troughton, conspire in giving a less expansion by heat to steel than iron, especially if the former is in an untempered state. Such specimens of steel wire and of watchspring as I possess expand less than iron. But this, as Sir W. Fairbairn observed to me, would in certain limits have the effect of strengthening rather than of weakening an iron wheel with a tire of steel.

The general conclusion is this—Frost does *not* make either iron (cast or wrought) or steel brittle, and accidents arise from neglect to submit wheels, axles, and all other parts of the rolling stock to a practical and sufficient test before using them.

Further Observations on the Strength of Garden Nails. *By* J. P. Joule, *D.C.L., F.R.S., &c.**

[Read before the Manchester Literary and Philosophical Society, Feb. 21, 1871. Proceedings of the Society, vol. x. p. 127.]

Since communicating the paper on the Alleged Influence of Cold in giving Brittleness to Iron, I have collated the results with cast-iron nails in order to show the range of strength in such specimens.

Height of Fall of Hammer.	Percentage of Fractures.
2 inches	0
2½ "	0
3 "	6·25
3½ "	23·5
4 "	30
4½ "	36·4
5½ "	37·5
6½ "	48
7 "	62·5
7½ "	64·3
8½ "	75
10 "	92·8

I chose the garden nails for experiment after some thought, as presenting a marked variety of metal in contrast with the iron and steel wire, tempered and untempered. I did not expect them to possess great strength; but having found them to require a heavier blow than I expected to fracture them, I have had the curiosity to make some experiments on them which may be interesting to the Society.

I took pairs of the nails, placed them head to point parallel to each other so that pressure applied in the middle by pincers sufficiently forcibly would fracture one of them. Paper slips were pasted on the edges of the nails, and their distances asunder measured by a microscope with micrometer-eyepiece divided by lines corresponding to $\frac{1}{800}$ of an inch. Weights were gradually added to the lever of one arm of the pincers until fracture took place, which was

* The experiments were made at Cliff Point.

always accompanied with a sharp report. The observed deflection or bending of the nails was taken continuously as the weights were laid on, and the calculation of what it would have been at the moment of rupture taken from the immediately preceding observations. The amount of deflection was almost exactly proportional to the weight laid on in each experiment.

No. of Experiment.	Length of Nail between Supports.	Breadth o Nail at Fracture.	Depth of Nail at Fracture.	Deflection.	Breaking Weight.
	in.	in.	in.	in.	lb.
1.	1·05	0·13	0·127	·0062	145·5
2.	1·1	0·114	0·125	·0067	141
3.	1·1	0·120	0·115	·0090	171
4.	1·08	0·111	0·106	·0073	142·5
5.	1·12	0·122	0·145	·0098	189
6.	1·06	0·138	0·120	·0087	184·5
7.	1·08	0·150	0·118	·0095	201
Average	1·084	0·1264	0·1223	·0082	167·8

If we compare the above with Mr. Brockbank's experiments we shall find, approximately, on reducing them to the dimensions he adopted, viz. 3 feet between supports and 1 inch section:—

	Breaking Weight.	Deflection.
Mr. Brockbank's, with large bars ..	860·7 lb.	·740 inch.
My own, with nails	2673	1·106

The metal, in the form I used it, was therefore more than three times as strong as that of the large bars to resist a compressing and tensile force, while its extent of spring at the breaking weight was half as much again. Therefore, so far from being of inferior quality, it would sustain a very much heavier blow without fracture.

Further Observations on the Strength of Garden Nails. *By* J. P. Joule, *LL.D., F.R.S.*

[Read before the Manchester Literary and Philosophical Society, March 7, 1871. Proceedings of the Society, vol. x. p. 131.]

The author thought it desirable to ascertain how far hardness had to do with the strength and elasticity of these small specimens of cast iron. For this purpose he plunged some of them at a heat near the melting-point into water; then selecting those which had been hardened sufficiently to resist the action of the file. Others he cooled slowly from a bright red heat. The experiments were conducted in the manner described in the previous communication.

	No. of Experiment.	Length of Nail between Supports.	Breadth of Nail at Fracture.	Depth of Nail at Fracture.	Deflection.	Breaking Weight.
		in.	in.	in.	in.	lb.
Hard nails.	1.	1·0	0·11	0·122	0·0067	129
	2.	1·04	0·12	0·12	0·0037	84
	3.	1·0	0·12	0·122	0·0028	81
	4.	1·02	0·143	0·102	0·0077	129
	5.	1·1	0·138	0·13	0·0071	203
	Average	1·032	0·1262	0·1192	0·0056	125·2
Soft nails.	6.	1·0	0·112	0·117	0·0088	141
	7.	1·05	0·139	0·114	0·0087	150
	8.	1·02	0·130	0·138	0·0051	176
	9.	1·04	0·117	0·090	0·0101	101
	10.	1·04	0·121	0·108	0·0073	113
	Average	1·03	0·1238	0·1134	0·008	136·2

Reducing to a length of 3 feet and 1 inch square section, and making a deduction of one sixth from the deflections on account of the taper of the nails, the above results, along with those in the last number of the Proceedings, become

	Breaking Weight.	Deflection.
Nails in the original state	2673 lb.	0·922 inch.
Hardened nails	2002	0·677
Softened nails	2448	0·924

Examples of the Performance of the Electro-Magnetic Engine. *By* J. P. Joule, *D.C.L., F.R.S., &c.*

[Read before the Manchester Literary and Philosophical Society, March 21, 1871. Proceedings of the Society, vol. x. p. 152.]

Some experiments and conclusions I arrived at a quarter of a century ago having been recently criticised, I have thought it might be useful to place the subject of work in connexion with electro-magnetism in a different and I hope clearer form than that in which I have hitherto placed it. The numbers given below are derived from recent experiments.

Suppose an electro-magnetic engine to be furnished with fixed permanent steel magnets, and a bar of iron made to revolve between the poles of the steel magnets by reversing the current in its coil of wire. Such an arrangement is perhaps the most efficient, as it is the most simple form of the apparatus. In considering it, we will first suppose the battery to consist of 5 large Daniell's cells in series, so large that their resistance may be neglected. We will also suppose that the coil of wire on the revolving bar is made of a copper wire 389 feet long, and $\frac{1}{16}$ of an inch diameter, or offering a resistance equal to one B.A. unit. Then, on connecting the terminals of this wire with the battery, and keeping the engine still, the current through the wire will be such as, with a horizontal force of earth's magnetism 3·678, would be able to deflect the small needle of a galvanometer furnished with a single circle of one foot diameter, to the angle of 54° 23′. Also this current, going through the above wire for one hour, will evolve heat that could raise 110·66 lb. of water 1°, a quantity equal to 85430 ft. lb. of work. In the meantime the zinc consumed in the battery will be 535·25 grains. Hence the work due to each grain of zinc is 159·6 ft. lb., and heat ·20674 of a unit.

I. In the condition of the engine being kept still we have, current being 1·396, as shown by a deflection of 54° 23′ :—

1. Heat evolved per hour by the wire, 110·66 units.
2. Consumption of zinc per hour, 535·25 grains.
3. Heat due to 535·25 grains, 110·66 units.
4. Therefore the work per hour will be (110·66 — 110·66)772=0.
5. And the work per grain of zinc will be $\frac{0}{535{\cdot}25}=0$.

II. If the engine be now started and kept by a proper load to a velocity which reduces the current to $\frac{2}{3}$, or ·9307, indicated by a deflection 42° 57′, we shall have

1. Heat evolved per hour by the wire, $110{\cdot}66 \times \left\{\frac{2}{3}\right\}^2 = 49{\cdot}18$ units.

2. Consumption of zinc per hour, $535{\cdot}25 \times \frac{2}{3} = 356{\cdot}83$ grains.

3. Heat due to 356·83 grains, $110{\cdot}66 \times \frac{2}{3} = 73{\cdot}77$ units.

4. Therefore the work per hour will be $(73{\cdot}77 - 49{\cdot}18)772 = 18983$ ft. lb.

5. And the work per grain of zinc will be $\frac{18983}{356{\cdot}83} = 53{\cdot}2$, or $\frac{1}{3}$ of the maximum duty.

III. If the load be lessened until the current is reduced to $\frac{1}{2}$ of the original amount, or to ·698, we shall have

1. Heat evolved per hour by the wire, $110{\cdot}66 \times \left(\frac{1}{2}\right)^2 = 27{\cdot}665$ units.

2. Consumption of zinc per hour, $535{\cdot}25 \times \frac{1}{2} = 267{\cdot}62$ grains.

3. Heat due to 267·62 grains, $110{\cdot}66 \times \frac{1}{2} = 55{\cdot}33$.

4. Therefore the work per hour will be $(55{\cdot}33 - 27{\cdot}665)772 = 21357$ ft. lb.

5. And the work per grain of zinc will be $\frac{21357}{267{\cdot}62} = 79{\cdot}8$, or $\frac{1}{2}$ of the maximum duty.

IV. If the load be still further reduced and velocity increased so as to bring down the current to $\frac{1}{3}$ of what it was when the engine was still, or to 4653, shown by a deflection of the galvanometer of 24° 57′, we shall have

1. Heat evolved per hour by the wire, $110{\cdot}66 \times \left(\frac{1}{3}\right)^2 = 12{\cdot}294$ units.

2. Consumption of zinc per hour, $535{\cdot}25 \times \frac{1}{3} = 178{\cdot}42$ grains.

3. Heat due to 178·42 grains, $110{\cdot}66 \times \frac{1}{3} = 36{\cdot}89$ units.

4. Therefore the work per hour will be $(36{\cdot}89 - 12{\cdot}294)772 = 18988$ ft. lb.

5. And the work per grain of zinc will be $\frac{18988}{178{\cdot}42} = 106{\cdot}4$, or $\frac{2}{3}$ of the maximum duty.

V. Remove the load still further until the velocity is increased so much that the current is brought down to $\frac{1}{100}$ of its quantity when the engine is still. Then we shall have

1. Heat evolved per hour by the wire, $110{\cdot}66 \times \left(\frac{1}{100}\right)^2 = {\cdot}011066$ of a unit.
2. Consumption of zinc per hour, $535{\cdot}25 \times \frac{1}{100} = 5{\cdot}3525$ grains.
3. Heat due to 5·3525 grains of zinc, $110{\cdot}66 \times \frac{1}{100} = 1{\cdot}1066$ units.
4. Therefore the work per hour will be $(1{\cdot}1066 - {\cdot}011066)772 = 845{\cdot}73$ ft. lb.
5. And the work per grain of zinc will be $\frac{845{\cdot}73}{5{\cdot}352} = 158$, or $\frac{99}{100}$ of the maximum duty.

When the velocity increases so that the current vanishes, the duty = 159·6 or the maximum.

I. Let us now improve the engine by giving it a coil of 4 times the conductivity, which will be done by using a copper wire 389 feet long and $\frac{1}{8}$th of an inch diameter, the same battery being used as before. Then, when the engine is kept still, we shall have a current $1{\cdot}396 \times 4 = 5{\cdot}584$, shown by a deflection of 79° 51′. Then we shall have

1. Heat evolved per hour by the wire, $110{\cdot}66 \times \frac{4^2}{4} = 442{\cdot}64$ units.
2. Consumption of zinc per hour, $535{\cdot}25 \times 4 = 2141$ grains.
3. Heat due to 2141 grains, 442·64 units.
4. Therefore the work per hour will be $(442{\cdot}64 - 442{\cdot}64)772 = 0$.
5. And the work per grain of zinc will be $\frac{0}{2141} = 0$.

II. Start the engine with such a load as shall reduce the current to $\frac{2}{3}$, or to 3·7227 (74° 58′); then we shall have

1. Heat evolved per hour by the wire, $442{\cdot}64 \times \left(\frac{2}{3}\right)^2 = 196{\cdot}73$ units.
2. Consumption of zinc per hour, $2141 \times \frac{2}{3} = 1427{\cdot}3$ grains.
3. Heat due to 1427·3 grains, $442{\cdot}64 \times \frac{2}{3} = 295{\cdot}09$ units.
4. Therefore the work per hour will be $(295{\cdot}09 - 196{\cdot}73)772 = 75934$.
5. And the work per grain of zinc will be $\frac{75934}{1427{\cdot}3} = 53{\cdot}2$, or $\frac{1}{3}$ of the maximum duty.

III. Lessen the load so that the velocity of the engine is increased until the current is reduced to one half its original amount, or 2·792, shown on the galvanometer by a deflection of 70° 18. Then we shall have

1. Heat evolved per hour by the wire, $442{\cdot}64 \times \left(\frac{1}{2}\right)^2 = 110{\cdot}66$ units.
2. Consumption of zinc per hour, $2141 \times \frac{1}{2} = 1070{\cdot}5$ grains.
3. Heat due to 1070·5 grains, $442{\cdot}64 \times \frac{1}{2} = 221{\cdot}32$ units.
4. Therefore the work per hour will be $(221{\cdot}32 - 110{\cdot}66)772 = 85430$ ft. lb.
5. And the work per grain of zinc will be $\frac{85429}{1070{\cdot}5} = 79{\cdot}8$, or $\frac{1}{2}$ the maximum duty.

IV. Let the load be further reduced until the velocity reduces the current to $\frac{1}{3}$, or to 1·8613, shown by a deflection of 61° 44′. Then we shall have

1. Heat evolved per hour by the wire, $442{\cdot}64 \times \left(\frac{1}{3}\right)^2 = 49{\cdot}182$ units.
2. Consumption of zinc per hour, $2141 \times \frac{1}{3} = 713{\cdot}66$ grains.
3. Heat due to 713·66 grains of zinc, $442{\cdot}64 \times \frac{1}{3} = 147{\cdot}55$ units.
4. Therefore the work per hour will be $(147{\cdot}55 - 49{\cdot}182)772 = 75940$ ft. lb.
5. And the work per grain of zinc will be $\frac{75940}{713{\cdot}66} = 106{\cdot}4$, or $\frac{2}{3}$ of the maximum duty.

V. Let the load be still further reduced until, with the increased velocity, the current becomes reduced to $\frac{1}{100}$, or to ·05584, showing a deflection of 3° 12′. Then we shall have

1. Heat evolved per hour by the wire, $442{\cdot}64 \times \left(\frac{1}{100}\right)^2 = {\cdot}044264$ of a unit.
2. Consumption of zinc per hour, $2141 \times \frac{1}{100} = 21{\cdot}41$ grains.
3. Heat due to 21·41 grains of zinc, $442{\cdot}64 \times \frac{1}{100} = 4{\cdot}4264$ units.
4. Therefore the work per hour will be $(4{\cdot}4264 - {\cdot}04426)772 = 3383$ ft. lb.
5. And the work per grain of zinc will be $\frac{3383}{21{\cdot}41} = 158$, or $\frac{99}{100}$ of the maximum duty.

Now suppose that we still further improve our engine by making the stationary magnets twice as powerful. In this case all the figures will remain exactly the same as before, the only difference being that the engine will only require to go at half the velocity in order to reduce the current to the same fraction of its first quantity. The attraction will be doubled, but the velocity being halved no change will take place in the amount of work given out.

In all cases the maximum amount of work per hour is obtained when the engine is going at such a velocity as reduces the current to one half of its amount when the engine is held stationary; and in this case the duty per grain of zinc is one half of the theoretical maximum.

The same principles apply equally well when, instead of employing the machine as an engine evolving work, we do work on it by forcibly reversing the direction of its motion. Suppose for instance we urge it with this reverse velocity until the quantity of current is quadrupled or becomes 22·336, indicated by a deflection of 87° 26′. Then we shall have

1. Heat evolved per hour by the wire, $442{\cdot}64 \times 4^2 = 7082{\cdot}2$ units.
2. Consumption of zinc per hour, $2141 \times 4 = 8564$ grains.
3. Heat due to 8564 grains of zinc, $442{\cdot}64 \times 4 = 1770{\cdot}56$ units.
4. Therefore the work per hour will be $(1770{\cdot}56 - 7082{\cdot}2)772 = -4100432$ ft. lb.
5. And the work per grain of zinc will be $\frac{-4100432}{8564} = -478{\cdot}8$, or -3 times the maximum working duty.

The principal reason why there has been greater scope for the improvement of the steam-engine than that of the electro-magnetic engine arises from the circumstance that in the formula $\frac{a-b}{a}$, applied to the steam-engine by Thomson, in which a and b are the highest and lowest temperatures, these values have been limited by practical difficulties. For a cannot easily be taken above $459° + 374° = 833°$ from absolute zero, since that temperature gives 12·425 atmospheres of pressure, nor can b be readily taken at less than the atmospheric temperature, or $459° + 60° = 519°$. Also there is much difficulty in preventing the escape of heat; whereas the insulation of electricity presents no difficulty.

I had arrived at the theory of the electro-magnetic engine in 1840, in which year I published a paper in the 4th volume of Sturgeon's 'Annals,' demonstrating that there is "no variation in economy, whatever the arrangement of the conducting metal, or whatever the size of the battery." The experiments of that paper indicate 36 foot lb. as the maximum duty for a grain of zinc in a Wollaston battery. Multiplying this by 4 to bring it to the intensity of a Daniell's battery, we obtain 144 foot lb. Here, as in the experiments in the paper on Mechanical Powers of Electro-Magnetism, Steam, and Horses, the actual duty is less than the theoretic; which is owing partly to the pulsatory nature of the current, and partly also to induced currents giving out heat in the substance of the iron cores of the electro-magnets; although these last were obviated as far as possible by using annealed tubes with slits down their sides.

On the Magnetic Storm of February 4, 1872.

[Read before the Manchester Literary and Philosophical Society, Feb. 6, 1872. Proceedings of the Society, vol. xi. p. 91.]

Dr. Joule, F.R.S., called attention to the very extraordinary magnetic disturbances on the afternoon of the 4th, and from which he anticipated the aurora which afterwards took place. The horizontally suspended needle was pretty steady in the forenoon of that day, but about 4 P.M. the north end was deflected strongly to the east of the magnetic meridian, and afterwards still more strongly to the west. The following were the observations he had made:—

Time.	Deflection from the Magnetic Meridian. ° ′	Time.	Deflection from the Magnetic Meridian. ° ′
4.0 P.M.	0 50 E.	6.10 P.M.	1 24 W.
4.30 „	0 47 W.	6.12 „	1 8 „
4.55 „	2 22 „	7.41 „	0 10 „
4.58 „	3 0 „	7.43 „	0 0 „
5.9 „	3 45 „	8.9 „	0 42 „
5.12 „	0 52 „	8.31 „	0 10 „
5.23 „	5 36 „	8.54 „	1 18 „
5.24 „	2 28 „	8.58 „	0 52 „
5.35 „	0 52 „	11.3 „	0 5 „
5.55 „	0 52 „		

Observation with the spectroscope of the aurora showed a bright and almost colourlesss line near the yellow part of the spectrum. This line appeared to whatever part of the heavens the instrument was directed, and could be plainly seen when the sky was covered with clouds and rain was falling. When looking at the most brilliant red light of the aurora, a faint red light was seen at the red end of the spectrum, and beyond the bright white line towards the violet end two broad bands of faint white light.

Mr. Sidebotham states that he also expected the magnificent aurora on account of the violet disturbance of the needle at Bowdon, amounting to at least 3°.

On the Polarization of Platina Plates by Frictional Electricity.

[Read before the Manchester Literary and Philosophical Society. Proceedings of the Society, vol. xi. p. 99.]

Dr. J. P. Joule described some experiments he had been making on the polarization by frictional electricity of platina plates, either immersed in water or rolled together with wet silk intervening. The charge was ascertained both in quality and quantity by transmitting it through the coil of a delicate galvanometer. He suggested that a condenser on this principle might be useful for the observation of atmospheric electricity.

Note added in 1883.

The experiments of which the above is a short notice were made in the commencement of 1872. The electrical machine had a cylinder of 9 inches diameter, and was furnished with insulated rubber and prime conductor in the usual manner. The galvanometer had a coil of half an inch diameter composed of a long and fine insulated wire, and the needle was astatic. An apparatus which we may call a condenser was formed by rolling up two sheets of platina foil and two sheets of silk or cotton fabric. The roll was bound tightly with silk thread, two fine platina wires being attached to the platina plates. It was then placed in a beaker filled with water. When, therefore, the wires were connected on one hand with the rubber, and on the other with the prime conductor of the electrical machine, and the latter was worked, a charge was accumulated in the condenser which could be measured by being transmitted through the galvanometer.

The charge, as indicated by the throw of the needle, was rather lower than in the simple proportion of the number of revolutions of the cylinder, which was owing to the time occupied allowing the dissipation of a portion of the charge. Thus I had in one instance (which will do as an example of the rest) :—

Turns.	Throw of Needle.
50	1° 52′
100	2 50
200	5 0
400	7 52

The condenser is not entirely discharged when its wires are placed in connexion with the galvanometer; on the contrary, a considerable residuary charge accumulates which, after being discharged, will again collect many times in successively decreasing quantities—a circumstance which makes it difficult to arrive at accurate quantitative results.

It is highly probable that if the condenser were immersed in perfectly pure water deprived of air, it would retain the charge much longer than it did in my experiments, although the town's water I used is one remarkably free from foreign ingredients. With this water the average time occupied in the reduction of the charge to one half was 40 minutes.

The addition of alcohol to the water in the proportion of one per cent. produced no sensible effect, the average time required to reduce the charge one half being 42 minutes.

Sugar, in a similar proportion, seemed to reduce the holding power, for a mean of two experiments showed that the charge was diminished to one half in 20 minutes.

With common salt in a similar proportion, the charge was reduced to one half in 15 minutes.

A small addition of sulphuric acid reduced the holding power very considerably; but there was little further reduction effected by an increase in its quantity—a proportion of 1 to 600 was sufficient to reduce the charge to one half in $6\frac{1}{2}$ minutes, a proportion of 1 to 300 had the same effect, while a still further increase of strength, viz. 1 to 150, only shortened the time required to reduce the charge to one half by half a minute more, or to 6 minutes.

As might have been anticipated, nitric acid had a much more powerful effect to destroy the holding power. One part in 600 caused the charge to be halved in $2\frac{1}{3}$ minutes, while 1 in 300 produced the same deterioration in $1\frac{3}{4}$ minutes.

On the Prevalence of Hydrophobia.

[Read before the Literary and Philosophical Society, Dec. 24, 1872. Proceedings of the Society, vol. xii. p. 41.]

THE President drew attention to the increasing number of cases of hydrophobia. There was every reason for believing that this dreadful disorder was communicated from one animal to another by a bite, and was seldom, if ever, spontaneously developed. Inasmuch, therefore, as the effects of a bite nearly always occurred within four months, it would only be necessary to isolate dogs for that period in order to stamp out the disease. That was the opinion of Dr. Bardsley, whose elaborate paper will be found in the 4th vol. of the

Memoirs of the Society, and probably gave rise to the practice of confining dogs at certain periods of the year; which, however, has unfortunately been rendered to a great extent nugatory in consequence of having been only partially adopted.

Note, 1883.—Having been told when a child by an old servant of my father's that he had himself assisted in smothering a victim to this horrid disease, I have ever since felt an anxious interest in the subject. There is no known preventive, except possibly that resorted to by my sister in a case which occurred in our own family—she kept the wound open a very long time. It is difficult to understand the intellectual and moral condition of a public which, by the neglect of the obvious precaution indicated by Dr. Bardsley, consigns to a frightful death probably as large a number of persons as those who by common custom are exclusively spoken of as murdered.

On a Mercurial Air-Pump.

[Read before the Manchester Literary and Philosophical Society, Feb. 18, 1873. Proceedings of the Society, vol. xii. p. 57.]

Dr. Joule, F.R.S., gave some further account of the improvements he had made in his air-exhausting apparatus. As stated in the last Proceedings, he had substituted a caoutchouc tube attached to the neck of a glass vessel for the original perpendicular pipe with its stop-cock. This is seen in the adjoining sketch, *c* and *d*. The two positions, viz. when *b* is being filled, and when it is being emptied, are shown by the full and the dotted drawing. It is convenient to introduce no air into *d* except that required to act as a cushion to avoid a shock when filled in the lower position. Sulphuric acid may be introduced into the receiver to be exhausted; but it is perhaps more convenient to place it over the mercury in *a*, whence it may occasionally be drawn into *b*, to effect the drying of the internal parts of the apparatus. Dr. Joule has met with some difficulty in using mercury-gauges to ascertain the residual pressure, inasmuch as he

Fig. 109.

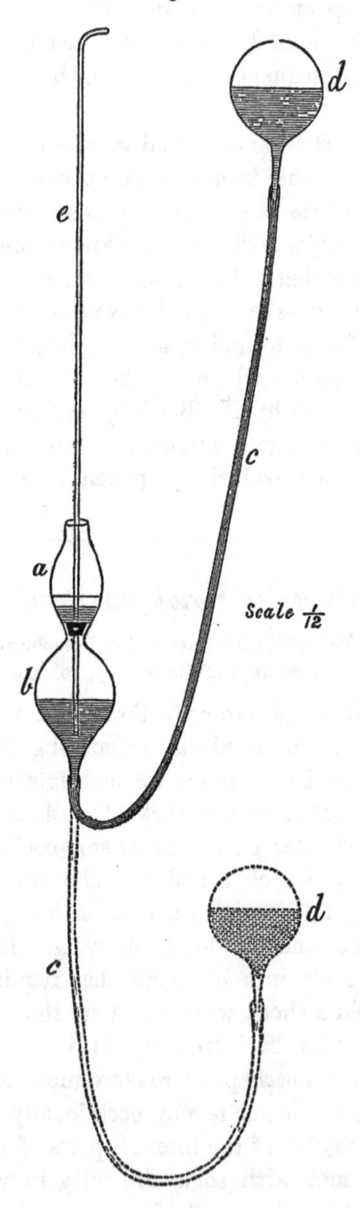

finds that mercury thoroughly boiled in clean glass tubes does not show a convex surface, but adheres strongly to the glass. However, he has confidence in giving the following results in working with his apparatus, with acid of various strength, obtained by successive dilutions by volume of sulphuric acid, of sp. gr. 1·845.

Sulphuric Acid.		Water.		Pressure in Inches of Mercury.
3	+	0	....	Inappreciable.
3	+	1	....	Inappreciable.
3	+	2	....	0·01 at 70°
1	+	1	....	0·03 at 63°
1	+	2	....	0·15 at 63°
1	+	4	....	0·30 at 55°
0	+	1	....	0·37 at 47°

[Read before the Manchester Literary and Philosophical Society, January 13, 1874. Proceedings of the Society, vol. xiii. p. 58.]

A DRAWING was shown representing some further improvements of Dr. Joule's mercurial air-exhauster described in the Proceedings of February 18, 1873.

In the section represented by fig. 110, I, W W is a wooden frame; P a pulley for raising or lowering a flask of mercury held in a wooden box, M, working in a slide; *s s s s* are india-rubber stoppers; E is the exhauster; *t*, *e* the entrance and exit tubes; *g* the gauge; *f* a funnel to admit sulphuric acid; B, B movable brackets to support any apparatus.

In fig. 110, 2, the exhauster is drawn to a larger scale. *t*, *e* are the entrance and exit tubes, fitting tightly in an india-rubber disk *a*, which disk is kept tightly pressed against the exhauster by means of the ring, *b*, *b*. The mercury is represented sunk below the entrance tube, as is the case when the movable flask is in its lower position. On raising the flask by means of the pulley, the mercury rises in the exhauster and forces any air it may contain into the upper part of the exhauster by raising the india-rubber plug. The air then makes its exit through the pipe *e*. This latter is also used for withdrawing the acid which gradually accumulates.

Fig 110.

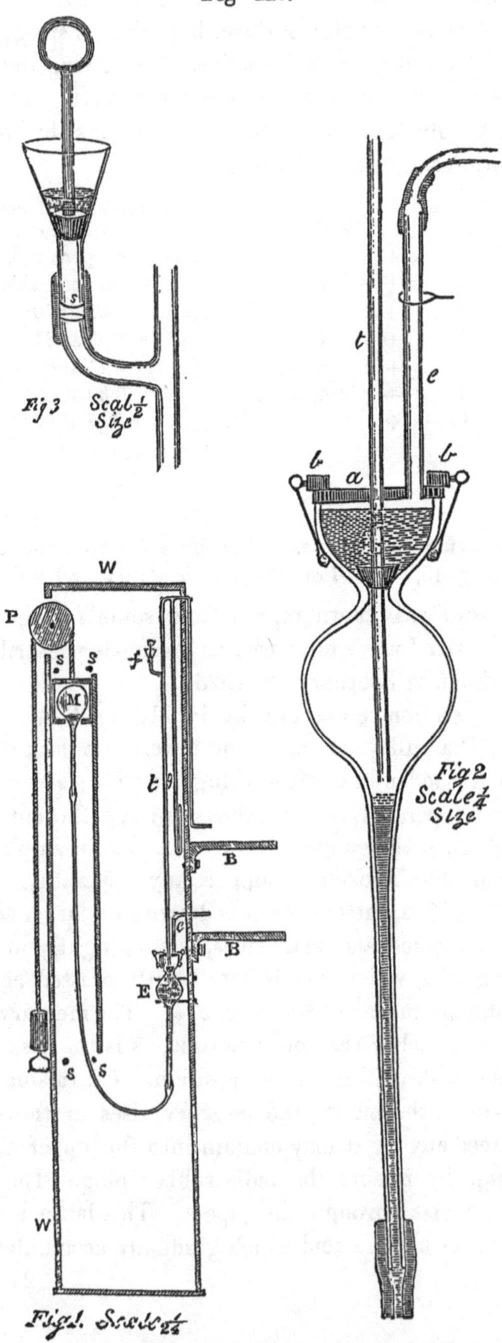
Fig 3 Scale ½ Size
s
t
e
b
b
a
Fig 2 Scale ¼ Size
W
P
s
s
f
M
t
B
c
B
E
s
s
W
Fig 1. Scale 1/24

Fig. 110, 3, also drawn to a large scale, represents a convenient means of introducing sulphuric acid for removing aqueous vapour, or to let air into the apparatus. The orifice at the bottom of the funnel is about $\frac{1}{100}$ of an inch diameter to prevent violent action.

It may be useful to mention that the junctions are made with black india-rubber tube fastened by softened iron wire.

Note, 1883.—In practice I find it more convenient to raise the globes with the hand than to work them with pulleys. I early thought of plungers consisting of iron cylinders working in mercury in order to alternate the level, which would obviate the necessity of using caoutchouc tubes. But this did not seem to offer any practical advantage.

[Proceedings, vol. xiv. p. 12.]

Fig. 111.

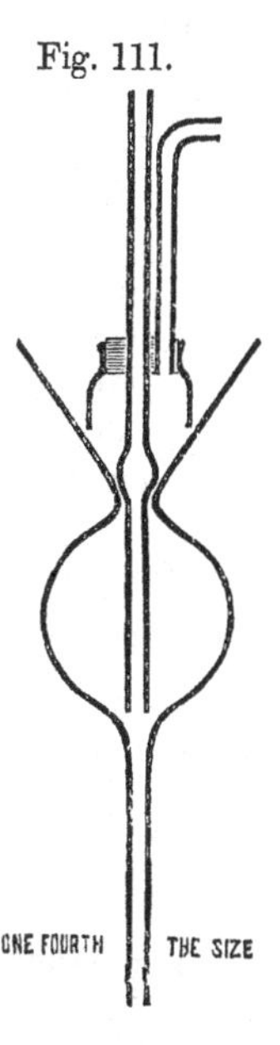

ON October 6, 1874, Dr. Joule made a further communication respecting his mercurial air-pump described in the Proceedings for Dec. 24, 1872, and Feb. 4, Feb. 18, and Dec. 30, 1873. He had successfully made use of the glass plug proposed in the Proceedings for Feb. 4, 1873. This he constructs by blowing out the entrance tube and grinding the bulb thus formed into the neck of the thistle-shaped glass vessel. To collect the pumped gases he now employs an inverted glass vessel attached to the entrance tube and dipping into the mercury in the upper part of the thistle-glass.

On a Glue-Battery. *By* Dr. J. P. Joule, *F.R.S.*

[Read before the Manchester Literary and Philosophical Society, October 5, 1875. Proceedings of the Society, vol. xv. p. 1.]

If sulphate of zinc or sulphate of copper be dissolved in solution of gelatine and then carefully dried, an elastic solid is produced holding the salt in conbination, which softens on the application of heat. I have taken advantage of this circumstance to form a voltaic couple which illustrates Faraday's discovery of the necessity of liquefaction for electrolysis, and which also may not be without some practical advantages.

I paint pieces of zinc with glue impregnated with salt of zinc, and pieces of copper with glue charged with sulphate of copper, dry, and lay them together in series. The pile thus formed is inert when cold, but capable of giving a good current when heated, as will be seen from the following results obtained with a single couple in connexion with a delicate galvanometer:—

Temp. Fahr.	Deflectio		Temp. Fahr.	Deflection.	
°	°		°	°	′
64	0		150	5	50
70	0	40	160	7	20
80	1	10	170	8	10
90	1	40	180	10	20
110	2	50	190	17	40
130	3	0	200	43	0
140	4	40			

After the lapse of three years I find that the above couple has retained its powers almost unimpaired.

On the Utilization of the Common Kite. *By* Dr. J. P. Joule.

[Read before the Manchester Literary and Philosophical Society, Dec. 28, 1875. Proceedings of the Society, vol. xv. p. 61.]

Unsuccessful attempts have recently been made for the purpose of utilizing a modification of the common kite as a means of obtaining a view of the surrounding country. The machine in each instance rose only to fall violently to the ground after remaining in the air a very short time. These trials have brought to my recollection some experiments I made more than six years ago, but of which I did not publish the results, imagining that all such matters must have been thoroughly elucidated by the Chinese, if not by our own more juvenile kite-flyers. The usual method of making the skeleton of a kite is to affix a rather slender bow to the top of a standard, tying the extremities of the bow to twine fastened to the bottom of the standard. The steadiness of the kite in the air depends on the fact that the wings yield with the wind. If the bow is too stiff and the surface nearly a plane, instability results. A kite ought to have a convex spherical surface for the wind to impinge upon. Such a surface I readily made by fixing two bows crosswise. The string was attached to a point a little above the centre of the upright bow, and a very light tail was fastened to the lower end. The kite stood in the air with almost absolute steadiness. I found that by pulling strings fastened to the right and left sides of the horizontal bow, the kite could be made to fly 30° or more from the direction of the wind, and hence that it would be possible to use it in bringing a vessel to windward. One great advantage of such a mode of propulsion over ordinary sails would be that the force, however great, could be applied low down, so as to produce no more careening than that desired by the seaman.

PS. 1883.—In 1827 Pocock yoked a pair of kites to a carriage in which he travelled from London to Bristol. He had an extra line to vary the angle of the surface of his kite, and so the direction of the tractile force.

On a Barometer. *By* Dr. J. P. Joule, *F.R.S. &c.*

[Read before the Manchester Literary and Philosophical Society, March 19, 1878. Proceedings of the Society, vol. xvii. p. 114.]

Some years ago I brought under the notice of the Society a

syphon barometer, the peculiarity of which consisted in the introduction of a small quantity of sulphuric acid. I hoped that the diminution of the capillary effect, and the extreme mobility of the mercury thus obtained, would present some advantages. I found, however, that the opinion expressed by Dr. R. A. Smith, viz. that the acid would act on the mercury, was fully justified in the event; for the barometer, after a few weeks had elapsed, stood at too low a level, and now, after an interval of 5 years, stands 3 inches below its proper height, while a plentiful crop of transparent prismatic crystals has been formed.

Fig. 112.

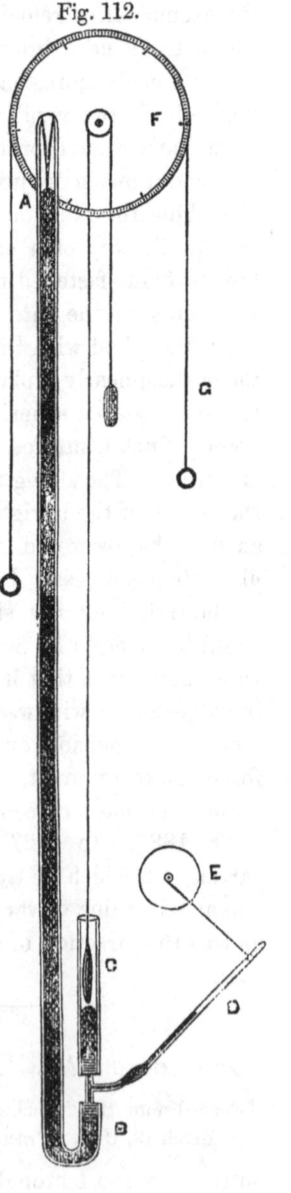

In the barometer which I now venture to submit, I have applied a principle which I have found very useful in the construction of the manometer. Sealed within to the top of the long leg of a syphon barometer is a piece of thin glass rod, with the extremity of which at A the mercurial column can be brought into contact by means of the adjunct to the lower short leg. C is a tube of the same diameter as the rest of the barometer: it is connected with the short leg by means of a T tube, to which also the narrow glass tube D is attached by a piece of rubber tubing. By raising this small tube by turning the axle E, the mercury can be brought to touch the glass point at A. Affixed to the upper part of the framework of the instrument is a graduated wheel F having a groove in its periphery and also a small V groove near its centre. This last holds a fine wire with a glass plummet at one end and an exact counterpoise at the other. To bring this plummet in contact with the mercury of the lower leg, the wheel is moved by means of the thread G. The contacts of the mercurial column are observed with microscopes, and the value of the graduations of the wheel by comparisons with a standard rule.

In order to facilitate the application of the needful temperature-corrections, it is desirable to secure the upper part of the barometer to the block of timber which supports the wheel, leaving the lower end free. There is also an advantage in introducing a small quantity of sulphuric acid into the small tube D, to promote the freedom of movement of the mercury in it.

The average time occupied by an observation is found to be $\frac{3}{4}$ of a minute, and the average error $\frac{1}{1560}$ of an inch. In the morning of yesterday the wind was high and gusty, causing the mercury to oscillate at short intervals of time through a space of about $\frac{1}{200}$ of an inch.

Method of Checking the Oscillations of a Telescope. *By* Dr. JOULE.

[Read before the Manchester Literary and Philosophical Society, Oct. 7, 1879. Proceedings of the Society, vol. xix. p. 4.]

Dr. JOULE described a simple means for checking the oscillations of his telescope. Leaden rings are placed centrally about the axis of the tube of the telescope, each being attached thereto by three or more elastic caoutchouc bands. He had usually employed two of these rings, one near the object-glass, the other near the eyepiece. Their united weight was only one quarter that of the telescope-tube; but nevertheless they diminished the time required for the cessation of vibration to one sixth of what it was before their application.

New Determination of the Mechanical Equivalent of Heat. By JAMES PRESCOTT JOULE, *D.C.L., LL.D., F.R.SS. L. and E., &c., President of the Literary and Philosophical Society of Manchester.*

[Philosophical Transactions, 1878, Part. II. Read January 24, 1878.]

[PLATE IV.]

THE Committee of the British Association on Standards of Electrical Resistance having judged it desirable that a fresh determination of the mechanical equivalent of heat should be made, by observing the thermal effects due to the transmission of electrical currents through resistances measured by the unit they had issued, I undertook experiments with that view, resulting in a larger figure (782·5) * than that which I had obtained from the friction of fluids (772·6) †.

The only way to account for this discrepancy was to admit the existence of error, either in my thermal experiments or in the unit of resistance. A committee, consisting of Sir Wm. Thomson, Professor P. G. Tait, Professor Clerk Maxwell, Professor B. Stewart, and myself, were appointed at the meeting of the British Association in 1870; and with the funds thus placed at my disposal I was charged with the present investigation, for the purpose of giving greater accuracy to the results of the direct method.

The plan I adopted was, in regard to the measurement of work, similar (as I afterwards found) to that used by Hirn, who has laboured so earnestly and successfully on this subject. He has described it as follows :—" L'appareil qui m'a servi pour cette étude consiste : 1°, en un cylindre en laiton de $0^{m}\cdot 3$ de diamètre, de 1^{m} de longeur, poli à sa périphérie externe, monté sur un axe solide en rapport avec un moteur d'un mouvement très régulier, et pouvant recevoir une vitesse

* Brit. Assoc. Report, Dundee, 1867, p. 522.

† Phil. Trans. 1850, p. 82.

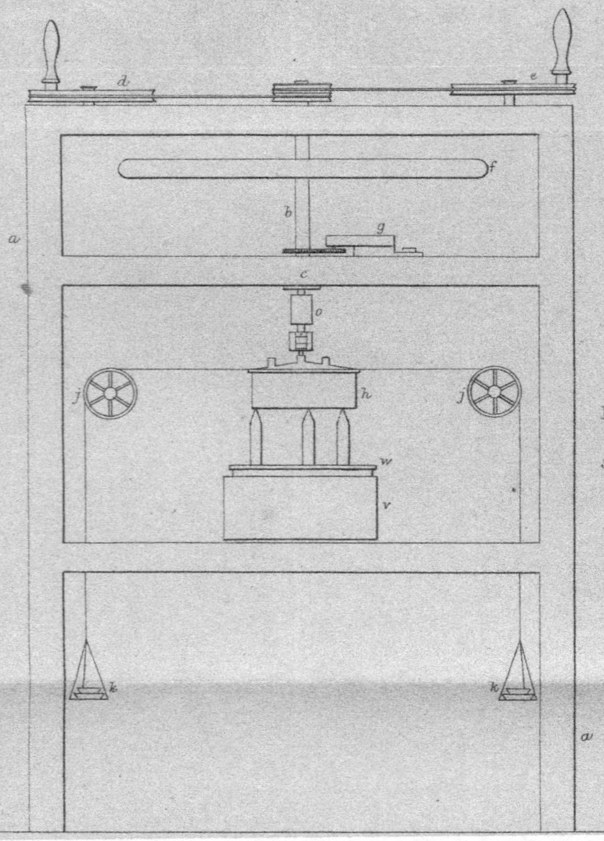

Fig. 113.

Scale $\frac{1}{18}$

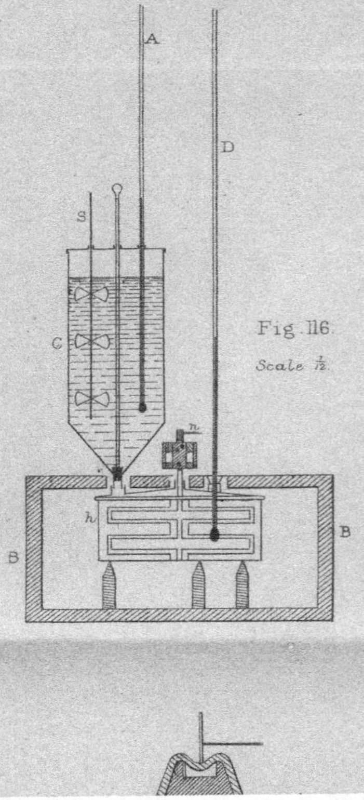

Fig. 116.

Scale $\frac{1}{12}$

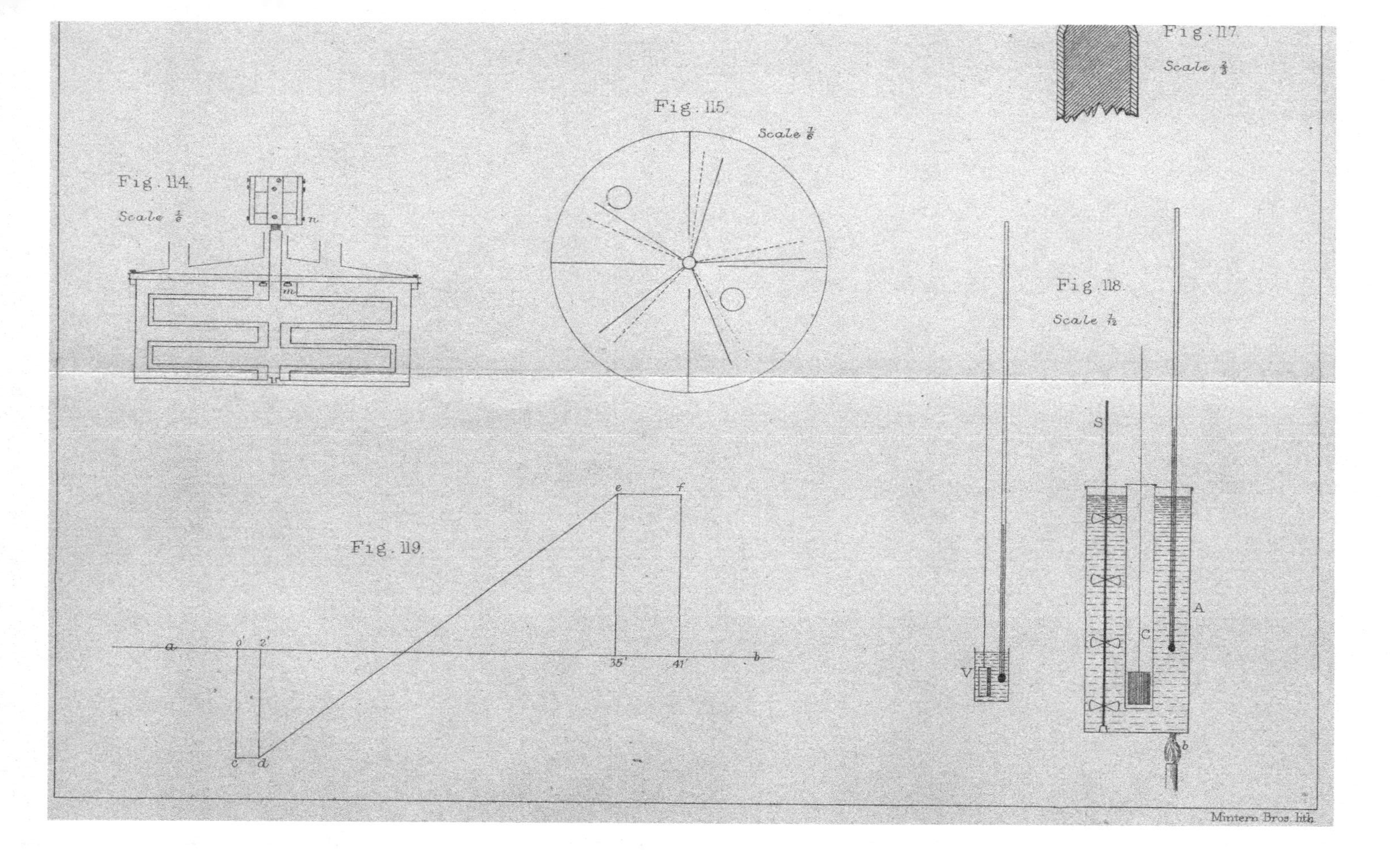
Fig. 114.
Scale 1/8
n
m
Fig. 115.
Scale 1/6
Fig. 117.
Scale 2/3
Fig. 118.
Scale 1/2
S
C
A
V
b
Fig. 119.
a
0'
2'
c
d
e
f
35'
41'
b
Mintern Bros. lith.

variant à volonté de 60 à 600^t par minute ; 2°, en un cylindre fixe, poli à son intérieur, concentrique au premier, éloigné partout de 0^m·03 de celui-ci. Les disques ou plateaux formant les extrémités de la cylindre étaient munis, à leur partie centrale, de boîtes à étoupes par où sortait l'axe du cylindre interne. Tout l'intervalle compris entre les deux cylindres pouvait être rempli ainsi d'un liquide quelconque que les boîtes à étoupes empêchaient de s'écouler par les centres.

"Lorsque le cylindre intérieur tournait, le frottement que sa surface externe exercait sur le liquide, et que le liquide, mis ainsi en mouvement lui-même, exercait à son tour sur la surface interne du cylindre externe, tendait à faire tourner celui-ci. Deux leviers parfaitement parallèles, adaptés aux deux extrémités, et portant des plateaux de balance, permettaient d'empêcher la rotation à l'aide de poids qui indiquaient ainsi la valeur du frottement. La tare des leviers, la valeur du frottement des boîtes à étoupes, etc., étaient déterminées aisément en faisant tourner très lentement le cylindre interne dans les deux sens alternativement. Deux tuyaux verticaux, soudés aux deux disques de fermeture, et aussi près que possible des boîtes à étoupes, permettaient d'établir dans l'appareil un courant continu et parfaitement régulier d'un liquide voulu. La température de ce liquide était prise à l'entrée et à la sortie. Autant que possible, la température à l'entrée était tenue à autant de degrés audessous de celle de l'appartement que celle du liquide sortant était supérieure. Du reste, la loi de refroidissement de l'appareil était soigneusement déterminée de manière à ce qu'il fût facile de faire les corrections nécessaires.

"Cet appareil, qui dans son ensemble constitue une véritable balance à frottement des liquides, pouvait très aisément servir à faire connaître, d'une part, le travail dépensé pour tel ou tel liquide, pour telle ou telle vitesse, et d'autre part, à l'aide des corrections convenables, à faire connaître le nombre de calories produit par ce frottement dans un liquide dont la capacité calorifique était connue.

"Les résultats obtenus ont été en général d'une régularité satisfaisante. Six expériences consécutives faites sur l'eau, et avec différentes vitesses, avec des quantités diverses de liquide introduites par seconde entre les deux tambours, m'ont donné 432[p]* pour le travail produisant une calorie, et par suite pour la valeur de l'équivalent." †

The method I adopted was to revolve a paddle in a suspended vessel of water, to find the heat thereby produced, measuring the work by the force required to hold the vessel from turning, and the distance run as referred to the point at which the force was applied. Fig. 113 (Plate IV.) represents the apparatus drawn one-eighteenth the actual size. A massive wooden framework, *a a*, resting on the asphalted floor of a cellar, is still further strengthened by means of timber abutting against the walls on every side. The perpendicular shaft *b* is supported by a conical collar ‡ turned on it at *c*. It is revolved, along with the fly-wheel *f*, weighing about 1 cwt., by means of the doubling hand-wheels, *d e*. A counter § is placed at *g*, for the purpose of reading off the number of revolutions. The calorimeter *h* has an accurately turned groove, from which silk threads pass over the light, accurately-turned pulleys *j j*, to the scales *k k*. The hydraulic supporter, *w v*, was not employed in the first two series of experiments, and will be described further on. Three sides of the frame are boxed in permanently; the fourth, or front, has shutters with windows which can be removed at pleasure. A delicate thermometer, suspended within the frame, is observed through a telescope, as is also the thermometer employed in reading the temperature of the calorimeter.

* This equals 787·4 in the measures I have adopted, viz., British feet, and degrees Fahr.

† 'Théorie Mécanique de la Chaleur,' 1865, p. 55. Maxwell has independently, in 1875, devised an apparatus of a similar description. He employs channelled cones, the revolution being on a vertical axis.

‡ Its surface, though only half a square inch, was found amply sufficient when castor-oil was employed as the lubricator. Other oils failed on trial.

§ In most of the experiments a second counter of my own construction was used to check the indications of the other. They were found in every instance to agree exactly.

Fig. 114 (Plate IV.) represents the section of the calorimeter, with its paddle, all of stout sheet brass; and fig. 115 (Plate IV.) gives a plan of the same. The dotted lines in the latter show the position of the fans in the upper part. The axle of the paddle works easily in the collar *m*, and is screwed into the boxwood piece *n*. There is another boxwood piece, *o*, fig. 113 (Plate IV.), placed to prevent any considerable quantity of heat arising from the friction of the shaft being conducted downwards. This friction was, however, so small that the precaution was afterwards found to be needless.

It will be seen in figs. 114 and 115 (Plate IV.) that there are four stationary vanes in the calorimeter, and two sets of rotating vanes, each of five arms, the upper set being fixed on the axis 9° behind the lower set. Hence no two vanes pass the fixed ones at the same moment; and inasmuch as the momentary alteration of resistance at crossing takes place 40 times in each revolution, the resistance may be considered as practically uniform.

The circumference of the groove of the calorimeter was found by measuring its diameter in various places, and also by measuring it directly with a fine wire, allowing for the thickness of the latter. The results, obtained with a rule verified by the Warden of Standards, are:—

Diameter in inches.	Circumference.
10·5850 . .	$\times\pi=$33·2538
10·5855 . .	$\times\pi=$33·2553
10·5855 . .	$\times\pi=$33·2553
Measured by wire $\frac{1}{120}$ in. diameter	=33·2538
" " $\frac{1}{55}$ "	=33·2563
Average	33·2549 inches=2·77124 feet.

The diameter of the silk cord, which was the finest that could be used with safety, was exactly $\frac{1}{100}$th inch. Hence the distance to be considered as run against the weights of the scales was, for each revolution, 2·77386 feet.

When a silk thread with a weight of 11,000 grains at each extremity was thrown over the small pulleys, 30 grains added to one of the weights was sufficient to keep both in motion. This friction, which includes the rigidity of the silk cord, taken with the distance traversed by the weights in their slight upward and downward motions during an experiment, gives the loss of work on the calorimeter from this cause. It did not amount to more than $\frac{1}{200000}$ subtractive from the equivalent, and could therefore be neglected.

The thermometer used to indicate the temperature of the calorimeter was the same which I employed in my former experiments. Those designated A* and D were calibrated with great care. I have recently compared them together at 50 different temperatures between 32° and 80° Fahr., the result being that, if the less sensitive was assumed to be correct, the other, or A, nowhere appeared more than 0°·023 in error; but taking the averages for each consecutive 10°, this error amounted to no more than 0°·008. I was anxious to compare these instruments with an air-pressure thermometer; and with that view have constructed an apparatus in which the height of the mercurial column is measured by a plummet hung over the axis of a graduated wheel—a method which I find capable of extreme accuracy, and which I purpose to apply to the construction of a new barometer. But owing to the use of caoutchouc in the connexion between the receiver and the rest of the apparatus, I fear that the zero point was subject to a slight displacement. The figures at which, after much labour, I have hitherto arrived, could not therefore be accepted as any improvement on Regnault's determinations of the expansion of air by heat.

The freezing-point of the standard D had risen from 13·3 divisions of its scale in 1844 to 15·14 in 1877. I think it probable that the boiling-point of this thermometer, if kept constantly at this temperature, would in the course of time fall as much. The five careful determinations of this boiling-point referred to 30 bar. and 60° are respectively 706, 706·4, 706, 705·9, and 706·15—mean 706·09. Subtracting 1·84, 704·25 will be the probable ultimate reading, from which if we take 15·14 we shall have 689·11 as the range between the fixed points cleared from the effects of imperfect elasti-

* Phil. Trans. 1850, p. 64.

city of the glass. Mr. E. Hodgkinson has pointed out* that the "set" of imperfectly elastic bodies is proportional to the square of the force applied. Therefore the effect of imperfect elasticity in the glass of the thermometers will be insensible for the small ranges used in the experiments; and the factor 3·3822 for reducing the indications of D to those of A may be confidently relied on.

We have therefore $\frac{180}{689{\cdot}11 \times 3{\cdot}3822} = 0^{\circ}{\cdot}07723$ as the most probable value of one division of A. In my former papers the number was taken as 0°·077214, which is so near that I shall continue to use it, trusting by long-continued observations of the fixed points to give it ultimately greater accuracy, and also by experiments above indicated to state it in terms of the absolute interval between these points.

The elevation of the mercurial column in A caused by the atmospheric pressure is five divisions; but inasmuch as in the limited time of an experiment the barometer never altered 0·1 inch, error from this cause was neglected. The depression occasioned by capillarity was 0·33 of a division.

A delicate calibrated thermometer E, each division of which indicated 0°·11195, was first used for taking the temperature of the air; but in consequence of a slight hitch in the motion of the mercury, an instrument called G was afterwards employed, each of whose divisions was equal to 0°·1911.

In registering the temperature of the air surrounding the calorimeter it was necessary to make allowance for the time which a thermometer takes in altering its temperature. I found that in a regularly rising or falling temperature, E was $3^{m}{\cdot}8$ behind time and G $3^{m}{\cdot}127$. This lagging of the thermometers was always carefully allowed for.

The capacity for heat of the calorimeter, calculated from the specific heat of brass given by Regnault, was equal to that of 5002 grains of water. But Regnault has shown how considerably the specific heat of metals of the same chemical composition is altered by changes in their hardness, and

* Brit. Assoc. Report, 1843, p. 23.

moreover there were the stoppers and other adjuncts to be taken into account. I therefore constructed the special apparatus represented by fig. 116 (Plate IV.), where B, B is a wooden box containing the calorimeter *h*; the projecting rim of the latter being supported by bits of string fastened at the top of three wooden legs, one of whch is shown in fig. 117 (Plate IV.). In the lid of the box are three holes which the tubulures of the calorimeter just enter without touching. The paddle of the calorimeter can be agitated by means of the boxwood piece *n*. C is a copper vessel covered with a non-conducting substance: its lid is perforated to admit a stirrer, a thermometer, and a rod furnished with a caoutchouc stopper.

In experimenting with this apparatus, the calorimeter was first weighed after the water which it might have contained was shaken out. It was then placed on its three supports, and left for three or more hours in an apartment of uniform temperature, until its thermometer ceased to show alteration. The vessel C, containing an adjusted quantity of hot distilled water, and placed at some distance, had its gradually descending temperature noted from minute to minute. At a given moment it was rapidly transferred to the position shown in the figure; and then on pulling the plug out, *h* was filled in a few seconds. C was then quickly removed, and the caoutchouc stopper belonging to the tubulure through which the water had entered having been replaced, the temperature of the water was noted again from minute to minute while *n* was constantly moved. These observations afforded the means of eliminating the effects of radiation. Finally the calorimeter, as filled with the water, was again weighed.

In the first half of the following Table, A was employed in determining the temperature of the water introduced into the calorimeter, and D was the thermometer plunged into the calorimeter. In the latter half their positions were reversed. The temperatures are all given in divisions of A. *w* includes the estimated value of the air displaced, reckoned at 8 grains of water.

EXPERIMENTS on Capacity for Heat of Calorimeter.

No.	Water already in calorimeter. w.	First temperature of calorimeter. T.	Grains of water poured in. W.	Temperature of water poured in. T′.	Corrected resulting temperature. T″.	Thermal capacity of calorimeter. $\frac{W(T'-T'')}{(T''-T)}-w$.
1	323·2	326·0	78887·6	457·63	449·47	4890·4
2	195·4	322·36	78996·6	464·62	456·5	4586·6
3	225·9	331·11	78984·6	475·84	467·4	4665·4
4	238·0	336·94	79042·6	505·36	495·35	4756·7
5	315·0	358·0	78916·8	504·04	495·4	4647·5
6	217·9	354·1	79029·6	512·36	502·13	5243·6
7	182·1	377·95	79127·7	514·1	506·18	4705·1
8	173·8	382·7	79044·7	534·42	525·29	4887·4
9	198·8	379·25	79059·2	613·83	599·82	4822·8
10	153·3	362·5	78959·2	614·04	598·56	5024·6
11	153·3	363·0	78920·2	673·03	654·57	4841·3
12	182·3	353·0	78914·0	641·11	623·77	4871·3
13	151·2	353·45	78789·2	658·06	640·11	4782·4
14	182·4	343·68	78574·0	668·01	649·02	4704·3
15	146·9	319·9	78817·7	654·83	634·85	4853·2
16	142·2	308·0	78933·2	640·97	621·13	4859·0
17	141·7	291·75	79021·7	668·07	646·07	4764·8
18	128·3	305·3	78946·7	647·78	627·8	4762·7
19	137·1	319·6	78895·2	681·64	660·31	4802·1
20	138·9	330·1	78821·7	654·51	635·38	4800·4
21	151·4	228·99	78774·7	624·35	600·3	4950·9
22	137·3	201·16	78996·2	638·52	612·66	4827·1
23	144·3	189·48	78965·7	637·71	611·18	4823·6
24	163·7	160·45	78884·7	636·91	607·95	4941·3
25	125·6	172·45	78801·2	643·44	615·12	4915·7
26	130·5	196·24	79026·7	613·78	589·46	4757·2
27	141·4	234·24	79094·2	665·48	640·53	4715·7
28	119·3	284·22	79031·2	669·39	647·19	4714·4
29	142·2	236·34	78913·2	669·26	643·56	4838·1
30	132·9	207·82	78976·7	662·14	635·38	4810·0
31	125·8	200·2	79019·7	674·56	646·63	4817·9
32	126·8	217·38	78915·7	667·95	641·0	4893·7
33	120·9	225·1	78824·7	672·3	646·14	4776·6
34	114·3	210·75	78986·2	673·03	646·03	4785·1
35	131·2	197·93	78806·7	648·93	622·51	4772·6
36	113·4	235·0	79024·2	684·54	658·32	4781·3
37	121·3	248·95	78882·2	675·41	650·4	4793·0
38	138·8	258·63	78865·7	682·38	657·24	4835·2
39	127·3	213·94	78696·0	682·26	654·53	4825·7
40	139·8	218·6	78811·7	681·67	654·57	4759·2
Average . .		278·91	. .	624·71	604·25	4815·15

The average temperatures T' and T'' are $78^{\circ}\cdot38$ and $76^{\circ}\cdot8$. Hence in order to express the foregoing result in terms of the capacity of a grain of water at 60°, we have from the experiments of Regnault, $4815\cdot15 \times 1\cdot00132 = 4821\cdot5$. Two further corrections were needed—one amounting, as was ascertained by means of experiments devised for the purpose, to 17·6, on account of the time allowed before the final reading of T'', limited to 8^{m}, not being sufficient to enable the caoutchouc stoppers and boxwood appendages to receive what would be their ultimate thermal distribution; the other, amounting to 3·3, arose from the thermal effect of the fall of water from one vessel to the other. Hence the final result for the capacity of the calorimeter, appendages, and thermometer, is 4842·4.

I thought it desirable to test this result by obtaining the sum of the capacities of the materials which composed the calorimeter. I had in my possession cuttings from the same sheets of brass that were used in the manufacture of the vessel and its paddle. These were formed into a compact bundle.

A copper vessel, A (Plate IV. fig. 118), filled with water, had a narrower vessel, C, immersed in it, to the bottom of which the material experimented on was let down by a fine wire. A Bunsen burner, *b*, kept the water at a constant temperature for not less than three hours, a continual agitation being given by revolving the stirrer *s*, formed on the principle of a screw-propeller. The temperature having been noted, the material was rapidly lifted by the thin wire, and transferred to a small copper vessel, V, filled with distilled water, and furnished with a thermometer and stirrer. After 5^{m}, which time was required for the equal distribution of temperature, the immersed thermometer was read off, and its observation was repeated each succeeding minute for some time, in order to obtain the cooling effect of the atmosphere *. The

* The method first employed was the opposite one of plunging the material at the atmospheric temperature into a small vessel filled with

following is a table of the results. The weight w of the bundle of brass was 2951·6 grains.

No.	Thermal capacity of small vessel of water. W.	Temperature to which the brass was heated. T.	Temperature in small vessel before immersion of brass. T′.	Corrected temperature after immersion of brass. T″.	Time occupied in transferring the brass to the small vessel, in seconds.	Specific heat uncorrected for transfer. $\frac{(T''-T')W}{(T-T'')w}$.
1	4733·2	901·25	174·78	213·24	5	0·08964
2	4762·8	900·58	175·5	213·8	4	0·08999
3	4747·0	947·27	213·26	252·29	3	0·09032
4	4727·7	952·13	158·0	200·56	4	0·09070
5	4750·3	1030·46	125·5	173·52	4	0·09018
6	4724·6	984·47	210·36	251·65	3	0·09019
7	4764·0	1069·6	197·95	244·3	4	0·09065
Average .		969·4 or 104°·92	179·34 or 44°·08	221·34 or 47°·32	3·86	0·09024
1	4756·0	894·49	104·35	145·72	30	0·08903
2	4794·2	985·82	154·5	197·44	30	0·08847
3	4717·4	994·44	165·33	208·97	30	0·08880
Average .		958·25 or 104°·06	141·39 or 41°·16	184·04 or 44°·44	30	0·08876

From the above we may estimate the correction arising from the time of transfer in the first seven experiments at ·00023, which, added to ·09024, gives ·09047 for the specific heat of brass at 76° compared with water at 46°. Regnault, in two trials, arrived at ·0939, but this appears to be in

hot water, and observing the temperature of mixture. The following specific heats were obtained by that method with brass and copper:—

	Brass.	Copper.
	·09200	·09516
	·08734	·09183
	·08945	·09295
	·09232	·08794
	·08734	..
Averages..	·08969	·09197

The wide discrepancy between the several results is owing to the great effect of the atmosphere on the small vessel, necessitating an absolute uniformity of stirring in order to give true temperatures.

reference to water taken as 1·008. When reduced to water taken as unity it becomes ·09315, which still differs considerably from my result. The method of cooling used by Regnault in this instance does not appear to me to be capable of as great accuracy as the method of mixtures used by the same physicist for other substances.

The interest I felt in this part of my subject induced me to try some experiments of a similar nature with copper sheet. It was tied in a bundle like the brass. Its weight w was 2777·9 grains.

No.	W.	T.	T'.	T''.	Time of transfer.	$\frac{(T''-T')\,W.}{(T-T'')\,w.}$
1	4734·5	855·26	199·3	232·62	6	0·09121
2	4772·7	900·58	172·4	209·57	4	0·09242
3	4738·5	946·26	218·85	255·49	5	0·09048
4	4732·2	948·07	166·18	206·18	5	0·09185
5	4786·3	1030·46	106·0	152·48	4	0·09121
6	4749·0	985·48	197·02	237·2	4	0·09180
7	4849·6	1069·0	103·35	208·46	4	0·09152
Average .		962·16 or 104°·36	166·16 or 43°·07	214·57 or 46°·80	4·57	0·09150
1	4749·9	891·11	106·6	145·2	30	0·08848
2	4815·1	985·14	161·46	201·14	30	0·08773
3	4768·0	997·47	170·35	210·99	30	0·08869
Average .		957·91 or 104°·03	146·14 or 41°·52	185·78 or 44°·58	30	0·08830

In the first seven the average time of transfer is 4^s·57 and the proximate specific heat 0·091497. In the last three we have 30^s and 0·088302. From these the specific heat of the sheet copper at 75° is determined at 0·092094.

The boxwood piece n, Plate IV. fig. 114, had a brass nut in its centre by which it was screwed on the axle of the brass stirrer. Being a bad conductor, and having nearly the whole of its surface in contact with the air, only a small portion of its capacity for heat could be counted in reckoning the whole capacity of the calorimeter. I determined this portion by ascertaining the heat communicated to a can of water when the boxwood piece was immersed in it after having been screwed on the calorimeter filled with hot water, for different periods of time. Calling the difference between the temperatures of the air and the calorimeter T, the gain of temperature in the small can t, the capacity of this can of water c, and C the modified or virtual capacity of the boxwood piece, we have $C=\frac{tc}{T}$. The following results were obtained showing the gradual approach of this virtual capacity to a certain limit :—

Time that the boxwood was screwed on the calorimeter.	Virtual capacity.
3^m	45·6
6	57·5
8	63·9
12	67·3
60	76·0

The virtual capacity of the caoutchouc stoppers was determined in the same manner :—

Time	Capacity.
3^m	15·35
8	21·8
30	27·45

The several capacities making up that of the calorimeter are therefore summed as follows* :—

Brass, 51979 grains × ·09047 =	4702·54
Caoutchouc stoppers	27·45
Boxwood piece	76·00
Thermometer	44·78
Total	4850·77

I had therefore great confidence in employing the value 4842·4, obtained, as already described, from experiments with the calorimeter itself.

In making an experiment for the equivalent, the weight of the calorimeter filled with distilled water was first carefully ascertained. It was then screwed on to the axis, and the

* The specific heat of boxwood, which I obtained by immersion in mercury, was 0·417 ; that of the caoutchouc, 0·29.

fine silk cords attached to the scales, *k k*, Plate IV. fig. 113, were adjusted. Thermometer A was then introduced into one of the tubulures, and after sufficient agitation of the water by means of the paddle itself, its indication was observed through a telescope. The thermometer was then removed and a caoutchouc stopper placed in the tubulure. The axle was then brought rapidly up to the velocity which produced friction sufficient to raise the weights about a foot from the ground. My son, Mr. B. A. Joule, who turned the wheel, could, by observing the position of the scales in a mirror, keep them very steadily at a constant height during the whole time of revolution. The wheel having been rapidly brought to a standstill, the temperature of the calorimeter was again ascertained.

In the experiments in Table I. the number of revolutions of the axis when the weights were off the ground was added to half the number occupied in the acts of starting from rest and returning to rest.

Previously to, and subsequently to, every such experiment others were made under similar conditions as to the observation of temperatures, &c., in order to ascertain the effect of the atmosphere on the temperature of the calorimeter. The indications of the thermometer for temperature of air are always reduced to the graduation of thermometer A.

Experience had already shown me that the thermal effect of the air on the calorimeter was not exactly proportional to the difference of their temperatures. This might arise from variations in the radiating powers of brass and glass from day to day. By making experiments for the air-effect immediately before and after one for the equivalent, I sought to neutralize any error arising from this circumstance. The last column but one of the first part of the following Tables gives the amount of correction required to be applied to the temperature of the air so as to make the effect proportional to the difference of temperatures. The figures in the last two columns are then used for calculating the corrected rise of temperature in the last column but one of Part 2 of the Tables.

Table I., Part 1.—Experiments to ascertain the Effect of Radiation, &c. Time occupied by each of the first fifteen, 50^m; by the last two, 41^m and 41^m 30^s.

No.	Mean temperature of calorimeter.	Mean temperature of air.	Difference.	Rise of temperature of calorimeter.	Correction to air-temperature.	Thermal effect of unit difference of temperature.
1*a* 1*b*	392·410 434·310	395·790 404·540	3·380+ 29·770−	0·20+ 3·40−	1·525−	0·1086
2*a* 2*b*	390·086 430·562	398·330 404·630	8·244+ 25·932−	0·62+ 2·96−	2·330+	0·1048
3*a* 3*b*	391·056 435·884	405·800 423·227	14·744+ 12·657−	1·52+ 1·35−	0·232−	0·1047
4*a* 4*b*	395·083 437·583	397·514 410·050	2·431+ 27·533−	0·26+ 3·06−	0·084−	0·1108
5*a* 5*b*	401·315 441·385	409·848 414·240	8·533+ 27·145−	0·61+ 3·13−	2·714−	0·1048
6*a* 6*b*	325·315 368·900	330·736 342·236	5·421+ 26·664−	0·49+ 2·96−	0·864−	0·1075
7*a* 7*b*	325·880 368·940	327·820 341·300	1·940+ 27·640−	0·16+ 2·88−	0·383−	0·1028
8*a* 8*b*	338·980 383·250	346·286 360·957	7·306+ 22·293−	0·74+ 2·30−	0·071−	0·1028
9*a* 9*b*	344·225 383·790	336·196 342·847	8·029− 40·943−	0·75+ 4·52−	1·460+	0·1145
10*a* 10*b*	327·350 373·705	344·610 361·645	17·260+ +12·060	2·00+ 1·29−	0·564+	0·1122
11*a* 11*b*	333·920 381·685	359·248 373·528	25·328+ 8·157−	3·12+ 0·67−	2·237+	0·1132
12*a* 12*b*	326·597 372·650	345·308 355·450	18·711+ 17·200−	2·26+ 1·74−	1·598+	0·1115
13*a* 13*b*	311·930 355·000	318·630 328·244	6·700+ 26·756−	0·66+ 2·96−	0·600−	0·1082
14*a* 14*b*	301·305 347·650	322·640 333·546	21·335+ 14·104−	2·37+ 1·30−	1·551+	0·1036
15*a* 15*b*	327·560 372·875	347·235 355·375	19·675+ 17·500−	1·92+ 1·85−	0·742−	0·1014
16*a* 16*b*	296·165 344·810	315·265 321·837	19·100+ 22·973−	1·71+ 2·38−	1·534−	0·0971
17*a* 17*b*	280·810 328·730	295·307 305·500	14·497+ 23·230−	1·15+ 2·20−	1·561−	0·0888

TABLE I., Part 2.—Experiments with Friction of Water and Brass. Weight, W, lifted in the first fifteen, 14619·5 grains; in the last two, 18122·9 grains. Average proportion of metallic to total friction, $\frac{1}{7\cdot 7}$. Time occupied by each of the first fifteen, 50^{m}; by the last two, 41^{m} and 41^{m} 30^{s}. Value, or V, of one division of the thermometer, $0^{\circ}\cdot077214$. Circumference of groove of calorimeter, P, 2·77386 feet.

No.	Number of revolutions. R.	Capacity of the calorimeter. C.	Mean temperature of the calorimeter.	Mean temperature of the atmosphere.	Difference.	Rise of temperature of calorimeter.	Ditto, corrected for radiation, &c. T.	Mechanical equivalent, or $\frac{RWP}{CTV}$
1	5545·0	84359·5	414·480	400·190	14·290−	42·761	44·478	776·15
2	5378·0	84413·4	412·570	399·130	13·440−	41·830	43·482	769·52
3	5522·6	84369·5	415·594	416·030	0·436+	44·606	44·585	771·06
4	5685·2	84309·1	418·020	402·390	15·630−	43·905	45·646	775·89
5	5321·3	84439·5	423·522	409·792	13·730−	41·347	43·071	768·43
6	5756·5	84429·4	349·325	335·160	14·165−	44·890	46·506	769·98
7	5753·5	84424·1	349·502	332·934	16·568−	44·509	46·251	773·87
8	5740·7	84429·1	363·325	351·585	11·740−	44·923	46·137	774·01
9	5714·0	84448·6	366·028	338·175	27·853−	42·682	45·703	777·55
10	5725·7	84415·0	352·730	351·466	1·264−	46·088	46·167	771·60
11	5695·0	84383·1	360·146	366·080	5·934+	46·631	45·706	775·51
12	5702·2	84370·1	351·998	349·282	2·716−	45·878	46·003	771·59
13	5681·2	84355·0	335·844	322·280	13·564−	44·295	45·827	771·80
14	5702·3	84353·0	326·922	325·294	1·628−	45·883	45·891	773·64
15	5670·8	84291·0	352·669	348·867	3·802−	45·368	45·829	770·97
16	4965·5	84282·1	324·334	316·810	7·524−	48·750	49·630	772·85
17	4953·3	84398·0	307·643	298·298	9·345−	48·535	49·503	771·87
Average			366·16 or 58°·46	..	..	..	..	772·72

The mean temperature of the atmosphere was derived from observations taken from minute to minute; but there were only two readings of the temperature of the calorimeter, viz. at the commencement and termination of an experiment, from which to determine its average temperature. Suppose *a b*, Plate IV. fig. 119, to represent the line of air-temperatures during an experiment lasting 41^m: the temperatures of the calorimeter will be represented by a line similar to *c d e f*. The wheel was set in motion 2^m after the first reading was taken. The temperature then rose until, at 35^m, the wheel was stopped. The temperature then declined slightly, until, at 41^m, the last reading was taken. The line is slightly curved; a few seconds are occupied in starting and stopping the wheel, and the thermometer reads a little backwards. Taking all these circumstances into account, I found that the average temperature for the whole time was very accurately represented by

$$\frac{37\frac{c+f}{2}+4f.}{41}$$

The mean temperature of the calorimeter for other times of experiment was estimated in a similar manner.

To obtain the corrected rise of temperature in the last column but one, the correction to the air-temperature indicated in the first part of the Table was supplied. For instance, in the first experiment the temperature of the air was virtually $14{\cdot}29+1{\cdot}525=15{\cdot}815$ lower than that of the calorimeter. Hence $15{\cdot}815\times0{\cdot}1086=1{\cdot}717$, which, added to $42{\cdot}761$, gives the value for T, $44{\cdot}478$.

TABLE II., Part 1.—Experiments to ascertain the Effect of Radiation, &c. Time occupied by each experiment, 41^m. Weight raised for an instant, 18229·0 grains.

No.	Revolutions. *r*.	Mean temperature of calorimeter.	Mean temperature of air.	Difference.	Rise of temperature of calorimeter.	Correction to air-temperature.	Thermal effect of unit difference of temperature.
1*a* 1*b*	25·5 23·5	259·220 305·925	268·008 277·775	8·788+ 28·150−	0·945+ 2·478−	1·410+	0·0926
2*a* 2*b*	26·0 22·92	261·740 308·925	273·313 286·733	11·573+ 22·192−	1·572+ 1·850−	3·938+	0·1013
3*a* 3*b*	23·60 24·12	267·270 318·620	291·597 307·351	24·327+ 11·269−	3·590+ 0·660−	5·741+	0·1194
4*a* 4*b*	22·75 23·17	298·510 344·910	308·706 316·151	10·196+ 28·759−	1·00+ 2·33−	1·502+	0·0855
5*a* 5*b*	22·0 21·08	311·560 355·710	311·188 317·040	0·372− 36·670−	0·12+ 3·38−	1·685+	0·0914
6*a* 6*b*	21·7 21·04	285·095 331·865	300·273 308·541	15·178+ 23·324−	1·59+ 1·77−	3·054+	0·0873
7*a* 7*b*	21·3 19·8	292·760 340·435	304·393 312·739	11·633+ 27·696−	1·28+ 2·23−	2·709+	0·0892
8*a* 8*b*	21·0 21·2	277·730 328·775	295·705 311·147	17·975+ 17·628−	2·46+ 1·21−	5·890+	0·1031
9*a* 9*b*	19·83 18·79	301·550 351·365	321·510 324·762	19·960+ 26·603−	1·90+ 2·33−	0·955+	0·0908
10*a* 10*b*	18·33 17·85	300·370 352·595	325·183 335·507	24·813+ 17·088−	2·34+ 1·29−	2·198+	0·0866
11*a* 11*b*	18·2 17·24	277·800 329·570	304·655 317·940	26·855+ 11·630−	2·80+ 0·86−	2·587+	0·0951
12*a* 12*b*	16·0 18·74	288·400 336·980	321·267 331·820	32·867+ 5·160−	3·20+ 0·40−	0·935+	0·0947
13*a* 13*b*	18·0 17·66	313·820 364·075	331·190 358·747	17·370+ 5·328−	3·73+ 0·15+	6·279+	0·1577
14*a* 14*b*	17·32 18·0	308·585 356·840	329·613 340·123	21·028+ 16·717−	2·33+ 1·28−	3·334+	0·0956
15*a* 15*b*	32·0 18·6	309·885 359·100	331·872 340·480	21·987+ 18·620−	2·00+ 1·46−	1·485+	0·0852

TABLE II., Part 2.—Experiments with Friction of Water and Brass. Weight, W, lifted, 18229·0 grains. Average proportion of metallic to total friction, $\frac{1}{8\cdot 3}$. Time occupied by each experiment 41m. $V=0^{\circ}\cdot 077214$; $P=2\cdot 77386$.

No.	Number of revolutions, $R+r$.	Capacity of the calorimeter. C.	Mean temperature of the calorimeter.	Mean temperature of the atmosphere.	Difference.	Rise of temperature of the calorimeter.	Ditto, corrected for radiation, &c. T.	Mechanical equivalent, or $\frac{R\,W\,P}{C\,T\,V}$.
1	4898·5	84349·7	286·242	271·743	14·499 −	47·686	48·898	773·94
2	4826·5	84242·7	288·94	277·634	11·306 −	47·477	48·223	774·09
3	4929·75	84191·7	296·43	298·242	1·812 +	50·046	49·144	776·48
4	4839·5	84324·0	324·9	311·823	13·077 −	47·288	48·278	774·80
5	4829·5	84270·0	336·865	312·924	23·941 −	46·119	48·153	775·93
6	4734·3	84256·0	311·387	303·190	8·197 −	46·893	47·342	773·74
7	4897·5	84234·5	320·1	308·110	11·990 −	48·168	48·996	773·84
8	5061·7	84295·7	306·684	303·337	3·347 −	50·640	50·378	777·30
9	5091·8	84294·0	330·182	321·677	8·505 −	50·237	50·922	773·88
10	5165·9	84250·0	330·3	328·046	2·254 −	51·800	51·805	772·38
11	5045·2	84302·0	307·173	309·280	2·107 +	50·820	50·374	775·28
12	4613·66	84292·0	315·95	324·655	8·705 +	47·250	46·337	770·62
13	4733·9	84304·0	342·25	343·864	1·614 +	48·326	47·081	778·11
14	4782·17	84284·0	335·99	333·819	2·171 −	47·771	47·660	776·73
15	4834· 0	84244·0	338·125	333·930	4·195 −	48·225	48·456	771·42
	Average	....	318·101 or 54°·76	..	..	..	..	774·57

Instead of reckoning one half of the revolutions which took place in the acts of starting and stopping the wheel, as was done in the case of Table I., I have eliminated them in the last and subsequent Tables by starting the wheel till the scales were raised for an instant and then immediately stopping it at some period in each experiment for determining radiation. The revolutions called r in the first part of the Table being subtracted from the revolutions called $R+r$ in the second part, give the numbers used in calculating the equivalent. This latter plan obviated some slight error to which the former method was possibly liable.

The irregularities in the values of R arise from the variations from time to time in the friction of the bearing which supports the calorimeter on the axis. In the subsequent experiments I adopted a method which removed nearly the whole of the metallic friction. In Plate IV. fig. 113, *v* and *w* represent two concentric vessels. The inner one has a lid surmounted by three uprights, such as that represented by fig. 5. When water is poured into the space between the vessels, the uprights are raised so as to press against the bottom rim of the calorimeter, thus relieving its weight on the axis. The arrangement was eminently successful in producing an almost absolute uniformity of motion.

TABLE III., Part 1.—Experiments to ascertain the Effect of Radiation, &c. Time occupied by each experiment, 41^{m}. Weight raised for an instant, 16477·4 grains.

No.	Revolutions. r.	Mean temperature of calorimeter.	Mean temperature of air.	Difference.	Rise of temperature of calorimeter.	Correction to air-temperature.	Thermal effect of unit difference of temperature.
1*a*	19·0	385·526	427·236	41·710+	4·05+	1·237+	0·1060
1*b*	19·5	432·690	424·879	7·811−	0·62−		
2*a*	18·58	366·510	376·830	10·320+	1·02+	0·683+	0·0927
2*b*	19·42	409·710	386·590	23·120−	2·08−		
3*a*	18·75	338·865	364·127	25·262+	2·325+	0·862+	0·0890
3*b*	20·33	386·535	368·066	18·469−	1·567−		
4*a*	19·5	339·675	368·504	28·829+	2·51+	0·429−	0·0884
4*b*	20·75	385·410	372·488	12·922−	1·18−		

TABLE III. Part 1 (*continued*).

No.	Revolutions. r.	Mean temperature of calorimeter.	Mean temperature of air.	Difference.	Rise of temperature of calorimeter.	Correction to air-temperature.	Thermal effect of unit difference of temperature.
5*a*	19·17	357·290	378·033	20·743+	1·78+	0·027 −	0·0859
5*b*	19·31	402·025	386·340	15·685−	1·35−		
6*a*	20·67	357·885	389·996	32·111+	2·97+	0·843+	0·0901
6*b*	20·93	404·650	397·150	7·500−	0·60−		
7*a*	19·33	355·780	383·923	28·143+	2·56+	0·600+	0·0891
7*b*	19·75	401·960	389·234	12·726−	1·08−		
8*a*	19·4	878·970	388·517	9·547+	0·90+	0·612+	0·0886
8*b*	21·2	421·730	392·446	29·284−	2·54−		
9*a*	19·46	342·575	361·843	19·268+	1·83+	0·503+	0·0926
9*b*	20·0	387·000	369·210	17·790−	1·60−		
10*a*	18·83	347·205	369·358	22·153+	1·75+	0·240−	0·0799
10*b*	20·43	391·930	373·889	18·041−	1·46−		
11*a*	19·0	349·530	378·721	29·191+	2·78+	1·341+	0·0911
11*b*	19·0	395·970	386·502	9·468−	0·74−		
12*a*	19·0	344·970	383·753	38·783+	3·54+	1·383+	0·0881
12*b*	19·0	393·040	390·749	2·291−	0·08−		
13*a*	18·0	348·125	372·988	24·863+	2·17+	0·627+	0·0851
13*b*	17·75	393·550	380·002	13·548−	1·10−		
14*a*	18·12	346·520	378·315	31·795+	3·00+	1·185+	0·0910
14*b*	17·0	393·310	386·409	6·901−	0·52−		
15*a*	17·73	356·395	390·069	33·674+	3·49+	0·688+	0·1016
15*b*	19·23	404·170	400·922	3·248−	0·26−		
16*a*	18·0	345·085	380·701	35·616+	3·17+	0·560+	0·0876
16*b*	16·72	392·365	386·442	5·923−	0·47−		
17*a*	19·0	366·715	388·706	21·991+	1·93+	0·086+	0·0874
17*b*	18·0	411·775	395·102	16·673−	1·45−		
18*a*	19·57	374·230	400·624	26·394+	2·46+	1·102+	0·0895
18*b*	18·97	420·020	408·859	11·161−	0·90−		
19*a*	18·72	396·755	412·830	16·075+	1·49+	0·501+	0·0899
19*b*	19·0	440·415	415·996	24·419−	2·15−		
20*a*	19·33	367·770	403·836	36·066+	3·46+	1·611+	0·0918
20*b*	17·77	415·070	411·934	3·136−	0·14−		
21*a*	17·0	386·805	415·792	28·987+	2·93+	1·158+	0·0972
21*b*	18·3	433·470	426·345	7·125−	0·58−		

TABLE III., Part 2.—Experiments with almost solely Friction of Water. Weight, W, lifted, 16477·4 grains. Average proportion of metallic to total friction, $\frac{1}{106}$. Time occupied by each experiment, 41ᵐ. V=0°·077214; P=2·77386.

No.	Number of revolutions. R+r.	Capacity of the calorimeter. C.	Mean temperature of the calorimeter.	Mean temperature of the air.	Difference.	Rise of temperature of the calorimeter.	Ditto, corrected for radiation, &c. T.	Mechanical equivalent, or $\frac{RWP}{CTV}$.
1	4904·95	84160·7	412·530	422·744	10·214+	45·509	44·295	775·78
2	4940·58	84124·2	390·866	381·225	9·641−	43·909	44·740	774·04
3	4925·28	84118·1	367·765	364·627	3·138−	44·564	44·767	771·14
4	4922·92	84071·2	365·695	369·159	3·464+	45·125	44·857	769·56
5	4938·0	84012·2	382·816	380·063	2·753−	44·747	44·986	770·39
6	4936·3	83930·2	384·452	391·613	7·161+	45·580	44·859	772·82
7	4958·0	83940·7	382·015	386·005	3·990+	45·530	45·121	771·82
8	4925·3	83882·2	403·368	390·360	13·008−	43·643	44·741	773·64
9	4925·9	83907·2	367·897	364·284	3·613−	44·564	44·852	771·68
10	4927·6	83884·2	372·565	370·435	2·130−	44·706	44·857	772·09
11	4934·3	83911·2	375·915	381·468	5·553+	45·438	44·810	773·81
12	4935·0	83889·2	372·285	386·270	13·985+	46·150	44·795	774·38
13	4929·0	83880·3	373·936	375·248	1·312+	44·930	44·765	774·22
14	4923·22	83888·7	373·092	380·689	7·597+	45·487	44·688	774·61
15	4923·0	83872·7	383·670	395·008	11·338+	45·850	44·629	775·60
16	4923·0	83897·2	372·033	382·095	10·062+	45·707	44·776	773·00
17	4928·0	83889·7	392·412	390·767	1·654−	44·652	44·788	773·47
18	4926·0	83904·7	400·189	402·495	2·306+	45·105	44·800	772·69
19	4925·75	83891·2	421·652	413·149	8·503−	44·148	44·967	771·68
20	4926·33	83944·2	394·541	406·037	11·496+	45·763	44·559	776·68
21	4924·0	83938·7	413·330	419·217	5·887+	45·459	44·774	772·76
	Average		385·86 or 59°·98	..	..	..	..	773·136

An error of four or five seconds in the time at which the wheel was started and stopped will account for the divergence of the revolutions in Nos. 1 and 7 from the average. For the rest, it will be seen with what great constancy the resistance of the paddle was kept up.

The weights were also so steady that the total distance run by them in their risings and fallings only amounted to about 30 feet in each experiment. This, taken with the friction of the pulleys = 30 grains, gives a quantity to be subtracted from the equivalent too small to require estimation.

TABLE IV., Part 1.—Experiments to ascertain the Effect of Radiation, &c. Time occupied by each experiment, 41^{m}. Weight raised for an instant, 7730·56 grains.

No.	Revolutions. *r*.	Mean temperature of calorimeter.	Mean temperature of air.	Difference.	Rise of temperature of calorimeter.	Correction to air-temperature.	Thermal effect of unit difference of temperature.
1*a* 1*b*	9·5 8·62	362·355 374·950	358·090 363·536	4·265— 11·414—	0·45— 1·10—	0·684—	0·0909
2*a* 2*b*	10·0 9·9	361·090 375·040	363·847 368·448	2·757+ 6·592—	0·18+ 0·60—	0·599—	0·0834
3*a* 3*b*	10·0 10·24	355·495 368·430	352·673 357·570	2·822— 10·860—	0·31— 1·02—	0·688—	0·0883
4*a* 4*b*	10·57 10·25	346·515 361·985	359·807 365·790	13·292+ 3·805+	1·01+ 0·15+	2·150—	0·0906
5*a* 5*b*	9·66 10·74	340·260 354·970	346·208 356·487	5·948+ 1·517+	0·52+ 0·08+	0·711—	0·0993
6*a* 6*b*	10·82 10·66	357·855 373·890	368·538 382·515	10·683+ 8·625+	1·01+ 0·78+	1·646—	0·1118

TABLE IV., Part 2.—Experiments with almost solely Friction of Water. Weight, W, lifted, 7730·56 grains. Average proportion of metallic to total friction, $\frac{1}{43}$. Time occupied by each experiment, 41^{m}. $V=0^{\circ}·077214$; $P=2·77386$.

No.	Number of revolutions. $R+r$.	Capacity of the calorimeter. C.	Mean temperature of the calorimeter.	Mean temperature of the air.	Difference.	Rise of temperature of the calorimeter.	Ditto, corrected for radiation, &c. T.	Mechanical equivalent, or $\frac{RWP}{CTV}$
1.	3336·66	83963·7	369·487	361·043	8·444−	13·495	14·325	768·32
2.	3344·2	83944·7	369·065	366·560	2·505−	14·078	14·337	769·39
3.	3343·42	83959·7	362·809	354·425	8·384−	13·646	14·447	763·18
4.	3341·23	83949·7	355·191	362·160	6·969+	14·927	14·490	760·44
5.	3330·75	83975·0	348·411	350·362	1·951+	14·416	14·293	768·30
6.	3335·33	83965·0	366·732	375·533	8·801+	15·040	14·240	772·19
	Average		361·949 or 58°·14	..	..	..	..	766·97

It will be obvious that, in the experiments of the above Table, where the heat evolved was able to raise the temperature of the calorimeter little more than 1°, great accuracy could not be expected without taking the average of a very large number of observations. In fact, the degree of accuracy will increase nearly with the square of the rise of temperature per unit of time*, and the square root of the number of observations.

Table V., Part 1.—Experiments to ascertain the Effect of Radiation, &c. Time occupied by each experiment, 41^{m}. Weight raised for an instant, 21729·56 grains.

No.	Revolutions. r.	Mean temperature of calorimeter.	Mean temperature of air.	Difference.	Rise of temperature of calorimeter.	Correction to air-temperature.	Thermal effect of unit difference of temperature.
1a	22·0	386·760	419·637	32·877+	3·52+	3·099+	0·09784
1b	20·64	454·455	427·134	27·321−	2·37−		
2a	21·16	390·200	433·283	43·083+	4·40+	2·444+	0·09665
2b	22·0	460·345	444·140	16·205−	1·33−		
3a	20·0	380·885	410·976	30·091+	2·77+	1·369+	0·08805
3b	21·22	448·250	418·488	29·762−	2·50−		
4a	22·0	386·310	420·428	34·118+	3·62+	1·965+	0·10032
4b	23·16	455·170	436·858	18·312−	1·64−		
5a	21·42	397·770	428·568	30·798+	3·46+	3·283+	0·10152
5b	22·0	465·440	441·275	24·165−	2·12−		
6a	19·0	396·300	438·413	42·113+	3·84+	2·100+	0·08685
6b	22·0	465·630	447·180	18·450−	1·42−		
7a	21·0	377·300	405·069	27·769+	2·60+	1·509+	0·08880
7b	22·0	444·240	414·354	29·886−	2·52−		

The average number of revolutions per minute in Table IV. pt. 2, and Table V. pt. 2, were 101·4 and 171·5. The fluid resistances, 7630·2 and 21548·5, were therefore almost exactly proportional to the squares of the velocities.

* *I. e.*, supposing the "Differences," for calculating the air-correction, increase with the values of T.

TABLE V., Part 2.—Experiments with almost solely Friction of Water. Weight, W, lifted, 21729·56 grains. Average proportion of metallic to total friction, $\frac{1}{108}$. Time occupied by each experiment, 41^m. $V=0^\circ{\cdot}077214$; $P=2{\cdot}77386$.

N°.	Number of revolutions. $R+r$.	Capacity of the calorimeter. C.	Mean temperature of the calorimeter.	Mean temperature of the air.	Difference.	Rise of temperature of the calorimeter.	Ditto, corrected for radiation, &c. T.	Mechanical equivalent, or $\frac{R\,W\,P.}{C\,T\,V}$
1.	5653·0	83924·2	425·354	419·585	5·769−	67·336	67·597	774·93
2.	5653·16	83960·2	430·165	436·986	6·821+	68·582	67·687	773·54
3.	5652·55	83970·5	419·268	413·363	5·905−	67·214	67·613	774·35
4.	5648·66	84001·7	425·460	426·763	1·303+	67·877	67·549	774·00
5.	5645·4	83986·2	436·432	433·830	2·602−	67·522	67·453	774·90
6.	5646·5	83973·7	435·661	439·140	3·479+	68·215	67·730	772·17
7.	5659·0	83970·7	415·580	408·720	6·860−	67·232	67·707	774·03
Average.............			426·846 or 63·14	..	..	..	..	773·99

The foregoing results are collected in the following Table :—

Table.	Number of experiments.	Proportion of metallic to total friction.	Mean rise of temperature per A.	Temperature of the calorimeter.	Mechanical equivalent of unit of heat.
1.	17	$\frac{1}{7 \cdot 7}$	45·907	58°·46	772·72
2.	15	$\frac{1}{8 \cdot 3}$	48·803	54°·76	774·57
3.	21	$\frac{1}{106}$	44·777	59°·98	773·136
4.	6	$\frac{1}{43}$	14·355	58°·14	766·97
5.	7	$\frac{1}{108}$	67·620	63°·14	773·99

The average of the first two gives 773·65 as the equivalent at a temperature of the calorimeter 56°·61 ; but inasmuch as the metallic friction is as much as $\frac{1}{8}$ of the whole, I prefer to use the last three, and to give each its due weight I will multiply the squares of the rise by the square root of the number of determinations :—

For the 3rd series $(44{\cdot}777)^2 \times \sqrt{21} = 9188$.
For the 4th series $(14{\cdot}355)^2 \times \sqrt{6} = 504{\cdot}76$.
For the 5th series $(67{\cdot}62)^2 \times \sqrt{7} = 12097{\cdot}7$.

Then—

$$\frac{733{\cdot}136 \times 9188 + 766{\cdot}97 \times 504{\cdot}76 + 773{\cdot}99 \times 12097{\cdot}7}{9188 + 504{\cdot}76 + 12097{\cdot}7} = 773{\cdot}467$$

is the equivalent at 61°·69 ; or, using Regnault's law of the increase of the specific heat of water with its temperature, 773·369 at 60°.

The latitude of the part of Higher Broughton, Manchester, where the experiments were made, is 53° 28½′ N. ; its elevation about 120 feet above the sea-level. The equivalent at the sea-level and the latitude at Greenwich will therefore be 773·492 foot lbs., defining the unit of heat to be that which a lb. of water, weighed by brass weights when the barometer stands at 30 inches, receives in passing from 60° to 61° Fahr. With water weighed in vacuo the equivalent is finally reduced to 772·55.

Printed in the United States
By Bookmasters